基金资助：① 绿色成形智能装备创新团队
② JAT190689
③ 2020S2002

方程式赛车动力学仿真

王孝鹏 ○ 著

西南交通大学出版社
·成 都·

图书在版编目（CIP）数据

方程式赛车动力学仿真 / 王孝鹏著. —成都：西南交通大学出版社，2021.8
ISBN 978-7-5643-8071-7

Ⅰ. ①方… Ⅱ. ①王… Ⅲ. ①赛车 – 车辆动力学 – 系统仿真 Ⅳ. ①U469.601

中国版本图书馆 CIP 数据核字（2021）第 123721 号

Fangchengshi Saiche Donglixue Fangzhen
方程式赛车动力学仿真
王孝鹏 著

责 任 编 辑	李 伟
封 面 设 计	何东琳设计工作室
出 版 发 行	西南交通大学出版社 （四川省成都市金牛区二环路北一段 111 号 西南交通大学创新大厦 21 楼）
发行部电话	028-87600564　028-87600533
邮 政 编 码	610031
网　　　址	http：//www.xnjdcbs.com
印　　　刷	成都蜀通印务有限责任公司
成 品 尺 寸	185 mm × 260 mm
印　　　张	37.25
字　　　数	930 千
版　　　次	2021 年 8 月第 1 版
印　　　次	2021 年 8 月第 1 次
书　　　号	ISBN 978-7-5643-8071-7
定　　　价	160.00 元

图书如有印装质量问题　本社负责退换
版权所有　盗版必究　举报电话：028-87600562

前　言

　　方程式赛车在运动过程中主要强调整车的稳定性，整车的稳定性核心在于底盘的设计与调试。与传统的乘用车相比，方程式赛车悬架系统的结构设计、弹簧刚度的设定、避震器的匹配较为复杂，同时需要考虑不同赛道、不同天气特征下使用的轮胎特性。采用多体系统动力学软件 ADAMS/CAR 模块可以快速构建不同悬架系统的结构及整车模型，分析其系统间的匹配特性。

　　近些年来，大学生方程式赛车在国内快速发展，但整车底盘的设计过于保守，且底盘设计同类化严重。针对此现象与问题，作者设计了不同系列的推杆式、拉杆式、扭杆弹簧式、变刚度式等新型悬架结构供参考与学习。

　　本书内容主要包含：(1) 系统性介绍了方程式赛车的转向系统、车身系统、动力系统、制动系统、路面模型等。(2) 系统性介绍了横向稳定杆模型，采用衬套特性模拟横向稳定杆的扭转刚度，模型整体简单，计算速度快，通过改变衬套刚度可以快速更改稳定杆特性；采用有限元柔性体 MNF 文件建立真实的横向稳定杆并模拟其特性，模型较为复杂，计算速度较慢。(3) 系统性介绍了不同种类的悬架模型（悬架模型是本书的核心）。需要强调的是，整车动态特性的优劣并不在于采用技术含量高的悬架，关键在于悬架与整车的匹配，这是底盘调试的精髓与核心。避震器纵置式推杆悬架通过增加辅助避震系统，可以改善摆臂的震动特性，同时提升赛车在起步加速时或制动时的"抬头"与"点头"现象。拉杆悬架模型的优势是可以在较大的裕度空间范围内布置拉杆，例如可以将拉杆布置到上下控制臂之外；对于推杆悬架来说，过长的推杆容易导致挠度变形过大，而拉杆不存在此问题。扭杆弹簧式推杆悬架可以将扭杆弹簧与推杆悬架的优势结合起来，对于支架摆臂的转动特性，可以考虑增加横向避震器，以限制或减小摆臂的旋转特性，进而减少车轮或车身上下振动的幅值特性，从而提升赛车的性能。空间斜置扭杆弹簧推杆式解耦悬架通过扭杆弹簧与螺旋弹簧两种不同刚度的设定，在起步、制动、高速行驶时采用扭转弹簧的大刚度特性，在转弯及低速行驶时采用螺旋弹簧，以提升整车的平顺性和弯道的稳定性。在静止、低速、高速时，可以通过滑阀避震器改变车身的高度；此种悬架设定与半主动悬架有本质的区别，半主动悬架是通过改变避震器的特性来改变悬架系统的特性，半主动悬架的弹簧刚度是不可调节的，此种悬架通过液压作用器改变扭杆弹簧与螺旋弹簧的工作特性，避震器可用滑阀特性改变车身高度，也可用变阻尼特性（如磁流变避震器）既改变高度，又改变阻尼特性。横置板簧悬架模型，包括板簧上横置和下横置等。（4）针对悬架前束角与外倾角，介绍其优化实验方法。（5）针对后轮随动转向、悬架变刚度特性、制动联合仿真案例，系统性介绍了问题的研究方案，同时分析模型并形成结论。

本书可作为高等院校高年级本科生、研究生及汽车工程研究院设计研发人员的学习用书，也可作为车辆系统动力学爱好者的参考资料。书中不同章节提供了相关模型，请扫码获取模型资源包。

王孝鹏

2021 年 4 月

仿真模型资源包

目 录 CONTENTS

第1章 中舵转向系统 ··· 1
   1.1 齿轮齿条式转向系统 ·· 2
   1.2 转向变量参数 ·· 22
   1.3 转向通信器 ·· 23
   1.4 转向仿真 ··· 33
   1.5 转向系统调试 ·· 35

第2章 车 身 ··· 39
   2.1 车身通信器 ·· 40
   2.2 车身参数变量 ·· 45
   2.3 车身测量函数 ·· 46

第3章 发动机 ·· 48
   3.1 发动机实验数据 ··· 49
   3.2 发动机扭矩图绘制程序 ··· 52
   3.3 发动机系统建模 ··· 53
   3.4 定半径转弯仿真 ··· 57

第4章 制动系统 ·· 59
   4.1 制动系统简介 ·· 60
   4.2 制动系统变量参数及通信器 ·· 60
   4.3 FSAE赛车Braking文件驱动仿真 ·· 61

第5章 轮胎模型 ·· 64
   5.1 轮胎模型适用性分类 ·· 65
   5.2 FSAE轮胎属性文件 ·· 66

第6章 路 面 ··· 71
   6.1 路面类型 ··· 71
   6.2 对开路面 ··· 73
   6.3 对接路面 ··· 75
   6.4 减速带路面 ·· 77
   6.5 连续障碍路面 ·· 79

  6.6 分离路面设置 ································· 81

第 7 章 横向稳定杆 I ································· 84
  7.1 横向稳定杆模型 ································· 85
  7.2 扭转弹簧 ································· 95
  7.3 单轮跳动仿真 ································· 99

第 8 章 横向稳定杆 II ································· 102
  8.1 横向稳定杆前处理 ································· 103
  8.2 柔体稳定杆模型 ································· 110
  8.3 车轮反向跳动仿真 ································· 115

第 9 章 非独立悬架 ································· 118
  9.1 板簧悬架模型 ································· 119
  9.2 钢板弹簧 ································· 132
  9.3 板簧悬架变量参数 ································· 142
  9.4 板簧悬架通信器 ································· 143
  9.5 板簧悬架通信器测试 ································· 145
  9.6 驱动轴显示组件 ································· 150
  9.7 双轮反向激振仿真 ································· 152
  9.8 整车模型 ································· 154
  9.9 Fish-Hook 仿真 ································· 155

第 10 章 避震器纵置式后推杆悬架 ································· 159
  10.1 后推杆悬架 ································· 160
  10.2 单轮激振测试验证模型 ································· 198
  10.3 悬架模型调试 ································· 200
  10.4 车轮同向跳动仿真 ································· 201

第 11 章 避震器纵置式前推杆悬架 ································· 203
  11.1 硬点信息 ································· 204
  11.2 FSAE 整车模型装配 ································· 205
  11.3 匀速直线仿真 ································· 206

第 12 章 避震器纵置式推杆悬架 ································· 209
  12.1 纵置式前推杆悬架模型 ································· 210
  12.2 纵置式后推杆悬架模型 ································· 226
  12.3 漂移仿真 ································· 228

第 13 章 避震器横置式推杆悬架 ································· 233
  13.1 刚度匹配 ································· 234

13.2	避震器横置式后推杆悬架	235
13.3	避震器横置式推杆悬架通信器	247
13.4	避震器横置式前推杆悬架	248
13.5	阶跃转向仿真	250

第 14 章　拉杆式悬架模型 Ⅰ ··· 254

14.1	前螺旋弹簧式拉杆悬架	255
14.2	后螺旋弹簧式拉杆悬架	270
14.3	单线移动超车仿真	271
14.4	螺旋弹簧式拉杆悬架调试	276
14.5	扭杆弹簧 MNF	277
14.6	刚度虚拟实验	278
14.7	前扭杆弹簧式拉杆悬架	280
14.8	后扭杆弹簧式拉杆悬架	287
14.9	转向盘角脉冲仿真	288
14.10	扭杆弹簧调试	294
14.11	谐波脉冲转向仿真	300
14.12	考虑横向避震器特性	303
14.13	正弦扫频转向仿真	305
14.14	稳定性参数对比	308

第 15 章　拉杆式悬架模型 Ⅱ ··· 310

15.1	拉杆悬架概述	311
15.2	拉杆悬架模型	311
15.3	收油门转弯仿真	345
15.4	纵置扭杆弹簧式拉杆悬架	348
15.5	弯道收油门仿真	348

第 16 章　扭力梁悬架 ··· 352

16.1	扭力梁悬架	353
16.2	FSAE 整车模型	377
16.3	定常半径转弯仿真	377

第 17 章　扭杆弹簧式推杆悬架 ··· 381

17.1	扭杆弹簧 MNF	382
17.2	扭杆弹簧垂置式推杆悬架	383
17.3	弯道制动仿真	396
17.4	扭杆弹簧纵置式推杆悬架	403
17.5	回正性仿真	408
17.6	考虑横向避震器	412

第18章 空间斜置扭杆弹簧推杆式悬架 416
18.1 前扭杆弹簧斜置式推杆悬架 417
18.2 四轮定位参数对标 441
18.3 刚度阻尼匹配 442
18.4 后扭杆弹簧斜置式推杆悬架 445
18.5 加速仿真 446

第19章 上横置板簧悬架模型 451
19.1 前上横置板簧悬架 452
19.2 后上横置板簧悬架 462
19.3 文件驱动仿真 464

第20章 下单横置板簧悬架模型 467
20.1 前下单横置板簧悬架 468
20.2 后下单横置板簧悬架 481
20.3 静平衡仿真 483
20.4 准静态定半径转弯 484
20.5 准静态定速转弯 486
20.6 准静态力矩仿真 487
20.7 准静态直线加速仿真 489

第21章 下双横置板簧悬架模型 490
21.1 双片板簧 MNF 491
21.2 多叶片板簧约束问题 492
21.3 前下双横置板簧悬架 493
21.4 后下双横置板簧悬架 508
21.5 收油门直线仿真 509

第22章 优化设计实验 513
22.1 双A臂悬架前束角优化数据库导入模型 514
22.2 推杆式悬架外倾角优化——ACAR 523

第23章 FSAE赛车后轮随动转向 528
23.1 随动转向数学模型 529
23.2 随动转向物理模型 531
23.3 柔性扭转梁 532
23.4 反向激振仿真 533
23.5 弯道仿真 534
23.6 扭力梁位置因素 536
23.7 衬套安装角度 538

23.8 总　结 ·· 540

第 24 章　变刚度悬架特性研究 ·· 541
24.1 横置板簧悬架 ··· 542
24.2 定常半径弯道仿真 ··· 545
24.3 板簧优化 ·· 547
24.4 总　结 ·· 550

第 25 章　弯道制动联合仿真 ·· 551
25.1 制动系统设置 ··· 552
25.2 函数编写 ·· 555
25.3 整车模型装配 ··· 558
25.4 ADAMS\Controls 设置 ·· 560
25.5 ADAMS 与 MATLAB 软件协同 ··· 561
25.6 双模糊理论 ·· 569
25.7 悬架辅助系统 ··· 571
25.8 制动联合仿真模型 ··· 572

第 26 章　整车平顺性仿真 ··· 575
26.1 整车装配 ·· 576
26.2 平顺性仿真 ·· 579

附　录 ··· 581

参考文献 ·· 585

第 1 章 中舵转向系统

汽车在行驶过程中，经常需要按照驾驶员的意志改变其行驶方向，即所谓汽车转向。就轮式汽车而言，实现汽车转向的方法是，驾驶员通过一套专设的机构，使汽车转向桥上的车轮相对于汽车纵轴线偏转一定角度。汽车在直线行驶时，往往转向轮也会受到路面侧向干扰力的作用，自动偏转而改变行驶方向。此时，驾驶员也可以利用这套机构使转向轮向相反的方向偏转，从而使汽车恢复原来的行驶方向。这一套用来改变或恢复汽车行驶方向的专设机构，即称为汽车转向系统。因此，汽车转向系统的功用是，保证汽车能按驾驶员的意志而进行转向行驶。动力转向系统是兼用驾驶员体力和发动机动力为转向能源的转向系统。在正常情况下，汽车转向所需能量，只有一小部分由驾驶员提供，而大部分是由发动机通过动力转向装置提供的。但在动力转向装置失效时，一般还应当能由驾驶员独立承担汽车转向任务。因此，动力转向系统是在机械转向系统的基础上加设一套动力转向装置而形成的。图 1-1 为 FSAE 赛车中舵（置）转向系统。

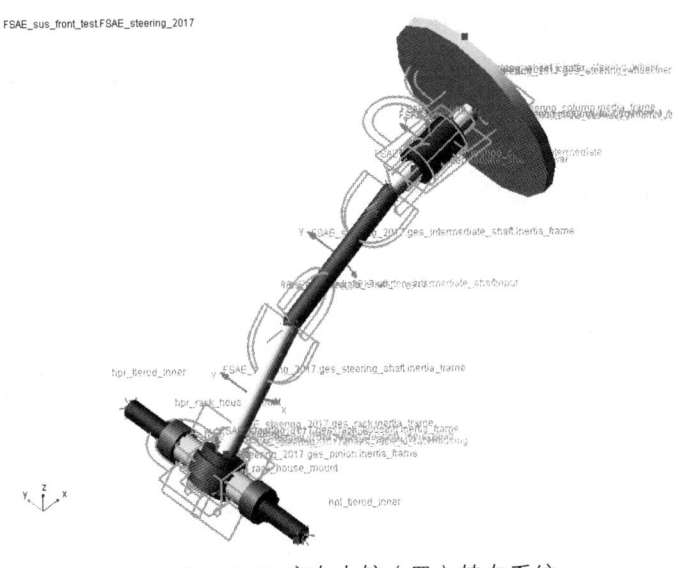

图 1-1　FSAE 赛车中舵（置）转向系统

ADAMS/CAR 中常用的转向系统包括齿轮齿条式转向系统和平行机构式转向系统，转向系统的通用性与可拓展性相对于悬架或其他系统较强，在建模过程直接在转向系统中修改硬点、传动比即可以满足相关要求。整车模型中除车身与转向系统外，其他子系统与实际模型相差较大，应尽量重新建立新模型，避免在共享数据库中修改模型。同时，对于四轮转向及多轮转向，也可以在基础的转向系统模板中进行拓展。

学习目标

（1）熟悉齿轮齿条式转向器。
（2）会转向器变量参数设定。
（3）会通信器匹配。
（4）会转向仿真分析。

1.1 齿轮齿条式转向系统

齿轮齿条式转向器属于可逆式转向器，其正效率与逆效率都很高，自动回正能力强。齿轮齿条式转向器结构简单、加工方便、工作可靠、使用寿命长，不需要调整齿轮齿条的间隙，因而得到了广泛的应用。与其他形式转向器相比，齿轮齿条式转向器结构简单、紧凑；壳体多采用铝合金或镁合金压铸而成，转向器质量比较小；采用齿轮齿条传动方式，传动效率较高；齿轮齿条之间因磨损产生间隙后，利用装在齿条背部、靠近主动小齿轮处的压紧力可以调节弹簧，从而自动消除齿间间隙，这不仅可以提高转向系统的刚度，还可以防止工作时产生冲击和噪声；转向器占用体积较小；没有转向摇臂和直拉杆，所以转向轮转角可以加大，制造成本较低。但其逆效率较高，汽车在不平路面上行驶时，发生在转向轮与路面之间的冲击力大部分能传至转向盘，造成驾驶员精神紧张，使驾驶员难以准确控制汽车行驶方向，同时转向盘突然转动又会造成打手，对驾驶员造成伤害。

模板设置步骤如下：
（1）启动 ADAMS/CAR，选择专家模块进入建模界面。
（2）单击 File > New 命令，弹出建模对话框，如图 1-2 所示。
（3）在模板名称中输入 my_steering，主特征选择 steering，单击 OK 按钮。

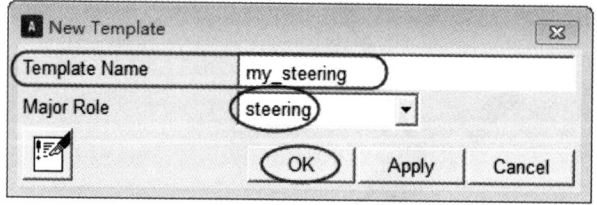

图 1-2 模板对话框

1.1.1 转向器硬点

转向器硬点设置步骤如下：
（1）单击 Build > Hardpoind > New 命令，弹出创建硬点对话框，如图 1-3 所示。
（2）在 Hardpoint Name（硬点名称）中输入 tierod_inner；类型选择 left；在位置文本框中输入 200，-425，300。

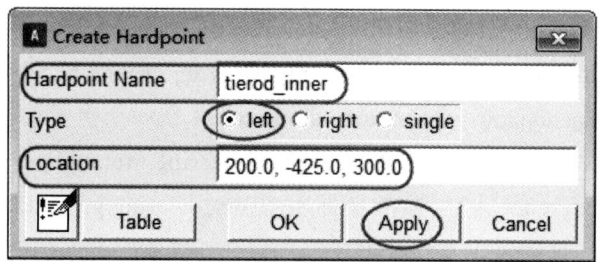

图 1-3 硬点创建对话框

（3）单击 Apply 按钮，完成 tierod_inner 硬点的创建。此时，屏幕上显示出左右对称的两个硬点。

（4）重复上述步骤完成图 1-4 中硬点的创建。

	loc x	loc y	loc z
hpl_rack_house_mount	200.0	-200.0	300.0
hpl_tierod_inner	200.0	-425.0	300.0
hps_intermediate_shaft_forward	400.0	-300.0	500.0
hps_intermediate_shaft_rearwar	550.0	-300.0	600.0
hps_pinion_pivot	200.0	-300.0	300.0
hps_steering_wheel_center	900.0	-300.0	700.0

图 1-4 转向器硬点数据

1.1.2 转向器结构框

（1）单击 Build > Construction Frame > New 命令，弹出创建结构框对话框，如图 1-5 所示。在下列对话框中输入相应的数据：

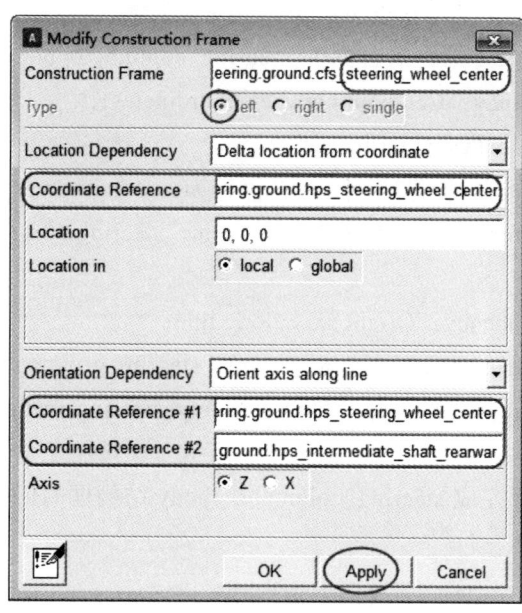

图 1-5 转向器硬点数据

- Construction Frame（结构框名称）：steering_wheel_center；
- Coordinate Reference（参考坐标）：._my_steering.ground.hps_steering_wheel_center；
- Orientation Dependency：Orient axis along line（在一条坐标轴 2 点连线）；
- Coordinate Reference #1（参考坐标）：._my_steering.ground.hps_steering_wheel_center；
- Coordinate Reference #2（参考坐标）：._my_steering.ground.hps_intermediate_shaft_rearwar。

（2）单击 Apply 按钮，完成 steering_wheel_center 结构框的创建。继续创建其他结构框：
- Construction Frame（结构框名称）：pinion_pivot；
- Coordinate Reference（参考坐标）：._my_steering.ground.hps_pinion_pivot；
- Orientation Dependency：Orient axis along line；
- Coordinate Reference #1（参考坐标）：._my_steering.ground.hps_pinion_pivot；
- Coordinate Reference #2（参考坐标）：._my_steering.ground.hps_intermediate_shaft_forward。

（3）单击 Apply 按钮，完成 pinion_pivot 结构框的创建。
- Construction Frame（结构框名称）：rack_mount；
- Location Dependency：Centered between coordinates；
- Centered between：Two Coordinates；
- Coordinate Reference #1（参考坐标）：._my_steering.ground.hpl_tierod_inner；
- Coordinate Reference #2（参考坐标）：._my_steering.ground.hpr_tierod_inner；
- Orientation Dependency：Orient axis along line（在一条坐标轴 2 点连线）；
- Coordinate Reference #1（参考坐标）：._my_steering.ground.hpl_tierod_inner；
- Coordinate Reference #2（参考坐标）：._my_steering.ground.hpr_tierod_inner。

（4）单击 Apply 按钮，完成 rack_mount 结构框的创建。
- Construction Frame（结构框名称）：steering_column_to_body；
- Location Dependency：Centered between coordinates；
- Centered between：Two Coordinates；
- Coordinate Reference #1（参考坐标）：._my_steering.ground.cfs_steering_wheel_center；
- Coordinate Reference #2（参考坐标）：._my_steering.ground.hps_intermediate_shaft_rearwar；
- Orientation Dependency：Orient axis along line；
- Coordinate Reference #1（参考坐标）：._my_steering.ground.cfs_steering_wheel_center；
- Coordinate Reference #2（参考坐标）：._my_steering.ground.hps_intermediate_shaft_rearwar。

（5）单击 OK 按钮，完成 steering_column_to_body 结构框的创建。创建完成的硬点与结构框图如图 1-6 所示。

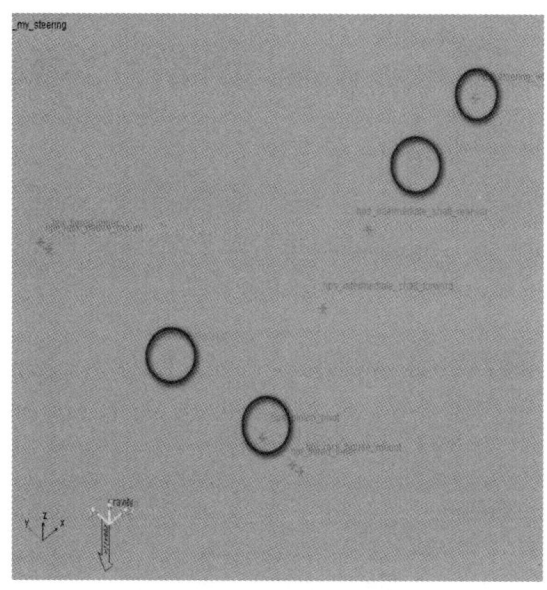

图 1-6　硬点与结构框图

1.1.3　转向轴部件

1. 转向轴部件

（1）单击 Build > Part > General Part > New 命令，弹出创建转向轴部件对话框，如图 1-7 所示。在下列对话框中输入相应的数据：

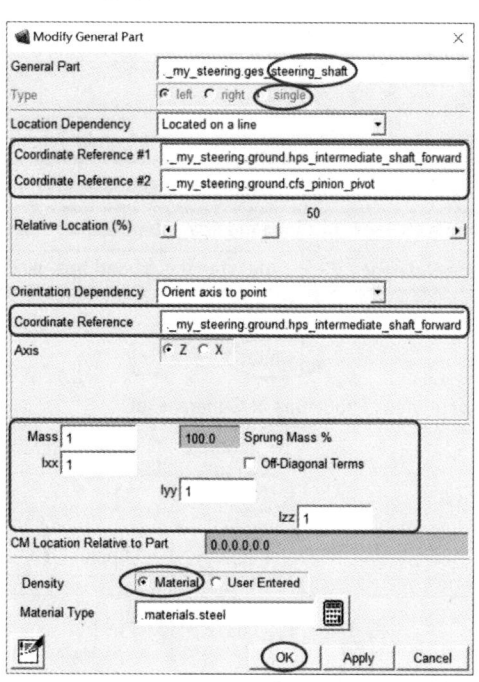

图 1-7　转向轴部件对话框

- General Part：steering_shaft；
- Location Dependency（定位）：Located on a line（在一条直线上）；
- Coordinate Reference #1（参考坐标）：._my_steering.ground.hps_intermediate_shaft_forward；
- Coordinate Reference #2（参考坐标）：._my_steering.ground.cfs_pinion_pivot；
- Relative Location（%）：50（部件参考点位于指定两点的连线上，相对位置百分比指相对于第一个参考点的位置，0%指的是第一个参考点，100%指的是第二个参考点，150%则位于第二点之外）；
- Orientation Dependency：Orient axis to point；
- Coordinate Reference（参考坐标）：._my_steering.ground.hps_intermediate_shaft_forward；
- Axis：Z；
- Mass：1；
- Ixx：1；
- Iyy：1；
- Izz：1；
- Density：Material；
- Material Type：.materials.steel。

（2）单击 OK 按钮，完成部件._my_steering.ges_steering_shaft 的创建。

2. 转向轴几何体

（1）单击 Build > Geometry > Link > New 命令，弹出转向轴几何体创建对话框，如图 1-8 所示。在下列对话框中输入相应的数据：

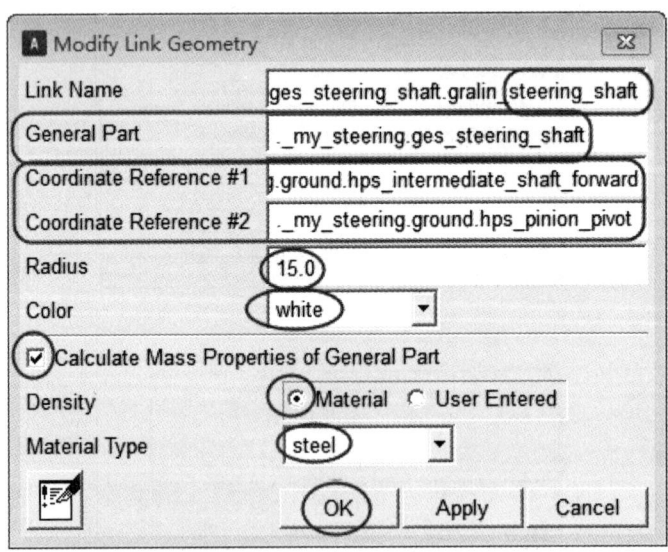

图 1-8 转向轴几何体对话框

- Link Name（连杆名称）：steering_shaft；
- General Part：._my_steering.ges_steering_shaft；

- Coordinate Reference #1（参考坐标）：._my_steering.ground.hps_intermediate_shaft_forward；
- Coordinate Reference #2（参考坐标）：._my_steering.ground.hps_pinion_pivot；
- Radius（半径）：15；
- Color：white；
- Density：Material；
- Material Type：Steel。

（2）选择 Calculate Mass Properties of General Part 复选框。

（3）单击 OK 按钮，完成 steering_shaft 几何体的创建。

1.1.4 齿条部件

1. 齿条部件

（1）单击 Build > Part > General Part > New 命令，创建齿条部件，参见图 1-7。在下列对话中输入相应的数据：

- General Part 输入：rack；
- Location Dependency（定位）：Located on a line（在一条直线上）；
- Coordinate Reference #1（参考坐标）：._my_steering.ground.hpl_tierod_inner；
- Coordinate Reference #2（参考坐标）：._my_steering.ground.hpr_tierod_inner；
- Relative Location（%）：50；
- Orientation Dependency：Orient axis to point；
- Coordinate Reference（参考坐标）：._my_steering.ground.hpl_tierod_inner；
- Axis：Z；
- Mass：1；
- Ixx：1；
- Iyy：1；
- Izz：1；
- Density：Material；
- Material Type：.materials.steel。

（2）单击 OK 按钮，完成部件._my_steering.ges_rack 的创建。

2. 齿条几何体

（1）单击 Build > Geometry > Link > New 命令，创建齿条几何体，参见图 1-8。在下列对话框中输入相应的数据：

- Link Name（连杆名称）：rack；
- General Part：._my_steering.ges_rack；
- Coordinate Reference #1（参考坐标）：._my_steering.ground.hpl_tierod_inner；
- Coordinate Reference #2（参考坐标）：._my_steering.ground.hpr_tierod_inner；

- Radius（半径）：10；
- Color：white；
- Density：Material；
- Material Type：Steel。

（2）选择 Calculate Mass Properties of General Part 复选框。

（3）单击 OK 按钮，完成._my_steering.ges_rack.gralin_rack 几何体的创建。

1.1.5 齿条箱部件

1. 齿条箱部件

（1）单击 Build > Part > General Part > New 命令，创建齿条箱部件，参见图 1-7。在下列对话中输入相应的数据：

- General Part：rack_housing；
- Location Dependency：Centered between coordinates；
- Centered between：Two Coordinates；
- Coordinate Reference #1（参考坐标）：._my_steering.ground.hpl_rack_house_mount；
- Coordinate Reference #2（参考坐标）：._my_steering.ground.hpr_rack_house_mount；
- Orientation Dependency：Orient axis to point；
- Coordinate Reference（参考坐标）：._my_steering.ground.hpl_rack_house_mount；
- Mass：1；
- Ixx：1；
- Iyy：1；
- Izz：1；
- Density：Material；
- Material Type：.materials.steel。

（2）单击 OK 按钮，完成部件._my_steering.ges_rack_housing 的创建。

2. 齿条箱几何体

（1）单击 Build > Geometry > Link > New 命令，创建齿条箱几何体，参见图 1-8。在下列对话中输入相应的数据：

- Link Name（连杆名称）：rack_housing；
- General Part：._my_steering.ges_rack_housing；
- Coordinate Reference #1（参考坐标）：._my_steering.ground.hpl_rack_house_mount；
- Coordinate Reference #2（参考坐标）：._my_steering.ground.hpr_rack_house_mount；
- Radius（半径）：25；
- Color：white；
- Density：Material；
- Material Type：Steel。

（2）选择 Calculate Mass Properties of General Part 复选框。

（3）单击 OK 按钮，完成._my_steering.ges_rack_housing.gralin_rack_housing 几何体的创建。

1.1.6 中间轴部件

1. 中间轴部件

(1) 单击 Build > Part > General Part > New 命令,创建转向中间轴部件,参见图 1-7。在下列对话框中输入相应的数据:
- General Part:intermediate_shaft;
- Location Dependency:Centered between coordinates;
- Centered between:Two Coordinates;
- Coordinate Reference #1(参考坐标):._my_steering.ground.hps_intermediate_shaft_rearwar;
- Coordinate Reference #2(参考坐标):._my_steering.ground.hps_intermediate_shaft_forward;
- Orientation Dependency:Orient axis to point;
- Coordinate Reference(参考坐标):._my_steering.ground.hps_intermediate_shaft_rearwar;
- Axis:Z;
- Mass:1;
- Ixx:1;
- Iyy:1;
- Izz:1;
- Density:Material;
- Material Type:.materials.steel。

(2) 单击 OK 按钮,完成部件._my_steering.ges_intermediate_shaft 的创建。

2. 中间轴几何体

(1) 单击 Build > Geometry > Link > New 命令,创建转向中间轴几何体,参见图 1-8。在下列对话框中输入相应的数据:
- Link Name(连杆名称):intermediate_shaft;
- General Part:._my_steering.ges_intermediate_shaft;
- Coordinate Reference #1(参考坐标):._my_steering.ground.hps_intermediate_shaft_rearwar;
- Coordinate Reference #2(参考坐标):._my_steering.ground.hps_intermediate_shaft_forward;
- Radius(半径):15;
- Color:white;
- Density:Material;
- Material Type:Steel。

(2) 选择 Calculate Mass Properties of General Part 复选框。

(3) 单击 OK 按钮,完成._my_steering.ges_intermediate_shaft.gralin_intermediate_shaft 几何体的创建。

1.1.7 转向柱部件

1. 转向柱部件

(1) 单击 Build > Part > General Part > New 命令，创建转向柱部件，参见图 1-7。在下列对话框中输入相应的数据：

- General Part：steering_column；
- Location Dependency：Centered between coordinates；
- Centered between：Two Coordinates；
- Coordinate Reference #1（参考坐标）：._my_steering.ground.hps_steering_wheel_center；
- Coordinate Reference #2（参考坐标）：._my_steering.ground.hps_intermediate_shaft_rearwar；
- Orientation Dependency：Orient axis to point；
- Coordinate Reference（参考坐标）：._my_steering.ground.hps_intermediate_shaft_rearwar；
- Axis：Z；
- Mass：1；
- Ixx：1；
- Iyy：1；
- Izz：1；
- Density：Material；
- Material Type：.materials.steel。

(2) 单击 OK 按钮，完成部件._my_steering.ges_steering_column_to_body 的创建。

2. 转向柱几何体

(1) 单击 Build > Geometry > Link > New 命令，创建转向柱几何体，参见图 1-8。在下列对话框中输入相应的数据：

- Link Name（连杆名称）：steering_column；
- General Part：._my_steering.ges_steering_column；
- Coordinate Reference #1（参考坐标）：._my_steering.ground.hps_intermediate_shaft_rearwar；
- Coordinate Reference #2（参考坐标）：._my_steering.ground.hps_steering_wheel_center；
- Radius（半径）：15；
- Color：White；
- Density：Material；
- Material Type：Steel。

(2) 选择 Calculate Mass Properties of General Part 复选框。

(3) 单击 OK 按钮，完成._my_steering.ges_steering_column.gralin_steering_column 几何体的创建。

1.1.8 转向柱支撑部件

1. 转向柱支撑部件

（1）单击 Build > Part > General Part > New 命令，创建转向柱支撑部件，参见图 1-7。在下列对话框中输入相应的数据。

- General Part：steering_column_to_body；
- Location Dependency：Centered between coordinates；
- Centered between：Two Coordinates；
- Coordinate Reference #1（参考坐标）：._my_steering.ground.hps_steering_wheel_center；
- Coordinate Reference #2（参考坐标）：._my_steering.ground.hps_intermediate_shaft_rearwar；
- Orientation Dependency：Orient axis to point；
- Coordinate Reference（参考坐标）：._my_steering.ground.cfs_steering_column_to_body；
- Axis：Z；
- Mass：1；
- Ixx：1；
- Iyy：1；
- Izz：1；
- Density：Material；
- Material Type：.materials.steel。

（2）单击 OK 按钮，完成部件._my_steering.ges_steering_column_to_body 的创建。

2. 转向柱支撑几何体

（1）单击 Build > Geometry > Cylinder（圆柱体）> New 命令，创建转向柱支撑几何体，如图 1-9 所示。在下列对话中输入相应的数据：

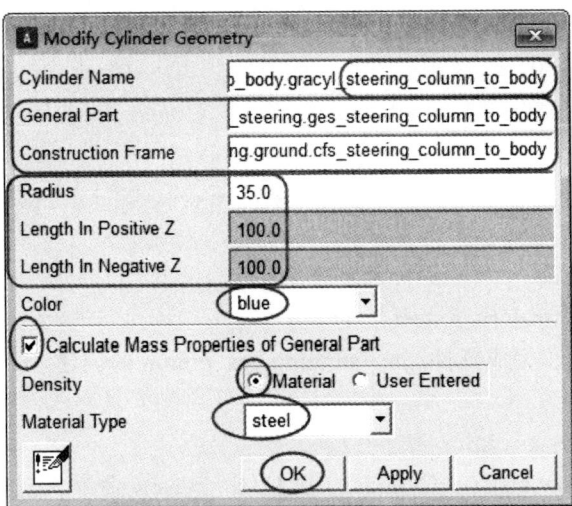

图 1-9 转向柱支撑几何体对话框

- Cylinder Name（连杆名称）：steering_column_to_body；
- General Part：._my_steering.ges_steering_column_to_body；
- Construction Frame（结构框）：._my_steering.ground.cfs_steering_column_to_body；
- Radius（半径）：35；
- Length In Positive Z（Z轴正方向长度）：100；
- Length In Negative Z（Z轴负方向长度）：100；
- Color：blue；
- Density：Material；
- Material Type：Steel。

（2）选择 Calculate Mass Properties of General Part 复选框。

（3）单击 OK 按钮，完成._my_steering.ges_steering_column_to_body.gracyl_steering_column_to_body 几何体的创建。

1.1.9 小齿轮部件

1. 小齿轮部件

（1）单击 Build > Part > General Part > New 命令，创建小齿轮部件，参见图 1-7。在下列对话框中输入相应的数据：

- General Part：pinion；
- Location Dependency（定位）：Delta location from coordinate；
- Coordinate Reference（参考坐标）：._my_steering.ground.hps_pinion_pivot；
- Location：0，0，0；
- Location in：local；
- Orientation Dependency：Delta orientation from coordinate；
- Construction Frame（结构框）：._my_steering.ground.cfs_pinion_pivot；
- Orientation：0，0，0；
- Mass：1；
- Ixx：1；
- Iyy：1；
- Izz：1；
- Density：Material；
- Material Type：.materials.steel。

（2）单击 OK 按钮，完成部件._my_steering.ges_pinion 的创建。

2. 小齿轮几何体

（1）单击 Build > Geometry > Cylinder（圆柱体）> New 命令，创建小齿轮几何体，参见图 1-9。在下列对话框中输入相应的数据：

- Cylinder Name（连杆名称）：pinion；
- General Part：._my_steering.ges_pinion；
- Construction Frame（结构框）：._my_steering.ground.cfs_pinion_pivot；
- Radius（半径）：25；
- Length In Positive Z（Z 轴正方向长度）：20；
- Length In Negative Z（Z 轴负方向长度）：20；
- Color：blue；
- Density：Material；
- Material Type：Steel。

（2）勾选 Calculate Mass Properties of General Part 复选框。

（3）单击 OK 按钮，完成 ._my_steering.ges_pinion.gracyl_pinion 几何体的创建。

1.1.10 方向盘部件

1. 方向盘部件

（1）单击 Build > Part > General Part > New 命令，创建方向盘部件，参见图 1-7。在下列对话框中输入相应的数据：

- General Part：steering_wheel；
- Location Dependency（定位）：Delta location from coordinate；
- Coordinate Reference（参考坐标）：._my_steering.ground.hps_steering_wheel_center；
- Location：0，0，0；
- Location in：local；
- Orientation Dependency：Delta orientation from coordinate；
- Construction Frame（结构框）：._my_steering.ground.cfs_steering_column_to_body；
- Orientation：0，0，0；
- Mass：1；
- Ixx：1；
- Iyy：1；
- Izz：1；
- Density：Material；
- Material Type：.materials.steel。

（2）单击 OK 按钮，完成部件 ._my_steering.ges_steering_wheel 的创建。

2. 方向盘几何体

（1）单击 Build > Geometry > Cylinder（圆柱体）> New 命令，创建方向盘几何体，参见图 1-9。在下列对话框中输入相应的数据：

- Cylinder Name（连杆名称）：steering_wheel；
- General Part：._my_steering.ges_steering_wheel；

- Construction Frame（结构框）：._my_steering.ground.cfs_steering_wheel_center；
- Radius（半径）：120；
- Length In Positive Z（Z 轴正方向长度）：6.35；
- Length In Negative Z（Z 轴负方向长度）：6.35；
- Color：blue；
- Density：Material；
- Material Type：Steel。

（2）勾选 Calculate Mass Properties of General Part 复选框。

（3）单击 OK 按钮，完成._my_steering.ges_steering_wheel.gracyl_steering_wheel 几何体的创建。

1.1.11　安装部件

1. 齿条与车体之间的安装部件 rack_to_body

（1）单击 Build > Part > Mount > New 命令，弹出创建安装部件对话框，如图 1-10 所示。在下列对话框中输入相应的数据：

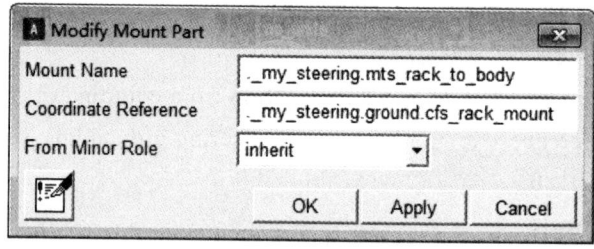

图 1-10　安装部件对话框

- Mount name（安装件名称）：rack_to_body；
- Coordinate Reference（参考坐标）：._my_steering.ground.cfs_rack_mount；
- From Minor Role（安装件特征选择）：inherit（继承特性）。

（2）单击 Apply 按钮，完成 rack_to_body 安装部件的创建。

2. 齿条箱与悬架副车架之间的安装部件 rack_housing_to_suspension_subframe

（1）在下列对话框中输入相应的数据：

- Mount name（安装件名称）：rack_housing_to_suspension_subframe；
- Coordinate Reference（参考坐标）：._my_steering.ground.cfs_rack_mount；
- From Minor Role（安装件特征选择）：inherit。

（2）单击 Apply 按钮，完成 rack_housing_to_suspension_subframe 安装部件的创建。

3. 转向柱与车体之间的安装部件 steering_column_to_body

（1）在下列对话框中输入相应的数据：

- Mount name（安装件名称）：steering_column_to_body；

- Coordinate Reference（参考坐标）：._my_steering.ground.hps_intermediate_shaft_rearwar；
- From Minor Role（安装件此特征选择）：inherit。

（2）单击 OK 按钮，完成 steering_column_to_body 安装部件的创建。

在创立安装件时，系统会自动创建同名的输入通信器，创建的通信器如下：

① ._my_steering.cis_steering_column_to_body；
② ._my_steering.cis_rack_housing_to_suspension_subframe；
③ ._my_steering.cis_rack_to_body。

1.1.12 Joint 连接

单击 Build > Attachments > Joint > New 命令，弹出创建约束件对话框，如图 1-11 所示。

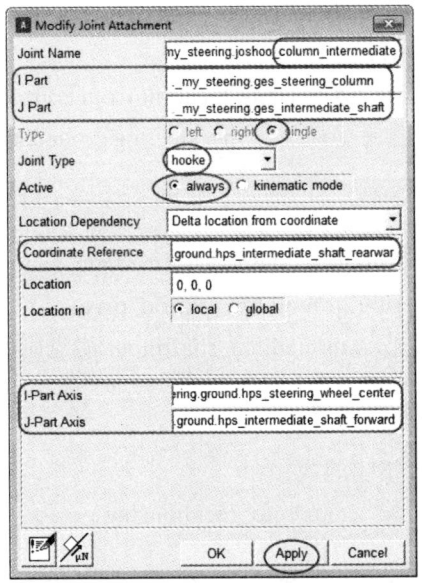

图 1-11 胡克副连接对话框

1. 转向柱与中间轴之间的胡克副

在下列对话框中输入相应的数据：

- Joint Name（约束副名称）：column_intermediate；
- I Part：._my_steering.ges_steering_column；
- J Part：._my_steering.ges_intermediate_shaft；
- Type：single；
- Joint Type（约束副类型）：hooke（胡克副，约束 3 个自由度）；
- Active（激活）：always；
- Location Dependency（定位）：Delta location from coordinate（坐标位置）；
- Coordinate Reference（参考坐标）：._my_steering.ground.hps_intermediate_shaft_ rearwar；

- Location：0，0，0；
- Location in：local；
- I-Part Axis：._my_steering.ground.hps_steering_wheel_center；
- J-Part Axis：._my_steering.ground.hps_intermediate_shaft_forward；

（2）单击 Apply 按钮，完成 ._my_steering.joshoo_column_intermediate 胡克副的创建。

2. 转向轴与中间轴之间的胡克副

（1）在下列对话框中输入相应的数据：

- Joint Name（约束副名称）：intermediate_shaftinput；
- I Part：._my_steering.ges_intermediate_shaft；
- J Part：._my_steering.ges_steering_shaft；
- Type：single；
- Joint Type（约束副类型）：hooke；
- Active（激活）：always；
- Location Dependency（定位）：Delta location from coordinate（坐标位置）；
- Coordinate Reference（参考坐标）：._my_steering.ground.hps_intermediate_shaft_ forward；
- Location：0，0，0；
- Location in：local；
- I-Part Axis：._my_steering.ground.hps_intermediate_shaft_rearwar；
- J-Part Axis：._my_steering.ground.hps_pinion_pivot。

（2）单击 Apply 按钮，完成 intermediate_shaftinput 胡克副的创建。

3. 齿条与齿条箱之间的移动副

（1）在下列对话框中输入相应的数据：

- Joint Name（约束副名称）：rack_to_rackhousing；
- I Part：._my_steering.ges_rack；
- J Part：._my_steering.ges_rack_housing；
- Type：single；
- Joint Type（约束副类型）：translational（移动副，约束 5 个自由度）；
- Active（激活）：always；
- Location Dependency（定位）：Delta location from coordinate（坐标位置）；
- Coordinate Reference（参考坐标）：._my_steering.ground.cfs_rack_mount。
- Location：0，0，0；
- Location in：local；
- Orientation Dependency：Orient axis along line；
- Coordinate Reference #1（参考坐标）：._my_steering.ground.hpl_tierod_inner；
- Coordinate Reference #2（参考坐标）：._my_steering.ground.hpr_tierod_inner。

（2）单击 Apply 按钮，完成 rack_to_rackhousing 移动副的创建。

4. 方向盘与转向柱之间的固定副

（1）在下列对话框中输入相应的数据：
- Joint Name（约束副名称）：steering_wheel；
- I Part：._my_steering.ges_steering_wheel；
- J Part：._my_steering.ges_steering_column；
- Type：single；
- Joint Type（约束副类型）：fix（固定副，约束 6 个自由度）；
- Active（激活）：always；
- Location Dependency（定位）：Delta location from coordinate（坐标位置）；
- Coordinate Reference（参考坐标）：._my_steering.ground.hps_steering_wheel_center；
- Location：0，0，0；
- Location in：local。

（2）单击 Apply 按钮，完成 steering_wheel 固定副的创建。

5. 转向柱支撑体与转向柱安装件之间的固定副

（1）在下列对话框中输入相应的数据：
- Joint Name（约束副名称）：steering_column_to_body；
- I Part：._my_steering.ges_steering_column_to_body；
- J Part：._my_steering.mts_steering_column_to_body；
- Type：single；
- Joint Type（约束副类型）：fix（固定副，约束 6 个自由度）；
- Active（激活）：always；
- Location Dependency（定位）：Delta location from coordinate（坐标位置）；
- Coordinate Reference（参考坐标）：._my_steering.ground.cfs_steering_column_to_body；
- Location：0，0，0；
- Location in：local。

（2）单击 Apply 按钮，完成 steering_column_to_body 固定副的创建。

6. 小齿轮与齿条箱之间的转动副

（1）在下列对话框中输入相应的数据：
- Joint Name（约束副名称）：pinion；
- I Part：._my_steering.ges_pinion；
- J Part：._my_steering.ges_rack_housing；
- Type：single；
- Joint Type（约束副类型）：revolute（转动副，约束 5 个自由度）；
- Active（激活）：always；
- Location Dependency（定位）：Delta location from coordinate（坐标位置）；
- Coordinate Reference（参考坐标）：._my_steering.ground.hps_pinion_pivot；

- Location：0，0，0；
- Location in：local。
- Orientation Dependency：Orient axis along line；
- Coordinate Reference #1（参考坐标）：._my_steering.ground.hps_intermediate_shaft_forward；
- Coordinate Reference #2（参考坐标）：._my_steering.ground.hps_pinion_pivot；
- Axis：Z。

（2）单击 Apply 按钮，完成 pinion 移动副的创建。

7. 齿条与齿条箱之间的移动副

（1）在下列对话框中输入相应的数据：
- Joint Name（约束副名称）：rack_to_rackhousing；
- I Part：._my_steering.ges_rack；
- J Part：._my_steering.ges_rack_housing；
- Type：single；
- Joint Type（约束副类型）：translational（移动副，约束5个自由度）；
- Active（激活）：always；
- Location Dependency（定位）：Delta location from coordinate（坐标位置）；
- Coordinate Reference（参考坐标）：._my_steering.ground.cfs_rack_mount；
- Location：0，0，0；
- Location in：local；
- Orientation Dependency：Orient axis along line；
- Coordinate Reference #1（参考坐标）：._my_steering.ground.hpl_tierod_inner；
- Coordinate Reference #2（参考坐标）：._my_steering.ground.hpr_tierod_inner；
- Axis：Z。

（2）单击 Apply 按钮，完成 rack_to_rackhousing 移动副的创建。

8. 转向输入轴部件与小齿轮部件之间的固定副

（1）在下列对话框中输入相应的数据：
- Joint Name（约束副名称）：steering_input_shaft；
- I Part：._my_steering.ges_steering_shaft；
- J Part：._my_steering.ges_pinion；
- Type：single；
- Joint Type（约束副类型）：fix（固定副，约束6个自由度）；
- Active（激活）：always；
- Location Dependency（定位）：Delta location from coordinate（坐标位置）；
- Coordinate Reference（参考坐标）：._my_steering.ground.hps_pinion_pivot；
- Location：0，0，0；
- Location in：local。

（2）单击 Apply 按钮，完成 steering_input_shaft 固定副的创建。

9. 齿条箱部件与齿条箱安装件之间的固定副

（1）在下列对话框中输入相应的数据：
- Joint Name（约束副名称）：rigid_rack_housing_mount；
- I Part：._my_steering.ges_rack_housing；
- J Part：._my_steering.sws_rack_house_mount；
- Type：single；
- Joint Type（约束副类型）：fix；
- Active（激活）：always；
- Location Dependency（定位）：Delta location from coordinate（坐标位置）；
- Coordinate Reference（参考坐标）：._my_steering.ground.cfs_rack_mount；
- Location：0，0，0；
- Location in：local。

（2）单击 Apply 按钮，完成 rigid_rack_housing_mount 固定副的创建。

10. 转向柱与转向柱支撑体圆柱副

（1）在下列对话框中输入相应的数据：
- Joint Name（约束副名称）：steering_column_to_body_1；
- I Part：._my_steering.ges_steering_column；
- J Part：._my_steering.ges_steering_column_to_body；
- Type：single；
- Joint Type（约束副类型）：cylindrical（圆柱副，约束 4 个自由度）；
- Active（激活）：always；
- Location Dependency（定位）：Delta location from coordinate（坐标位置）；
- Coordinate Reference（参考坐标）：._my_steering.ground.cfs_steering_column_to_body；
- Location：0，0，0；
- Location in：local；
- Orientation Dependency：Delta location from coordinate；
- Construction Frame（结构框）：._my_steering.ground.cfs_steering_column_to_body；
- Orientation（方向）：90，180，0（按 313 原则旋转，此处一定要填写正确，否则会导致方向向左转，车轮向右转，方向盘和车轮的转向不同向）。

（2）单击 OK 按钮，完成 steering_column_to_body_1 圆柱副的创建。至此，转向系统中的所有约束副创建完成。

11. 减速齿轮（耦合副）

（1）单击 Build > Gear > Reduction Gear > New 命令，创建减速齿轮对话框，如图 1-12 所示；减速齿轮本质上是一对耦合副，需要指定输入输出约束及传动比。在下列对话框中输入相应的数据：

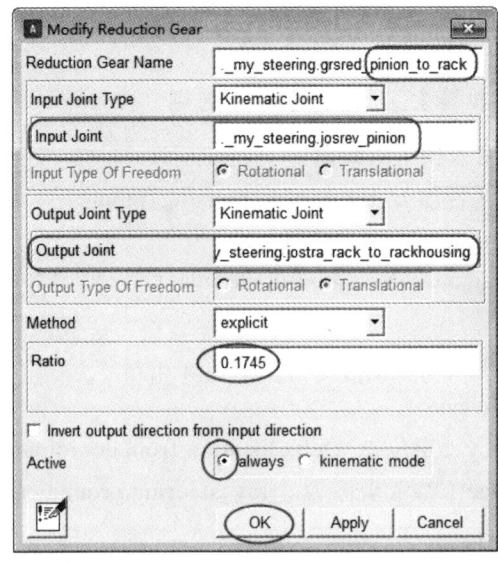

图 1-12 减速齿轮对话框

- Reduction Gear Name（减速器齿轮名称）：pinion_to_rack；
- Input Joint（输入约束名称）：._my_steering.josrev_pinion；
- Output Joint（输出约束名称）：._my_steering.jostra_rack_to_rackhousing；
- Reduction Ratio（减速比）0.1745；
- Active（激活）：always。

（2）单击 OK 按钮，完成 pinion_to_rack 减速齿轮的创建。

1.1.13 bushing 连接

轴套定义 6 个自由度状态连接（包括 X、Y、Z 3 个方向的平移与旋转）。轴套通过属性文件定义其 3 个方向的线性刚度和 3 个方向旋转的扭转刚度，刚度可以是线性数据，也可以是非线性数据。需要强调的是，轴套属性文件中数据曲线参考的是局部坐标系，而非全局坐标系。

单击 Build > Attachments > Bushing > New 命令，弹出创建衬套对话框，如图 1-13 所示。

1. 齿条箱与齿条箱安装件之间轴套

（1）在下列对话框中输入相应的数据：
- Bushing Name（约束副名称）：rack_housing_bushing；
- I Part：._my_steering.ges_rack_housing；
- J Part：._my_steering.sws_rack_house_mount；
- Inactive（抑制）：kinematic mode（运动学模式）；
- Geometry Length（几何长度）：20；
- Geometry Radius（几何半径）：40；

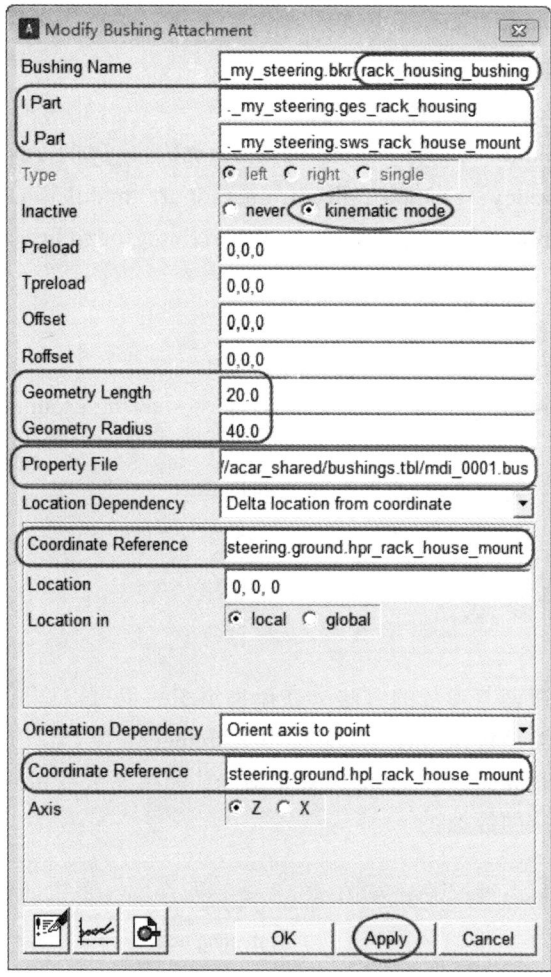

图 1-13 轴套创建对话框

- Property File：mdids：//acar_shared/bushings.tbl/mdi_0001.bus；
- Location Dependency：Delta location from coordinate；
- Coordinate Reference（参考坐标）：._my_steering.ground.hpl_rack_house_mount；
- Location：0，0，0；
- Location in：local；
- Orientation Dependency：Orient axis to point；
- Coordinate Reference（参考坐标）：._my_steering.ground.hpr_rack_house_mount。

（2）单击 Apply 按钮，完成 rack_housing_bushing 轴套的创建。

2. 转向轴与小齿轮之间轴套

（1）在下列对话框中输入相应的数据：
- Bushing Name（约束副名称）：torsion_bar；
- I Part：._my_steering.ges_steering_shaft；
- J Part：._my_steering.ges_pinion；

- Inactive（抑制）：never；
- Geometry Length（几何长度）：20；
- Geometry Radius（几何半径）：30；
- Property File：mdids：//acar_shared/bushings.tbl/mdi_0001.bus；
- Location Dependency：Delta location from coor arb_middledinate；
- Coordinate Reference（参考坐标）：._my_steering.ground.hps_pinion_pivot；
- Location：0，0，0；
- Location in：local；
- Orientation Dependency：Orient axis to point；
- Coordinate Reference（参考坐标）：._my_steering.ground.hps_intermediate_shaft_forward。

（2）单击 OK 按钮，完成 torsion_bar 轴套的创建。

1.2 转向变量参数

参数变量提供了快速调节参数的方法，在模型组织后仍可以进行修改，主要用于系统的优化设计。参数变量的类型包括 String（字符串）、Integer（整数）、Real（实数）三种类型。

（1）单击 Build > Parameter Variable > New 命令，弹出参数变量对话框，如图 1-14 所示。在下列对话框中输入相应的数据：

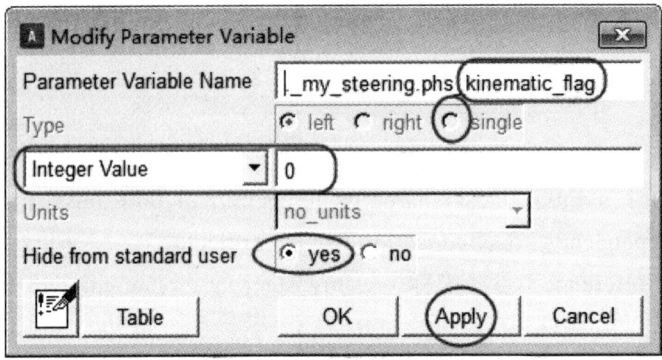

图 1-14 参数变量对话框

- Parameter Variable Name（参数名称）：kinematic_flag；
- 参数类型：Integer Value（整数），数值为 0；
- Hide from standard user（是否从标准界面隐藏）：yes。

（2）单击 Apply 按钮，完成变量 kinematic_flag 的创建。

（3）按同样的方法创建其他变量。

- Parameter Variable Name（参数名称）：steering_assist_active；
- 参数类型：Integer Value（整数），数值为 0；
- Hide from standard user（是否从标准界面隐藏）：no。

（4）单击 Apply 按钮，完成变量 steering_assist_active 的创建；此变量参数不建立不影响转向系统的正确仿真，此变量为转向助力在 ADAMS 中的开关，在 ADAMS 软件中，转向助力涉及的是液压助力转向。现在轿车大多数为电子助力转向，转向助力的模型建议在 ADAMS 与 MATLAB 软件的联合仿真中模拟，在 MATLAB 软件中建立详细的电子助力参数及控制策略。联合仿真模型参考后续章节案例。

- Parameter Variable Name（参数名称）：max_steering_angle；
- 参数类型：Real Value（整数），数值为 540；
- Units（单位）：angle（度）；
- Hide from standard user（是否从标准界面隐藏）：no；
- 单击 Apply 按钮，完成变量 max_steering_angle 的创建。
- Parameter Variable Name（参数名称）：max_rack_displacement；
- 参数类型：Real Value（整数），数值为 100；
- Units（单位）：length（长度）；
- Hide from standard user（是否从标准界面隐藏）：no；
- 单击 Apply 按钮，完成变量 max_rack_displacement 的创建。
- Parameter Variable Name（参数名称）：max_rack_force；
- 参数类型：Real Value（整数），数值为 500；
- Units（单位）：force（力）；
- Hide from standard user（是否从标准界面隐藏）：no；
- 单击 Apply 按钮，完成变量 max_rack_force 的创建。
- Parameter Variable Name（参数名称）：max_steering_torque；
- 参数类型：Real Value（整数），数值为 500；
- Units（单位）：torque（力矩）；
- Hide from standard user（是否从标准界面隐藏）：no；
- 单击 OK 按钮，完成变量 max_steering_torque 的创建。

1.3 转向通信器

1. 输入输出通信器的建立

转向系统包含 3 个输入通信器：① ._my_steering.cis_steering_column_to_body；② ._my_steering.cis_rack_housing_to_suspension_subframe；③ ._my_steering.cis_rack_to_ body；在创建安装的同时，输入通信器同时创建完成。输出通信器的创建如下：

（1）单击 Build > Communicator > Output >New 命令，弹出输出通信器对话框，如图 1-15 所示。在下列对话框中输入相应的数据：

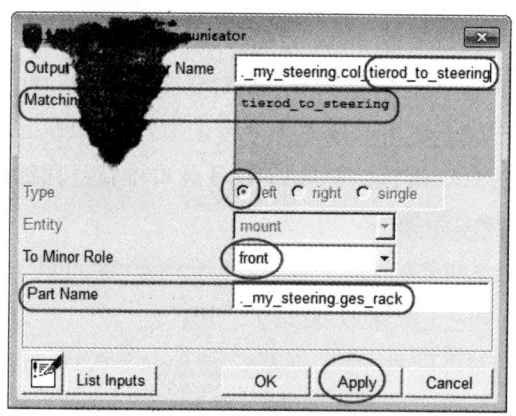

图 1-15 输出通信器对话框

- Output Communicator Name（输出通信器名称）：tierod_to_steering；
- Matching Name（s）：tierod_to_steering；
- Type：left；
- Entity：mount；
- To Minor Role：front；
- Part Name：._my_steering.ges_rack。

（2）单击 Apply 按钮，完成通信器 col_tierod_to_steering 的创建。

（3）按同样的方法创建其他输出通信器。

- Output Communicator Name（输出通信器名称）：steering_wheel_joint；
- Matching Name（s）：steering_wheel_joint；
- Type：single；
- Entity：joint for motion；
- To Minor Role：inherit；
- Joint Name：._my_steering.joscyl_steering_column_to_body_1；
- 单击 Apply 按钮，完成通信器 col_ steering_wheel_joint 的创建。
- Output Communicator Name（输出通信器名称）：steering_rack_joint；
- Matching Name（s）：steering_rack_joint；
- Type：single；
- Entity：joint for motion；
- To Minor Role：inherit；
- Joint Name：._my_steering.jostra_rack_to_rackhousing；
- 单击 Apply 按钮，完成通信器 col_steering_rack_joint 的创建。
- Output Communicator Name（输出通信器名称）：max_steering_angle；
- Matching Name（s）：max_steering_angle；
- Type：single；
- Entity：parameter real；
- To Minor Role：inherit；

- Parameter Variable Name：._my_steering.pvs_max_steering_angle；
- 单击 Apply 按钮，完成通信器 col_max_steering_angle 的创建。
- Output Communicator Name（输出通信器名称）：max_rack_displacement；
- Matching Name（s）：max_rack_displacement；
- Type：single；
- Entity：parameter real；
- To Minor Role：inherit；
- Parameter Variable Name：._my_steering.pvs_max_rack_displacement；
- 单击 Apply 按钮，完成通信器 col_ max_rack_displacement 的创建。
- Output Communicator Name（输出通信器名称）：max_rack_force；
- Matching Name（s）：max_rack_force；
- Type：single；
- Entity：parameter real；
- To Minor Role：inherit；
- Parameter Variable Name：._my_steering.pvs_max_rack_force；
- 单击 Apply 按钮，完成通信器 col_max_rack_force 的创建。
- Output Communicator Name（输出通信器名称）：max_steering_torque；
- Matching Name（s）：max_steering_torque；
- Type：single；
- Entity：parameter real；
- To Minor Role：inherit；
- Parameter Variable Name：._my_steering.pvs_max_steering_torque；
- 单击 OK 按钮，完成通信器 col_max_steering_torque 的创建。

至此齿轮齿条式转向系统的输入输出通信器建立完成，选项系统在装配过程中，主要与悬架系统及车身进行匹配，因此对其3个部件进行通信器测试，查看其通信器的匹配特性，保证通信器建立的正确性。

2. 通信器测试

（1）单击 Build > Communicator > Test 命令，弹出输出通信器测试对话框，如图1-16所示。在下列对话框中输入相应的数据：

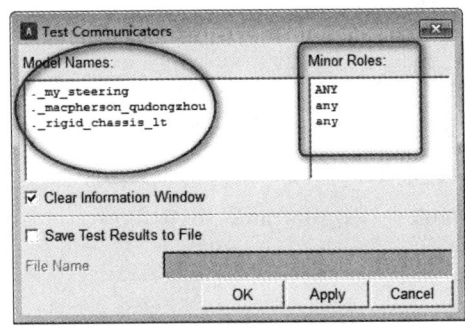

图1-16 通信器测试对话框

- Model Names：①._my_steering.；② _macpherson(共享数据模型)；③ _rigid_chassis_lt（共享数据库模型）；
- Minor Roles：any。

（2）单击 OK 按钮，完成麦弗逊悬架和悬架试验台：① ._my_steering.；② _macpherson（共享数据模型）；③ _rigid_chassis_lt（共享数据库模型）的匹配测试。

!---------------- Matched communicators：----------------!匹配的通信器

Communicator Matching Name：rack_to_body
Input Communicator Name：cis_rack_to_body
Located in：_my_steering
Output Communicator Name：cos_rack_to_body
Output from：_rigid_chassis_lt

Communicator Matching Name：rack_housing_to_suspension_subframe
Input Communicator Name：cis_rack_housing_to_suspension_subframe
Located in：_my_steering
Output Communicator Name：cos_rack_housing_to_suspension_subframe
Output from：_macpherson_qudongzhou

Communicator Matching Name：steering_column_to_body
Input Communicator Name：cis_steering_column_to_body
Located in：_my_steering
Output Communicator Name：cos_steering_column_to_body
Output from：_rigid_chassis_lt

Communicator Matching Name：strut_to_body
Input Communicator Name：ci[lr]_strut_to_body
Located in：_macpherson_qudongzhou
Output Communicator Name：co[lr]_strut_to_body
Output from：_rigid_chassis_lt

Communicator Matching Name：subframe_to_body
Input Communicator Name：cis_subframe_to_body
Located in：_macpherson_qudongzhou
Output Communicator Name：cos_subframe_to_body
Output from：_rigid_chassis_lt

Communicator Matching Name：tierod_to_steering

Input Communicator Name：ci[lr]_tierod_to_steering

Located in：_macpherson_qudongzhou

Output Communicator Name：co[lr]_tierod_to_steering，co[lr]_tierod_to_steering

Output from：_my_steering，_rigid_chassis_lt

!------------ Unmatched input communicators：--------------! 不匹配的输入通信器

Input Communicator Name：ci[lr]_tripot_to_differential

Class：mount

From Minor Role：any

Matching Name（s）：tripot_to_differential

In Template：_macpherson_qudongzhou

Input Communicator Name：cis_std_tire_ref

Class：location

From Minor Role：any

Matching Name（s）：std_tire_ref

In Template：_rigid_chassis_lt

!------------- Unmatched output communicators：------------!不匹配的输出通信器

Output Communicator Name：cos_steering_wheel_joint

Class：joint_for_motion

To Minor Role：any

Matching Name（s）：steering_wheel_joint

In Template：_my_steering

Output Communicator Name：cos_steering_rack_joint

Class：joint_for_motion

To Minor Role：any

Matching Name（s）：steering_rack_joint

In Template：_my_steering

不匹配的输出通信器省略。

通过通信器测试发现，不匹配的通信器大多在车身上，主要是因为车身是承载体，几乎所有的子系统都会与其有装配关系，因此车身结构体虽较为简单，但其承载的通信器较多，车身模板的建立主要集中在参数变量和通信器的建立上。

3．转换部件

转换部件可以实现多个部件的拓扑连接，同时它也是一种无质量的部件。在装配过程中，转换部件会自动寻找系统的存在部件，随着系统组装的完成，转接件会自动删除。

（1）单击 Build > Part > Switch > New 命令，弹出转向中间轴部件对话框，如图 1-17 所示。在下列对话框中输入相应的数据：

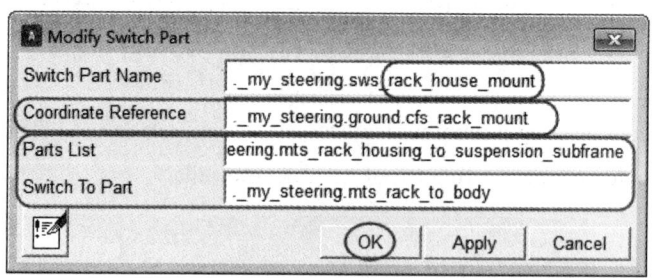

图 1-17 转换部件对话框

- Switch Part Name（转换部件名称）：rack_house_mount；
- Coordinate Reference（参考坐标）：._my_steering.ground.cfs_rack_mount；
- Parts List（部件清单，可以输入一系列可能连接的部件，最少为两个部件）：①._my_steering.mts_rack_to_body，② ._my_steering.mts_rack_housing_to_suspension_ subframe；
- Switch To Part（转接到部件，在部件清单中选择一个被转接件当前固定的部件）：._my_steering.mts_rack_to_body。

（2）单击 OK 按钮，完成转接件 sws_rack_house_mount 的创建。

（3）单击 File > Save As 命令，弹出保存模板对话框，如图 1-18 所示。在下列对话框中输出相应的参数：

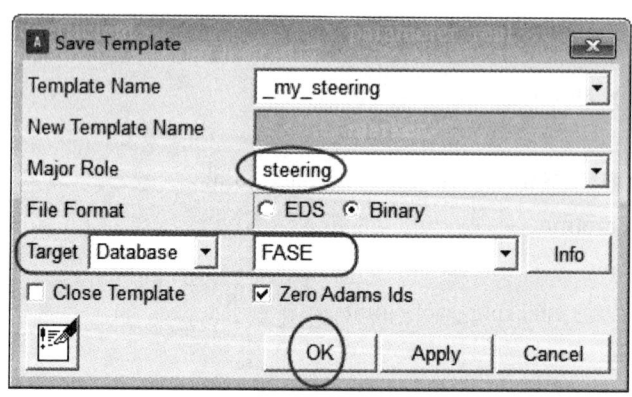

图 1-18 保存转向系统模型对话框

- Major Role（主特征）：steering；
- Database（保存数据库）：FASE。

（4）单击 OK 按钮，完成左舵麦弗逊悬架模板的保存，此转向系统在我国内地应用较多，方向盘在左边，靠右行驶。左舵转向模型如图 1-19 所示。

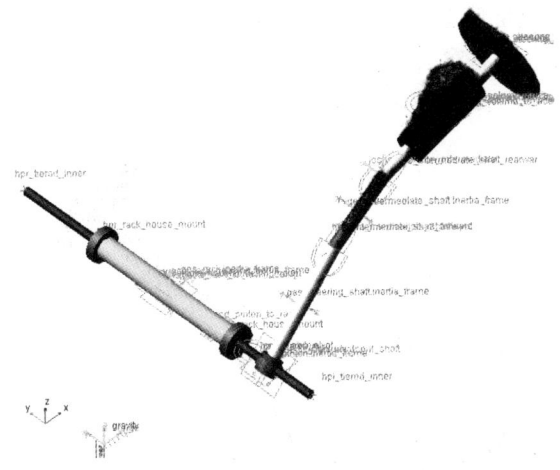

图 1-19　左舵转向系统

（5）单击 Build > Hardpoind > Table 命令，弹出硬点修改对话框，如图 1-20 所示。

	loc x	loc y	loc z
hpl_rack_house_mount	200.0	-200.0	300.0
hpl_tierod_inner	200.0	-425.0	300.0
hps_intermediate_shaft_forward	400.0	-300.0	500.0
hps_intermediate_shaft_rearwar	550.0	-300.0	600.0
hps_pinion_pivot	200.0	-300.0	300.0
hps_steering_wheel_center	900.0	-300.0	700.0

图 1-20　转向系统硬点对话框

- 将方框中的参数值全部修改为正 300，单击 Apply 按钮，此时转变为右舵转向系统，如图 1-21 所示。此转向系统在英国、日本、东南亚国家及中国香港和中国澳门地区应用较多，方向盘在右边，靠左行驶。

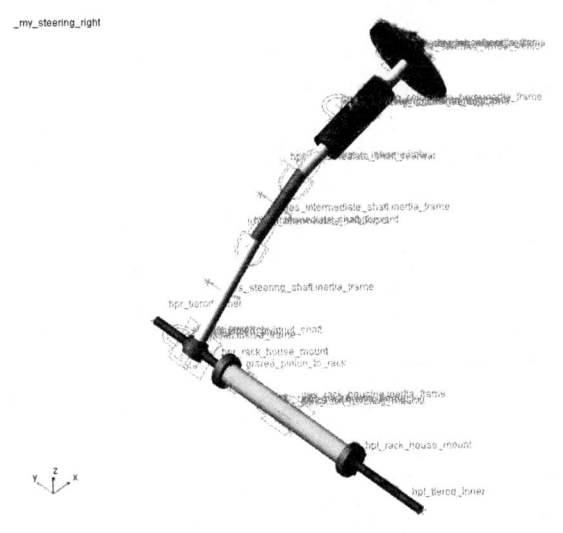

图 1-21　右舵转向系统

（6）单击 File > Save As 命令，保存为 my_steering_right 模板。

• 将方框中的参数值全部修改为 0，单击 Apply 按钮，此时转变为中置转向系统，如图 1-22 所示。此转向系统在一些赛车上用得较多，如卡丁车、F3 赛车（三级方程式赛车）和 F1 赛车（一级方程式赛车）及一些比较特殊的超级跑车上。

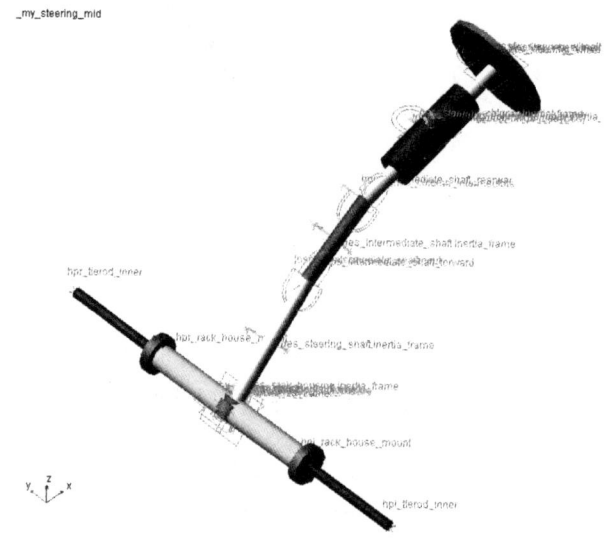

图 1-22　中置转向系统

（7）单击 File > Save As 命令，保存为 my_steering_mid 模板。

至此，齿轮齿条式转向系统建立基本完成。此模型没有创建助力转向特性，系统共享数据库模型的转向系统助力特性是基于液压助力转向特性建立的。在 ADMAS 中建立助力特性的使用性不强，建议建立转向系统的联合仿真模型，控制策略在 MATLAB 软件中实现，这样其控制算法及控制细节可以更加精确。

4. 齿轮齿条转向子系统的建立

（1）将模板转换到标准模式，单击 File > New > Subsystem 命令，弹出子系统对话框，如图 1-23 所示。在下列对话框中输入相应的数据：

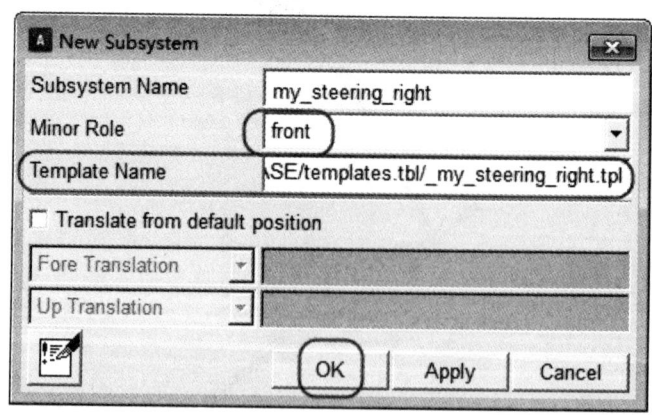

图 1-23　转向系统的创建

- Subsystem Name（系统名称）：my_steering_right；
- Minor Role（副特征）：front（指前轮为转向轮）；
- Template Name（模板路径）：mdids：//my_adams/templates.tbl/_my_steering_right.tpl。

（2）单击 OK 按钮，完成齿轮齿条式转向子系统 my_steering_right 的创建。

5. 转向系统与麦弗逊悬架系统装配

- 单击 File > New > Suspension Assembly 命令，弹出转向装配对话框，如图 1-24 所示。在下列对话框中输入相应的数据：

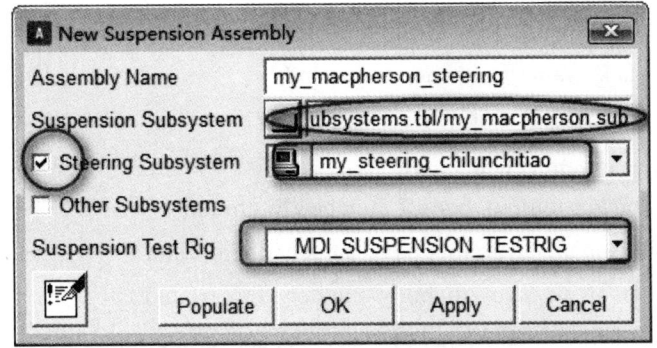

图 1-24　转向系统与麦弗逊悬架装配对话框

- Assembly Name（系统名称）：my_macpherson_steering；
- Suspension Subsystem（悬架子系统）：mdids：//my_adams/subsystems.tbl/my_macpherson.sub；
- 勾选 Steering Subsystem，在悬架仿真中加入转向系统；
- Steering Subsystem（转向子系统）：mdids：//my_adams/subsystems.tblmy_steering_right.sub；
- Suspension Test Rig：MDI_SUSPENSION_TESTRIG（悬架试验台）。

（2）单击 OK 按钮，完成麦弗逊悬架、转向系统、试验台架的装配创建，如图 1-25 所示。

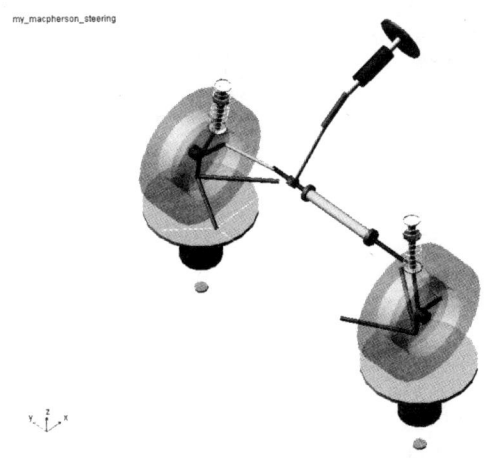

图 1-25　转向系统与麦弗逊悬架装配模型

在创建悬架与齿轮齿条式转向系统及试验台的装配过程中，系统会弹出如下信息，信息包括装配模型含有的子系统已经不匹配的通信器，系统间不匹配的通信器会自动与大地连接。装配信息如下：

Creating the suspension assembly：'my_macpherson_steering'...

Opening the front suspension subsystem：'my_macpherson'...（麦弗逊悬架系统）

Opening the front steering subsystem：'my_steering_right'...（右舵转向系统）

Assembling subsystems...

Assigning communicators...

WARNING：The following input communicators were not assigned during assembly：

 testrig.cil_jack_frame（attached to ground） （连接到大地中的通信器）

 testrig.cir_jack_frame（attached to ground）

 testrig.cis_leaf_adjustment_steps

 testrig.cis_powertrain_to_body（attached to ground）

 my_macpherson.cil_strut_to_body（attached to ground）

 my_macpherson.cir_strut_to_body（attached to ground）

 my_macpherson.cis_subframe_to_body（attached to ground）

 my_steering_right.cis_rack_to_body（attached to ground）

 my_steering_right.cis_steering_column_to_body（attached to ground）

Assignment of communicators completed.

Assembly of subsystems completed.

 Suspension assembly ready. （模型装配完成）

6. 悬架载荷设定

（1）在悬架装配仿真模型上，右击弹簧选择修改参数，弹出弹簧修改对话框，如图 1-26 所示。

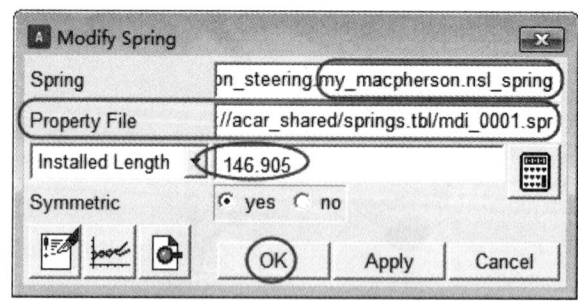

图 1-26 弹簧修改对话框

（2）在弹出的菜单中输入预载荷 2 940 N，弹簧安装长度会自动更新为 146.905。

（3）单击 OK 按钮，麦弗逊悬挂载荷定义完成。

7. 悬架参数设置

（1）单击 Simulate > Suspension Analysis > Set Suspension Parameters 命令，弹出悬架参数设置对话框，如图 1-27 所示。在下列对话框中输入相应的数据：

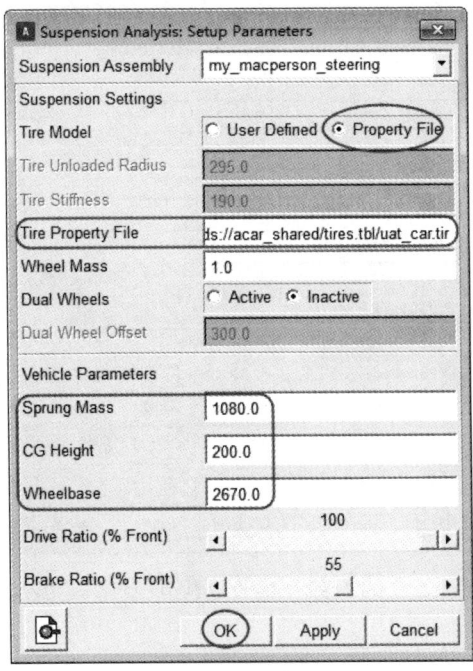

图 1-27　悬架参数对话框

- Tire Model（轮胎模型）：Property File（轮胎以属性文件方式给出）；
- Tire Property File（轮胎属性文件）：mdids: //acar_shared/tires.tbl/uat_car.tir；轮胎的相关参数可以在属性文件 uat_car.tir 修改后保存再引入；
- Sprung Mass：1080；
- CG Height：200；
- Wheelbase：2670；
- Drive Ratio（% Front）：拖动滚动条为 100，车辆为前轮前驱动；如果为后轮驱动，则拖动滚动条为 0；如果为四轮驱动，具体驱动力分配根据实际的驱动力进行分配。

（2）单击 OK 按钮，完成悬架转向装配系统参数相关设置。

至此，齿轮齿条式转向系统与麦弗逊悬架系统装配模型的相关参数全部设置完成，之后可以进行装配系统的各种性能仿真实验。

1.4　转向仿真

传统的转向机构在设计过程中忽略转向机构与悬架系统之间的干涉问题，造成实际工况下的前轮定位参数变化过大，使操作稳定性下降。为提高实际工况下的转向性能，针对原地转向及车辆跳动工况进行相关仿真，检验阿克曼转角是否符合要求。

（1）单击 Simulate > Suspension Analysis > Steering 命令，弹出转向仿真对话框，如图 1-28 所示。在下列对话框中输入相应的数据：

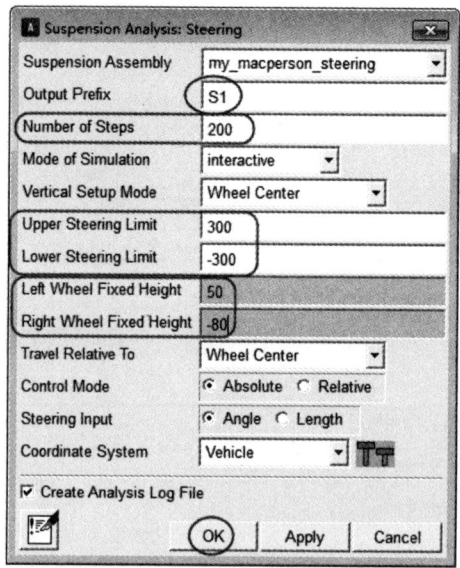

图 1-28　转向仿真对话框

- Output Prefix（输出别名）：s1；
- Number of Steps（仿真步数）：200；
- Upper Steering Limit：300，单位为度（°）；
- Lower Steering Limit：-300，单位为度（°）；
- Left Wheel Fixed Height：50；
- Right Wheel Fixed Height：-80，单位为毫米。

（2）单击 OK 按钮，完成转向系统与麦弗逊悬架装配体在 C 模式下的仿真。

（3）按 F8 键，此时从标准进入后处理模块。

- Simulation（仿真结果）：s1_steering；
- Source（输出）：Result Set（结果设定）；
- Component：同时选择 left\right。

（4）点击 Date（数据）按钮，在弹出的对话框中选择 Request >steering_wheel_input；Component > steering_wheel_input。

（5）单击 OK 按钮，将 steering_wheel_input 转向盘输入角度作为横坐标。

- Result Set（结果设定）：ackerman、ackerman_angle、ackerman_error；
- 勾选 Surf，点击 Add Curves；
- 绘制左右车轮的阿克曼角，阿克曼角误差、车轮外倾角、主销后倾角，如图 1-29 ~ 图 1-32 所示。

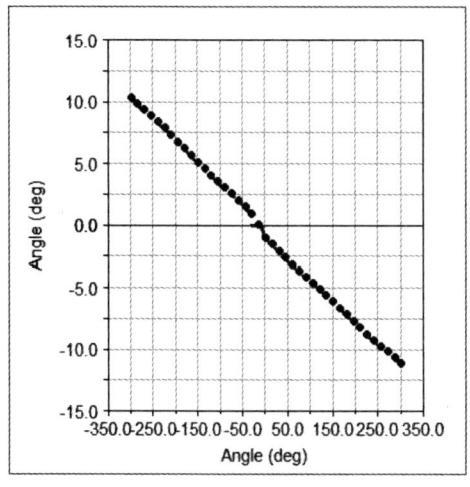

图 1-29 左右阿克曼角

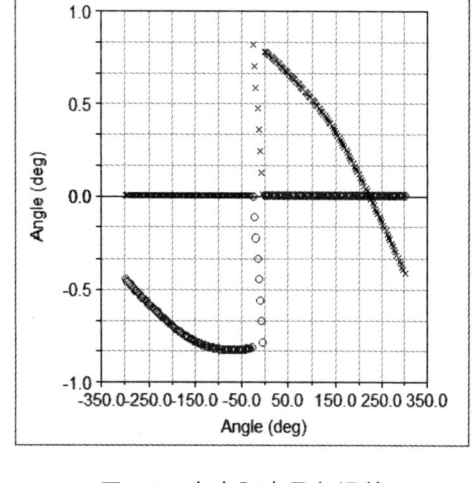

图 1-30 左右阿克曼角误差

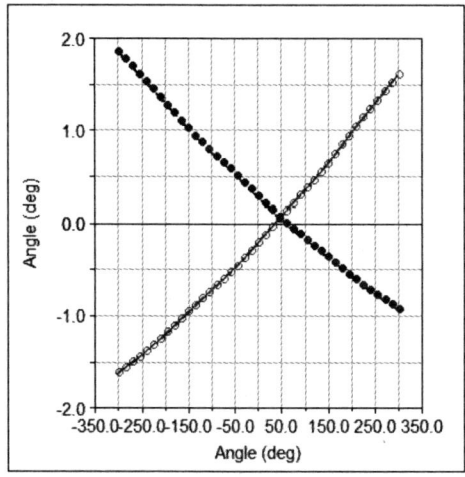

图 1-31 左右车轮外倾角

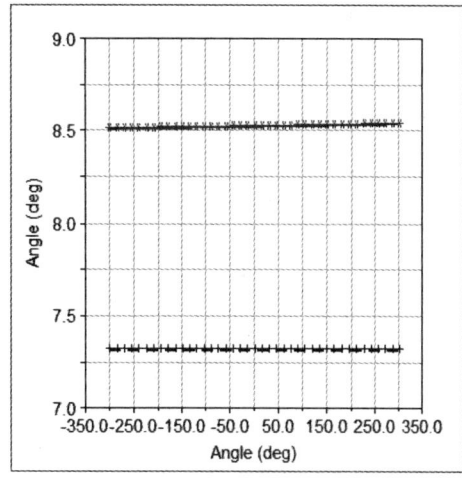

图 1-32 左右主销后倾角

1.5 转向系统调试

FSAE 赛车转向系统为中置转向系统，模型物理结构与上述转向系统完全相同，区别是硬点位置不同。因此可以通过在相关界面中修改硬点的位置完成 FSAE 赛车中置悬架模型的建立。FSAE 赛车中置转向模型硬点如图 1-33 所示，硬点修改完成后中置转向系统模型如图 1-1 所示。

	loc_x	loc_y	loc_z
hpl_rack_house_mount	50.8	-60.0	152.4
hpl_tierod_inner	50.8	-127.0	152.4
hps_intermediate_shaft_forward	164.3	0.0	279.4
hps_intermediate_shaft_rearwar	316.7	0.0	355.6
hps_pinion_pivot	50.8	0.0	152.4
hps_steering_wheel_center	443.7	0.0	381.0

图 1-33　FSAE 赛车中置转向系统硬点

（1）单击 File > Save As 命令，保存为 FSAE_steering_mid 模板。

（2）将模板转换到标准模式，单击 File > New > Subsystem 命令，弹出子系统对话框，如图 1-34 所示。在下列对话框中输入相应的数据。

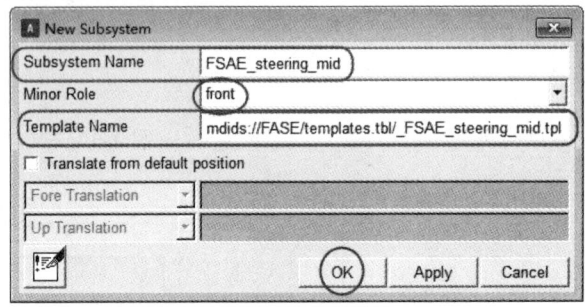

图 1-34　FSAE 赛车中置转向子系统对话框

- Subsystem Name（系统名称）：FSAE_steering_mid；
- Minor Role（副特征）：front；
- Template Name（模板路径）：mdids：//my_adams/templates.tbl/_FSAE_steering_mid.tpl。

（3）单击 OK 按钮，完成齿轮齿条式转向子系统 FSAE_steering_mid 的创建。

（4）单击 File > Save As > Subsystem 命令。

（5）单击 OK 按钮，完成子系统 FSAE_steering_mid 的保存。

（6）单击 Open > Assembly 命令，在 Assembly Name 中输入 mdids：//FASE/subsystems.tbl/FSAE_steering_mid.sub；

（7）单击 OK 按钮，打开推杆式前悬架与实验台装配模型。

（8）单击 File > Manage Assemblies > Add Subsystems 命令，弹出添加转向子系统对话框，如图 1-35 所示。在下列对话框中输入相应的数据：

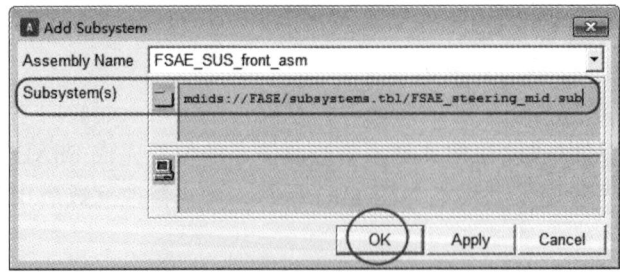

图 1-35　添加转向子系统对话框

- Subsystem（s）：mdids：//FASE/subsystems.tbl/FSAE_steering_mid.sub。

（9）单击 OK 按钮，完成中置转向系统的添加。

（10）右击轮胎，点击修改参数。

- Tire Model（轮胎模型）：User Defined；
- Tire Unload Radius：200。

（11）单击 OK 按钮，完成参数设置后的悬架与转向系统参数如图 1-36 所示。

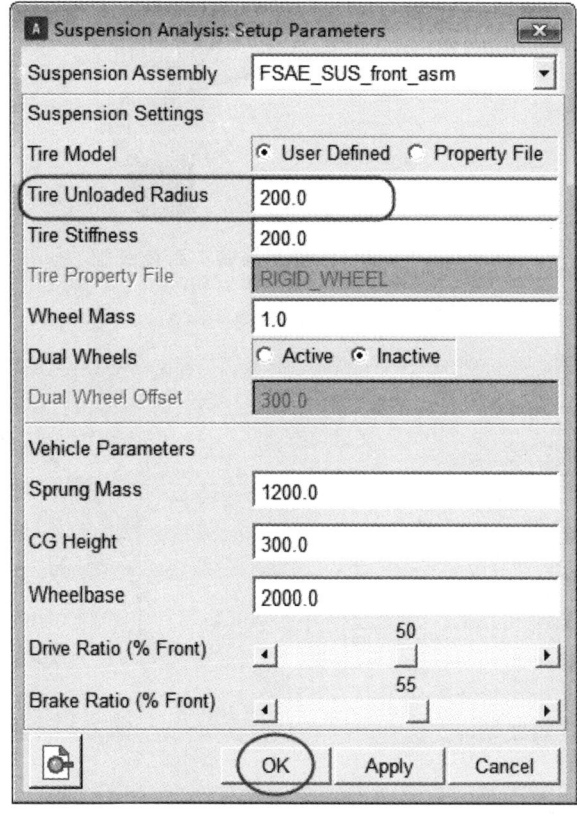

图 1-36　悬架参数设定对话框

（12）单击 Simulate > Suspension Analysis > Steering 命令，转向仿真对话框设置如图 1-28 所示。在下列对话框中输入相应的数据：

- Output Prefix（输出别名）：s2；
- Number of Steps（仿真步数）：200；
- Upper Steering Limit：200，单位为度（°）；
- Lower Steering Limit：-200，单位为度（°）；
- Left Wheel Fixed Height：0；
- Right Wheel Fixed Height：0，单位为毫米。

（13）单击 OK 按钮，完成转向系统与麦弗逊悬架装配体在 C 模式下的仿真，如图 1-37 所示。车轮定位参数与转向盘输入角度的关系如图 1-38 ~ 图 1-41 所示。

图 1-37　推杆式悬架与转向系统装配

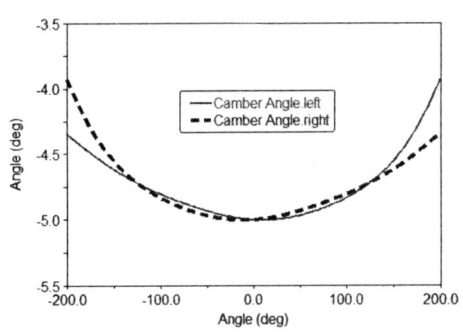

图 1-38　车轮外倾角

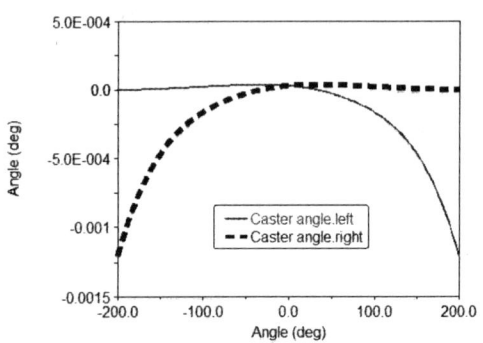

图 1-39　主销后倾角

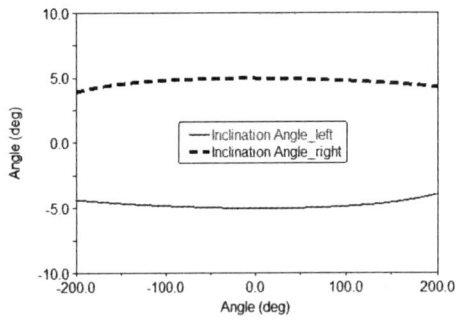

图 1-40　主销内角

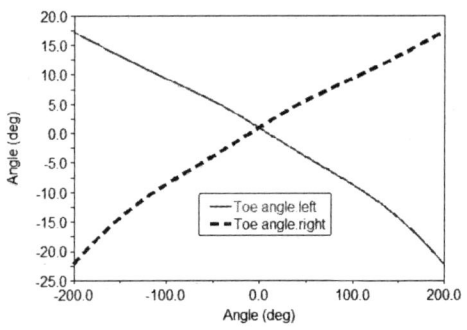

图 1-41　车轮前束角

第 2 章 车 身

车身为整车的承载部分，同时与各个子系统之间存在装配关系及数据交换。ADAMS 车身模型较为简单，主要是一些变量参数的设置及输出通信器。输出通信器的主要作用是与其他子系统之间建立虚拟意义上的装配，建立好的简化车身模型如图 2-1 所示。车身建模过程如下，将软件界面切换到专家模式界面。

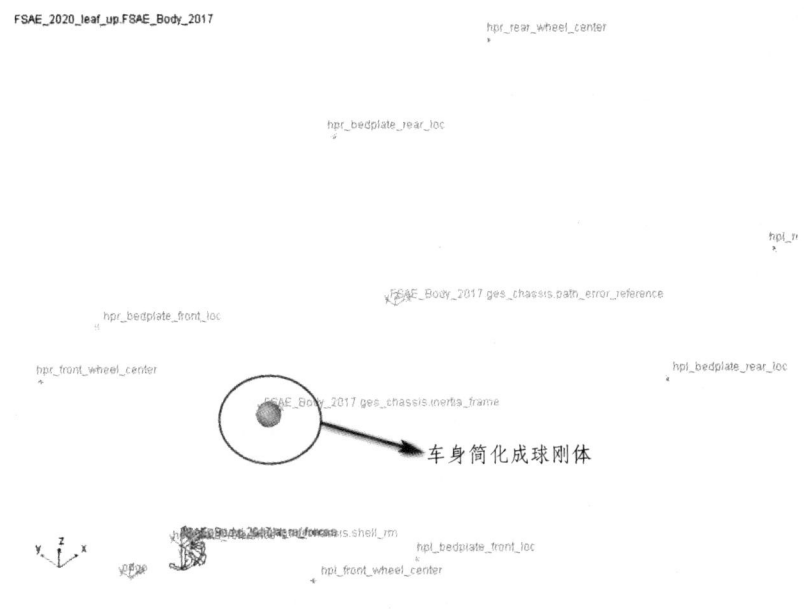

图 2-1　车身简化刚体模型

学习目标

（1）熟悉输出通信器。
（2）会变量参数设定。
（3）会测量函数。

2.1 车身通信器

（1）单击 File > New 命令，弹出创建模板对话框，如图 2-2 所示。在下列对话框中输入相应的数据：

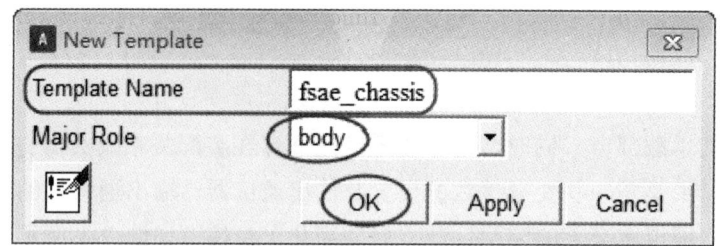

图 2-2 车身模板创建对话框

- Template Name：fsae_chassis；
- Major Role：body。

（2）单击 OK 按钮，完成车身模板的创建。

1. 车身硬点参数

（1）单击 Build > Hardpoind > New 命令，弹出创建硬点对话框，如图 2-3 所示。在下列对话框中输入相应的数据：

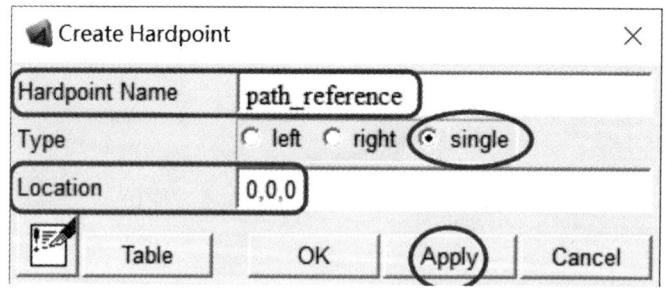

图 2-3 硬点创建对话框

- Hardpoind Name：path_reference；
- Type：single；
- Location：0.0，0.0，0.0。

（2）单击 Apply 按钮，完成 path_reference 硬点的创建。创建另外一个硬点：

- Hardpoind Name：ground_height_reference；
- Type：single；
- Location：0.0，0.0，0.0。

（3）单击 Apply 按钮，完成 ground_height_reference 硬点的创建。

2. 结构框参数

（1）单击 Build > Construction Frame > New 命令，创建定向结构框，如图 2-4 所示。在

下列对话框中输入相应的数据：

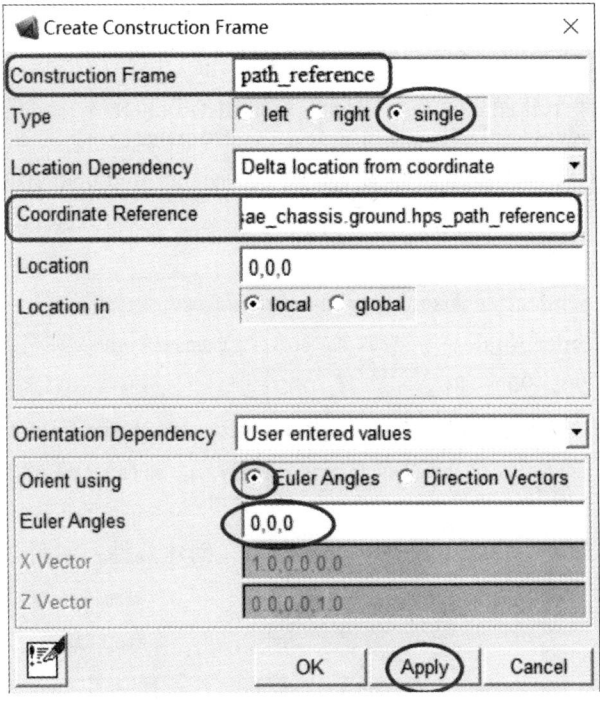

图 2-4　结构框 path_reference

- Construction Frame（结构框名称）：path_reference；
- Type：single；
- Location Dependency：Delta location from coordinate；
- Coordinate Reference（参考坐标）：._fsae_chassis.ground.hps_path_reference；
- Location：0，0，0；
- Location in：local；
- Orientation Dependency：User entered values；
- Orient using：Euler Angles；
- Euler Angles：0，0，0。

（2）单击 Apply 按钮，完成 ._fsae_chassis.ground.cfs_path_reference 结构框的创建。继续创建其他结构框：

- Construction Frame（结构框名称）：driver_reference；
- Type：single；
- Location Dependency：Delta location from coordinate；
- Coordinate Reference（参考坐标）：._fsae_chassis.ground.hps_path_reference；
- Location：0，0，0；
- Location in：local；
- Orientation Dependency：User entered values；
- Orient using：Euler Angles；

- Euler Angles：180，0，0。

（3）单击 Apply 按钮，完成._fsae_chassis.ground.cfs_driver_reference 结构框的创建。
- Construction Frame（结构框名称）：aero_force_reference；
- Type：single；
- Location Dependency：Delta location from coordinate；
- Coordinate Reference（参考坐标）：._fsae_chassis.ground.hps_ground_height_reference；
- Location：0，0，0；
- Location in：local；
- Orientation Dependency：User entered values；
- Orient using：Euler Angles；
- Euler Angles：90，90，180。

（4）单击 OK 按钮，完成._fsae_chassis.ground.cfs_aero_force_reference 结构框的创建。

3．车身部件与几何体

（1）单击 Build > Part > General Part > New 命令，创建部件，如图 2-5 所示。在下列对话框中输入相应的数据：

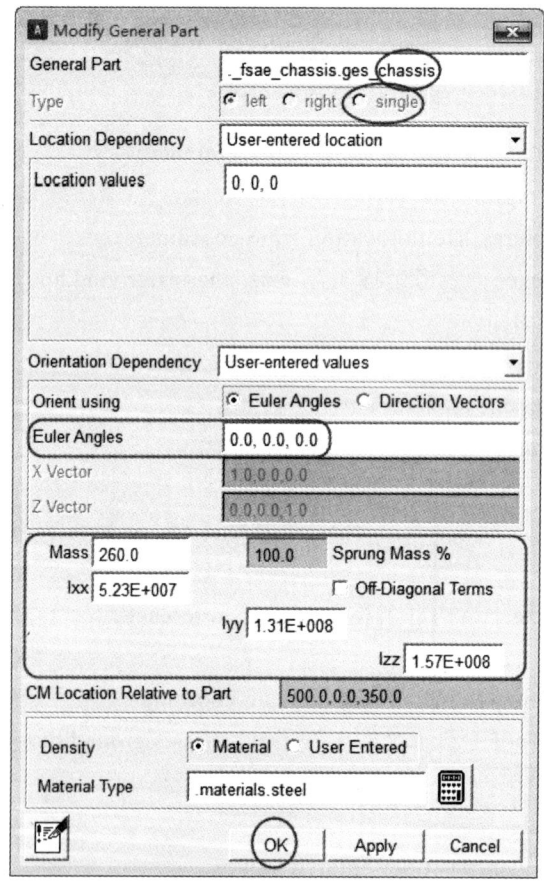

图 2-5　车身部件对话框

- General Part：chassis；
- Location Dependency：User-entered location；
- Location values：0，0，0；
- Orientation Dependency：User-entered values；
- Orient using：Euler Angles；
- Euler Angles：0，0，0；
- Mass：260；
- Ixx：5.23e7；
- Iyy：1.31e8；
- Izz：1.57e8；
- Density：Material；
- Material Type：.materials.steel。

（2）单击 OK 按钮，完成车身部件._fsae_chassis.ges_chassis 的创建。

（3）单击 Build > Geometry > Ellipsoid > New 命令，创建椭圆几何体，如图 2-6 所示。在下列对话框中输入相应的数据：

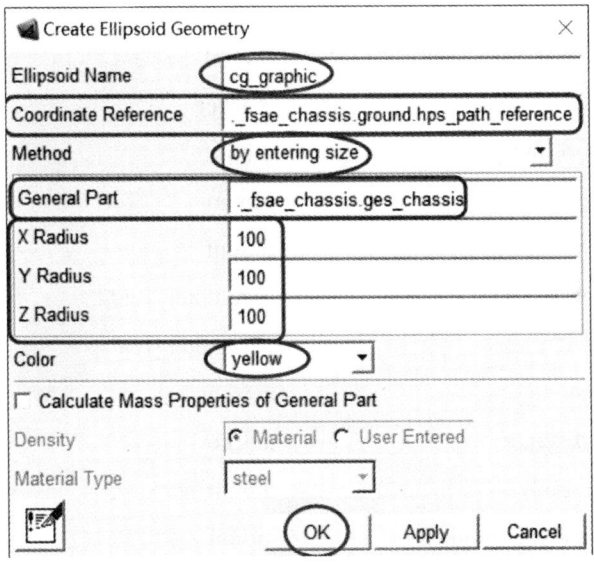

图 2-6　椭圆几何体对话框

- Ellipsoid Name（椭圆体名称）：cg_graphic；
- Coordinate Reference（参考坐标）：._fsae_chassis.ground.hps_path_reference；
- Method：by entering size；
- General Part：._fsae_chassis.ges_chassis；
- X Radius：100；
- Y Radius：100；
- Z Radius：100；
- Color（杆件几何体颜色）：yellow；
- Density：Material；

- Material Type：steel。

（4）其余保持默认设置，单击 OK 按钮，完成 cg_graphic 几何体的创建。

4. 输出通信器

FSAE 简化车身共包含 26 个输出通信器，具体见表 2-1。

表 2-1　车身通信器

Communicator Name	Entity Class	To Minor Role
co[lr]_tierod_to_steering	mount	rear
co[lr]_trod	mount	inherit
co[lr]_tv_link	mount	inherit
cos_aero_drag_force	solver_variable	inherit
cos_aero_frontal_area	parameter_real	inherit
cos_air_density	parameter_real	inherit
cos_body	mount	inherit
cos_body_subsystem	mount	inherit
cos_chassis_path_reference	marker	inherit
cos_column_support_mount	mount	inherit
cos_concept_to_body	mount	inherit
cos_diff_housing_to_body	mount	rear
cos_downforce_coefficient	parameter_real	inherit
cos_drag_coefficient	parameter_real	inherit
cos_driver_reference	marker	inherit
cos_measure_for_distance	marker	inherit
cos_powertrain_to_body	mount	inherit
cos_propshaft_support_to_body	mount	rear
cos_rack_housing_mount	mount	inherit
cos_rack_to_body	mount	inherit
cos_steering_column_to_body	mount	inherit
cos_subframe_to_body	mount	inherit
cos_suspension_to_chassis	mount	inherit

（1）单击 Build > Communicator > Output >New 命令，弹出输出通信器对话框，如图 2-7 所示。在下列对话框中输入相应的数据：

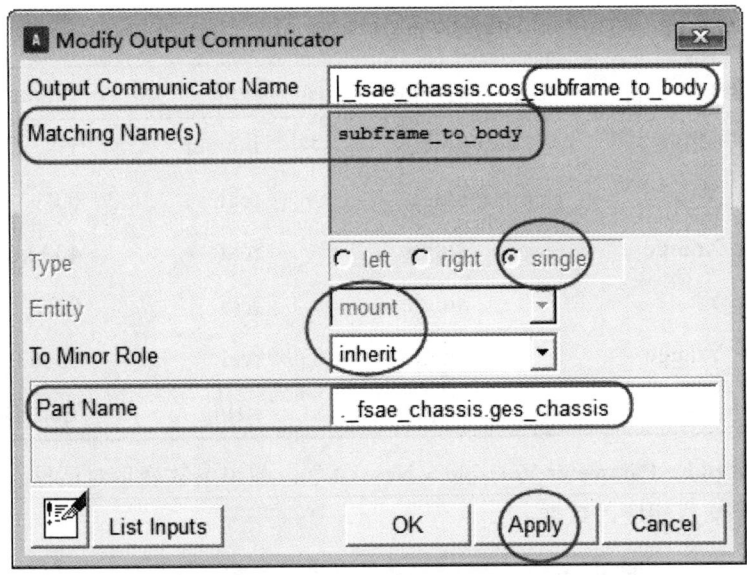

图 2-7 输出通信器对话框

- Output Communicator Name（输出通信器名称）：subframe_to_body；
- Matching Name（s）：subframe_to_body；
- Type：single；
- Entity：mount；
- To Minor Role：inherit；
- Part Name：._fsae_chassis.ges_chassis。

（2）单击 Apply 按钮，完成通信器._fsae_chassis.cos_subframe_to_body 的创建。
（3）重复上述步骤，按顺序完成表格中对应输出通信器的创建。

2.2 车身参数变量

车身参数变量如表 2-2 所示。

表 2-2 车身变量参数

parameter name	symmetry	type	value
kinematic_flag	single	integer	0
aero_drag_active	single	integer	1
aero_frontal_area	single	real	1.8
air_density	single	real	1.22
downforce_coefficient	single	real	0.0
drag_coefficient	single	real	0.36

续表

parameter name	symmetry	type	value
lap_beacon_active	single	integer	1
lap_beacon_X	single	real	0.0
lap_beacon_Xrange	single	real	4000.0
lap_beacon_Y	single	real	0.0
lap_beacon_Yrange	single	real	8000.0
lap_info_file	single	string	'Lap_Data'

（1）单击 Build > Parameter Variable > New 命令，弹出参数变量对话框，如图 2-8 所示。在下列对话框中输入相应的数据。

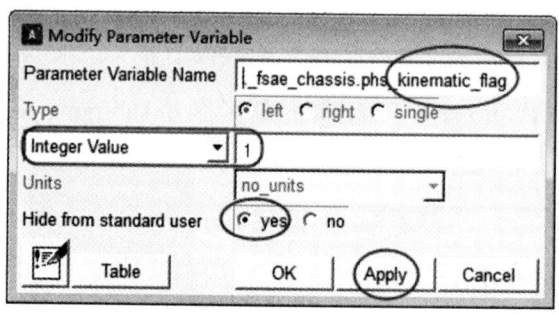

图 2-8 变量参数对话框

- Parameter Variable Name：kinematic_flag；
- Integer Value（实数值）：1；
- Units：no_units；
- Hide from standard user（是否从标准界面隐藏）：yes。

（2）单击 Apply 按钮，完成变量 ._fsae_chassis.phs_kinematic_flag 的创建。

（3）重复上述步骤，按顺序完成表格中对应参数变量的创建。

2.3 车身测量函数

1. 车身测量函数

车身测量函数主要包括 X、Y、Z 方向上的加速度及转动角加速度，这六个参数反映整车在运行过程中的车身状态，可以判定整车的稳定性及平顺性。

① 车身 X 方向加速度 ._fsae_chassis.av_x：ACCX（._fsae_chassis.ges_chassis.inertia_frame）；

② 车身 Y 方向加速度 ._fsae_chassis.av_y：ACCY（._fsae_chassis.ges_chassis.inertia_frame）；

③ 车身 Z 方向加速度：._fsae_chassis.av_z：ACCZ（._fsae_chassis.ges_chassis.inertia_frame）;

④ 绕车身 X 方向转动角加速度：._fsae_chassis.WDTX：WDTX（._fsae_chassis.ges_chassis.inertia_frame）;

⑤ 绕车身 Y 方向转动角加速度：._fsae_chassis.WDTY：WDTY（._fsae_chassis.ges_chassis.inertia_frame）;

⑥ 绕车身 Z 方向转动角加速度：._fsae_chassis.WDTZ：WDTZ（._fsae_chassis.ges_chassis.inertia_frame）。

输出通信器建立完成后，车身模型建立完成。

（1）单击 File > Save as 命令，在下列对话框中输入相应的数据：
- Major Role：body；
- File Format：Binary。

（2）其余设置保持默认，单击 OK 按钮，完成车身 fsae_chassis 的存储。

2. 车身子系统

（1）按 F9 键切换到标准模板，单击 File > New > Subsystem 命令，弹出子系统对话框，参见图 1-23。在下列对话框中输入相应的数据：
- Subsystem Name：FSAE_Body_2017；
- Minor Role：any；
- Template Name：mdids：//FASE/templates.tbl/_fsae_chassis.tpl。

（2）单击 OK 按钮，完成车身子系统 FSAE_Body_2017.sub 的创建。

（3）单击 File > Save as 命令，在下列对话框中输入相应的数据：
- Subsystem Name：FSAE_Body_2017；
- Minor Role：any；
- File Format：Teim Orbit；
- Target：Database，FSAE；

（4）单击 OK 按钮，完成轮胎子系统 FSAE_Body_2017.sub 的存储。

第 3 章　发动机

发动机模型是整车模型中的一部分。整车不包含发动机模型，依然可以进行整车部分相关工况的仿真，但在仿真过程中不能运行准静态平衡，在仿真刚开始会伴随有较大的振动。发动机模型的建立关键在于获取精确的发动数据，发动机数据需要通过发动机台架试验获取。同时，发动机数据还需要和其他变量参数相匹配，如换挡转速、变速器驱动轴参数等。本章提供了一个 600cc 排量（0.6 L 排量）四缸发动机参数，将此参数编排成一个发动机属性文件：600cc_engine_map.pwr，并替换掉通用数据单元数据曲线中发动机输出扭矩的系统默认属性文件，建模过程中参考共享数据库中的发动机模型，删除原有发动机的造型参数，重新构建发动机几何外形；并提供了发动机与车身安装需要对应的支撑点，此支撑点衬套的参数对发动机振动影响很大，建模中采用与悬架衬套相同的参数（此参数不准确，但可适用），也可以在完整的整车架构下研究发动机的振动特性，对发动机的支撑架轻量化、频率及支撑点衬套的精确刚度等参数进行优化。建立好发动机模型并添加到整车模型中，如图 3-1 所示，经过相关调试，整车所有的工况基本上都可以准确仿真，同时仿真前可以运行准静态平衡。

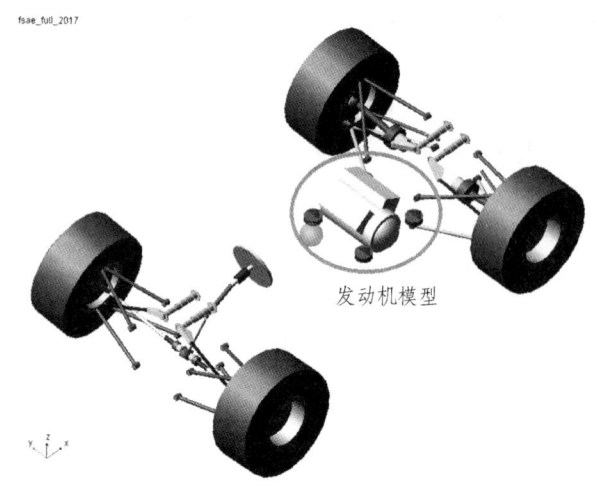

图 3-1　发动机系统模型

学习目标

（1）熟悉发动机实验数据。
（2）掌握发动机扭矩图绘制程序。
（3）会发动机系统建模。
（4）会定半径转弯仿真。

3.1 发动机实验数据

发动机的输出扭矩与制动系统相互关联,方程式赛车采用中置 600cc 排量四缸发动机,点火顺序为 1—3—4—2。整车在制动过程中,由发动机经过变速器及传动系统输出到车轮上的力矩不同,尤其是在弯道制动中,内外侧车轮的力矩及所承受的载荷也不一样,因此,精确的整车模型必须考虑发动机输出扭矩及传动系统和差速器等子系统的影响。发动机外特性实验采用 ET2000 发动机自动测控系统,测试过程中,发动机水温系统、燃油恒温系统正常,发动机实验条件按国家标准《汽车发动机性能实验方法》(GB/T 18297—2001)的规定进行控制;扭矩测量精度为 ±0.4%FS,转速测量精度为 ±1 r/min,转动惯量测量精度为 0.231 kg·m^2;测试中每增加 500 转,记录一次输出扭矩及制动力矩。发动机在 9 900 r/min 时发动机输出最大扭矩为 56 004.1 N·mm。

方程式赛车发动机实验数据如下所示,根据发动机文件编排规则编排文件,保存重命名为:600cc_engine_map.pwr。

```
发动机信息,根据实验数据编排发动机文件如下:
$-------------------------------------------------------------MDI_HEADER
[MDI_HEADER]
 FILE_TYPE      =  'pwr'
 FILE_VERSION   =  1.0
 FILE_FORMAT    =  'ASCII'
$-------------------------------------------------------------------UNITS
[UNITS]
(BASE)
{length      force        angle       mass        time}
 'mm'        'newton'     'degrees'   'kg'        'sec'
(USER)
{unit_type   length   force   angle   mass   time   conversion}
 'rpm'         0        0       1       0     -1      6.0
 'torque'      1        1       0       0      0      1.0
$-----------------------------------------------------------------ENGINE
[ENGINE]
(Z_DATA)
{throttle <no_units>}
0.0
1.00
(XY_DATA)
{engine_speed <rpm>    torque@throttle <torque>}    %以下为实验参数
0                      0
1000                   -67.85                        10856
```

2000	-67.85	10856
2700	-135.7	15334.1
2800	-203.55	14519.9
2900	-271.4	13977.1
3000	-339.25	13705.7
3100	-407.1	13705.7
3200	-474.95	13841.4
3300	-542.8	14112.8
3400	-610.65	14248.5
3500	-678.5	14384.2
3600	-746.35	14519.9
3700	-814.2	14927
3800	-882.05	15605.5
3900	-949.9	16555.4
4000	-1017.75	17505.3
4100	-1085.6	18455.2
4200	-1153.45	19812.2
4300	-1221.3	21304.9
4400	-1289.15	23204.7
4500	-1357	25375.9
4600	-1424.85	30490.7
4700	-1492.7	31304.9
4800	-1560.55	32119.1
4900	-1628.4	33476.1
5000	-1696.25	34968.8
5100	-1764.1	36868.6
5200	-1831.95	39175.5
5300	-1899.8	42609.8
5400	-1967.65	43424
5500	-2035.5	44102.5
5600	-2103.35	45188.1
5700	-2171.2	45323.8
5800	-2239.05	46138
5900	-2306.9	46545.1
6000	-2374.75	46680.8
6100	-2442.6	46002.3
6200	-2510.45	44916.7
6300	-2578.3	43831.1

6400	-2646.15	42474.1
6500	-2714	41117.1
6600	-2781.85	40981.4
6700	-2849.7	40302.9
6800	-2917.55	40710
6900	-2985.4	41659.9
7000	-3053.25	42474.1
7100	-3121.1	44781
7200	-3188.95	45866.6
7300	-3256.8	47223.6
7400	-3324.65	48987.7
7500	-3392.5	50887.5
7600	-3460.35	51023.2
7700	-3528.2	52651.6
7800	-3596.05	52380.2
7900	-3663.9	51973.1
8000	-3731.75	51294.6
8100	-3799.6	50073.3
8200	-3867.45	48444.9
8300	-3935.3	47630.7
8400	-4003.15	47223.6
8500	-4071	47223.6
8600	-4138.85	47766.4
8700	-4206.7	49259.1
8800	-4274.55	49259.1
8900	-4342.4	51158.9
9000	-4410.25	51701.7
9100	-4478.1	52108.8
9200	-4545.95	52651.6
9300	-4613.8	52787.3
9400	-4681.65	53737.2
9500	-4749.5	54687.1
9600	-4817.35	54687.1
9700	-4885.2	55501.3
9800	-4953.05	55501.3
9900	-5020.9	56044.1	%9900转是发动机输出最大扭矩
10000	-5088.75	55772.7
10100	-5156.6	55772.7

10200	-5224.45	55637
10300	-5292.3	55365.6
10400	-5360.15	55772.7
10500	-5428	55501.3
11000	-5500	55000
12000	-5500	53000
13000	-5500	51000
14000	-5500	51000
15000	-5500	51000
16000	-5500	51000
17000	-5500	51000
18000	-5500	51000
19000	-5500	51000
20000	-5500	51000

3.2　发动机扭矩图绘制程序

发动机实验数据按属性文件编排好之后，对发动机的输出扭矩及制动力矩进行图形绘制，可以更加明确地观察发动机参数的变化趋势，如图 3-2 所示，绘图采用 MATLAB 软件编写程序，具体信息如下：

```
发动机输出及制动扭矩绘图程序：
a=[2000 2700 2800 2900 3000 3500 4000 4500 5000 5500 6000 6500 7000 7500 8000 8500 9000 9500 10000 10500 11000 12000 13000 14000 15000];
b=[-67.85 -135.7 -203.55 -271.4 -339.25 -678.5 -1017.75 -1357 -1696.25 -2035.5 -2374.75 -2714 -3053.25 -3392.5 -3731.75 -4071 -4410.25 -4749.5 -5088.75 -5428 -5500 -5500 -5500 -5500 -5500];
c=[10856 15334.1 14519.9 13977.1 13705.7 14384.2 17505.3 25375.9 34968.8 44102.5 46680.8 41117.1 42474.1 50887.5 51294.6 47223.6 51701.7 54687.1 55772.7 55501.3 55000 53000 51000 51000 51000];
plot(a,b,'r-',a,c,'b-')
xlabel('转速/rpm')
ylabel('扭矩/N*mm')
```

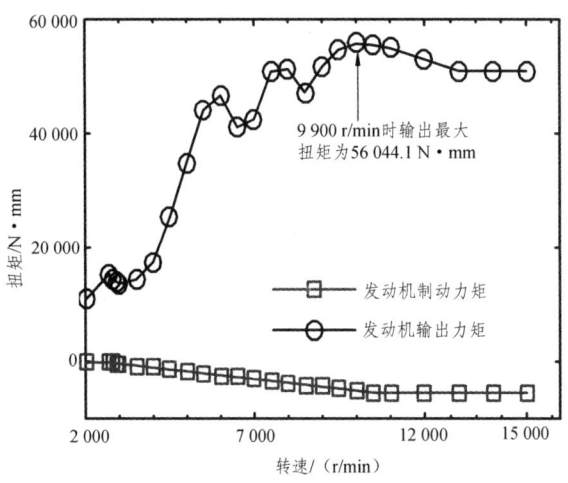

图 3-2 外特性与制动力矩曲线

3.3 发动机系统建模

系统信息包含相关硬点、部件、转换部件、变量参数、通信器等。建模过程中，难点在于 Adams Arrays、Requests 等参数的设置以及彼此之间的关系特性较为复杂，建议采用共享数据库中的发动机系统模型，通过发动机实验数据编制属性文件进行重新匹配。发动机外形几何体可以修改（几何体修改不会对发动机的性能产生任何影响，只是与整车外形大小等视觉搭配效果有关），还可以通过数据对话框对外形进行删除，然后把界面转换到 ADAMS/View 通用界面，用基本的几何体构造发动机的简单外形。FSAE 赛车发动机模型通过几何体重新构造、发动机属性文件替换等最终建立好的子系统如图 3-3 所示。发动机子系统建立完成后系统信息如下：

```
File Name         :   <FASE>/subsystems.tbl/powertrain_fsae_2017.sub
Template          :   mdids://FASE/templates.tbl/_powertrain.tpl
Comments          :
Template          :   Example of a non-spinning powertrain
Subsystem         :   *no subsystem comments found*
Major Role        :   powertrain
Minor Role        :   rear

HARDPOINTS:

hardpoint name          symmetry        x_value     y_value     z_value
------------------      ---------       -------     -------     -------
graphics_reference      single          0.0         0.0         0.0
front_engine_mount      left/right      950.0       -150.0      200.0
rear_engine_mount       left/right      1250.0      -150.0      200.0
```

```
PARTS:

    powertrain
      symmetry                    :  single
      mass                        :  5.0
      location (dependent)        :  1100.0,  0.0,  200.0
      orientation                 :  zp_vector=0.0, 0.0, 1.0
                                  :  xp_vector=1.0, 0.0, 0.0
      cm_location_from_part       :  0.0, 0.0, 0.0
      Ixx, Iyy, Izz               :  1.0, 1.0, 1.0
      Ixy, Izx, Iyz               :  0.0, 0.0, 0.0

    diff_output
      symmetry                    :  left/right
      mass                        :  2.0
      location (dependent)        :  1500.0, -200.0, 225.0
      orientation (dependent)     :  zp_vector=0.0, -1.0, 0.0
                                  :  xp_vector=1.0, 0.0, 0.0
      cm_location_from_part       :  0.0, 0.0, 0.0
      Ixx, Iyy, Izz               :  1.0, 1.0, 1.0
      Ixy, Izx, Iyz               :  0.0, 0.0, 0.0

SWITCH PARTS:

    engine_mount_option
      symmetry      :  single
      switched to   :  chassis (general part)
      part list     :  chassis (general part)
                    :  chassis (general part)

GENERAL SPLINES:

    differential
      symmetry       :  single
      type           :  'two_dimensional'
      property file  :  mdids://FASE/differentials.tbl/MSC_viscous.dif
      curve_name     :  'DIFFERENTIAL'
    engine_torque
      symmetry       :  single
      type           :  'three_dimensional'
      property file  :  mdids://FASE/powertrains.tbl/600cc_engine_map.pwr
      curve_name     :  'ENGINE'
```

PARAMETERS:

parameter name	symmetry	type	value
kinematic_flag	single	integer	0
clutch_capacity	single	real	1.00E+06
clutch_close	single	real	0.25
clutch_damping	single	real	10000.0
clutch_open	single	real	0.75
clutch_stiffness	single	real	1.00E+06
clutch_tau	single	real	0.05
ems_gain	single	real	0.005
ems_max_throttle	single	real	100.0
ems_throttle_off	single	real	1.0
engine_idle_speed	single	real	10.0
engine_inertia	single	real	70000.0
engine_rev_limit	single	real	14000.0
final_drive	single	real	3.28
gear_1	single	real	3.231
gear_2	single	real	2.571
gear_3	single	real	2.125
gear_4	single	real	1.789
gear_5	single	real	1.55
gear_6	single	real	1.0
gear_r	single	real	-3.0
graphics_flag	single	integer	1
max_gears	single	integer	6
max_throttle	single	real	100.0

Listing of input communicators in '_fsae_powertrain'

Communicator Name:	Entity Class:	From Minor Role:
ci[lr]_diff_tripot	location	inherit
ci[lr]_tire_force	force	inherit
cis_clutch_demand	solver_variable	inherit
cis_engine_to_subframe	mount	inherit
cis_initial_engine_rpm	parameter_real	any
cis_powertrain_to_body	mount	inherit
cis_sse_diff1	diff	inherit
cis_throttle_demand	solver_variable	inherit
cis_transmission_demand	solver_variable	inherit

```
Listing of input communicators in '_fsae_powertrain'
------------------------------------------------------------------
Communicator Name:              Entity Class:            To Minor Role:
    co[lr]_output_torque            force                    inherit
    co[lr]_tripot_to_differential   mount                    inherit
    cos_clutch_displacement_ic      solver_variable          inherit
    cos_default_downshift_rpm       parameter_real           inherit
    cos_default_upshift_rpm         parameter_real           inherit
    cos_diff_ratio                  parameter_real           inherit
    cos_engine_idle_rpm             parameter_real           inherit
    cos_engine_map                  spline                   inherit
    cos_engine_max_rpm              parameter_real           inherit
    cos_engine_rpm                  solver_variable          inherit
    cos_engine_speed                solver_variable          inherit
    cos_max_engine_braking_torque   solver_variable          inherit
    cos_max_engine_driving_torque   solver_variable          inherit
    cos_max_gears                   parameter_integer        inherit
    cos_max_throttle                parameter_real           inherit
    cos_powertrain_gse              general_state_equation   inherit
    cos_transmission_input_omega    solver_variable          inherit
    cos_transmission_spline         spline                   inherit
```

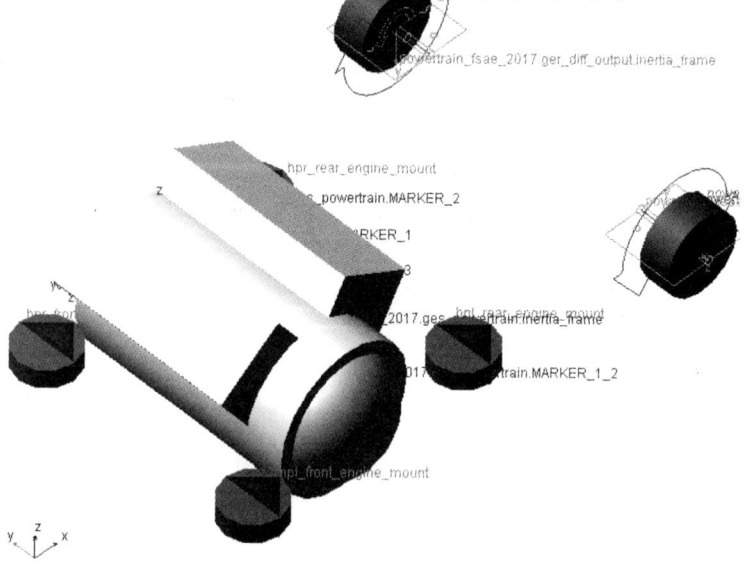

图 3-3 发动机系统模型

3.4 定半径转弯仿真

FSAE 赛车包含发动机模型后，定半径转弯工况才可以正常仿真。同时在仿真过程中，可以设置挡位参数，准静态平衡可以在仿真计算前先计算系统的静平衡，勾选自动换挡"Shift Gears"，整车可以在运行中自动换挡。

（1）单击 Simulate > Full-Vehicle Analysis > Cornering Events > Constant Radius Cornering 命令，弹出定半径转弯仿真对话框，如图 3-4 所示。在下列对话框中输入相应的数据：

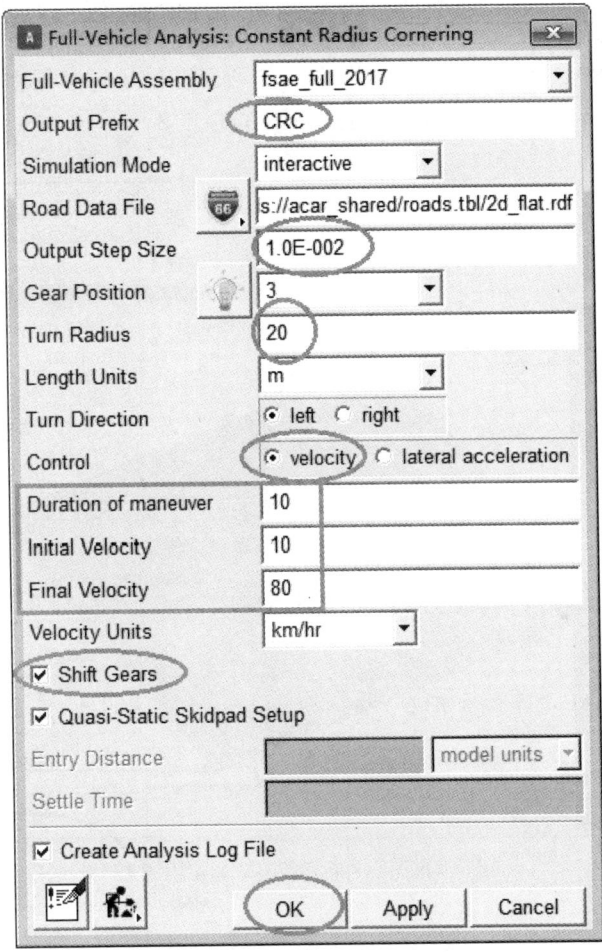

图 3-4 定半径转弯仿真对话框

- Output Prefix：CRC；
- Simulation Mode：interactive；
- Road Date File：mdids：//acar_shared/roads.tbl/2d_flat.rdf；
- Output Step Size（仿真步数）：0.01；
- Gear Position：3；
- Turn Radius：20；
- Length Units：m；

- Turn Direction：left；
- Control：velocity；
- Duration of maneuver：10；
- Initial Velocity：10；
- Final Velocity：80；
- Velocity Units：km/hr。

（2）勾选"Shift Gears"，计算机在运行过程中，FSAE 赛车可以自行换挡运行。

（3）勾选 Quasi-Static Straight-Line Setup，整车模型包含发动模型后可以运行准静态平衡。

（4）单击 OK 按钮，完成定半径转向 Constant Radius Cornering 仿真设置并提交运算。

仿真结束后，FSAE 赛车的运行轨迹如图 3-5 所示，在运行过程中，整车的侧向加速度、车身横摆角加速度及后驱动半轴的输出力矩如图 3-6～3-8 所示。

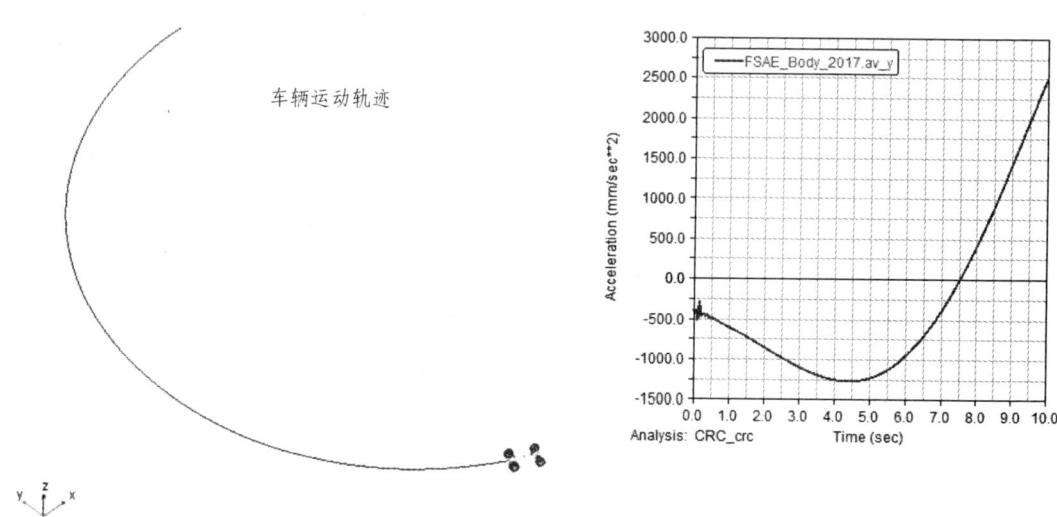

图 3-5　FSAE 赛车运行轨迹　　　　　图 3-6　车身侧向加速度

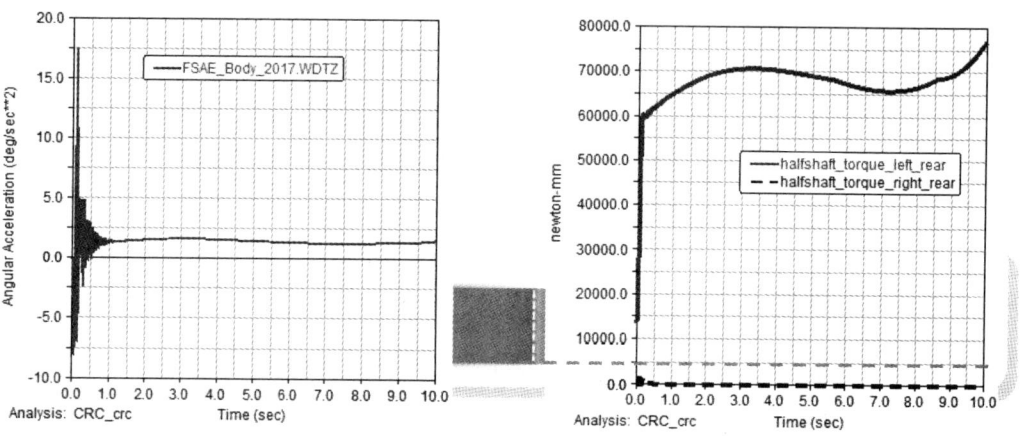

图 3-7　车身横摆角加速度　　　　　图 3-8　后驱动半轴输出力矩

第4章 制动系统

制动系统的好坏直接关系到整车的安全特性。整车在制动过程中的制动力减速度与制动距离、制动时方向的稳定性以及制动盘的抗热衰退性能是衡量制动器系统的三个重要指标。制动盘的抗热衰退性能需要借助有限元软件进行模拟；制动减速度及制动距离、制动时方向的稳定性可以采用 ADAMS 多体动力学软件下的整车模型进行模拟。ABS（防抱死制动系统）是现在乘用车与商用车的标准配置之一，制动系统多体模型与 MATLAB 控制软件结合可以模拟不同控制算法下制动系统的制动效能。制动系统模型如图 4-1 所示。制动系统中制动力矩的关键在于制动力矩函数的构造，可以在原有函数的基础上根据设计的要求增加或者减少状态变量项，即考虑最终制动力矩由哪些参数决定。同时，制动盘的直径大小、接触面积、摩擦系数等参数可以通过变量参数直接修改，以影响制动力矩的大小。制动系统建模也推荐采用共享数据库中的制动模板，根据实际需求对制动模型中的有关参数进行修改。

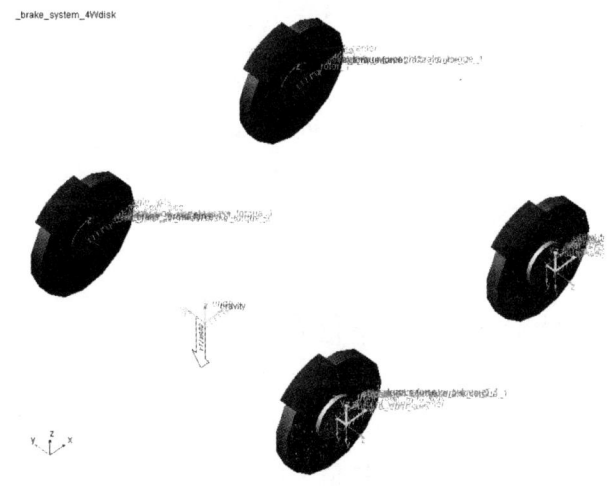

图 4-1　制动系统模型

学习目标

（1）了解制动系统。
（2）熟悉制动系统变量参数及通信器。
（3）会 FSAE 赛车 Braking 文件驱动仿真。
（4）会客车 Braking 仿真。
（5）会牵引车 Braking 仿真。

4.1 制动系统简介

基于 ADAMS 整车环境模式下对制动系统进行研究可以取得较好的效果，其仿真结果可以作为设计制造制动器的依据，同时也可以验证不同制动控制算法的优劣。对制动系统建模的关键是要充分考虑影响制动力矩的因素，ADAMS/CAR 中四轮制动系统（左前轮）制动力矩函数如下：

2.0*._brake_system_4Wdisk.pvs_front_piston_area*._brake_system_4Wdisk.pvs_front_brake_bias*VARVAL（._brake_system_4Wdisk.cis_brake_demand_adams_id）*._brake_system_4Wdisk.force_to_pressure_cnvt*._brake_system_4Wdisk.pvs_front_brake_mu*._brake_system_4Wdisk.pvs_front_effective_piston_radius*STEP（VARVAL（._brake_system_4Wdisk.left_front_wheel_omega），-10D，1，10D，-1）

同理，以左前轮制动力矩函数为例，式中：

（1）._brake_ABS.pvs_front_piston_area 为制动缸活塞有效面积；

（2）._brake_ABS.pvs_front_brake_bias 为前轴系制动力分配系数；

（3）VARVAL（._brake_ABS.cis_brake_demand_adams_id）为制动踏板力；

（4）._brake_ABS.force_to_pressure_cnvt 为换算系数，将制动踏板力直接转化为制动总管液体介质压强，默认为 0.1；

（5）._brake_ABS.pvs_front_brake_mu 为制动器摩擦系数；

（6）._brake_ABS.pvs_front_effective_piston_radius 为制动油缸在制动盘上的作用半径；

（7）STEP（VARVAL（._brake_ABS.left_front_wheel_omega），-10D，1，10D，-1）为阶跃函数，确保制动力矩与车轮旋转方向相反。

4.2 制动系统变量参数及通信器

制动系统的变量参数及输入输出通信器如表 4-1 和表 4-2 所示。在研究制动系统时，可以根据真实的制动系统数据更改变量的参数值（包含制动系统的几何参数、摩擦系数等）。

表 4-1 制动系统变量参数

parameter name	symmetry	type	value
kinematic_flag	single	integer	0
front_brake_bias	single	real	0.6
front_brake_mu	single	real	0.4
front_effective_piston_radius	single	real	135.0
front_piston_area	single	real	2500.0
front_rotor_hub_wheel_offset	single	real	25.0

parameter name	symmetry	type	value
front_rotor_hub_width	single	real	40.0
front_rotor_width	single	real	−25.0
max_brake_value	single	real	100.0
rear_brake_mu	single	real	0.4
rear_effective_piston_radius	single	real	120.0
rear_piston_area	single	real	2500.0
rear_rotor_hub_wheel_offset	single	real	25.0
rear_rotor_hub_width	single	real	40.0
rear_rotor_width	single	real	−25.0

表 4-2 制动系统输入输出通信器

Communicator Name	Entity Class	From Minor Role
ci[lr]_front_camber_angle	parameter_real	front
ci[lr]_front_rotor_to_wheel	mount	front
ci[lr]_front_suspension_upright	mount	front
ci[lr]_front_tire_force	force	front
ci[lr]_front_toe_angle	parameter_real	front
ci[lr]_front_wheel_center	location	front
ci[lr]_rear_camber_angle	parameter_real	rear
ci[lr]_rear_rotor_to_wheel	mount	rear
ci[lr]_rear_suspension_upright	mount	rear
ci[lr]_rear_tire_force	force	rear
ci[lr]_rear_toe_angle	parameter_real	rear
ci[lr]_rear_wheel_center	location	rear
cis_brake_demand	solver_variable	any
cos_max_brake_value	parameter_real	inherit

4.3 FSAE 赛车 Braking 文件驱动仿真

（1）启动 ADAMS/CAR，选择 Standard 标准模块进入界面。

（2）单击 File > Open > Assembly 命令，弹出装配打开对话框，在 Assembly Name 中输入：mdids：//FASE/assemblies.tbl/fsae_full_2017.asy。

（3）单击 OK 按钮，完成方程式赛车整车模型的打开。

（4）单击 Simulate > Full-Vehicle Analysis > Straight-Line Event > Braking 命令，弹出制动

仿真对话框，在下列对话框中输入相应的数据：
- Output Prefix（输出别名）：B_line；
- End Time：10；
- Number Of Steps：1000；
- Simulation Mode（仿真类型）：interactive；
- Road Date File：mdids：//FASE/roads.tbl/2d_flat.rdf；
- Steering Input（转向输入）：lock，转向时保持转向锁定；
- Start Time：4；
- 选择闭环制动模式：Closed-Loop Brake；
- Longitudinal Decel（G's）（制动时侧向加速度）：0.63；
- Gear Position（挡位）：4 挡。

（5）单击 OK 按钮，完成直线 B_line 制动仿真设置并提交运算。

仿真完成后，在计算目录存放一个文件：B_line_brake.xml，路径为：file：//C：/Users/Administrator/B_line_brake.xml；此文件可以用来作为驱动控制文件进行驱动文件控制仿真。仿真结果如图 4-2 ~ 图 4-5 所示。

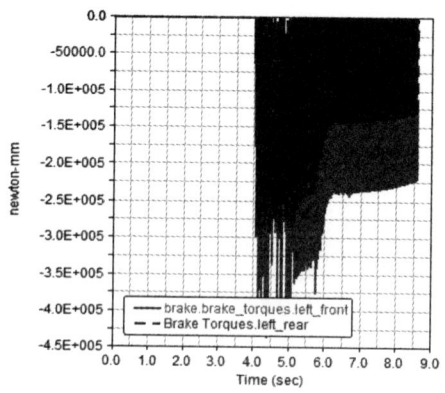

图 4-2　左前轮与左后轮制动力矩

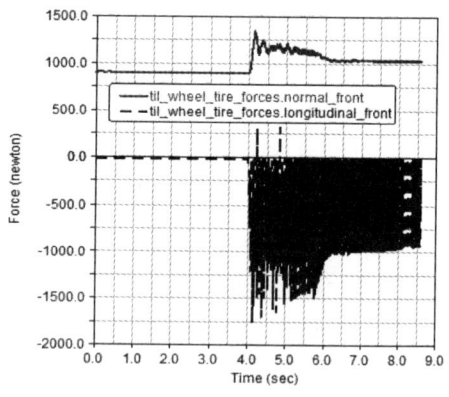

图 4-3　左前轮胎法向力与纵向力

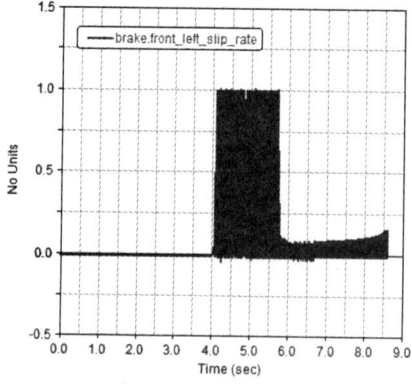

图 4-4　左前轮滑移率

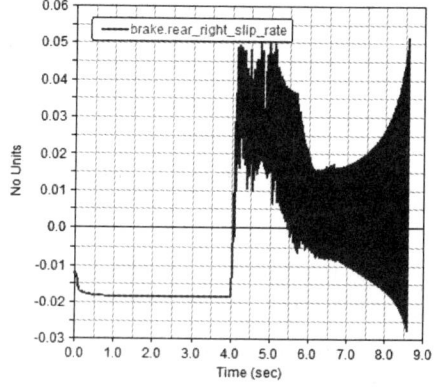

图 4-5　右后轮滑移率

（6）单击 Simulate > Full-Vehicle Analysis > File Driven Event 命令，弹出驱动控制文件仿真对话框，如图 4-6 所示。在下列对话框中输入相应的数据：

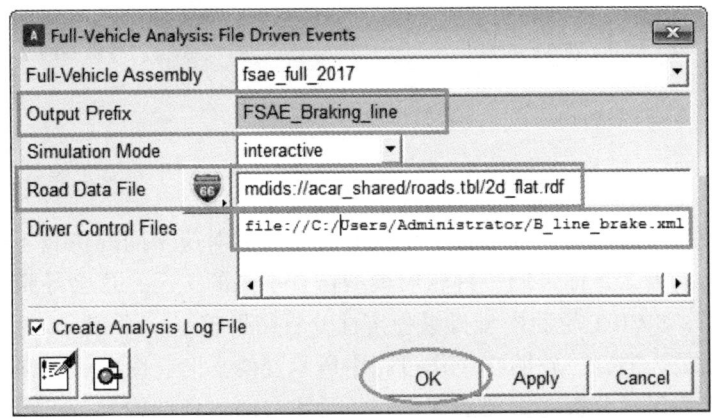

图 4-6　Braking 驱动控制仿真设置对话框

- Output Prefix（输出别名）：FSAE_Braking_line；
- Simulation Mode（仿真类型）：interactive；
- Road Date File：mdids：//acar_shared/roads.tbl/2d_flat.rdf；
- Driver Control Files：file：//C：/Users/Administrator/B_line_brake.xml；此文件为上述 B_line 制动仿真在目录文件夹中的存根。

（7）单击 OK 按钮，完成直线制动 FSAE_Braking_line 驱动控制仿真设置并提交运算，B_line 制动仿真与直线制动 FSAE_Braking_line 驱动控制仿真计算结果完全一样，在此主要为了对驱动控制文件仿真进行说明。

第 5 章　轮胎模型

轮胎是整车模型中必不可少的部件，不同的轮胎模型对应不同的仿真工况。轮胎模型是整车中最容易忽略的因素，即采用一种轮胎模型仿真不同的工况，这会导致有些仿真工况偏离实际较大。例如，SWIFT 轮胎模型操纵稳定性分析结果较好，而 FTire 轮胎模型在平顺性及耐久性分析方面较为准确。轮胎在工作过程中所承受的垂向、纵向、侧向及回正力矩对整车的平顺性、操纵稳定性分析有重要的作用；轮胎本身的物理结构复杂，力学特性为高度非线性，目前为止没有一种轮胎模型可以适用于整车所有的仿真工况。有限元轮胎模型是目前为止对轮胎结构最为详细的描述，能够准确地计算出轮胎的稳态与动态响应，但其计算经济成本太高，主要用于轮胎的设计制造分析。ADAMS/CAR 中建立的比较好的 FSAE 轮胎模型如图 5-1 所示。

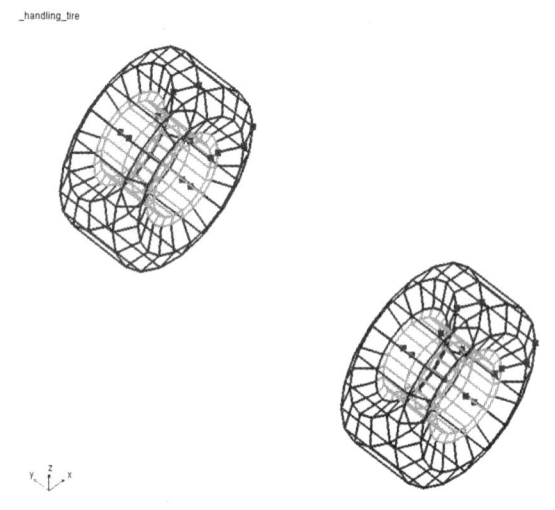

图 5-1　轮胎模型

学习目标

（1）了解轮胎类型。
（2）会轮胎属性文件编制。

5.1 轮胎模型适用性分类

ADAMS/CAR 共享数据库中包含 55 个轮胎模型属性文件，需要注意的是每个轮胎都有特定的实用性范围，不能乱用，否则会导致错误的分析结果。表 5-1 为每种轮胎的正确使用领域；表 5-2 为不同的轮胎模型使用的最佳状况。

表 5-1 轮胎使用领域

	Tire model	transient	gyroscopic effect	scaling factors	combined slip	camber effects	coordinate system	max valid frequency /Hz	tire enveloping effects
ADAMS /Handling Tire	PAC2002	Y	Y	Y	Y	Y	ISO	8	N
	PAC89	N	N	limited	ellips	limited	SAE/ISO	0.5	N
	PAC94	N	N	limited	ellips	Y	SAE	0.5	N
	FLALA	N	N	N	N	N	SAE	0.5	N
	5.2.1	N	N	N	ellips	Y	SAE	0.5	N
	UA Tire	Y	Y	N	ellips	Y	SAE	8	N
Specific Models	PAC_MC	Y	Y	Y	Y	Y	ISO	8	N
	SWIFT-Tire	Y	Y	Y	Y	Y	ISO	60	Y
	FTire	Y	Y	N	Y	Y	ISO	120	Y

表 5-2 轮胎模型使用的最佳状况

Simulation event	ADAMS/Handling Tire						Specific Models		
	PAC2002	PAC89	PAC94	FIALA	5.2.1	UA Tire	PAC_MC	SWIFT-Tire	FTire
standstill	○	–	–	–	–	–	○	○	○
steady-state cornering	+	○	+	○	○	○	+	+	○
cornering over bumpy road	○	–	–	–	–	○	○	+	○
lane change	+	○	○	○	○	○	+	+	○
ABS braking	○	–	–	–	–	–	○	○	+
braking/power-off in a turn	+	○	○	○	○	○	+	+	+
same on bumpy road	○	–	–	–	–	–	○	+	+
shimmy	○	–	–	–	–	○	○	+	+
ride/comfort	–	–	–	–	–	–	–	–	+
chassis control system > 8 Hz	–	–	–	–	–	–	–	+	+
chassis control with ride	–	–	–	–	–	–	–	+	+
durability	–	–	–	–	○	○	–	○	○
– : not possible, not realistic; ○: possible; +: best									

5.2 FSAE 轮胎属性文件

1. 轮胎属性文件

FSAE 赛车轮胎采用 260 干胎品牌，型号、规格分别是：Hossier 43120、19.5×7.5-10；无载荷车轮半径：247.65 mm；轮胎断面宽度：190.5 mm；扁平率：63.3%。

打开 D:\fsae_MD_2010.cdb\tires.tbl 文件夹中的轮胎属性文件 mdi_tire01.tir，参数可在 DIMENSION 数据块中进行更改，更改完成后将轮胎属性文件另存为：fsae_tire_front.tir 及 fsae_tire_rear.tir。

```
轮胎属性文件信息如下：
$---------------------------------------------------------MDI_HEADER
[MDI_HEADER]
 FILE_TYPE       = 'tir'
 FILE_VERSION    = 2.0
 FILE_FORMAT     = 'ASCII'
(COMMENTS)
{comment_string}
'Tire      - XXXXXX'
'Pressure  - XXXXXX'
'Test Date - XXXXXX'
'Test tire'
'New File Format v2.1'
$---------------------------------------------------------UNITS
[UNITS]
 LENGTH              = 'mm'
 FORCE               = 'newton'
 ANGLE               = 'radians'
 MASS                = 'kg'
 TIME                = 'sec'
$---------------------------------------------------------MODEL
[MODEL]
! use mode    1  2  3  4      11  12  13   14
! ------------------------------------------------
! smoothing      X  X          X   X
! combined          X   X  X   X
! transient                    X   X   X    X
 PROPERTY_FILE_FORMAT  = 'PAC89'
 USE_MODE              = 12.0
$---------------------------------------------------------DIMENSION
```

```
[DIMENSION]
 UNLOADED_RADIUS        = 247.65
 WIDTH                  = 190.5
 ASPECT_RATIO           = 0.633
$--------------------------------------------------PARAMETER
[PARAMETER]
 VERTICAL_STIFFNESS     = 150
 VERTICAL_DAMPING       = 9.0
 LATERAL_STIFFNESS      = 190.0
 ROLLING_RESISTANCE     = 0.01
$--------------------------------------------------shape
[SHAPE]
{radial width}
 1.0    0.0
 1.0    0.2
 1.0    0.4
 1.0    0.5
 1.0    0.6
 1.0    0.7
 1.0    0.8
 1.0    0.85
 1.0    0.9
 0.9    1.0
$--------------------------------------------------LATERAL_COEFFICIENTS
[LATERAL_COEFFICIENTS]
 a0  =   2.0
 a1  =  -34.0
 a2  =   1250.00
 a3  =   3036.00
 a4  =   12.80
 a5  =   0.00501
 a6  =  -0.02103
 a7  =   0.77394
 a8  =   0.0022890
 a9  =   0.013442
 a10 =   0.003709
```

```
   a11 =   19.1656
   a12 =   1.21356
   a13 =   6.26206
$------------------------------------------------longitudinal
[LONGITUDINAL_COEFFICIENTS]
   b0  =   2.3
   b1  =   -10
   b2  =   1400
   b3  =   0
   b4  =   175
   b5  =   0.1
   b6  =   0.005
   b7  =   -0.1
   b8  =   1
   b9  =   0
   b10 =   0
$------------------------------------------------aligning
[ALIGNING_COEFFICIENTS]
   c0  =   2.34000
   c1  =   1.4950
   c2  =   6.416654
   c3  =   -3.57403
   c4  =   -0.087737
   c5  =   0.098410
   c6  =   0.0027699
   c7  =   -0.0001151
   c8  =   0.1000
   c9  =   -1.33329
   c10 =   0.025501
   c11 =   -0.02357
   c12 =   0.03027
   c13 =   -0.0647
   c14 =   0.0211329
   c15 =   0.89469
   c16 =   -0.099443
   c17 =   0.0
```

转换到专家模板，单击 File > Open 命令，弹出模板打开对话框，如图 5-2 所示。

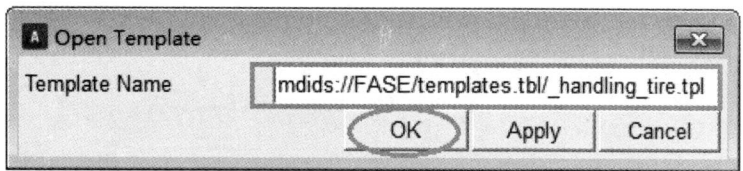

图 5-2　轮胎模板打开对话框

在 Template Name 中输入：mdids：//FASE/templates.tbl/_handling_tire.tpl；单击 OK 按钮，轮胎._handling_tire 显示在窗口中。选择轮胎（左右轮胎均可）右击 Wheel：whr_wheel > Modify，弹出轮胎修改对话框，如图 5-3 所示。

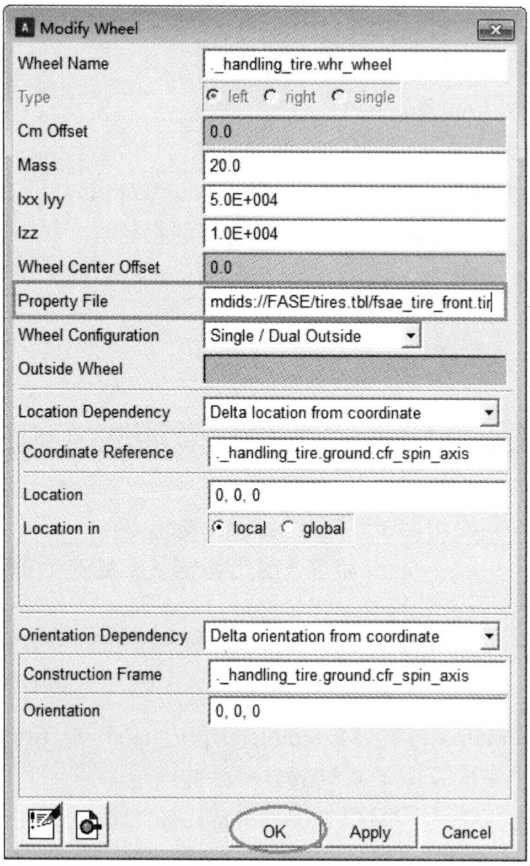

图 5-3　轮胎修改对话框

在 Property File 中输入：mdids：//FASE/tires.tbl/fsae_tire_front.tir，并输入上述修改好的轮胎属性文件 fsae_tire_front.tir 或者 fsae_tire_rear.tir 均可，其余参数均保持默认设置；单击 OK 按钮，完成轮胎._handling_tire.whr_wheel 的修改，此时轮胎在窗口中的尺寸发生了变化。

单击 File > Save as 命令，弹出 Save Template 对话框，保持默认状态，单击 OK 按钮，完成轮胎._handling_tire.whr_wheel 的存储。

2. 轮胎前后子系统

（1）按 F9 键切换到标准模板，单击 File > New > Subsystem 命令，弹出新建子系统对话框，如图 5-4 所示。在下列对话框中输入相应的数据：

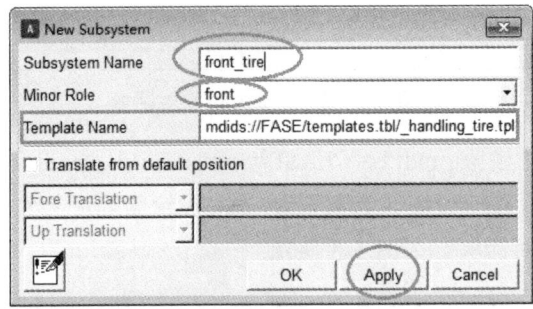

图 5-4 轮胎子系统创建对话框

- Subsystem Name：front_tire；
- Minor Role：front；
- Template Name：mdids://FASE/templates.tbl/_handling_tire.tpl。

（2）单击 OK 按钮，完成轮胎子系统 front_tire 的创建。

（3）单击 File > Save as 命令，弹出 Save Subsystem 对话框，在下列对话框中输入相应的数据：

- Subsystem Name：front_tire；
- Minor Role：front；
- File Format：TeimOrbit；
- Target：Database，FSAE。

（4）单击 OK 按钮，完成轮胎子系统 front_tire 的存储。

（5）单击 File > New > Subsystem 命令，弹出新建子系统对话框，在下列对话框中输入相应的数据：

- Subsystem Name：rear_tire；
- Minor Role：rear；
- Template Name：mdids://FASE/templates.tbl/_handling_tire.tpl。

（6）单击 OK 按钮，完成轮胎子系统 rear_tire 的创建。

（7）单击 File > Save as 命令，弹出 Save Subsystem 对话框，在下列对话框中输入相应的数据：

- Subsystem Name：rear_tire；
- Minor Role：rear；
- File Format：TeimOrbit；
- Target：Database，FSAE。

（8）单击 OK 按钮，完成轮胎子系统 rear_tire 的存储。

第 6 章　路　面

整车模型计算仿真的前提是必须在路面上进行。路面的状态类型较为繁多，以适应不同计算工况的需要。在对整车制动系统评估时，需要设置对开及对接路面；对整车的平顺性计算仿真时，需要不同等级的路面及通过减速带、连续坑洼路面等。ADAMS\CAR 模块共享数据库中 ROAD 文件夹提供的路面文件足以满足日常所需的工况仿真要求，但对于一些特殊工况需要的路面仍需要读者自己建立。

学习目标

（1）了解路面类型。
（2）熟悉对开路面。
（3）熟悉对接路面。
（4）熟悉减速带路面。
（5）熟悉连续障碍路面。
（6）会分离轮胎路面设置。

6.1　路面类型

6.1.1　路面类型简介

路面模型可以分为 2D（二维）与 3D（三维）路面模型。2D 路面接触通常采用点式跟踪法；3D 路面模型为三维轮胎-路面接触模型，用来计算路面和轮胎之间交叉的体积，路面采用一系列离散的三角形片表示，而轮胎用一系列圆柱表示。采用 3D 路面模型（或者称 3D 等效体积路面模型），可以模拟车辆在运动过程中碰到路边台阶、凹坑、粗糙路面及不规则路面运动的情形。3D 等效体积路面模型如图 6-1 所示，此路面由 6 个节点构成 4 个三角形面单元，每个三角形单元的向外单位法向矢量如图 6-1 所示，与有限元网格中定义较为相似。ADAMS/Tire 在定义路面时需要首先指定每个节点在路面参考坐标系下的坐标，再按顺序指定三个节点构成三角形单元，对应每个单元，可以指定不同的摩擦系数。除此之外还有 3D 光滑路面，用于定义停车场、赛道路面等。3D 光滑路面的路面曲率小于轮胎的曲率。

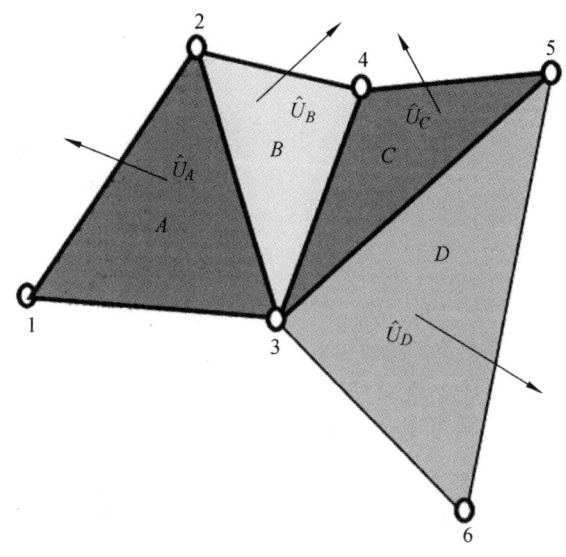

图 6-1 3D 等效体积路面模型

6.1.2 2D 路面类型

路面模型存储于共享数据库的文件夹中，路径为：D：\MSC.Software\Adams_x64\2014\acar\shared_car_database.cdb\roads.tbl。2D 路面模型除平整路面 FLAT 外，其他路面在仿真时均不能显示几何图形。2D 路面类型如下：

① DRUM：测试轮胎用转鼓试验台；
② FLAT：平整路面；
③ PLANK：矩形凸块路面；
④ POLY_LINE：折线路面；
⑤ POT_HOLE：凹坑路面；
⑥ RAMP：斜坡路面；
⑦ ROOF：三角形凸块路面；
⑧ SINE：正弦波路面；
⑨ SINE_SWEEP：正弦波波纹路面；
⑩ STOCHASTIC_UNEVEN：随机不平路面。

（1）单击 Simulate > Component Analysis > cosin/tiretlls 命令，弹出 cosin2014-3 插件对话框，如图 6-2 所示。

（2）单击 File > Open road 命令，弹出选择路面文件对话框，选择正弦波波纹路面 2d_sine_sweep.rdf。

（3）单击"打开"按钮，弹出 roadtools 工具对话框，如图 6-3 所示。

（4）单击显示按钮快捷方式，显示正弦波波纹路面，如图 6-4 所示，其余不同类型路面形状读者可自行尝试打开观看。

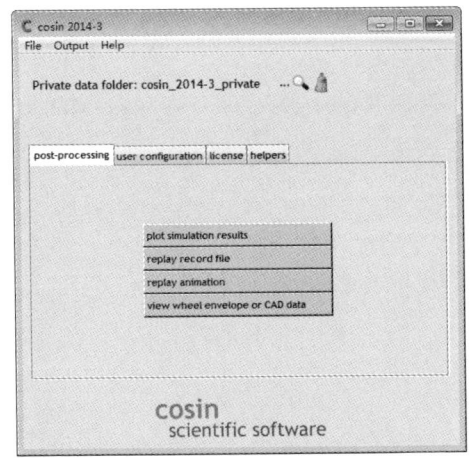

图 6-2 cosin2014-3 插件

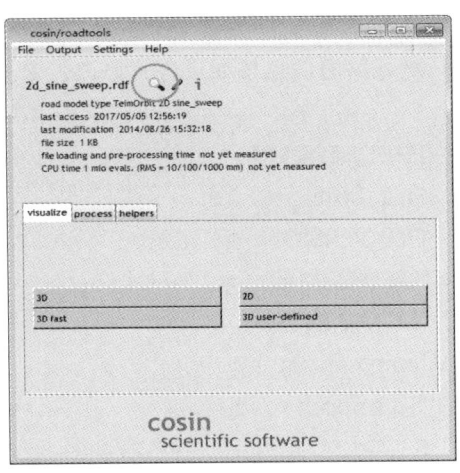

图 6-3 roadtools 工具对话框

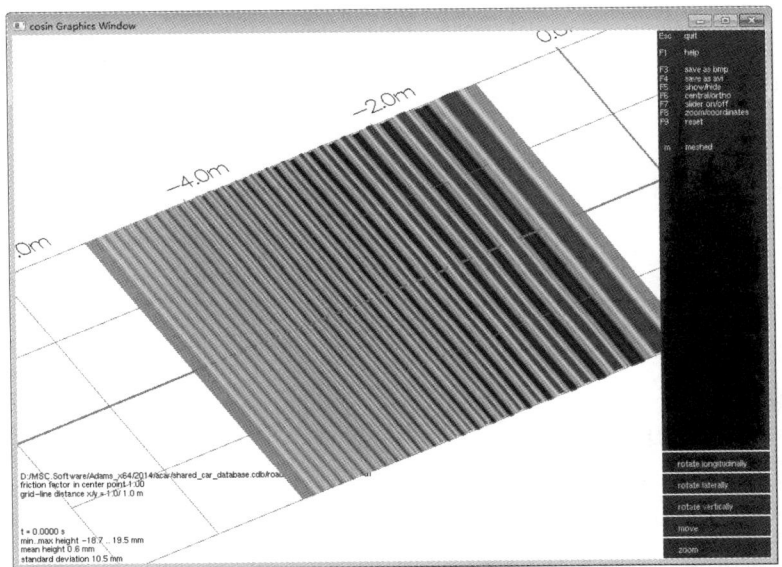

图 6-4 正弦波波纹路面图

6.2 对开路面

对开路面主要用于车辆 ABS 制动状态下系统的仿真，以路面中轴线为界，左右两侧的路面摩擦系数不用，或者模拟车辆在失控状态下整车的稳定性能。真实车辆在制动过程中，左右两侧车轮可能处在不同的路面上。对开路面编辑以 3D 样条路面：mdi_3d_smooth_road.rdf 为模板，对路面左侧摩擦系数 MU_LEFT 与右侧路面摩擦系数 MU_RIGHT 进行更改；高低附路面以摩擦系数 0.5 为中间值，大于 0.5 为高附路面，小于 0.5 为低附路面，同时要求高低附路面摩擦系数比值大于等于 2。对 3D 样条路面 mdi_3d_smooth_road.rdf 进行局部修改，修改部分用斜体加下划线标注。修改好的路面另存为 mdi_3d_smooth_road_DK.rdf，文件存放于章节文件夹中。

对开路面信息按如下方式修改：

```
$---------------------------------------------------------------MDI_HEADER
[MDI_HEADER]
FILE_TYPE        =   'rdf'
FILE_VERSION     =   5.00
FILE_FORMAT      =   'ASCII'
(COMMENTS)
{comment_string}
'3d smooth road'
$---------------------------------------------------------------UNITS
[UNITS]
 LENGTH                = 'meter'
 FORCE                 = 'newton'
 ANGLE                 = 'radians'
 MASS                  = 'kg'
 TIME                  = 'sec'
$---------------------------------------------------------------DEFINITION
[MODEL]
 METHOD                = '3D_SPLINE'
 FUNCTION_NAME         = 'ARC903'
 VERSION               = 1.00
$---------------------------------------------------------------ROAD_PARAMETERS
[GLOBAL_PARAMETERS]
 CLOSED_ROAD           = 'no'
 SEARCH_ALGORITHM      = 'FaSt'
 ROAD_VERTICAL         = '0.0 0.0 1.0'
 FORWARD_DIR           =  'NORMAL'
 MU_LEFT               = 1.0
 MU_RIGHT              = 1.0
 WIDTH                 = 7.000
 BANK                  = 0.0

$---------------------------------------------------------------DATA_POINTS
[DATA_POINTS]
{  X              Y              Z           WIDTH  BANK  MU_LEFT  MU_RIGHT }
 12.50000E+00  0.00000E-00  0.00000E-00   7.000  0.000   _**0.800**_   _**0.400**_
```

```
 10.50000E+00   0.00000E-00   0.00000E-00   7.000 0.000   0.800   0.400
  5.50000E+00   0.00000E-00   0.00000E-00   7.000 0.000   0.800   0.400
  0.50000E+00   0.00000E-00   0.00000E-00   7.000 0.000   0.800   0.400
  0.00000E+00   0.00000E-00   0.00000E-00   7.000 0.000   0.800   0.400
 -2.50000E+00   0.00000E-00   0.00000E-00   7.000 0.000   0.800   0.400
 -5.00000E+00   0.00000E-00   0.00000E-00   7.000 0.000   0.800   0.400
 -1.00000E+01   0.00000E-00   0.00000E-00   7.000 0.000   0.800   0.400
 -2.00000E+01   0.00000E-00   0.10000E-00   7.000 0.000   0.800   0.400
 -3.00000E+01   0.00000E-00   0.20000E-00   7.000 0.000   0.800   0.400
 -4.00000E+01   0.00000E-00   0.30000E-00   7.000 0.000   0.800   0.400
 -5.00000E+01   0.00000E-00   0.40000E-00   7.000 0.000   0.800   0.400
 -6.00000E+01   0.00000E-00   0.50000E-00   7.000 0.000   0.800   0.400
 -7.00000E+01   0.00000E-00   0.60000E-00   7.000 0.000   0.800   0.400
 -8.00000E+01   0.00000E-00   0.70000E-00   7.000 0.000   0.800   0.400
 -9.00000E+01   0.00000E-00   0.80000E-00   7.000 0.000   0.800   0.400
 -1.00000E+02   0.00000E-00   0.90000E-00   7.000 0.000   0.800   0.400
 -1.10000E+02   0.00000E-00   1.00000E+00   7.000 0.000   0.800   0.400
 -1.20000E+02   0.00000E-00   1.10000E-00   7.000 0.000   0.800   0.400
 -1.30000E+02   0.00000E-00   1.20000E-00   7.000 0.000   0.800   0.400
$-----------------------------------------------------------END_DATA_POINTS
```

6.3 对接路面

对接路面同样用于车辆 ABS 制动状态下系统的仿真,对接路面以长度为单位作为一个整体,每个整体路面摩擦系数不同,以路面中轴线为界。对接路面编辑以 3D 样条路面:mdi_3d_smooth_road.rdf 为模板,经过某一个长度后(长度的大小可以对整车进行直线制动仿真估计),路面左右侧的摩擦系数同时变更,一般情况下变小;高低附路面以摩擦系数 0.5 为中间值,大于 0.5 为高附路面,小于 0.5 为低附路面,同时要求高低附路面摩擦系数比值大于等于 2。对 3D 样条路面 mdi_3d_smooth_road.rdf 进行局部修改,修改部分用斜体加下划线标注。修改好的路面另存为 mdi_3d_smooth_road_DJ.rdf,文件存放于章节文件夹中。

```
对接路面信息按如下方式修改:
$-----------------------------------------------------------------MDI_HEADER
[MDI_HEADER]
FILE_TYPE     =  'rdf'
FILE_VERSION  =   5.00
FILE_FORMAT   =  'ASCII'
(COMMENTS)
```

```
{comment_string}
'3d smooth road'
$--------------------------------------------------------------UNITS
[UNITS]
 LENGTH                = 'meter'
 FORCE                 = 'newton'
 ANGLE                 = 'radians'
 MASS                  = 'kg'
 TIME                  = 'sec'
$----------------------------------------------------------DEFINITION
[MODEL]
 METHOD                = '3D_SPLINE'
 FUNCTION_NAME         = 'ARC903'
 VERSION               = 1.00
$-----------------------------------------------------ROAD_PARAMETERS
[GLOBAL_PARAMETERS]
 CLOSED_ROAD           = 'no'
 SEARCH_ALGORITHM      = 'FaSt'
 ROAD_VERTICAL         = '0.0 0.0 1.0'
 FORWARD_DIR           =  'NORMAL'
 MU_LEFT               =  1.0
 MU_RIGHT              =  1.0
 WIDTH                 =  7.000
 BANK                  =  0.0
$---------------------------------------------------------DATA_POINTS
[DATA_POINTS]
{     X              Y              Z         WIDTH  BANK  MU_LEFT  MU_RIGHT }
 12.50000E+00   0.00000E-00    0.00000E-00    3.000  0.000  0.900   0.900
 10.50000E+00   0.00000E-00    0.00000E-00    3.000  0.000  0.900   0.900
  5.50000E+00   0.00000E-00    0.00000E-00    3.000  0.000  0.900   0.900
  0.50000E+00   0.00000E-00    0.00000E-00    3.000  0.000  0.900   0.900
  0.00000E+00   0.00000E-00    0.00000E-00    3.000  0.000  0.900   0.900
 -2.50000E+00   0.00000E-00    0.00000E-00    3.000  0.000  0.900   0.900
 -5.00000E+00   0.00000E-00    0.00000E-00    3.000  0.000  0.900   0.900
 -1.00000E+01   0.00000E-00    0.00000E-00    3.000  0.000  *0.300*  *0.300*
 -2.00000E+01   0.00000E-00    0.10000E-00    3.000  0.000  *0.300*  *0.300*
 -3.00000E+01   0.00000E-00    0.20000E-00    3.000  0.000  *0.300*  *0.300*
 -4.00000E+01   0.00000E-00    0.30000E-00    3.000  0.000  *0.300*  *0.300*
```

-5.00000E+01	0.00000E-00	0.40000E-00	3.000	0.000	*0.300*	*0.300*
-6.00000E+01	0.00000E-00	0.50000E-00	3.000	0.000	*0.300*	*0.300*
-7.00000E+01	0.00000E-00	0.60000E-00	3.000	0.000	*0.300*	*0.300*
-8.00000E+01	0.00000E-00	0.70000E-00	3.000	0.000	*0.300*	*0.300*
-9.00000E+01	0.00000E-00	0.80000E-00	3.000	0.000	*0.300*	*0.300*
-1.00000E+02	0.00000E-00	0.90000E-00	3.000	0.000	*0.300*	*0.300*
-1.10000E+02	0.00000E-00	1.00000E+00	3.000	0.000	*0.300*	*0.300*
-1.20000E+02	0.00000E-00	1.10000E-00	3.000	0.000	*0.300*	*0.300*
-1.30000E+02	0.00000E-00	1.20000E-00	3.000	0.000	*0.300*	*0.300*

$------------------------------------END_DATA_POINTS

6.4 减速带路面

减速带主要设置在路口、学校、小区门口等车流量较多、人口较为密集的地方，提示车辆减速慢行，注意安全。减速带规格类型较多，此案例采用的减速带规格为 250 mm × 350 mm × 50 mm（长、宽、高），其中减速带断面参数为 350 mm × 50 mm；通过 ADAMS\CAR 建立减速带模型，模拟 FSAE 赛车通过减速带时整车的运动状态。

（1）单击 Simulate > Full-Vehicle Analyses > Road builder 命令，弹出路面构建对话框，如图 6-5 所示；对话框主要包含四部分：路面文件、标题栏、路面文件版本信息、路面单位信息。

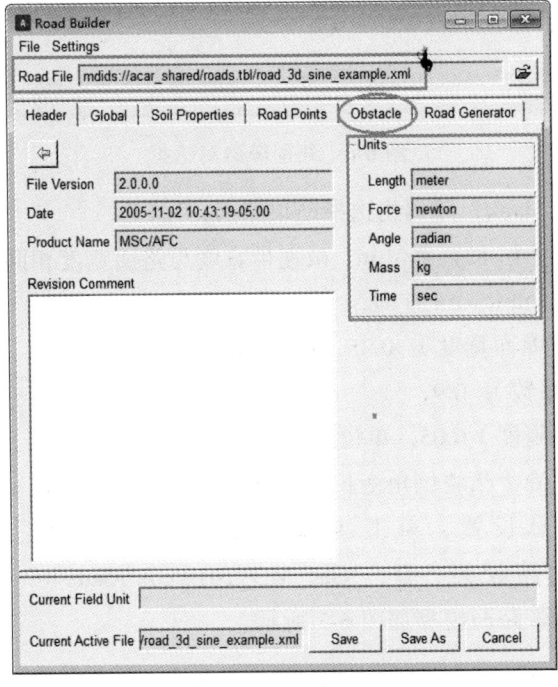

图 6-5 路面构建对话框

在 Road File 中输入路面文件：D：\MSC.Software\Adams_x64\2014\acar\shared_car_database.cdb\roads.tbl\ road_3d_sine_example.xml；路面建模器打开后默认存在，也可以点击后面的文件快捷方式输入其他路面文件；界面其余设置均保持默认。

（2）单击 Obstacle（障碍物，包括凸块路面、凹坑路面、三角形凸台路面等），此时图 6-5 转换成障碍物路面设置界面，如图 6-6 所示。在下列数据框中输入相应的数据：

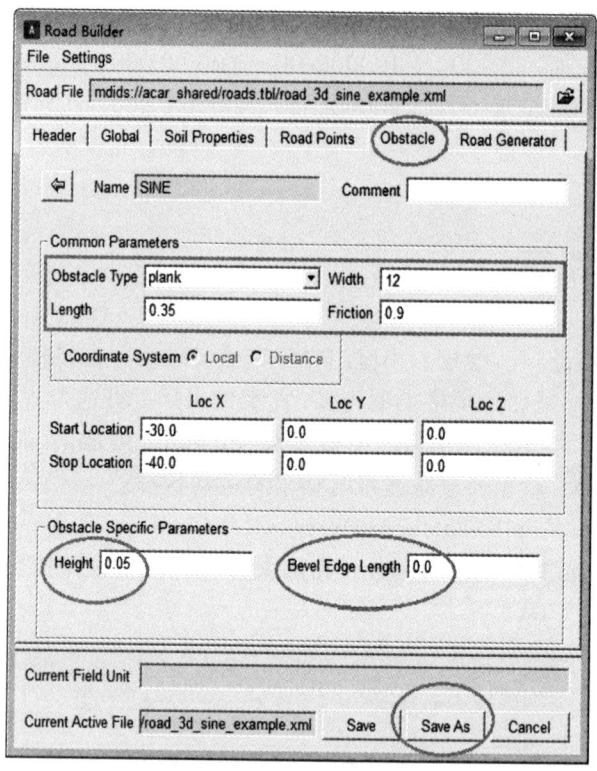

图 6-6　路面障碍对话框

- Obstacle Type：plank，障碍物选择凸块路面；
- Width（路面宽度）：12，单位 m，减速带宽度与路面宽度相同，路面宽度可以用记事本打开 road_3d_sine_example.xml 查询；
- Length（减速带断面宽度）：0.35，单位 m；
- Friction（摩擦系数）：0.9；
- Height（减速带高度）0.05，单位 m；
- Bevel Edge Length（凸块倒角变长度，默认角度为 45°）：0，单位 m。

（3）其余保持默认设置，单击 Save As 标签，另存为：road_3d_sine_example_JIANSUDAI.xml；存储路径为：D：\fsae_MD_2010.cdb\roads.tbl\ road_3d_sine_example_JIANSUDAI.xml。减速带路面模型如图 6-7 所示。

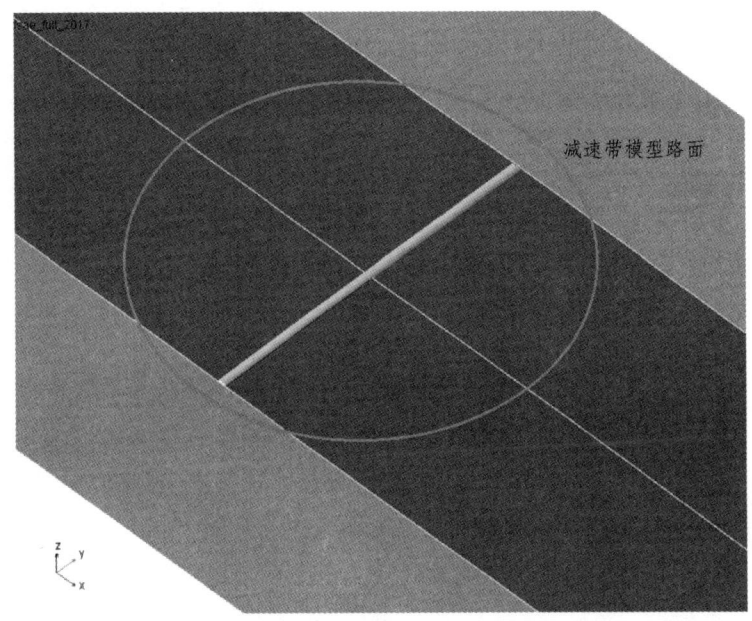

图 6-7 减速带路面模型

6.5 连续障碍路面

整车在高速路上行驶时，会遇到多个连续减速带，提示驾驶员与前车保持合适的车距；在整车设计量产之前，需要对整车的性能进行评估，也需要整车在随机不平路面上或者连续障碍路面上行驶。连续 3 个减速带路面创建如下，其他障碍路面创建也可参考：

（1）单击 Simulate > Full-Vehicle Analyses > Road builder 命令，弹出路面构建对话框，如图 6-5 所示。在 Road File 中输入：D：\fsae_MD_2010.cdb\roads.tbl\ road_3d_sine_example_JIANSUDAI.xml。

（2）单击 Obstacle，弹出路面障碍对话框。

（3）单击 Display table view，显示出连续障碍路面设置对话框，如图 6-8 所示。在 Name 中输入：sine_1。

（4）单击 Add 按钮，双击列表中的 sine_1，界面转换成图 6-6 所示。在下列对话框中输入相应的数据：

- Obstacle Type：plank，障碍物选择凸块路面；
- Width（路面宽度）：12，单位 m，减速带宽度与路面宽度相同，路面宽度可以用记事本打开 road_3d_sine_example.xml 查询；

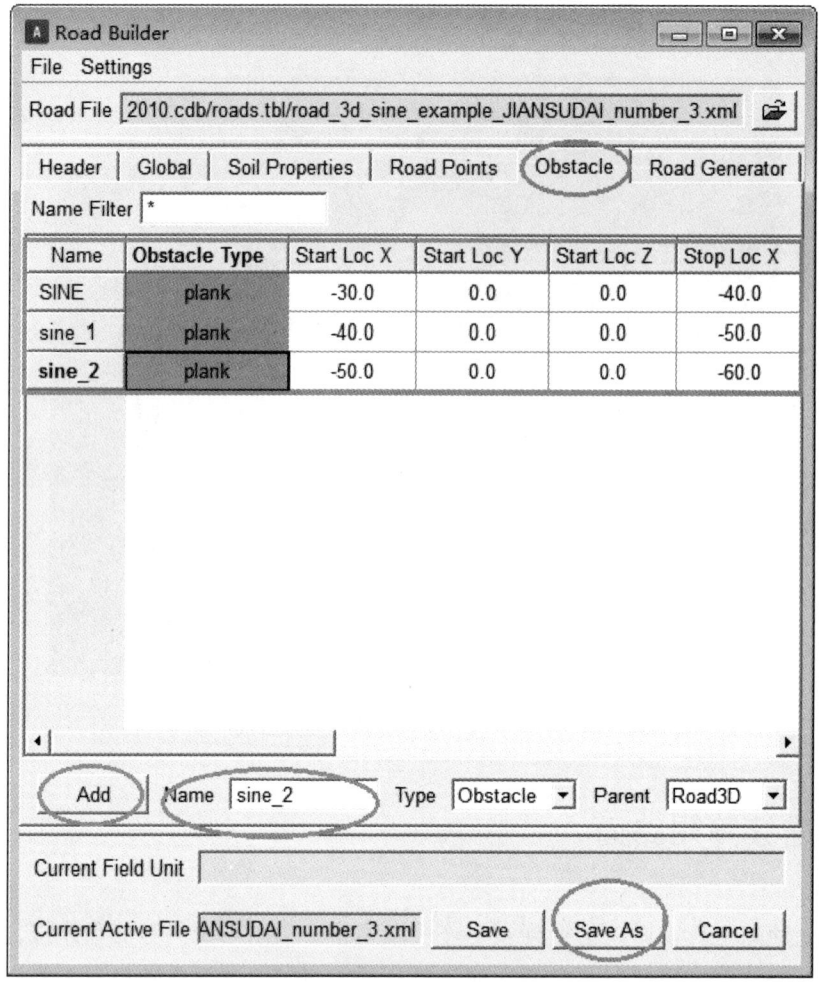

图 6-8　路面连续障碍设置对话框

- Length（减速带断面宽度）：0.35，单位 m；
- Friction（摩擦系数）：0.9；
- Height（减速带高度）0.05，单位 m；
- Start Loc X ：-40；
- Stop Loc X ：-50。

（5）单击 Display table view，重复一次上述过程；在 Name 中输入：sine_2。同样，输入相应的数据：

- Start Loc X ：-50；
- Stop Loc X ：-60；

（6）连续减速带路面如图 6-9 所示，其余保持默认设置，单击 Save As 标签，另存为：road_3d_sine_example_JIANSUDAI_number_3.xml；存储路径为：D：\fsae_MD_2010.cdb\roads.tbl\ road_3d_sine_example_JIANSUDAI_number_3.xml。

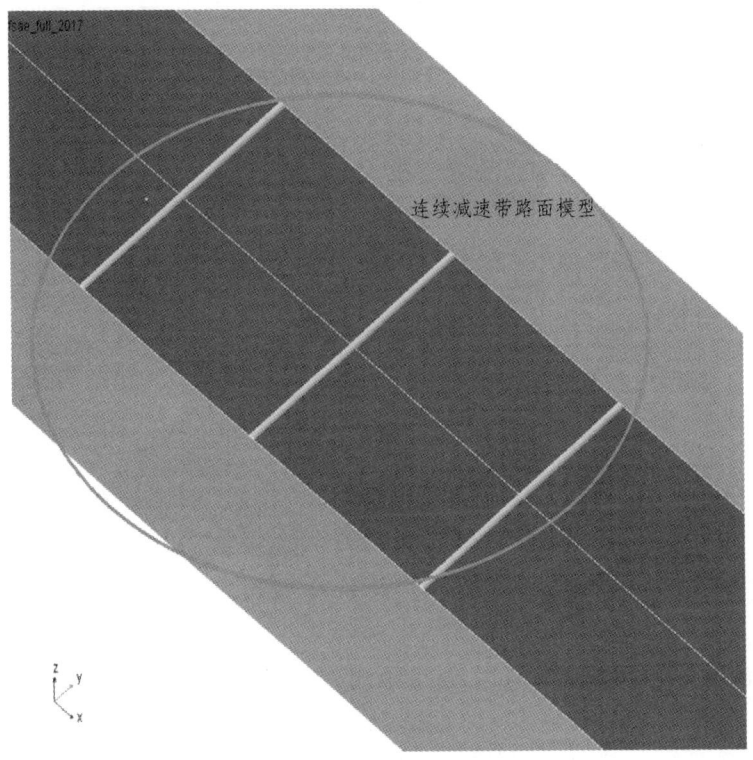

图 6-9 连续减速带路面模型图

6.6 分离路面设置

整车在行驶过程中,四个轮胎接触的路面不可能完全相同,即使是在良好的一级路面上,也会存在微小差异。针对整车的制动特性,在一些特殊路面,如雨地、雪地、坑洼泥泞路面,四个车轮(或者多个车轮)与路面接触不可能具有相同的摩擦系数。因此,有必要在虚拟仿真时设置分离路面,左右车轮或者四个车轮设置不同的摩擦系数。

根据文件夹路径 D:\fsae_MD_2010.cdb\roads.tbl,用记事本格式打开平整路面文件 2d_flat.rdf,如下信息所示,在 PARAMETERS 栏修改 MU=0.5,保存文件重命名为:2d_flat_mu_0.5.rdf。

```
平整路面信息如下:
$---------------------------------------------------------MDI_HEADER
[MDI_HEADER]
FILE_TYPE       =   'rdf'
FILE_VERSION    =   5.00
FILE_FORMAT     =   'ASCII'
(COMMENTS)
{comment_string}
'flat 2d contact road for testing purposes'
```

```
$------------------------------------------------------------UNITS
[UNITS]
 LENGTH              = 'mm'
 FORCE               = 'newton'
 ANGLE               = 'radians'
 MASS                = 'kg'
 TIME                = 'sec'
$------------------------------------------------------------MODEL
[MODEL]
 METHOD              = '2D'
 FUNCTION_NAME       = 'ARC901'
 ROAD_TYPE           = 'flat'
$------------------------------------------------------------GRAPHICS
[GRAPHICS]
 LENGTH              = 160000.0
 WIDTH               = 80000.0
 NUM_LENGTH_GRIDS    = 16
 NUM_WIDTH_GRIDS     = 8
 LENGTH_SHIFT        = 10000.0
 WIDTH_SHIFT         = 0.0       %此栏参数也可以修改,用以改变路面的大小
$------------------------------------------------------------PARAMETERS
[PARAMETERS]
 MU                  = 0.5       %可修改的轮胎与路面的接触摩擦系数,范围在0到1之间;
$------------------------------------------------------------REFSYS
[REFSYS]
 OFFSET                  = 0.0 0.0 0.0
 ROTATION_ANGLE_XY_PLANE = 0.0
```

在标准界面下打开整车模型:fsae_full_2017,也可用本书其他章节的整车模型。

(1)单击 Simulate > Full-Vehicle Analyses > Vehicle Set-Up > Set Road for individual Tires 命令,弹出分离轮胎路面数据文件对话框,如图 6-10 所示。在下列对话框中输入相应的数据:

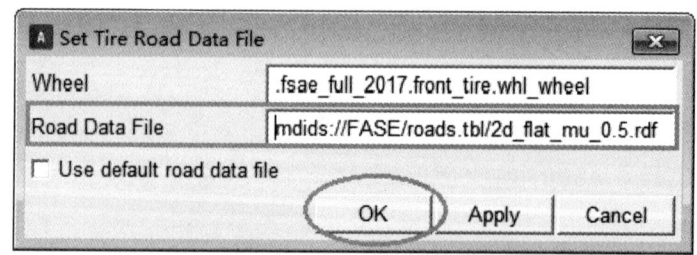

图 6-10 分离轮胎路面设置对话框

- Wheel：.fsae_full_2017.front_tire.whl_wheel（在方框中右击 Wheel > Pick 选择）；
- 不勾选使用路面默认文件：Use default road date file；
- Road Date File：mdids：//FASE/roads.tbl/2d_flat_mu_0.5.rdf。

（2）单击 Apply 按钮，完成左前轮轮胎路面的设置，用同样的方法设置左后轮胎路面。

- Wheel：.fsae_full_2017.rear_tire.whl_wheel（在方框中右击 Wheel > Pick 选择）；
- 不勾选使用路面默认文件：Use default road date file；
- Road Date File：mdids：//FASE/roads.tbl/2d_flat_mu_0.5.rdf。

（3）单击 OK 按钮，完成左后轮轮胎路面的设置。

第 7 章　横向稳定杆 I

横向稳定杆的功用是防止车身在转弯时发生过大的横向侧倾，尽量使车身保持平衡。其目的是减少汽车横向侧倾程度和改善平顺性。横向稳定杆实际上是一个横置的扭杆弹簧，在功能上可以看成是一种特殊的弹性元件。当车身只做垂直运动时，两侧悬架变形相同，横向稳定杆不起作用。当汽车转弯时，车身侧倾，两侧悬架跳动不一致，外侧悬架会压向稳定杆，稳定杆就会发生扭曲，杆身的弹力会阻止车轮抬起，从而使车身尽量保持平衡，起到横向稳定的作用。图 7-1 为 FSAE 赛车前横向稳定杆，该横向稳定杆采用简化模型。

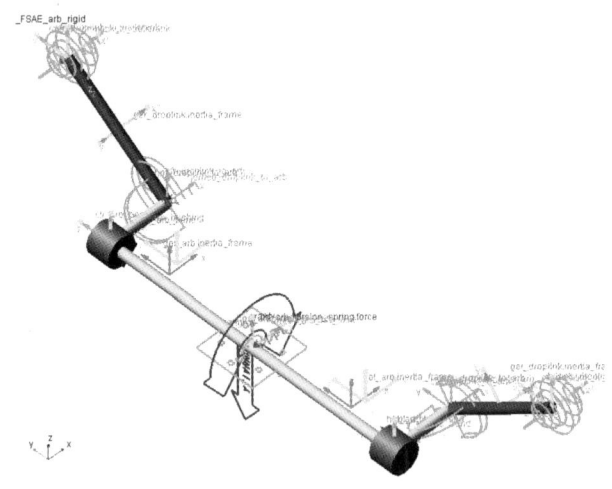

图 7-1　横向稳定杆

学习目标

（1）了解横向稳定杆简化思路。

（2）会横向稳定杆约束。

（3）了解横向稳定杆通信器。

（4）了解扭转弹簧刚度参数。

（5）会模型装配。

（6）会单轮跳动仿真。

（7）会横向稳定杆匹配。

7.1 横向稳定杆模型

（1）启动 ADAMS/CAR，选择专家模块进入建模界面。

（2）单击 File > New 命令，弹出建模对话框，如图 7-2 所示。在下列对话框中输入相应的数据：

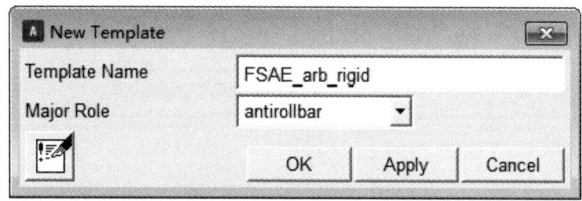

图 7-2 横向稳定杆模板框

- Template Name（模板名称）：FSAE_arb_rigid；
- Major Role（主特征）：antirollbar；

（3）单击 OK 按钮，完成横向稳定杆模板的创建。

7.1.1 横向稳定杆硬件数据

（1）单击 Build > Hardpoind > New 命令，弹出创建硬点对话框，如图 7-3 所示。在下列对话框中输入相应的数据：

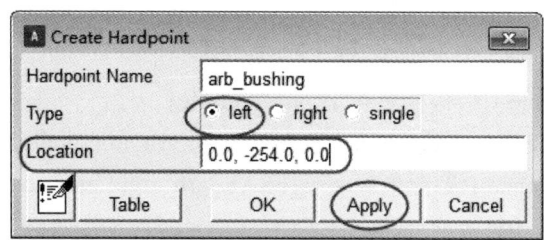

图 7-3 硬点创建对话框

- Hardpoint Name：arb_bushing；
- Type：left；
- Location（置文本框）：0.0，-254.0，0.0。

（2）单击 Apply 按钮，完成._FSAE_arb_rigid.ground.hpl_arb_bushing 硬点的创建；此时在屏幕上显示出左右对称的两个硬点；以此类推，重复上述步骤完成图 7-4 中硬点的创建。

	loc x	loc y	loc z
hpl_arb_bend	0.0	-254.0	0.0
hpl_arb_bushing	0.0	-254.0	0.0
hpl_droplink_to_arb	101.6	-254.0	0.0
hps_arb_middle	0.0	0.0	0.0

图 7-4 硬点数据

（3）单击 Build > Construction Frame > New 命令，创建结构框，如图 7-5 所示。在下列对话框中输入相应的数据：

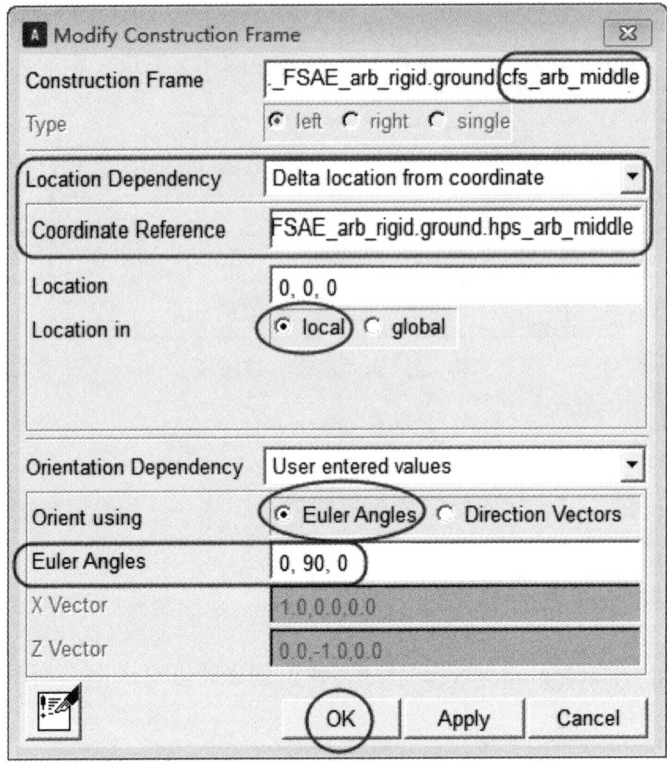

图 7-5　结构框

- Construction Frame（结构框名称）：arb_middle；
- Location Dependency：Delta location from coor arb_middledinate；
- Coordinate Reference（参考坐标）：._FSAE_arb_rigid.ground.hps_arb_middle；
- Location：0，0，0；
- Location in：local；
- Orientation Dependency：User entered values；
- Orient using：Euler Angles；
- Euler Angles：0，90，0。

（4）单击 Apply 按钮，完成._FSAE_arb_rigid.ground.cfs_arb_middle 结构框的创建。同理，创建其他结构框。

- Construction Frame（结构框名称）：arb_bend；
- Location Dependency：Delta location from coor arb_middledinate；
- Coordinate Reference（参考坐标）：._FSAE_arb_rigid.ground.hpl_arb_bend；
- Location：0，0，0。
- Location in：local；
- Orientation Dependency：User entered values；
- Orient using：Euler Angles；

- Euler Angles：0，0，0。

（5）单击 Apply 按钮，完成._FSAE_arb_rigid.ground.cfl_arb_bend 结构框的创建。
- Construction Frame（结构框名称）：droplink_to_arb；
- Location Dependency：Delta location from coor arb_middledinate；
- Coordinate Reference（参考坐标）：._FSAE_arb_rigid.ground.hpl_droplink_to_arb；
- Location：0，0，0；
- Location in：local；
- Orientation Dependency：User entered values；
- Orient using：Euler Angles；
- Euler Angles：0，0，0。

（6）单击 OK 按钮，完成._FSAE_arb_rigid.ground.cfl_droplink_to_arb 结构框的创建。

（7）单击 Build > Part > General Part > New 命令，弹出创建扭杆部件对话框，如图 7-6 所示。在下列对话框中输入相应的数据：

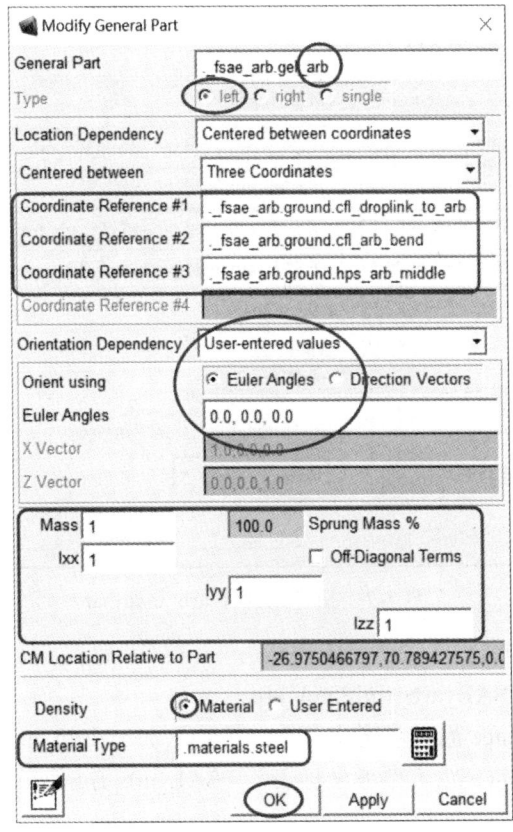

图 7-6　扭杆部件_arb

- General Part：arb；
- Location Dependency：Centered between coordinates；
- Centered between：Three Coordinates；
- Coordinate Reference #1（参考坐标）：._FSAE_arb_rigid.ground.cfl_droplink_to_arb；

- Coordinate Reference #2（参考坐标）：._FSAE_arb_rigid.ground.cfl_arb_bend；
- Coordinate Reference #3（参考坐标）：._FSAE_arb_rigid.ground.cfs_arb_middle；
- Orientation Dependency：User entered values；
- Orient using：Euler Angles；
- Euler Angles：0，0，0；
- Mass：1；
- Ixx：1；
- Iyy：1；
- Izz：1；
- Density：Material；
- Material Type：.materials.steel。

（8）单击 OK 按钮，完成部件._FSAE_arb_rigid.gel_arb 的创建。

（9）单击 Build > Geometry > Link > New 命令，弹出扭杆几何体对话框，如图 7-7 所示。在下列对话框中输入相应的数据：

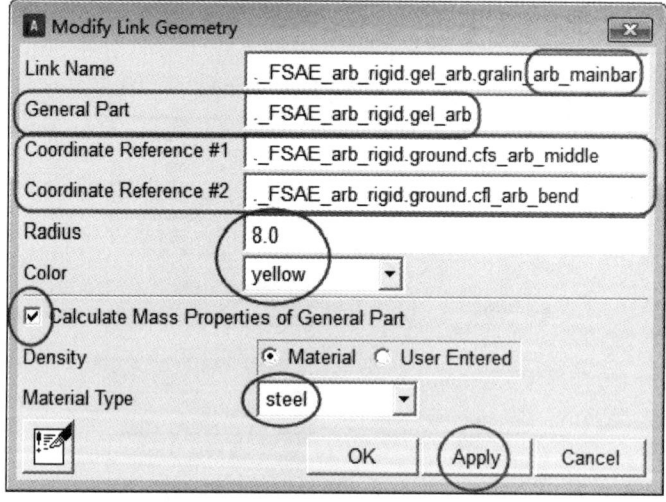

图 7-7 扭杆几何体 arb_mainbar

- Link Name（连杆名称）：arb_mainbar；
- General Part：._FSAE_arb_rigid.gel_arb；
- Coordinate Reference #1（参考坐标）：._FSAE_arb_rigid.ground.cfs_arb_middle；
- Coordinate Reference #2（参考坐标）：._FSAE_arb_rigid.ground.cfl_arb_bend；
- Radius（半径）：8；
- Color（杆件几何体颜色）：yellow；

• 选择 Calculate Mass Properties of General Part 复选框，当几何体建立好之后，会更新对应部件的质量和惯量参数；

- Density：Material；
- Material Type：steel。

（10）单击 Apply 按钮，完成 arb_mainbar 几何体的创建。同理，创建其他几何体。
- Link Name（连杆名称）：arb_arm；
- General Part：._FSAE_arb_rigid.gel_arb；
- Coordinate Reference #1（参考坐标）：._FSAE_arb_rigid.ground.cfl_arb_bend；
- Coordinate Reference #2（参考坐标）：._FSAE_arb_rigid.ground.cfl_droplink_to_arb；
- Radius（半径）：8；
- Color（杆件几何体颜色）：yellow；
- 选择 Calculate Mass Properties of General Part 复选框，当几何体建立好之后，会更新对应部件的质量和惯量参数；
- Density：Material；
- Material Type：steel。

（11）单击 OK 按钮，完成 arb_arm 几何体的创建。

7.1.2 稳定杆通信器

（1）单击 Build > Communicator > Input >New 命令，弹出通信器参数对话框，如图 7-8 所示。在下列对话框中输入相应的数据：

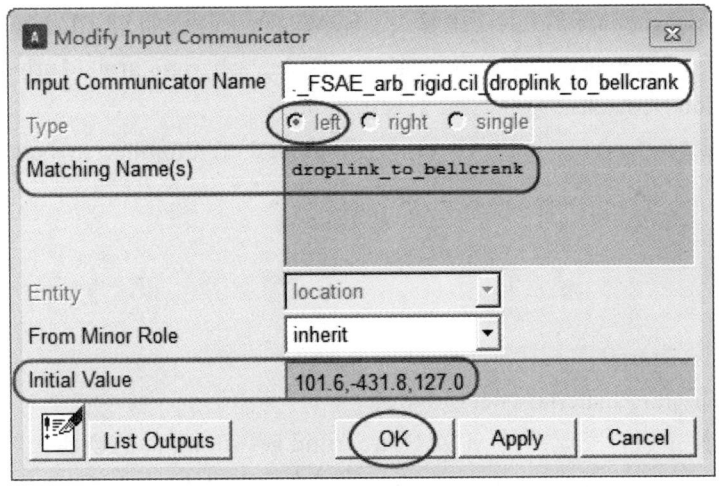

图 7-8 通信器对话框

- Output Communicator Name（输出通信器名称）：droplink_to_bellcrank；
- Matching Name（s）：droplink_to_bellcrank；
- Type：left；
- Entity：location；
- From Minor Role：inherit；
- Initial Value：101.6，-431.8，127.0。

（2）单击 OK 按钮，完成通信器._FSAE_arb_rigid.cil_droplink_to_bellcrank 的创建。

7.1.3 扭杆臂连杆部件

（1）单击 Build > Construction Frame > New 命令，创建结构框，如图 7-5 所示。在下列对话框中输入相应的数据：
- Construction Frame（结构框名称）：droplink_to_bellcrank；
- Location Dependency：Location input communicator；
- Input communicator（输入通信器）：._FSAE_arb_rigid.cil_droplink_to_bellcrank；
- Orientation Dependency：User entered values；
- Orient using：Euler Angles；
- Euler Angles：0，0，0。

（2）单击 OK 按钮，完成._FSAE_arb_rigid.ground.cfl_droplink_to_bellcrank 结构框的创建。在通信器基础上建立参考坐标的优势是，在子系统或者整车状态时，可以快速对接不同部件的连接，同时对应的几何体参数也会根据位置响应调整。

（3）单击 Build > Part > General Part > New 命令，创建扭杆臂连接部件，如图 7-6 所示。在下列对话框中输入相应的数据：
- General Part：droplink；
- Location Dependency：Centered between coordinates；
- Centered between：Two Coordinates；
- Coordinate Reference #1（参考坐标）：._FSAE_arb_rigid.ground.cfl_droplink_to_bellcrank；
- Coordinate Reference #2（参考坐标）：._FSAE_arb_rigid.ground.cfl_droplink_to_arb；
- Orientation Dependency：Orient axis to point；
- Coordinate Reference：._FSAE_arb_rigid.ground.cfl_droplink_to_bellcrank；
- Mass：1；
- Ixx：1；
- Iyy：1；
- Izz：1；
- Density：Material；
- Material Type：.materials.steel。

（4）单击 OK 按钮，完成部件._FSAE_arb_rigid.gel_droplink 的创建。然后创建几何体。

（5）单击 Build > Geometry > Link > New，创建梁几何体。
- Link Name（连杆名称）：droplink；
- General Part：._FSAE_arb_rigid.gel_droplink；
- Coordinate Reference #1（参考坐标）：._FSAE_arb_rigid.ground.cfl_droplink_to_bellcrank；
- Coordinate Reference #2（参考坐标）：._FSAE_arb_rigid.ground.cfl_droplink_to_arb；
- Radius（半径）：8；
- Color（杆件几何体颜色）：red；
- 选择 Calculate Mass Properties of General Part 复选框，当几何体建立好之后，会更新对应部件的质量和惯量参数；
- Density：Material；

- Material Type：steel。

（6）单击 OK 按钮，完成._FSAE_arb_rigid.gel_droplink.gralin_droplink 几何体的创建。

7.1.4 稳定杆安装部件

（1）单击 Build > Part > Mount > New 命令，弹出创建安装部件对话框，如图 7-9 所示。在下列对话框中输入相应的数据：

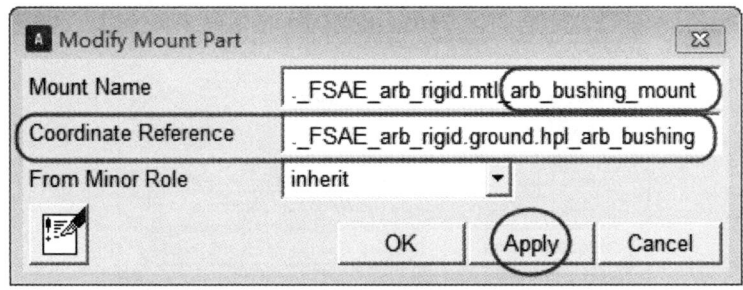

图 7-9 安装部件对话框

- Mount name（安装件名称）：arb_bushing_mount；
- Coordinate Reference（参考坐标）：._FSAE_arb_rigid.ground.hpl_arb_bushing；
- From Minor Role（安装件特征选择）：inherit（继承特性）。

（2）单击 Apply 按钮，完成._FSAE_arb_rigid.mtl_arb_bushing_mount 安装部件的创建。同理，创建其他安装部件。

- Mount name（安装件名称）：ARB_pickup；
- Coordinate Reference（参考坐标）：._FSAE_arb_rigid.ground.cfl_droplink_to_bellcrank；
- From Minor Role（安装件特征选择）：inherit（继承特性）。

（3）单击 OK 按钮，完成._FSAE_arb_rigid.mtl_ARB_pickup 安装部件的创建。

7.1.5 稳定杆约束

1. 部件 gel_arb 与 ger_arb 之间的 revolute 约束

（1）单击 Build > Attachments > Joint > New 命令，弹出刚性约束件对话框，如图 7-10 所示。在下列对话框中输入相应的数据：

- Joint Name（约束副名称）：arb_rev_joint；
- I Part：._FSAE_arb_rigid.gel_arb；
- J Part：._FSAE_arb_rigid.ger_arb；
- Type：single；
- Joint Type（约束副类型）：revolute，转动副，约束 5 个自由度；
- Active（激活）：always；

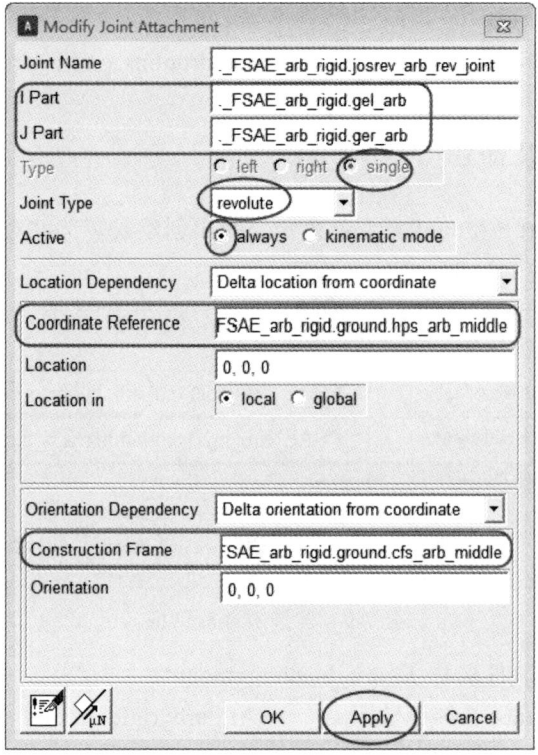

图 7-10 刚性约束-revolute

- Location Dependency：Delta location from coordinate；
- Coordinate Reference（参考坐标）：._FSAE_arb_rigid.ground.hps_arb_middle；
- Location：0，0，0；
- Location in：local；
- Orientation Dependency：Delta location from coordinate；
- Coordinate Reference（参考坐标）：._FSAE_arb_rigid.ground.cfs_arb_middle；
- Orientation：0，0，0。

（2）单击 Apply 按钮，完成._FSAE_arb_rigid.josrev_arb_rev_joint 转动副的创建。

2. 部件 mtl_ARB_pickup 与 gel_droplink 之间的 spherical 约束

（1）单击 Build > Attachments > Joint > New 命令，弹出刚性约束件对话框，如图 7-10 所示。在下列对话框中输入相应的数据：

- Joint Name（约束副名称）：droplink_to_bellcrank；
- I Part：._FSAE_arb_rigid.mtl_ARB_pickup；
- J Part：._FSAE_arb_rigid.gel_droplink；
- Joint Type（约束副类型）：spherical，转动副；
- Active（激活）：always；
- Location Dependency：Delta location from coordinate；
- Coordinate Reference（参考坐标）：._FSAE_arb_rigid.ground.cfl_droplink_to_bellcrank；

- Orientation Dependency：Delta location from coordinate；
- Coordinate Reference（参考坐标）：._FSAE_arb_rigid.ground.cfs_arb_middle；
- Location：0，0，0；
- Location in：local；
- Orientation：None。

（2）单击 Apply 按钮，完成球副._FSAE_arb_rigid.jolsph_droplink_to_bellcrank 的创建。

3. 部件 gel_arb 与 gel_droplink 之间的 hook 约束

（1）单击 Build > Attachments > Joint > New 命令，弹出刚性约束对话框，如图 7-10 所示。在下列对话框中输入相应的数据：
- Joint Name（约束副名称）：droplink_to_arb；
- I Part：._FSAE_arb_rigid.gel_arb；
- J Part：._FSAE_arb_rigid.gel_droplink；
- Joint Type（约束副类型）：hook；
- Active（激活）：always；
- Location Dependency：Delta location from coordinate；
- Coordinate Reference（参考坐标）：._FSAE_arb_rigid.ground.cfl_droplink_to_arb；
- Location：0，0，0；
- Location in：local；
- I-Part Axis：._FSAE_arb_rigid.ground.cfl_arb_bend；
- J-Part Axis：._FSAE_arb_rigid.ground.cfl_droplink_to_bellcrank。

（2）单击 OK 按钮，完成胡克副._FSAE_arb_rigid.jolhoo_droplink_to_arb 的创建。

4. 部件 arb 与 arb_bushing_mount 之间的 bushing 约束

（1）单击 Build > Attachments > Bushing > New 命令，弹出创建衬套件对话框，如图 7-11 所示。在下列对话框中输入相应的数据：
- Bushing Name（约束副名称）：arb_bushing；
- I Part：._FSAE_arb_rigid.gel_arb；
- J Part：._FSAE_arb_rigid.mtl_arb_bushing_mount；
- Inactive（抑制）：never；
- Preload：0，0，0；
- Tpreload：0，0，0；
- Offset：0，0，0；
- Roffset：0，0，0；
- Geometry Length：40；
- Geometry Radius：25；
- Property File：mdids: //FASE/bushings.tbl/MSC_antiroll_bar.bus；

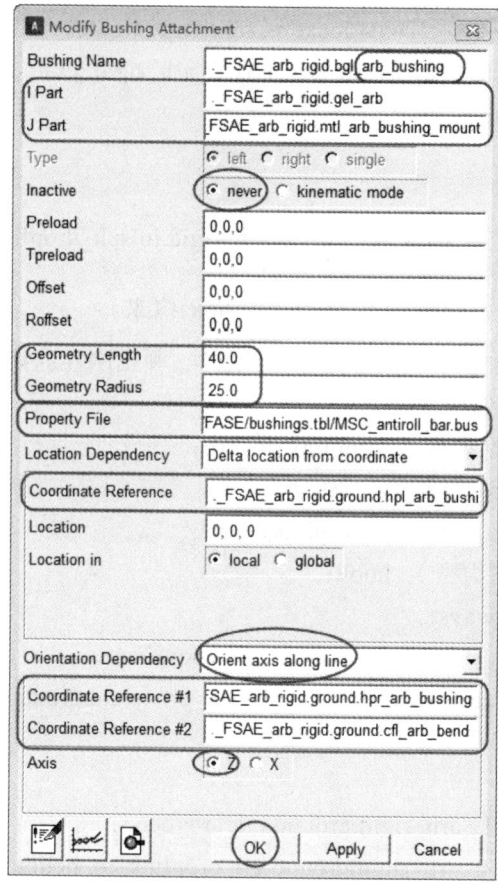

图 7-11 衬套约束-bushing

- Location Dependency：Delta location from coordinate；
- Coordinate Reference（参考坐标）：._FSAE_arb_rigid.ground.hpl_arb_bushing；
- Location：0，0，0；
- Location in：local；
- Orientation Dependency：Orient axis along line；
- Coordinate Reference #1（参考坐标）：._FSAE_arb_rigid.ground.hpr_arb_bushing；
- Coordinate Reference #2（参考坐标）：._FSAE_arb_rigid.ground.cfl_arb_bend；
- Axis：Z。

（2）单击 OK 按钮，完成轴套._FSAE_arb_rigid.bgl_arb_bushing 的创建。

7.1.6 参数变量

（1）单击 Build > Parameter Variable > New 命令，弹出参数变量对话框，如图 7-12 所示。在下列对话框中输入相应的数据：

- Parameter Variable Name：：torsional_spring_stiffness；
- Real Value（实数值）：3572；

- Units：torsion_stiffness；
- Hide from standard user（是否从标准界面隐藏）：no。

（2）单击 OK 按钮，完成变量 ._FSAE_arb_rigid.pvs_torsional_spring_stiffness 的创建。

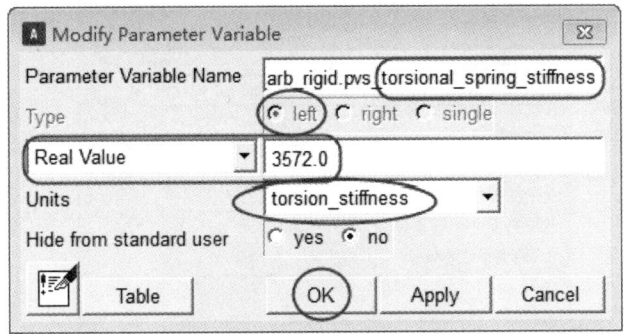

图 7-12　参数变量对话框

7.2　扭转弹簧

横向稳定杆本质上是一个扭杆弹簧，以上建立的模型均为刚性体部件，不存在弹性体。通过简化的方法，在对称点旋转约束上施加扭转弹簧可以近似地模拟横向稳定杆，也可以通过柔性体的方法把整个扭杆弹簧用有限元的方法做成模态中性文件。此处采用简化方法，用扭转刚度与扭转角度的乘积模拟横向稳定杆的总体刚度。采用该方法的优势是扭转刚度是一个可变量，可以通过修改参数快速设置刚度，从而快速调整整体悬架子系统或者整车的性能。如果采用柔性体，稳定杆的扭转刚度是定值，刚度只与扭转弹簧的材料及界面直径的大小有关。另外，因为模型规模较大，计算速度较慢。

扭转弹簧的建模可以采用多种方法：① 采用函数的方法，即采用扭转刚度（采用参数变量定义）与扭转角度乘积建立模型；② 直接在 View 模块中建立扭转弹簧，输入刚度即可。

7.2.1　方案 A

（1）单击 Build > Actuator > Joint Force > New 命令，弹出约束副驱动力对话框，如图 7-13 所示。在下列对话框中输入相应的数据：
- Actuator Name：arb_torsion_spring；
- Joint：._FSAE_arb_rigid.josrev_arb_rev_joint；
- Function：-（._FSAE_arb_rigid.pvs_torsional_spring_stiffness）*AZ（ger_arb.jxs_joint_j_1，gel_arb.jxs_joint_i_1），可以通过单击函数框后面的图标，弹出函数构建对话框，如图 7-14 所示，函数构建遵循扭转胡克定律，构建完成后，可以通过 Verify 菜单判断构造函数的正确性。

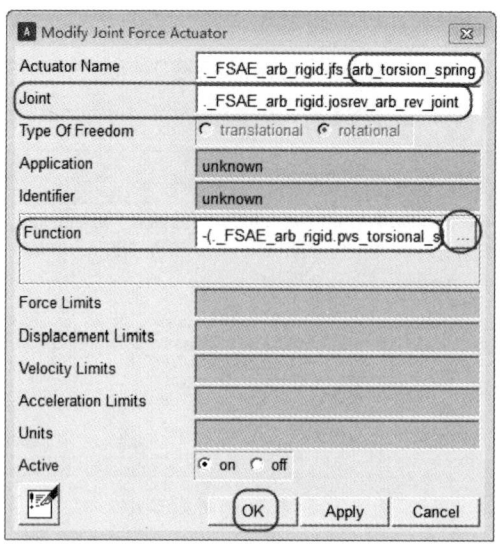

图 7-13 约束副驱动力对话框

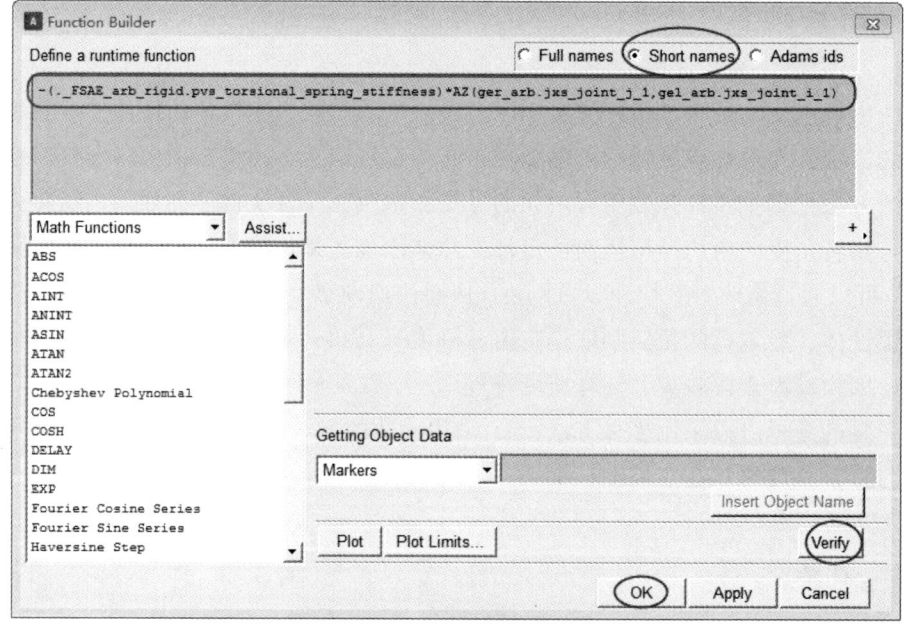

图 7-14 函数构建对话框

（2）单击 OK 按钮，完成函数-（._FSAE_arb_rigid.pvs_torsional_spring_stiffness）*AZ（ger_arb.jxs_joint_j_1，gel_arb.jxs_joint_i_1）的构建，然后返回到约束副驱动力对话框。

（3）单击 OK 按钮，完成约束力._FSAE_arb_rigid.jfs_arb_torsion_spring 的创建，此时扭杆弹簧如图 7-1 所示。

7.2.2 方案 B

（1）单击 Tools > Adams/View interface，切换到 View 通用模块界面，如图 7-15 所示。

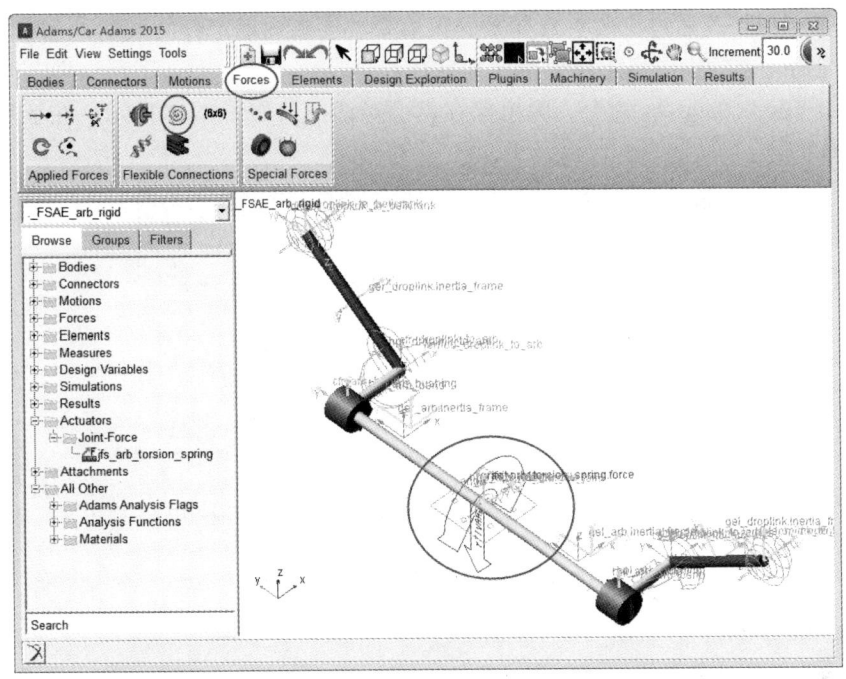

图 7-15　Adams/View 界面

（2）单击 Forces 按钮，在出现的 Flexible Connections 菜单框中单击 Create a Rotational spring-damper（旋转弹簧及阻尼器）图标，扭转弹簧阻尼器设置参数如图 7-16 所示。

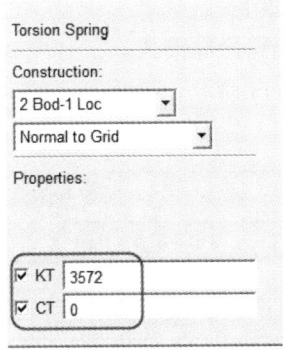

图 7-16　扭转弹簧设置参数

（3）勾选 KT，输入 3572，此处的数值与上述所建的参数变量中的数值一样，即扭转弹簧的刚度。

（4）勾选 CT，输入 0，CT 为扭转阻尼，输入 0 即扭转阻尼不起作用；也可以不勾选，即默认没有阻尼。

（5）调整网格线为 XZ 平面，根据窗口下方命令提示，先后选择部件 gel_arb 与 ger_arb，位置选择点 ._FSAE_arb_rigid.ground.hps_arb_middle，完成扭转弹簧的添加，如图 7-1 所示。

（6）单击 Tools > Select mode > Switch to A/car Template mode Builder，切换到专家模式界面。

方案 A 与方案 B 建立的横向稳定杆本质是一样的，即稳定杆都通过简化的扭转弹簧来模拟；采用方案 B 建模时，可以不用建立扭转弹簧的参数变量，如果已经建立，可以删除；方案 A 的优势是可以快速通过更改参数变量的数值来改变横向稳定杆的刚度,同时作为优化设计变量。至此简化式的横向稳定杆模型建立完成。

(7)单击 File > Save As 命令,弹出保存模型对话框,如图 7-17 所示。在下列对话框中输入相应的数据:

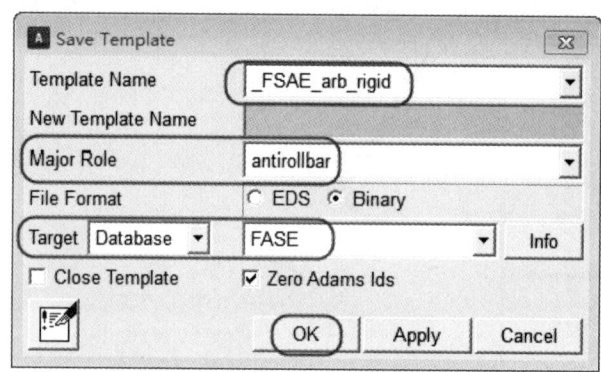

图 7-17 稳定杆模型保存对话框

- Template Name:_FSAE_arb_rigid;
- Major Role(主特征):antirollbar;
- File Format:Binary;
- Target:Datebase/FASE。

(8)单击 OK 按钮,完成横向稳定杆模型模板 FSAE_arb_rigid 的保存。

横向稳定杆子系统的创建步骤如下:

(1)按 F9 键,把专家模板转换到标准模式。

(2)单击 File > New > Subsystem 命令,弹出子系统对话框,如图 7-18 所示。在下列对话框中输入相应的数据:

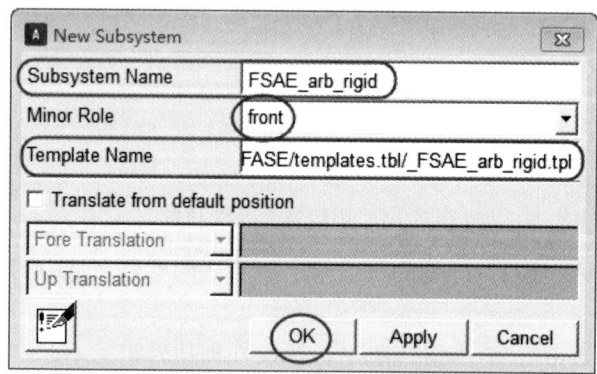

图 7-18 横向稳定杆子系统

- Subsystem Name:(系统名称):FSAE_arb_rigid;
- Minor Role(副特征):front(指前横向稳定杆);
- Template Name(模板路径):mdids://FASE/templates.tbl/_FSAE_arb_rigid.tpl。

(3)单击 OK 按钮,完成推杆式悬架子系统 FSAE_arb_rigid 的创建。

（4）单击 File > Save As 命令，保持默认设置。
（5）单击 OK 按钮，完成横向稳定杆子系统 FSAE_arb_rigid 的保存。

7.3 单轮跳动仿真

横向稳定杆与悬架装配步骤如下：

（1）单击 File > Open > Assembly 命令，启动悬架装配，如图 7-19 所示。在 Assembly Name 中输入：mdids：//FASE/assemblies.tbl/FSAE_sus_front_test.asy。

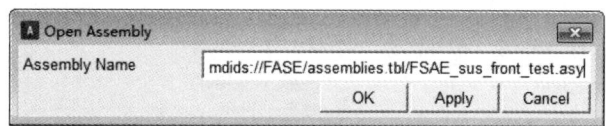

图 7-19 悬架装配对话框

（2）单击 OK 按钮，打开悬架与试验台的装配。
（3）单击 File > Manage > Assemblies > Add Subsystem 命令，启动添加子系统对话框，如图 7-20 所示。在 Subsystem(s)中输入：mdids：//FASE/subsystems.tbl/FSAE_arb_rigid_front.sub。
（4）单击 OK 按钮，添加子系统后的悬架如图 7-21 所示。

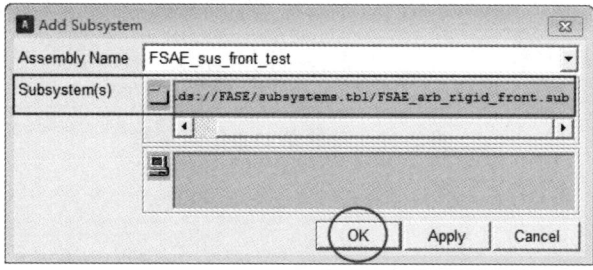

图 7-20 添加子系统横向稳定杆

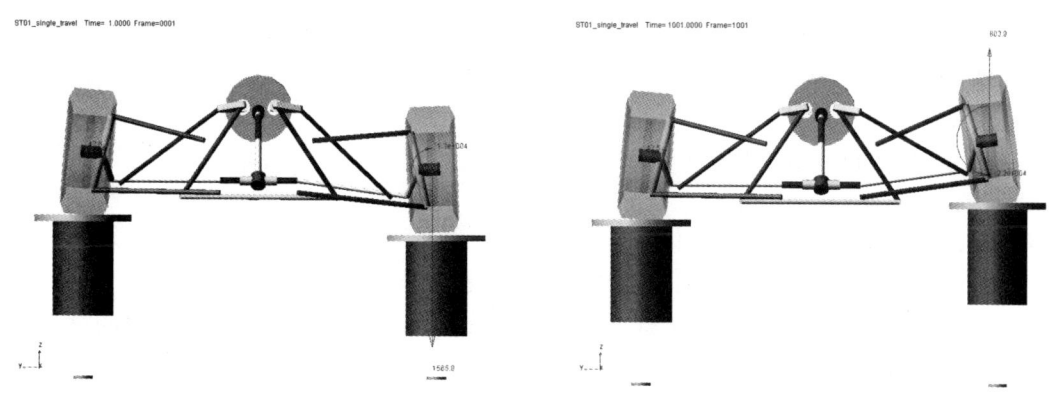

（a）左车轮下跳　　　　　　　　　　　（b）左车轮上跳

图 7-21 悬　架

（5）单击 Simulate > Suspension Analysis > Single Travel 命令，弹出双轮单向激振对话框，如图 7-22 所示。在下列对话框中输入相应的数据。

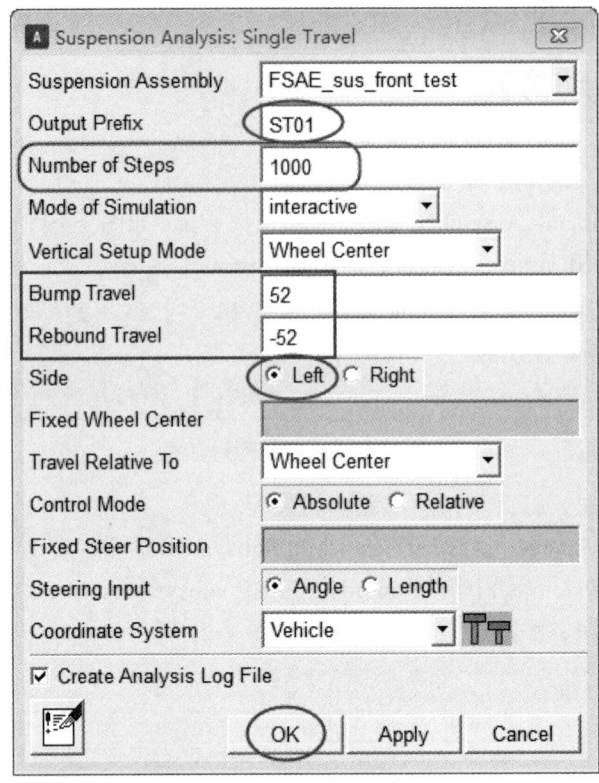

图 7-22 双轮单向激振对话框

- Output Prefix：ST01；
- Number of Steps（仿真步数）：1000；
- Mode of Simulation：interactive；
- Vertical Setup Mode：Wheel Center；
- Bump Travel：52；
- Rebound Travel：-52；
- Side：Left；
- Travel Relative To：Wheel Center；
- Control Mode：Absolute；
- Coordinate System：Vehicle。

（6）单击 Apply 按钮，完成推杆式双横臂悬架在 C 模式下的仿真。

（7）按 F8 键，界面转换到后处理模块。点击 Independent Axis / date，在弹出的对话框中输入相应的数据：

- Result Set：wheel travel；
- Component：vertical left（左侧车轮垂向跳动的距离，即仿真设置的上下跳动 52 mm，数值 52 是 FSAE 赛车设计的一个要求，实际可以根据情况做适当调整）。

(8)单击 OK 按钮,返回到后处理主界面,在下列对话框中输入相应的数据:
- Filter:user defined;
- Request:选择 bushing(衬套各方向)与 torsion_spring(扭力弹簧各方向)。

(9)绘制参数曲线,如图 7-23~图 7-27 所示。从图中可以看出,对于衬套来说,垂向力在 Y、Z 方向受力较大,扭转力在 X 方向受力较大;对于稳定杆来说,在 Y 方向(绝对坐标系下)受扭转力最大,通过改变刚度值或是参数变量的刚度值,可以改变其在 Y 方向下的力矩大小,也可以通过优化实验的方法,在设定目标后,逆向求解横向稳定杆的最佳刚度值。

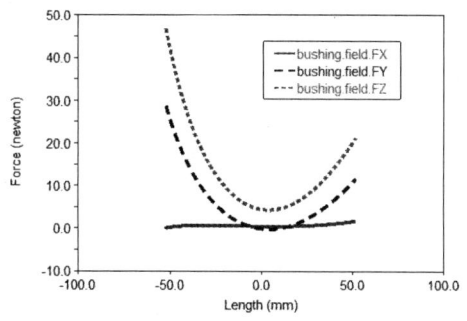

图 7-23 衬套 X/Y/Z 方向的垂向力

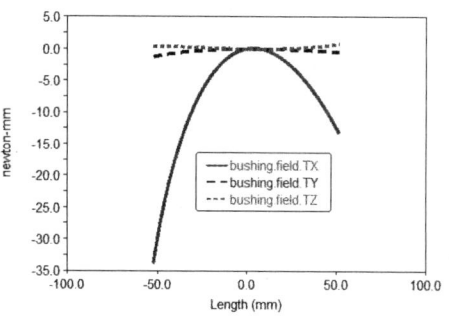

图 7-24 衬套 X/Y/Z 方向的扭转力

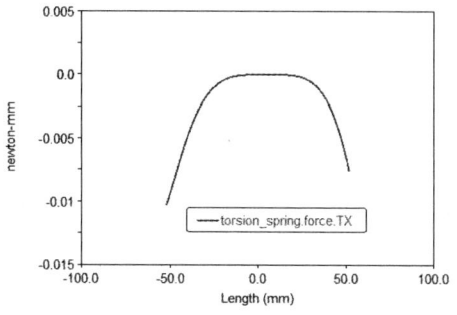

图 7-25 稳定杆 X 方向的扭转力

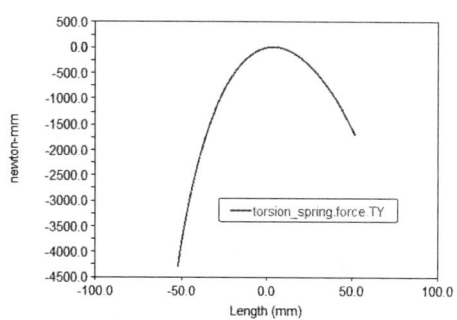

图 7-26 稳定杆 Y 方向的扭转力

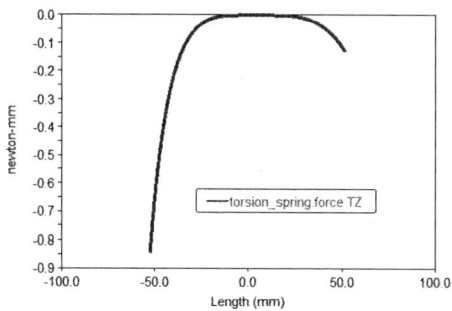

图 7-27 稳定杆 Z 方向的扭转力

第 8 章 横向稳定杆 Ⅱ

在 ADAMS 多体系统动力软件中,横向稳定杆一般采用简化模型建立。所谓的建模模型,即稳定杆采用扭转弹簧的方式来模拟横向稳定杆的刚度特性,该方法的优势是可以在一定程度上近似,同时计算效率较高,但其精确性和准确性并不完美。采用模态中性文件 MNF,即采用有限元的方法可以真实地模拟横向稳定杆的刚度特性,但其缺点是有限元模型相对简化模型计算规模较大,建模过程也较难,本章采用多柔体动力学方法建立横向稳定杆模型,分析其在车轮跳动时期的特性。图 8.1 为 FSAE 赛车前多柔体横向稳定杆模型,蓝色部分为采用有限元方法建立的柔性体。

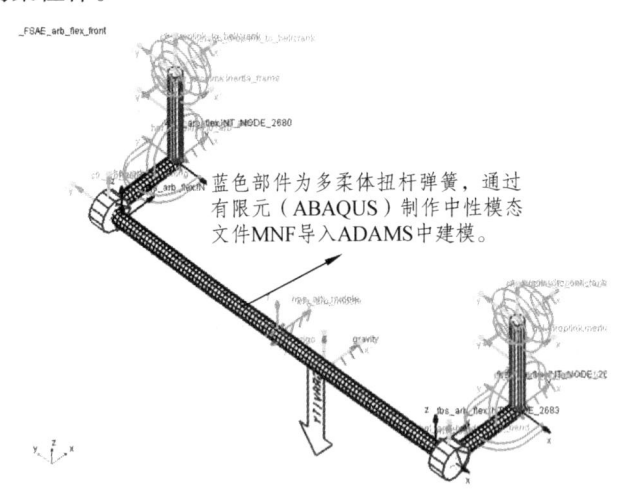

图 8-1 横向稳定杆

学习目标

(1)会创建稳定杆有限元模型。
(2)会 MNF 参数设置。
(3)会约束设置。
(4)会修改程序。
(5)会将柔性体导入 ADAMS。
(6)会创建多柔体稳定杆模型。
(7)会横向稳定杆匹配。

8.1 横向稳定杆前处理

（1）启动 Abaqus/CAE，切换到 Part 模块，通过扫掠方法建立横向稳定杆几何体模型，横向稳定杆的扫掠路径草图如图 8-2 所示，横向稳定杆横截半径为 8 mm。

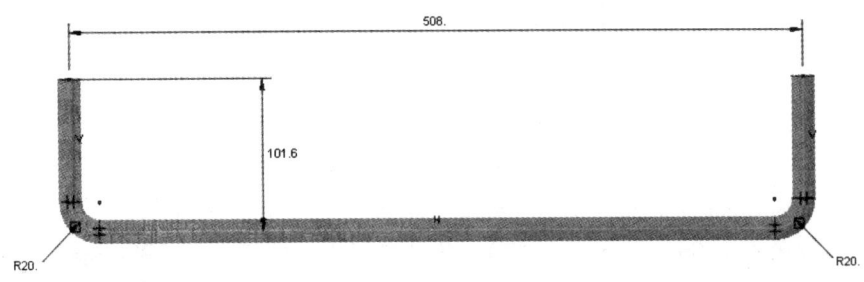

图 8-2　横向稳定杆

（2）切换到 Property 界面，创建材料属性。弹性模量：2.06×10^5；泊松比：0.29；密度：7.74×10^{-9}。不同有限元软件的单位制不同，材料属性参数一定要保证正确，否则会导致计算出的模态出错。材料属性界面输入参数如图 8-3 和图 8-4 所示；横向稳定杆采用的材料为 60Si2Mn。

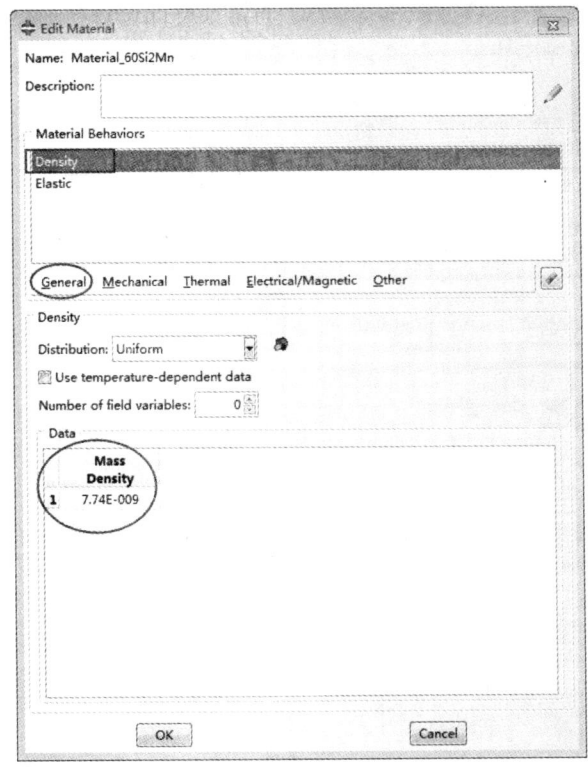

图 8-3　密度参数

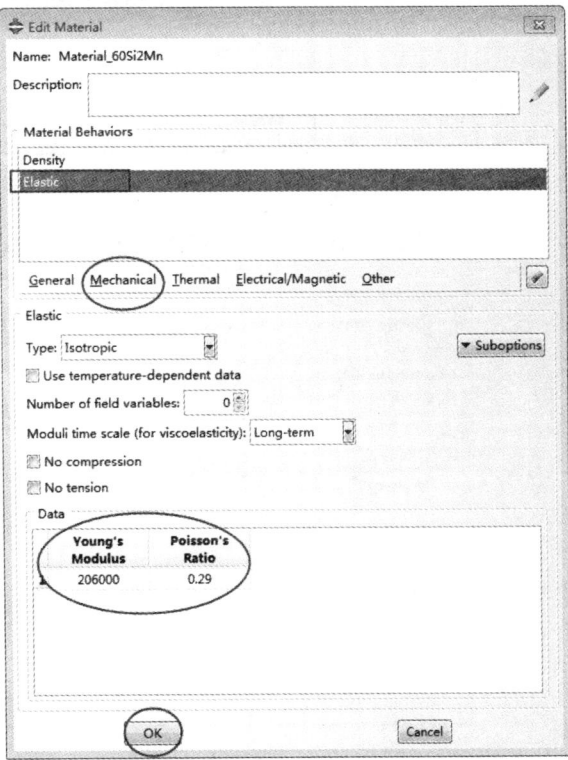

图 8-4　杨氏模量与泊松比

（3）材料通过截面属性赋予到三维模型上，赋予成功后，横向稳定杆几何体颜色变为浅绿色；截面属性如图 8-5 所示。

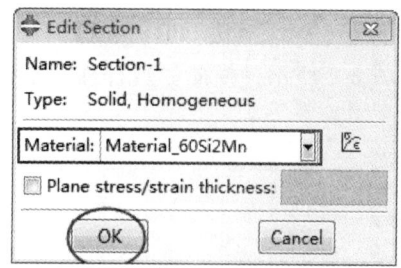

图 8-5　材料界面属性_60Si2Mn

（4）切换到 Assembly 界面，单击装配完成单体部件装配。

（5）切换到分析步 Step 界面，完成两个分析步创建，如图 8-6 所示。Step-1 为模态分析步，设置提取前 20 阶模态；Step-2 为子结构生成，子结构即把整个连杆作为一个单一部件。在 Basic 选项卡设置 Step-2 子结构标识（Substructure identifier：Z101），点选 Whole model，在后续方框中选择整个模型；切换到 Options 选项卡，勾选 Specify retained eigenmodes by，点选 Mode range，在 Date 方框中输入 1，20，1。Step-2 设置如图 8-7 和图 8-8 所示。

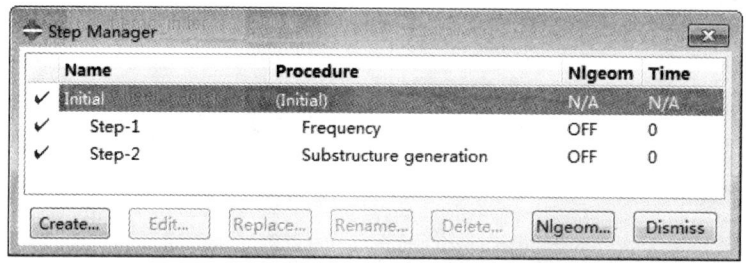

图 8-6　分析步设置

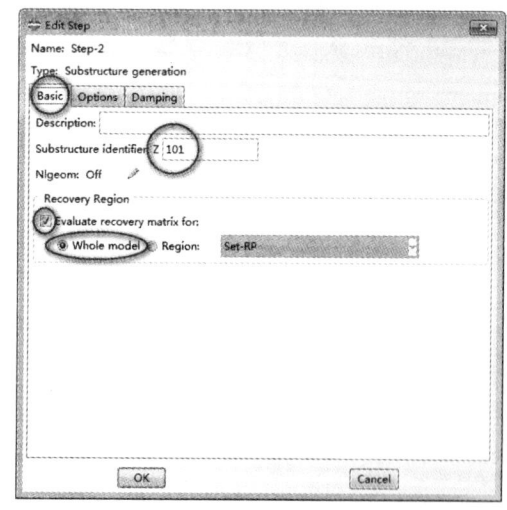

图 8-7　Basic 选项卡设置

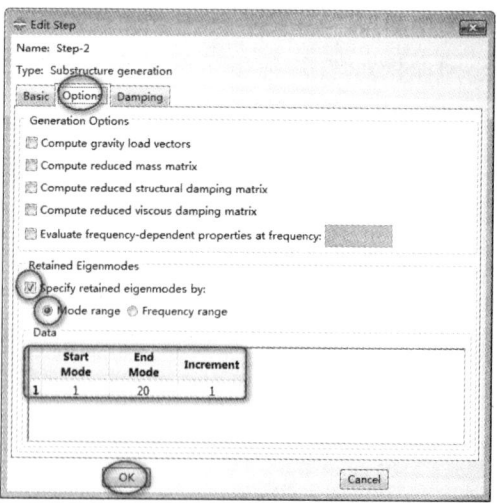

图 8-8　Options 选项卡设置

（6）切换到相互作用 Interaction 界面，在横向稳定杆与稳定杆拉杆处创建 RP-1、RP-2 两个约束参考点，在横向稳定杆与车身衬套连接处创建 RP-3、RP-4 两个约束参考点，如图 8-9 所示。

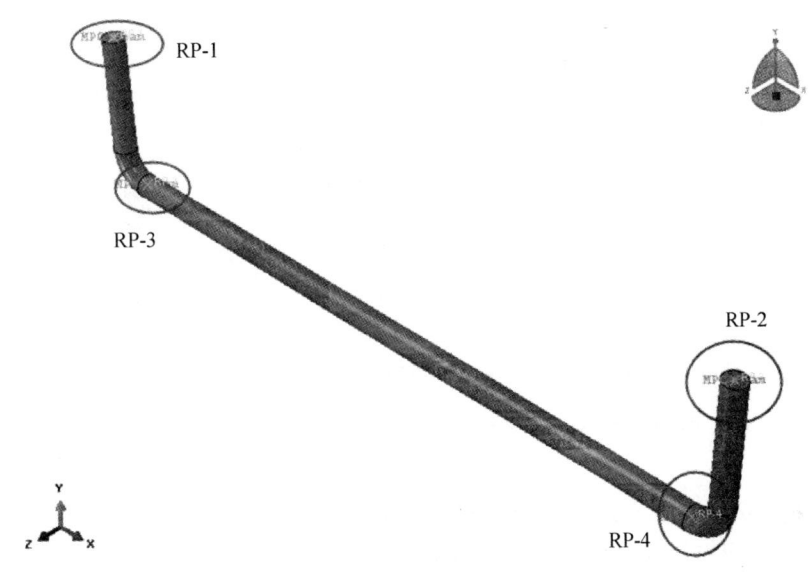

图 8-9　RP 参考点

（7）建立 RP 点与端面的 MPC 多点约束，如图 8-10 所示。多点约束为参考点 RP 与对应的断面边线通过梁单元建立连接，具体参考模型文件。

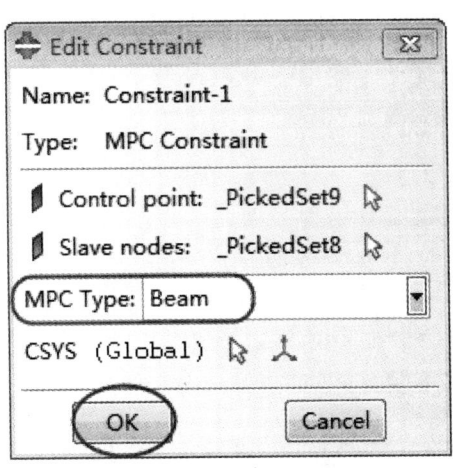

图 8-10　MPC 多点约束

（8）切换到网格划分 Mesh 界面，设置网格全局尺寸为 5 mm，网格划分完成后如图 8-11 所示，其中共包含 1 820 个六面体单元，经检查，网格全部符合要求。

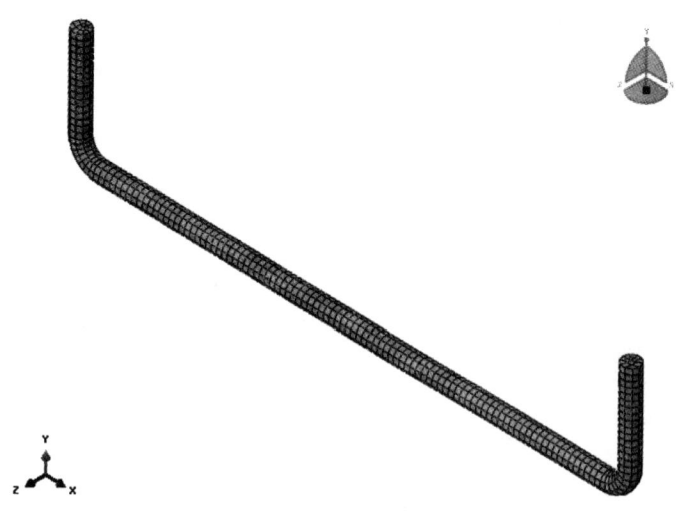

图 8-11 横向稳定杆网格划分-C3D8R 单元

（9）切换到 Load 界面，在 Step-1 分析步下约束 RP-1、RP-2、RP-3、RP-4 四个参考点完全固定。

（10）在 Step-2 分析步下选择 Retained nodal dofs，点击继续 RP-1、RP-2、RP-3、RP-4 四个参考点，弹出编辑界面对话框，如图 8-12 所示，勾选全部约束。

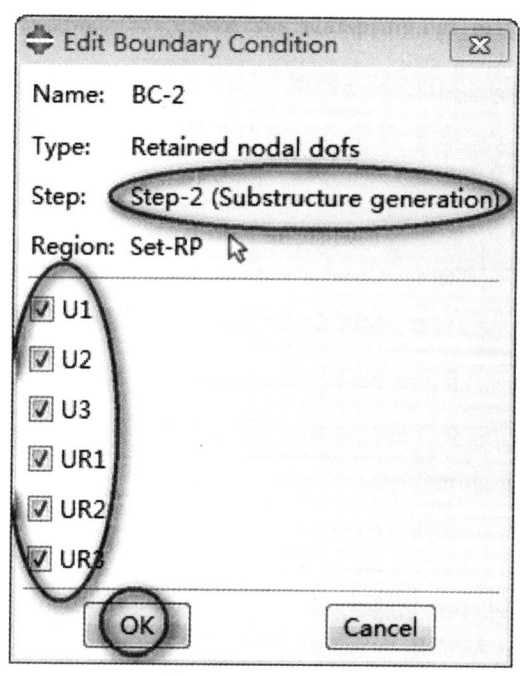

图 8-12 约束设置

（11）切换到 Job 界面，需要添加以下关键字符：

关键字符如下：

MASS MATRIX=YES   %质量矩阵

*FLEXIBLE BODY, TYPE=ADAMS   %转换为 ADAMS 关键字；

*ELEMENT RECOVERY MATRIX, POSITION=AVERAGED AT NODES   %计算结果中显示应力应变

S,   %   在 ADAMS 中显示应力大小信息

E,   %   在 ADAMS 中显示应变大小信息

（12）整个关键字信息及上述需要添加关键字符的位置如下（下划线处）：

```
以下为关键字信息，在 Step-2 中需要添加对应的关键字才能完成 MNF 文件的制作：
** ----------------------------------------------------------------
**
** PART INSTANCE: Part-1-1
**
*Element, type=C3D8R
*Nset, nset=Part-1-1__PickedSet2, generate
*Elset, elset=Part-1-1__PickedSet2, generate
** Section: Section-1
*Solid Section, elset=Part-1-1__PickedSet2, material=si2mn
,
*Surface, type=NODE, name=_PickedSet8_CNS_
*Surface, type=NODE, name=_PickedSet10_CNS_
*Surface, type=NODE, name=_PickedSet12_CNS_
*Surface, type=NODE, name=_PickedSet14_CNS_
** Constraint: Constraint-1
*MPC
BEAM, _PickedSet8, _PickedSet9
** Constraint: Constraint-2
*MPC
BEAM, _PickedSet10, _PickedSet11
** Constraint: Constraint-3
*MPC
BEAM, _PickedSet12, _PickedSet13
** Constraint: Constraint-4
*MPC
BEAM, _PickedSet14, _PickedSet15
**
** MATERIALS
**
*Material, name=si2mn
*Density
```

```
       7.74e-09,
      *Elastic
       206000., 0.29
      ** ----------------------------------------------------------------
      **
      ** STEP: Step-1
      **
      *Step, name=Step-1, nlgeom=NO, perturbation
      *Frequency, eigensolver=Lanczos, acoustic coupling=on,
normalization=displacement
       20, , , , ,
      **
      ** BOUNDARY CONDITIONS
      **
      ** Name: BC-1 Type: Symmetry/Antisymmetry/Encastre
      *Boundary
      _PickedSet16, ENCASTRE
      **
      ** OUTPUT REQUESTS
      **
      *Restart, write, frequency=0
      **
      ** FIELD OUTPUT: F-Output-1
      **
      *Output, field, variable=PRESELECT
      *End Step
      ** ----------------------------------------------------------------
      **
      ** STEP: Step-2
      **
      *Step, name=Step-2, nlgeom=NO
      *Substructure Generate, overwrite, type=Z101, recovery matrix=YES, MASS
MATRIX=YES
      *FLEXIBLE BODY, TYPE=ADAMS
      *ELEMENT RECOVERY MATRIX, POSITION=AVERAGED AT NODES
      S,
      E,
      *Select Eigenmodes, generate
       1, 20, 1
      *Damping Controls, structural=COMBINED, viscous=COMBINED
      *Retained Nodal Dofs
      _PickedSet17, 1, 6
      *End Step
```

（13）创建 fsae_arb_flex 分析作业并提交运算，运算完成后可以在后处理模块中显示连杆的模态变形及对应的频率。前 6 阶模态变形如图 8-13～图 8-18 所示。

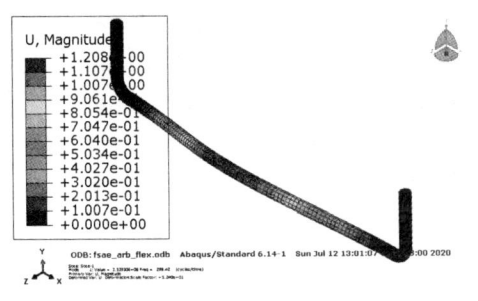

图 8-13　横向稳定杆一阶模态

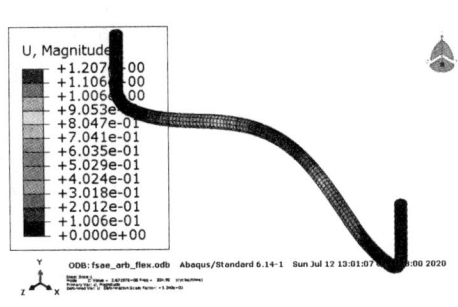

图 8-14　横向稳定杆二阶模态

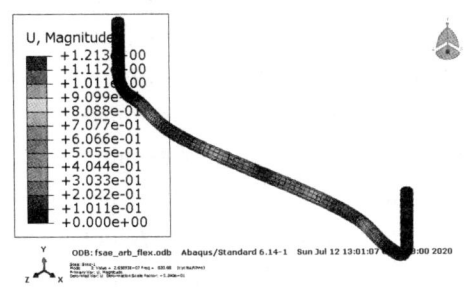

图 8-15　横向稳定杆三阶模态

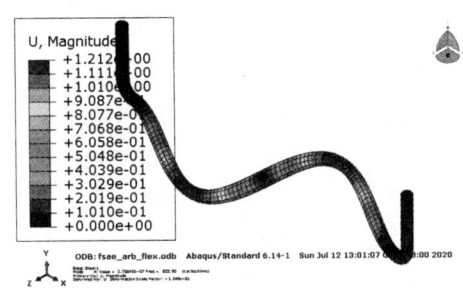

图 8-16　横向稳定杆四阶模态

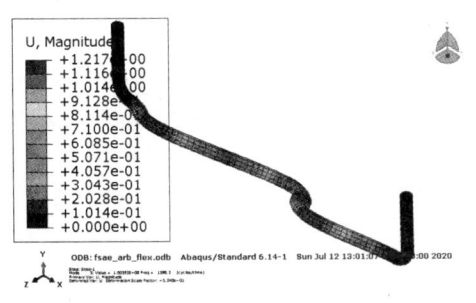

图 8-17　横向稳定杆五阶模态

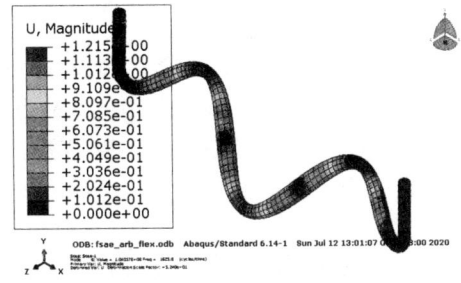

图 8-18　横向稳定杆六阶模态

（14）横向稳定杆前 20 阶频率如图 8-19 所示。在进行模态分析时施加对应的约束，因此为约束模态。如果为自由模态分析，前 1 到 6 阶模态的刚体模态数值较小，可以忽略，真实的模态从第 7 阶开始。一阶频率为 299.42 Hz，根据经验建立的刚柔耦合稳定杆模型在低频状态下不会发生共振（有时候频率过低会导致在低频状态下发生共振，此时刚柔模型不能正常仿真，当然可以根据模态的贡献率对相应的模态进行抑制）。

1	Mode	1: Value = 3.53930E+06 Freq =	299.42	(cycles/time)
2	Mode	2: Value = 3.67197E+06 Freq =	304.98	(cycles/time)
3	Mode	3: Value = 2.65893E+07 Freq =	820.68	(cycles/time)
4	Mode	4: Value = 2.75845E+07 Freq =	835.90	(cycles/time)
5	Mode	5: Value = 1.00593E+08 Freq =	1596.3	(cycles/time)
6	Mode	6: Value = 1.04357E+08 Freq =	1625.8	(cycles/time)
7	Mode	7: Value = 2.69394E+08 Freq =	2612.2	(cycles/time)
8	Mode	8: Value = 2.79401E+08 Freq =	2660.3	(cycles/time)
9	Mode	9: Value = 3.48742E+08 Freq =	2972.2	(cycles/time)
10	Mode	10: Value = 5.86513E+08 Freq =	3854.4	(cycles/time)
11	Mode	11: Value = 6.08042E+08 Freq =	3924.5	(cycles/time)
12	Mode	12: Value = 9.21691E+08 Freq =	4831.8	(cycles/time)
13	Mode	13: Value = 9.27846E+08 Freq =	4848.0	(cycles/time)
14	Mode	14: Value = 1.04377E+09 Freq =	5141.9	(cycles/time)
15	Mode	15: Value = 1.11648E+09 Freq =	5318.0	(cycles/time)
16	Mode	16: Value = 1.15868E+09 Freq =	5417.5	(cycles/time)
17	Mode	17: Value = 1.39556E+09 Freq =	5945.6	(cycles/time)
18	Mode	18: Value = 1.44874E+09 Freq =	6057.8	(cycles/time)
19	Mode	19: Value = 1.46409E+09 Freq =	6089.8	(cycles/time)
20	Mode	20: Value = 1.91595E+09 Freq =	6966.5	(cycles/time)

图 8-19 横向稳定杆前 20 阶频率

（15）打开 Abaqus Command，输入 cd D：\ADAMS_MNF，切换命令至 ADAMS_MNF 文件夹。

（16）继续输入以下命令：abaqus adams job=fsae_arb_flex substructure_sim=fsae_arb_flex_Z101 model_odb= fsae_arb_flex length=mm mass=tonne time=sec force=N，命令输入完成后，Abaqus Command 完成提交并运算后产生 fsae_arb_flex.mnf 中性文件。

8.2 柔体稳定杆模型

为避免重复建模，在横向稳定杆简化模型中删除扭杆部件对应的约束，用有限元柔性体扭杆替代的方法建模，具体如下：

（1）启动 ADAMS/CAR，选择专家模块进入建模界面。

（2）单击 File > Open 命令，打开简化横向稳定杆，如图 8-20 所示。在 Template Name 中输入：mdids：//FASE/templates.tbl/_FSAE_arb_rigid.tpl。

（3）单击 OK 按钮，完成横向稳定杆的打开操作。

（4）删除稳定杆部件 arb，此时对应的刚性约束：①._FSAE_arb_rigid.josrev_arb_rev_joint，②._FSAE_arb_rigid.jolhoo_droplink_to_arb；柔性约束：._FSAE_arb_rigid.bgl_arb_bushing；约束力：._FSAE_arb_rigid.jfs_arb_torsion_spring 同时被删除。

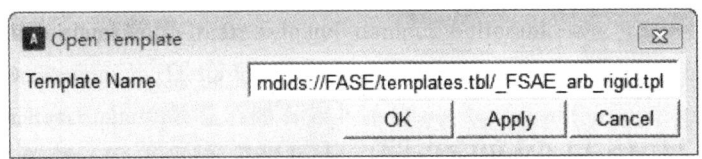

图 8-20 横向稳定杆（简化模型）

（5）删除参数变量：._FSAE_arb_rigid.pvs_torsional_spring_stiffness（不删除也不会影响柔性体稳定杆模型）。

（6）单击 File > Save As 命令，弹出保存模型对话框，如图 8-21 所示。在下列对话框中输入相应的数据：

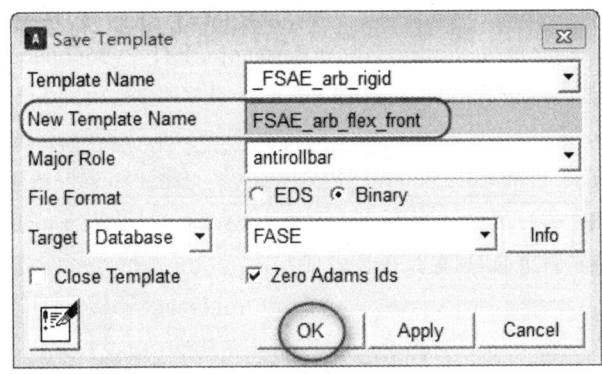

图 8-21 模型保存对话框

- New Template Name：FSAE_arb_flex_front；
- Major Role（主特征）：antirollbar；
- File Format：Binary；
- Target：Datebase/FASE。

（7）单击 OK 按钮，完成横向稳定杆模型模板 FSAE_arb_flex_front 的保存。

1. 柔性体部件 arb_flex

（1）单击 Build > Part > Flexible Body > New 命令，弹出创建柔性扭杆部件对话框，如图 8-22 所示。在下列对话框中输入相应的数据：

- General Part：arb_flex；
- Location Dependency：Delta location from coordinates；
- Coordinate Reference（参考坐标）：._FSAE_arb_flex_front.ground.hps_arb_middle；
- Location：0，0，0；
- Location in：local；
- Orientation Dependency：User entered values；
- Orient using：Euler Angles；
- Euler Angles：-90，0，0；
- MNF File：file：//D：/ADAMS_MNF/fsae_arb_flex.mnf；
- Color：white。

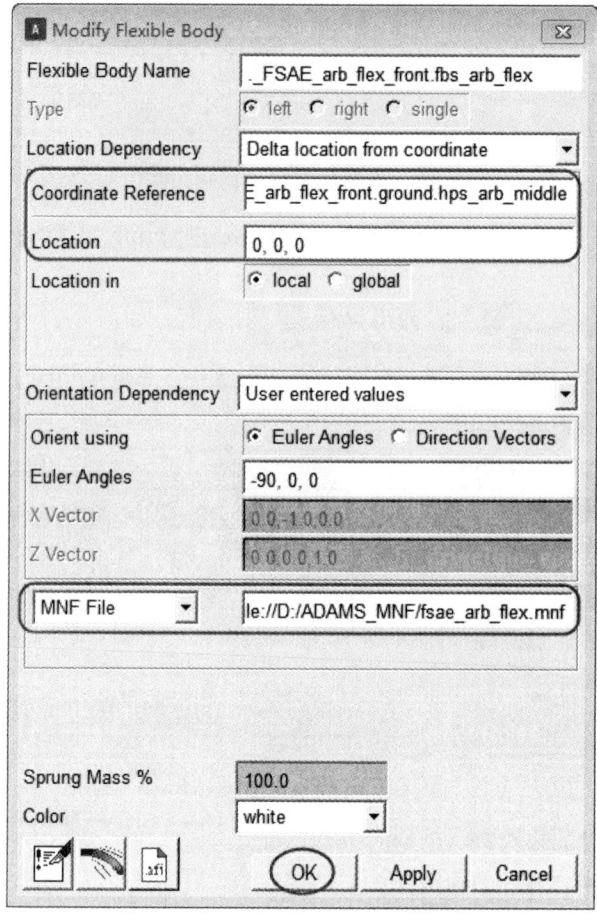

图 8-22　柔性体部件对话框

（2）单击 OK 按钮，完成柔性体部件._FSAE_arb_flex_front.fbs_arb_flex 的创建。

2. 部件 arb_flex 与 arb_bushing_mount 之间的 bushing 约束

（1）单击 Build > Attachments > Bushing > New 命令，弹出创建衬套件对话框，如图 8-23 所示。在下列对话框中输入相应的数据：

- Bushing Name（约束副名称）：arb_bushing；
- I Part：._FSAE_arb_flex_front.fbs_arb_flex；
- J Part：._FSAE_arb_flex_front.mtl_arb_bushing_mount；
- Inactive（抑制）：never；
- Preload：0，0，0；
- Tpreload：0，0，0；
- Offset：0，0，0；
- Roffset：0，0，0；
- Geometry Length：20；
- Geometry Radius：20；

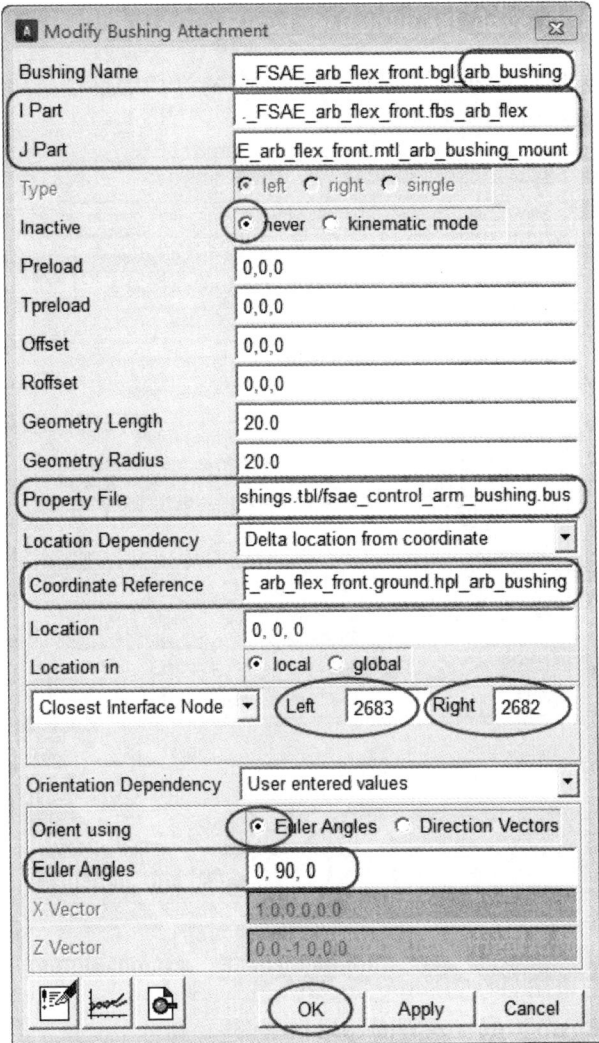

图 8-23 衬套约束

- Property File：mdids：//FASE/bushings.tbl/fsae_control_arm_bushing.bus；
- Location Dependency：Delta location from coordinate；
- Coordinate Reference（参考坐标）：._FSAE_arb_flex_front.ground.hpl_arb_bushing，当选中此硬点后，计算机会自动搜索附近的接口节点并自动填入下面的接口对话框；
- Location：0，0，0；
- Location in：local；
- Closest Interface Node（附近接口节点）：left 为 2683/Right 为 2682，这两个点是在 ABAQUS 有限元模型中对应的 RP-3\RP-4；
- Orientation Dependency：User entered values；
- Orient using：Euler Angles；
- Euler Angles：0，90，0。

（2）单击 OK 按钮，完成轴套._FSAE_arb_flex_front.bgl_arb_bushing 的创建。

3. 部件 fbs_arb_flex 与 gel_droplink 之间的 hook 约束

（1）单击 Build > Attachments > Joint > New 命令，弹出刚性约束件对话框，如图 8-24 所示。在下列对话框中输入相应的数据：

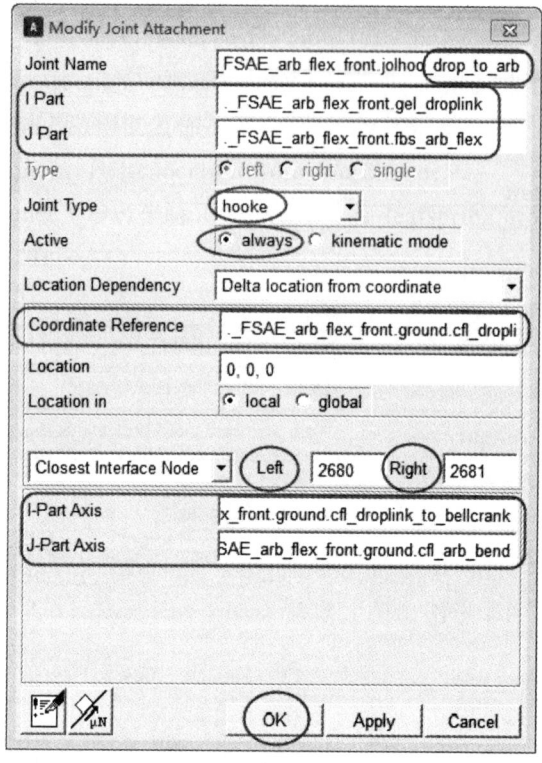

图 8-24　胡克副约束对话框

- Joint Name（约束副名称）：drop_to_arb；
- I Part：._FSAE_arb_flex_front.gel_droplink；
- J Part：._FSAE_arb_flex_front.fbs_arb_flex；
- Joint Type（约束副类型）：hook；
- Active（激活）：always；
- Location Dependency：Delta location from coordinate；
- Coordinate Reference（参考坐标）：._FSAE_arb_flex_front.ground.cfl_droplink_to_arb；
- Location：0，0，0；
- Location in：local；
- Closest Interface Node（附近接口节点）：left 为 2680/Right 为 2681；
- I-Part Axis：._FSAE_arb_flex_front.ground.cfl_droplink_to_bellcrank；
- J-Part Axis：._FSAE_arb_flex_front.ground.cfl_arb_bend。

（2）单击 OK 按钮，完成胡克副._FSAE_arb_flex_front.jolhoo_drop_to_arb 的创建。

（3）单击 File > Save As 命令，弹出保存模型对话框，如图 8-21 所示，保持默认设置。

（4）单击 OK 按钮，完成横向稳定杆模型模板 FSAE_arb_flex_front 的保存。

8.3 车轮反向跳动仿真

（1）单击 Simulate > Suspension Analysis > Opposite Travel 命令，弹出双轮反向激振对话框，如图 8-25 所示。在下列对话框中输入相应的数据：

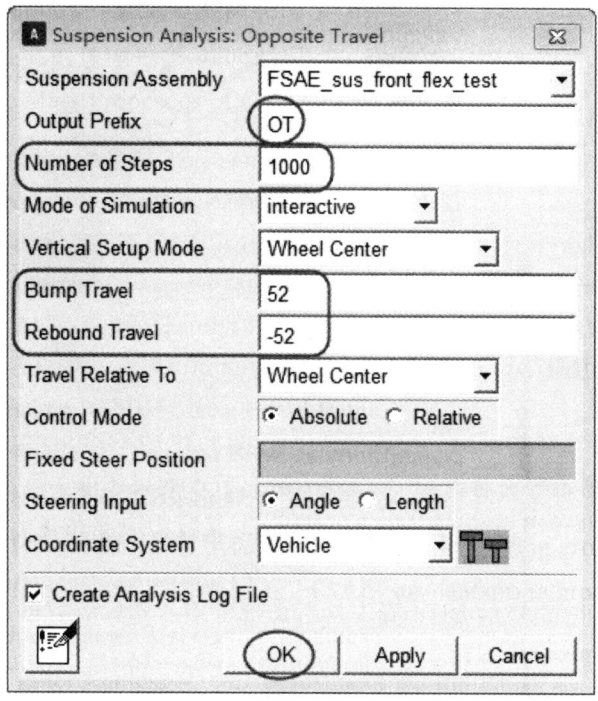

图 8-25 双轮反向激振对话框

- Output Prefix：OT；
- Number of Steps（仿真步数）：1000；
- Mode of Simulation：interactive；
- Vertical Setup Mode：Wheel Center；
- Bump Travel：52；
- Rebound Travel：-52；
- Travel Relative To：Wheel Center；
- Control Mode：Absolute；
- Steeling input：Angle；
- Coordinate System：Vehicle。

（2）单击 Apply 按钮，提交运算并完成仿真，仿真完成后车轮跳动如图 8-26 所示。从图中可以看出，柔性体横向稳定杆已经产生云图。

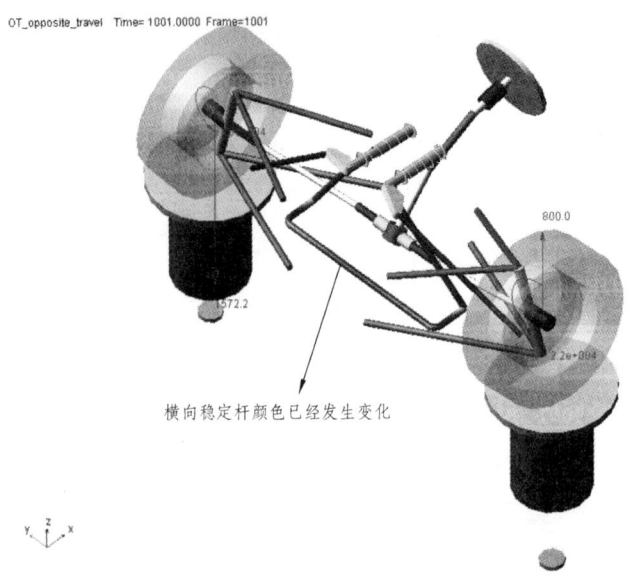

图 8-26 车轮反向跳动

（3）按 F8 键，界面转换到后处理模块。

（4）随机在横向稳定杆上选取 10 个节点（所有节点均可选取，篇幅有限，仅选取 10 个，读者可在模型中查看其他节点应力应变变化情况），其应力应变在各个方向的变化如图 8-27 和图 8-28 所示。

（5）点击 Independent Axis/date，在下列对话框中输入相应的数据：

- Result Set：wheel travel；
- Component：vertical left（左侧车轮垂向跳动的距离，即仿真设置的上下跳动 52 mm，数值 52 是 FSAE 赛车设计的一个要求，实际可以根据情况做适当调整）。

（6）单击 OK 按钮，返回到后处理主界面。在下列对话框中输入相应的数据：

- Source：Object；
- Filter：body；

VON MISES Hot Spots for fbs_arb_flex Date= 2020-07-15 11:41:22						
Model=.FSAE_sus_front_flex_test.FSAE_arb_flex_front		Analysis= OT_opposite_travel		Time = 1 to 1001 sec	
Top 10 Hot Spots		Abs	Radius= 0.0 mm		
Hot Spot	Stress	Node	Time	Location wrt LPRF (mm)	
#	(newton/mm**2)	id	(sec)	X	Y	Z
1	26.071	2267	1001	-239.781	7.99495	4.40897
2	26.071	405	1	239.781	7.99495	4.40897
3	25.7115	2248	1001	-236.965	7.00948	4.40897
4	25.7115	424	1	236.965	7.00948	4.40897
5	25.6788	2286	1001	-242.308	9.58241	4.40897
6	25.6788	386	1	242.308	9.58241	4.40897
7	25.0505	2305	1001	-244.418	11.6922	4.40897
8	25.0505	367	1	244.418	11.6922	4.40897
9	24.7032	2249	1001	-236.672	8.2931	-0.356773
10	24.7031	425	1	236.672	8.29311	-0.356773

图 8-27 节点应力变化

VON MISES Hot Spots for fbs_arb_flex Date= 2020-07-15 11:41:22							
Model= .FSAE_sus_front_flex_test.FSAE_arb_flex_front			Analysis= OT_opposite_travel		Time = 1 to 1001 sec		
Top 10 Hot Spots			Abs	Radius= 0.0 mm		
Hot Spot	Strain	Node	Time	Location wrt LPRF (mm)		
#	(mm/mm)	id	(sec)	X	Y	Z
1	8.91449e-005	2343	1001	-246.991	17.035	4.40897
2	8.91449e-005	329	1	246.991	17.035	4.40897
3	8.58844e-005	348	1	246.005	14.2187	4.40897
4	8.58844e-005	2324	1001	-246.005	14.2187	4.40897
5	8.46022e-005	2362	1001	-247.325	20	4.40897
6	8.46021e-005	310	1	247.325	20	4.40897
7	8.18707e-005	333	1	256.237	14.9245	-7.49063
8	8.18706e-005	2347	1001	-256.237	14.9245	-7.49063
9	8.14816e-005	314	1	256.809	20	-7.49063
10	8.14816e-005	2366	1001	-256.809	20	-7.49063

图 8-28 节点应变变化

- Object：FSAE_arb_flex_front > fbs_arb_flex > INT_NODE_2680 / INT_NODE_2681；
- Characteristic：Total_Force_At_Location；
- Component：X/Y/Z。

（7）单击 Add Curves，完成 2680、2681 节点在 X、Y、Z 三个方向的受力分析如图 8-29 和图 8-30 所示。

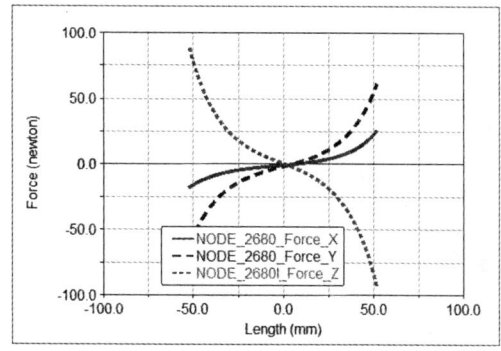

图 8-29 节点 2680 受力

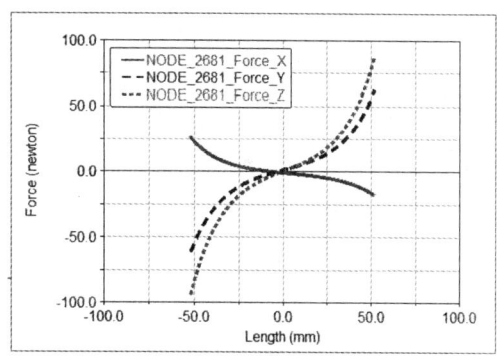

图 8-30 节点 2681 受力

第 9 章　非独立悬架

钢板弹簧由于其特殊性，在工、农等行业均有应用，农用三轮车后悬架，商用牵引车、挂车、工程车辆等均采用钢板弹簧。悬架系统的优劣不在于采用所谓的悬架类型，关键在于悬架与车身等参数的匹配与调试，调试参数包含连接位置、衬套刚度、弹簧与阻尼器参数等。采用同一个底盘的不同车辆并不能得到相同性能的原因主要在此。例如，克尔维特跑车、沃尔沃依然采用横置板簧悬架系统，横置板簧可以节省安装空间，同时起到拉杆的作用，其特性表现均衡。牵引车及工程车辆后驱动轴采用平衡悬架在较差的路面依然极具优势。本章介绍创建 FSAE 赛车尺寸架构下的非独立悬架系统，建立好悬架模型后并将其装配到整车上验证动态特性。

图 9-1　板簧悬架

学习目标

（1）了解板簧悬架模型。
（2）了解驱动轴。
（3）了解变量参数。
（4）会通信器匹配。
（5）会半车模型装配。
（6）会车轮激振分析。
（7）会整车模型装配。
（8）会 Fish-Hook 仿真。

9.1 板簧悬架模型

（1）启动 ADAMS/CAR，选择专家模块 Expert 进入建模界面。
（2）单击 File > New 命令，弹出建模对话框，如图 9-2 所示。

图 9-2 模板框

在模板名称中输入：FSAE_sus_leafspringr，主特征选择 suspension，单击 OK 按钮。

9.1.1 板簧悬架硬点

（1）单击 Build > Hardpoind > New 命令，弹出创建硬点对话框，如图 9-3 所示。

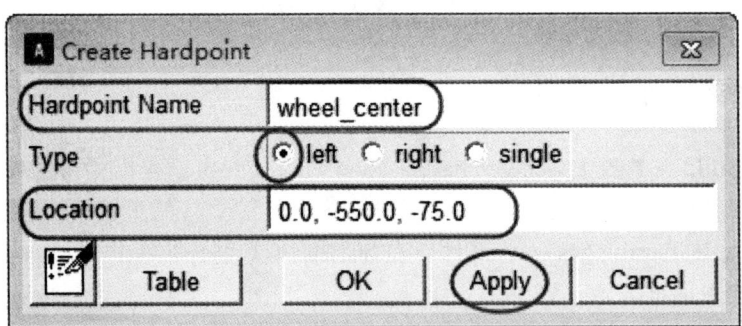

图 9-3 硬点对话框

（2）在硬点名称中输入：wheel_center，类型选择 left；在位置文本框输入：0.0，-750.0，-75.0。
（3）单击 Apply 按钮，完成 wheel_center 硬点的创建。此时，在屏幕上显示出左右对称的两个硬点。
（4）重复上述步骤，完成图 9-4 中硬点的创建。

	loc x	loc y	loc z
hpl_drive_shaft_in	0.0	-250.0	-70.0
hpl_hub_down	0.0	-450.0	-150.0
hpl_hub_up	0.0	-400.0	75.0
hpl_p7_down	160.0	-300.0	-135.0
hpl_pt1	-300.0	-300.0	-130.0
hpl_pt2	-240.0	-300.0	-130.0
hpl_pt3	-160.0	-300.0	-130.0
hpl_pt3_down	-160.0	-300.0	-135.0
hpl_pt4	-80.0	-300.0	-130.0
hpl_pt4_down	-80.0	-300.0	-135.0
hpl_pt5	0.0	-300.0	-130.0
hpl_pt5_down	0.0	-300.0	-135.0
hpl_pt6	80.0	-300.0	-130.0
hpl_pt6_down	80.0	-300.0	-135.0
hpl_pt7	160.0	-300.0	-130.0
hpl_pt8	240.0	-300.0	-130.0
hpl_pt9	300.0	-300.0	-130.0
hpl_pt9_up	300.0	-300.0	-80.0
hpl_tierod_inner	200.0	-150.0	-75.0
hpl_tierod_outer	200.0	-500.0	-75.0
hpl_wheel_center	0.0	-550.0	-75.0

图 9-4 板簧硬点参数

9.1.2 横梁部件

(1)单击 Build > Part > General Part > New 命令,弹出创建横梁部件对话框,如图 9-5 所示。在下列对话框中输入相应的数据(图 9-5 为已经建立好的板簧非独立悬架模型,通过右击._FSAE_sus_leafspring.ges_axis 部件,在弹出的快捷菜单中单击 Modify 修改对话框,修改对话框与新建对话框完全一致):

- General Part:axis;
- Location Dependency:Centered between coordinates;
- Centered between:two Coordinates;
- Coordinate Reference #1(参考坐标):._FSAE_sus_leafspring.ground.hpl_wheel_center;
- Coordinate Reference #2(参考坐标):._FSAE_sus_leafspring.ground.hpr_wheel_center;
- Orient using:Euler Angles,部件定向采用欧拉角模式;
- Euler Angles:0,0,0;
- Mass:1;
- Ixx:1;
- Iyy:1;
- Izz:1;

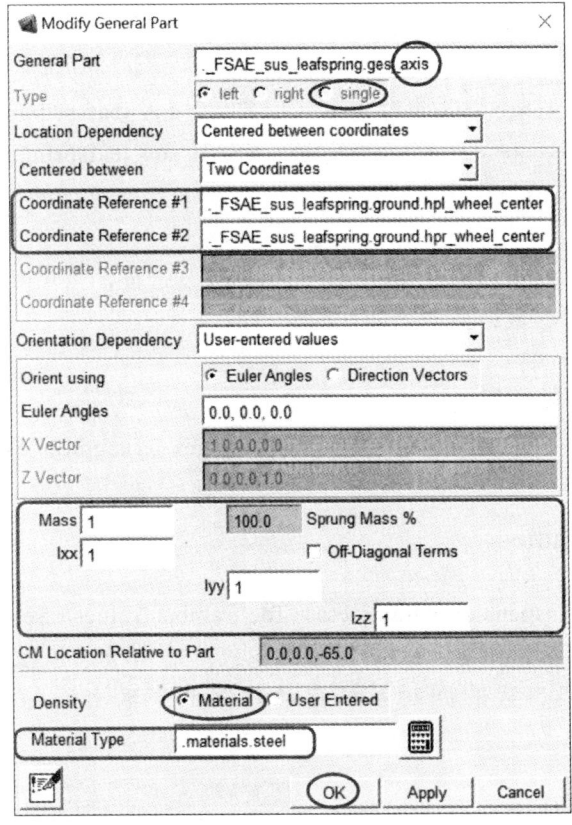

图 9-5　横梁部件对话框

- Density：Material；
- Material Type：.materials.steel。

（2）单击 OK 按钮，完成部件._FSAE_sus_leafspring.ges_axis 的创建。

（3）单击 Build > Geometry > Link > New 命令，弹出横梁部件连杆几何体创建对话框，如图 9-6 所示。在下列对话框中输入相应的数据：

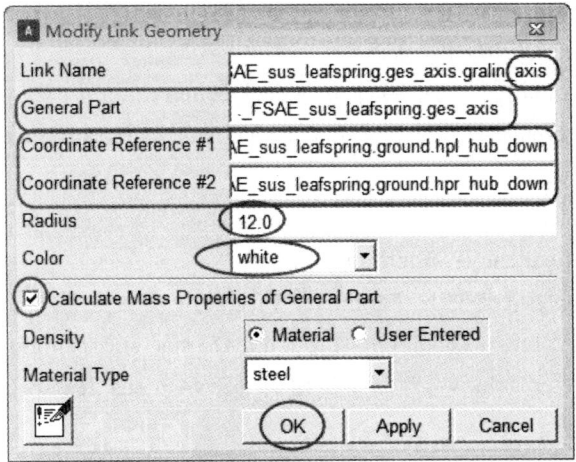

图 9-6　横梁部件连杆几何体创建对话框

- Link Name（连杆名称）：axis；
- General Part：._FSAE_sus_leafspring.ges_axis；
- Coordinate Reference #1（参考坐标）：._FSAE_sus_leafspring.ground.hpl_hub_down；
- Coordinate Reference #2（参考坐标）：._FSAE_sus_leafspring.ground.hpr_hub_down；
- Radius（半径）：12；
- Color：white；
- 勾选 Calculate Mass Properties of General Part 复选框，当几何体建立好之后，会更新对应部件的质量和惯量参数；
- Density：Material；
- Material Type：steel。

（4）单击 OK 按钮，完成._FSAE_sus_leafspring.ges_axis.gralin_axis 几何体的创建。

9.1.3 轮毂 spindle

（1）单击 Build > Suspension Parameters > Toe/Camber Values > Set 命令，弹出悬架参数对话框，如图 9-7 所示。前束角输入：0；外倾角输入：0；单击 OK 按钮，完成前束与外倾参数的创建。创建悬架参数的同时系统自动建立两个输出通信器：col[r]_toe_angle、col[r]_camber_angle。

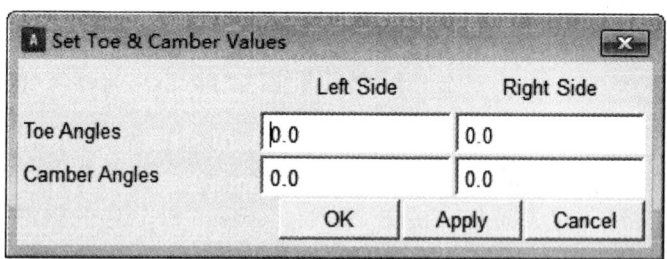

图 9-7 悬架参数

（2）单击 Build > Construction Frame > New 命令，弹出创建结构框对话框，如图 9-8 所示。在下列对话框中输入相应的数据：

- Construction Frame（结构框名称）：wheel_center；
- Coordinate Reference（参考坐标）：._FSAE_sus_leafspring.ground.hpl_wheel_center；
- Location：0，0，0；
- Location in：local；
- Orientation Dependency：Toe/Camber；
- Variable Type（变量类型）：Parameter Variable（参数变量）；
- Toe Parameter Values（前束变量值）：._FSAE_sus_leafspring.pvl_toe_angle；
- Camber Parameter Values（外倾变量值）：._FSAE_sus_leafspring.pvl_camber_angle。

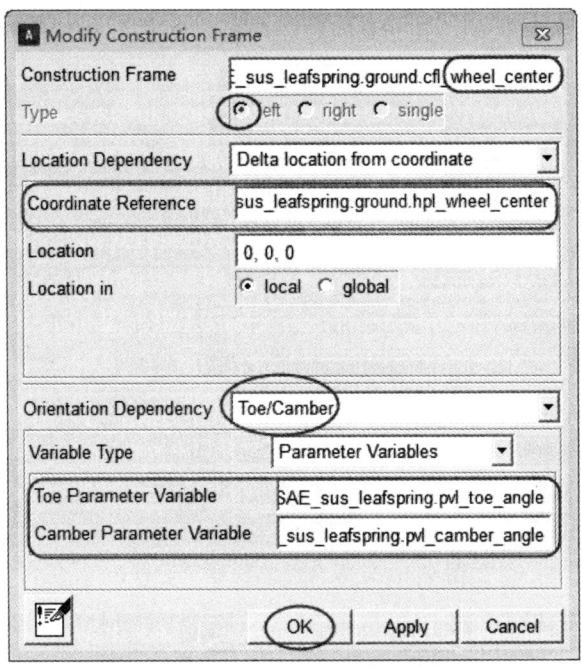

图 9-8 wheel_center 结构框对话框

（3）单击 OK 按钮，完成._FSAE_sus_leafspring.ground.cfl_wheel_center 结构框的创建。

（4）单击 Build > Part > General Part > New 命令，弹出创建部件对话框，如图 9-5 所示。在下列对话框中输入相应的数据：

- General Part：spindle；
- Location Dependency：Delta location from coordinate；
- Coordinate Reference（参考坐标）：._FSAE_sus_leafspring.ground.hpl_wheel_center；
- Location：0, 0, 0；
- Location in：local；
- Orientation Dependency：Delta orientation from coordinate；
- Construction Frame：._FSAE_sus_leafspring.ground.cfl_wheel_center；
- Orientation：0, 0, 0；
- Mass：1；
- Ixx：1；
- Iyy：1；
- Izz：1；
- Density：Material；
- Material Type：.materials.steel。

（5）单击 OK 按钮，完成部件._FSAE_sus_leafspring.gel_spindle 的创建。

（6）单击 Build > Geometry > Cylinder（圆柱体）> New 命令，弹出创建部件对话框，如图 9-9 所示。在下列对话框中输入相应的数据：

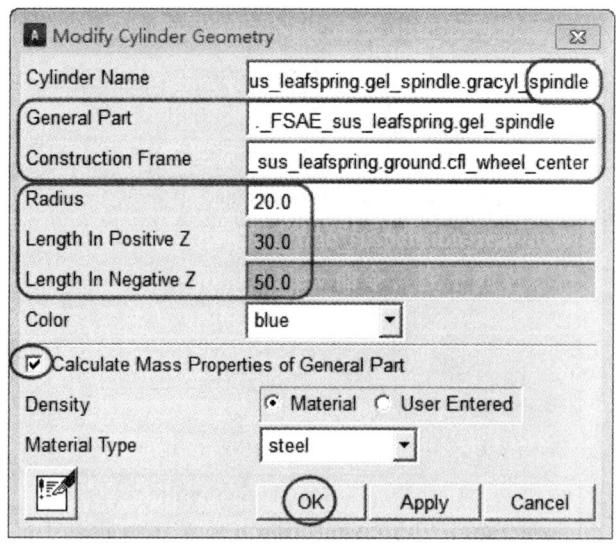

图 9-9 轮毂几何体创建对话框

- Cylinder Name（连杆名称）：spindle；
- General Part：._FSAE_sus_leafspring.gel_spindle；
- Construction Frame：._FSAE_sus_leafspring.ground.cfl_wheel_center；
- Radius（半径）：20；
- Length In Positive Z（Z轴正方向长度）：30；
- Length In Negative Z（Z轴负方向长度）：50；
- Color（圆柱体几何体颜色）：blue；
- 选择 Calculate Mass Properties of General Part 复选框。

（7）单击 OK 按钮，完成轮毂圆柱体._FSAE_sus_leafspring.gel_spindle.gracyl_spindle 几何体的创建。

9.1.4 转向节 upright

（1）单击 Build > Part > General Part > New 命令，弹出创建部件对话框，如图 9-5 所示。在下列对话框中输入相应的数据：

- General Part：upright；
- Location Dependency：Delta location from coordinate；
- Coordinate Reference（参考坐标）：._FSAE_sus_leafspring.ground.cfl_wheel_center；
- Location：0，0，0；
- Location in：local；
- Orient using：Euler Angles；
- Euler Angles：0，0，0；
- Mass：1；
- Ixx：1；

- Iyy：1；
- Izz：1；
- Density：Material；
- Material Type：.materials.steel。

（2）单击 OK 按钮，完成部件._FSAE_sus_leafspring.gel_upright 的创建。

（3）单击 Build > Geometry > Link > New 命令，在下列对话框中输入相应的数据：
- Link Name（连杆名称）：link1；
- General Part：._FSAE_sus_leafspring.gel_upright；
- Coordinate Reference #1（参考坐标）：._FSAE_sus_leafspring.ground.hpl_hub_down；
- Coordinate Reference #2（参考坐标）：._FSAE_sus_leafspring.ground.cfl_wheel_center；
- Radius（半径）：15；
- Color（杆件几何体颜色）：green；
- 选择 Calculate Mass Properties of General Part 复选框，当几何体建立好之后，会更新对应部件的质量和惯量参数；
- Density：Material；
- Material Type：steel。

（4）单击 Apply 按钮，完成._FSAE_sus_leafspring.gel_upright.gralin_link1 几何体的创建。同理，创建其他几何体。
- Link Name（连杆名称）：link2；
- General Part：._FSAE_sus_leafspring.gel_upright；
- Coordinate Reference #1（参考坐标）：._FSAE_sus_leafspring.ground.hpl_hub_up；
- Coordinate Reference #2（参考坐标）：._FSAE_sus_leafspring.ground.cfl_wheel_center；
- Radius（半径）：15；
- Color（杆件几何体颜色）：green；
- 选择 Calculate Mass Properties of General Part 复选框，当几何体建立好之后，会更新对应部件的质量和惯量参数；
- Density：Material；
- Material Type：steel。

（5）单击 Apply 按钮，完成._FSAE_sus_leafspring.gel_upright.gralin_link2 几何体的创建。
- Link Name（连杆名称）：link3；
- General Part：._FSAE_sus_leafspring.gel_upright；
- Coordinate Reference #1（参考坐标）：._FSAE_sus_leafspring.ground.hpl_tierod_outer；
- Coordinate Reference #2（参考坐标）：._FSAE_sus_leafspring.ground.cfl_wheel_center；
- Radius（半径）：15；
- Color（杆件几何体颜色）：green；
- 选择 Calculate Mass Properties of General Part 复选框，当几何体建立好之后，会更新对应部件的质量和惯量参数；
- Density：Material；
- Material Type：steel。

（6）单击 OK 按钮，完成 ._FSAE_sus_leafspring.gel_upright.gralin_link3 几何体的创建。

9.1.5 转向横拉杆

（1）单击 Build > Part > General Part > New 命令，弹出创建部件对话框，如图 9-5 所示。在下列对话框中输入相应的数据：

- General Part：tierod；
- Location Dependency：Centered between coordinates；
- Centered between：Two Coordinates；
- Coordinate Reference #1（参考坐标）：._FSAE_sus_leafspring.ground.hpl_tierod_outer；
- Coordinate Reference #2（参考坐标）：._FSAE_sus_leafspring.ground.hpl_tierod_inner；
- Orientation Dependency：User-entered values；
- Orient using：Euler Angles；
- Euler Angles：0，0，0；
- Mass：1；
- Ixx：1；
- Iyy：1；
- Izz：1；
- Density：Material；
- Material Type：.materials.steel。

（2）单击 OK 按钮，完成部件 ._FSAE_sus_leafspring.gel_tierod 的创建。

（3）单击 Build > Geometry > Link > New 命令，在下列对话框中输入相应的数据：

- Link Name（连杆名称）：tierod；
- General Part：._FSAE_sus_leafspring.gel_tierod；
- Coordinate Reference #1（参考坐标）：._FSAE_sus_leafspring.ground.hpl_tierod_outer；
- Coordinate Reference #2（参考坐标）：._FSAE_sus_leafspring.ground.hpl_tierod_inner；
- Radius（半径）：10；
- Color（杆件几何体颜色）：red；
- 选择 Calculate Mass Properties of General Part 复选框，当几何体建立好之后，会更新对应部件的质量和惯量参数；
- Density：Material；
- Material Type：steel。

（4）单击 OK 按钮，完成横拉杆 ._FSAE_sus_leafspring.gel_tierod.gralin_tierod 几何体的创建。

9.1.6 吊　耳

（1）单击 Build > Part > General Part > New 命令，弹出创建部件对话框，如图 9-5 所示。在下列对话框中输入相应的数据：

- General Part：shackle；
- Location Dependency：Delta location from coordinate；
- Coordinate Reference（参考坐标）：._FSAE_sus_leafspring.ground.hpl_p9；
- Location：0，0，0；
- Location in：local；
- Orientation Dependency：User-entered values；
- Orient using：Euler Angles；
- Euler Angles：0，0，0；
- Mass：1；
- Ixx：1；
- Iyy：1；
- Izz：1；
- Density：Material；
- Material Type：materials.steel。

（2）单击 OK 按钮，完成部件._FSAE_sus_leafspring.gel_shackle 的创建。

（3）单击 Build > Geometry > Link > New 命令，在下列对话框中输入相应的数据：
- Link Name（连杆名称）：shackle；
- General Part：._FSAE_sus_leafspring.gel_shackle；
- Coordinate Reference #1（参考坐标）：._FSAE_sus_leafspring.ground.hpl_p6；
- Coordinate Reference #2（参考坐标）：._FSAE_sus_leafspring.ground.hpl_p9；
- Radius（半径）：10；
- Color（杆件几何体颜色）：red；
- 选择 Calculate Mass Properties of General Part 复选框，当几何体建立好之后，会更新对应部件的质量和惯量参数；
- Density：Material；
- Material Type：steel。

（4）单击 OK 按钮，完成吊耳._FSAE_sus_leafspring.gel_shackle.gralin_shackle 几何体的创建；

9.1.7 驱动轴 drive_shaft 部件

（1）单击 Build > Parameter Variable > New 命令，弹出参数变量对话框，如图 9-10 所示。在下列对话框中输入相应的数据：
- Parameter Variable Name：drive_shaft_offset；
- Real Value（实数值）：50；
- Units：no units；
- Hide from standard user（是否从标准界面隐藏）：no。

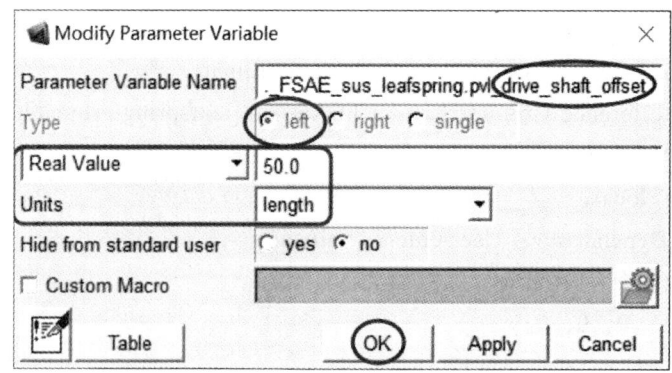

图 9-10 drive_shaft_offset 变量对话框

（2）单击 OK 按钮，完成变量._FSAE_sus_leafspring.pvl_drive_shaft_offset 的创建。

（3）单击 Build > Construction Frame > New 命令，在下列对话框中输入相应的数据：

- Construction Frame（结构框名称）：drive_shaft_otr；
- Location Dependency：Delta location from coordinate；
- Coordinate Reference（参考坐标）：._FSAE_sus_leafspring.ground.cfl_wheel_center；
- Location：0.0，0.0，(-1.0 * ._FSAE_sus_leafspring.pvl_drive_shaft_offset)；
- Location in：local；
- Orientation Dependency：Orient axis to point；
- Coordinate Reference（参考坐标）：._FSAE_sus_leafspring.ground.hpl_wheel_center；
- Axis：Z。

（4）单击 OK 按钮，完成._FSAE_sus_leafspring.ground.cfl_drive_shaft_otr 结构框的创建。

（5）单击 Build > Part > General Part > New 命令，在下列结构框中输入相应的数据：

- General Part：drive_shaft；
- Location Dependency：Delta location from coordinate；
- Coordinate Reference（参考坐标）：._FSAE_sus_leafspring.ground.hpl_drive_shaft_inr；
- Location：0，0，0；
- Location in：local；
- Orientation Dependency：Oriented in plane；
- Coordinate Reference #1（参考坐标）：._FSAE_sus_leafspring.ground.cfl_drive_shaft_otr；
- Coordinate Reference #2（参考坐标）：._FSAE_sus_leafspring.ground.hpl_drive_shaft_inr；
- Coordinate Reference#3（参考坐标）：._FSAE_sus_leafspring.ground.hpl_wheel_center；
- Axis：ZX；
- Mass：1；
- Ixx：1；
- Iyy：1；
- Izz：1；
- Density：Material；
- Material Type：.materials.steel。

（6）单击 OK 按钮，完成部件._FSAE_sus_leafspring.gel_drive_shaft 的创建。

（7）单击 Build > Geometry > Link > New 命令，在下列对话框中输入相应的数据：
- Link Name（连杆名称）：drive_shaft；
- General Part：._FSAE_sus_leafspring.gel_drive_shaft；
- Coordinate Reference #1（参考坐标）：._FSAE_sus_leafspring.ground.hpl_drive_shaft_inr；
- Coordinate Reference #2（参考坐标）：._FSAE_sus_leafspring.ground.cfl_drive_shaft_otr；
- Radius（半径）：15；
- Color（杆件几何体颜色）：yellow；
- 选择 Calculate Mass Properties of General Part 复选框，当几何体建立好之后，会更新对应部件的质量和惯量参数；
- Density：Material；
- Material Type：steel。

（8）单击 OK 按钮，完成._FSAE_sus_leafspring.gel_drive_shaft.gralin_drive_shaft 几何体的创建。

（9）单击 Build > Geometry > Ellipsoid > New 命令，在下列对话框中输入相应的数据：
- Ellipsoid Name（连杆名称）：cv_housing；
- Coordinate Reference（参考坐标）：._FSAE_sus_leafspring.ground.cfl_drive_shaft_otr；
- Link：._FSAE_sus_leafspring.gel_drive_shaft.gralin_drive_shaft；
- X Scale：2；
- Y Scale：2；
- Z Scale：2；
- Color（杆件几何体颜色）：yellow；
- 选择 Calculate Mass Properties of General Part 复选框，当几何体建立好之后，会更新对应部件的质量和惯量参数；
- Density：Material；
- Material Type：steel。

（10）单击 Apply 按钮，完成._FSAE_sus_leafspring.gel_drive_shaft.graell_cv_housing 几何体的创建。

（11）单击 Build > Geometry > Ellipsoid > New 命令，在下列对话框中输入相应的数据：
- Ellipsoid Name（连杆名称）：tripot_housing；
- Coordinate Reference（参考坐标）：._FSAE_sus_leafspring.ground.hpl_drive_shaft_inr；
- Link：._fsae_suspension_rear_axle.gel_drive_shaft.gralin_drive_shaft；
- X Scale：2；
- Y Scale：2；
- Z Scale：2；
- Color（杆件几何体颜色）：yellow；
- 选择 Calculate Mass Properties of General Part 复选框；
- Density：Material；
- Material Type：steel。

（12）单击 OK 按钮，完成 ._FSAE_sus_leafspring.gel_drive_shaft.graell_tripot_housing 几何体的创建。

9.1.8 等速万向节 tripot 部件

（1）单击 Build > Construction Frame > New 命令，在下列对话框中输入相应的数据：
- Construction Frame（结构框名称）：drive_shaft_inr；
- Location Dependency：Delta location from coordinate；
- Coordinate Reference（参考坐标）：._FSAE_sus_leafspring.ground.hpl_drive_shaft_inr；
- Location：0，0，0；
- Location in：local；
- Orientation Dependency：Orient in plane；
- Coordinate Reference #1（参考坐标）：._FSAE_sus_leafspring.ground.hpl_drive_shaft_inr；
- Coordinate Reference #2（参考坐标）：._FSAE_sus_leafspring.ground.hpr_drive_shaft_inr；
- Coordinate Reference#3（参考坐标）：._FSAE_sus_leafspring.ground.cfl_drive_shaft_otr；
- Axis：ZX。

（2）单击 OK 按钮，完成 ._FSAE_sus_leafspring.ground.cfl_drive_shaft_inr 结构框的创建。

（3）单击 Build > Part > General Part > New 命令，弹出创建部件对话框，如图 9-5 所示。在下列对话框中输入相应的数据：
- General Part：tripot；
- Location Dependency：Delta location from coordinate；
- Coordinate Reference（参考坐标）：._FSAE_sus_leafspring.ground.hpl_drive_shaft_inr；
- Location：0，0，0；
- Location in：local；
- Orientation Dependency：Orient to zpoint-xpoint；
- Coordinate Reference #1（参考坐标）：._FSAE_sus_leafspring.ground.hpl_drive_shaft_inr；
- Coordinate Reference #2（参考坐标）：._FSAE_sus_leafspring.ground.cfl_drive_shaft_otr；
- Axis：ZX；
- Mass：1；
- Ixx：1；
- Iyy：1；
- Izz：1；
- Density：Material；
- Material Type：.materials.steel。

（4）单击 OK 按钮，完成部件 ._FSAE_sus_leafspring.gel_tripot 的创建。

（5）单击 Build > Geometry > Cylinder（圆柱体）> New 命令，弹出创建部件对话框，如图 9-5 所示。在下列对话框中输入相应的数据：

- Cylinder Name（连杆名称）：tripot_housing_extention；
- General Part：._FSAE_sus_leafspring.gel_tripot；
- Radius（半径）：30；
- Length In Positive Z（Z轴正方向长度）：50；
- Length In Negative Z（Z轴负方向长度）：0；
- Color（圆柱体几何体颜色）：blue；
- 选择 Calculate Mass Properties of General Part 复选框。

（6）单击 OK 按钮，完成圆柱体._FSAE_sus_leafspring.gel_tripot.gracyl_tripot_housing_extention 几何体的创建。

9.1.9 安装部件

（1）单击 Build > Construction Frame > New 命令，在下列对话框中输入相应的数据：
- Construction Frame（结构框名称）：subframe；
- Location Dependency：Centered between coordinates；
- Centered between：Two Coordinates；
- Coordinate Reference #1（参考坐标）：._FSAE_sus_leafspring.ground.hpl_wheel_center；
- Coordinate Reference #2（参考坐标）：._FSAE_sus_leafspring.ground.hpr_wheel_center；
- Orientation Dependency：User-entered values；
- Orient using：Euler Angles；
- Euler Angles：0，0，0。

（2）单击 OK 按钮，完成._FSAE_sus_leafspring.ground.cfs_subframe 结构框的创建。

（3）单击 Build > Part > Mount > New 命令，弹出创建部件对话框，如图 9-11 所示。在下列对话框中输入相应的数据：

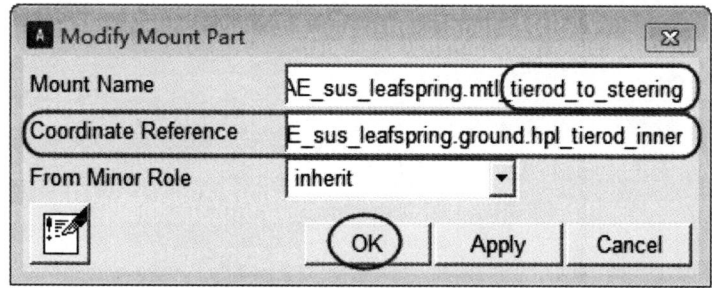

图 9-11 安装部件对话框

- Mount name（安装件名称）：tierod_to_steering；
- Coordinate Reference（参考坐标）：._FSAE_sus_leafspring.ground.hpl_tierod_inner；
- From Minor Role：inherit（继承特性）。

（4）单击 Apply 按钮，完成 ._FSAE_sus_leafspring.mtl_tierod_to_steering 安装部件的创建。同理，完成其他安装部件的创建。

- Mount name（安装件名称）：tripot_to_differential；
- Coordinate Reference（参考坐标）：._FSAE_sus_leafspring.ground.hpl_drive_shaft_inr；
- From Minor Role：inherit（继承特性）。

（5）单击 Apply 按钮，完成 ._FSAE_sus_leafspring.mtl_tripot_to_differential 安装部件的创建。

- Mount name（安装件名称）：leafspring_to_body；
- Coordinate Reference（参考坐标）：._FSAE_sus_leafspring.ground.cfs_subframe；
- From Minor Role：inherit（继承特性）。

（6）单击 OK 按钮，完成 ._FSAE_sus_leafspring.mts_leafspring_to_body 安装部件的创建。

9.2 钢板弹簧

（1）单击 Build > Part > Nonlinear Beam > New 命令，弹出创建非线性梁对话框，如图 9-12 所示。在下列对话框中输入相应的数据：

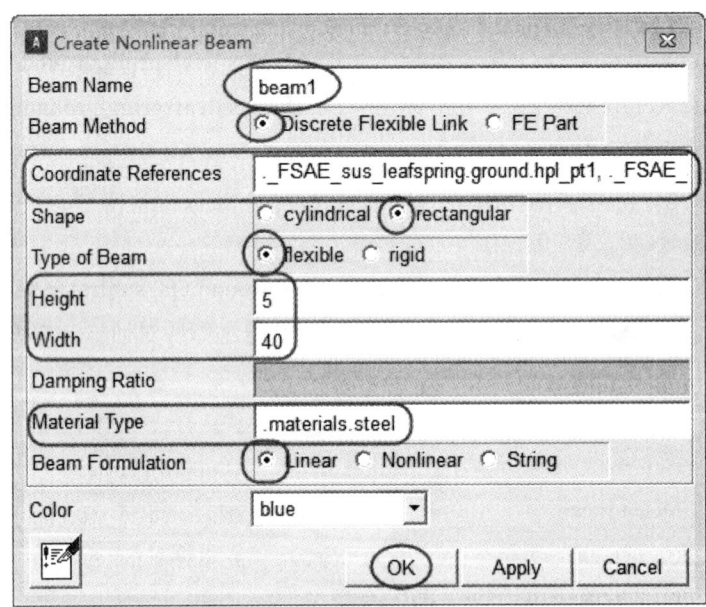

图 9-12 非线性梁部件 Beam1

- General Part：beam1；
- Coordinate Reference（参考坐标，依次输入如下硬点信息，硬点信息属性不能错乱，右击鼠标选择 Pick 选取）：

[1] ._FSAE_sus_leafspring.ground.hpl_pt1；

[2] ._FSAE_sus_leafspring.ground.hpl_pt2；

[3] ._FSAE_sus_leafspring.ground.hpl_pt3；

- [4] ._FSAE_sus_leafspring.ground.hpl_pt4；
- [5] ._FSAE_sus_leafspring.ground.hpl_pt5；
- [6] ._FSAE_sus_leafspring.ground.hpl_pt6；
- [7] ._FSAE_sus_leafspring.ground.hpl_pt7；
- [8] ._FSAE_sus_leafspring.ground.hpl_pt8；
- [9] ._FSAE_sus_leafspring.ground.hpl_pt9；
- Shape（非线性梁形状，包括圆形和矩形两种）：rectangular；
- Height：5；
- Width：10；
- Material Type：steel；
- Type of Beam：flexible；
- Beam Formulation：linear。

（2）单击 Apply 按钮，完成._FSAE_sus_leafspring.nrl_1_beam1 部件的创建。同理，创建 beam2 部件。

- General Part：beam2；
- Coordinate Reference（参考坐标，依次输入如下硬点信息，硬点信息属性不能错乱，右击鼠标选择 Pick 选取）：

- [1] ._FSAE_sus_leafspring.ground.hpl_pt3_down；
- [2] ._FSAE_sus_leafspring.ground.hpl_pt4_down；
- [3] ._FSAE_sus_leafspring.ground.hpl_pt5_down；
- [4] ._FSAE_sus_leafspring.ground.hpl_pt6_down；
- [5] ._FSAE_sus_leafspring.ground.hpl_pt7_down；
- Shape（非线性梁形状，包括圆形和矩形两种）：rectangular；
- Height：5；
- Width：40；
- Material Type：steel；
- Type of Beam：flexible；
- Beam Formulation：linear。

（3）单击 OK 按钮，完成._FSAE_sus_leafspring.nrl_1_beam2 部件的创建。

9.2.1 簧片接触力

（1）单击 Tools > Adams/View Interface 命令，切换到 View 通用界面，如图 9-13 所示。

（2）单击 Forces > Create a Contact 命令，弹出创建接触对话框，如图 9-14 所示。在下列对话框中输入相应的数据：

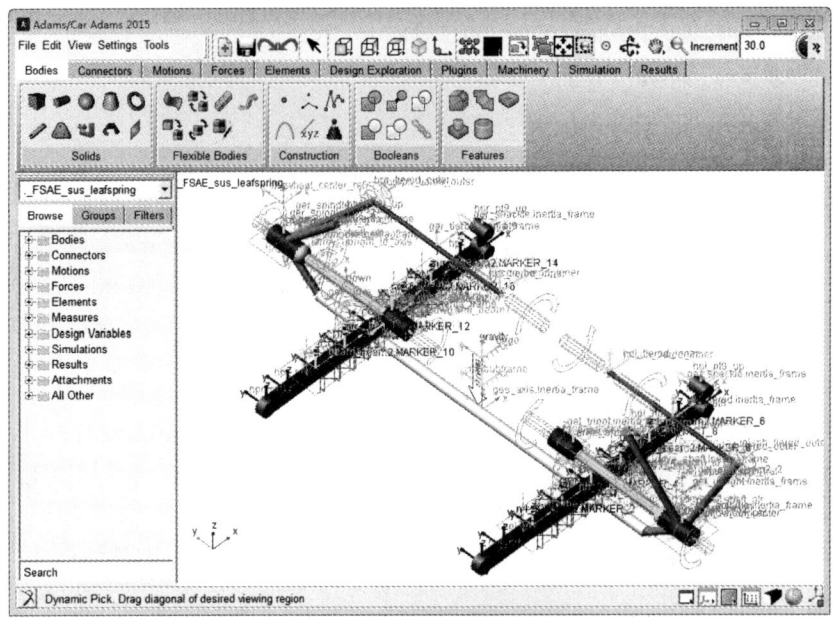

图 9-13 View 通用界面

图 9-14 接触力创建对话框

- Contact Type：Solid to Solid；
- I Solid（s）：._FSAE_sus_leafspring.nrl_1_beam1.nrl_gra_i_1；
- J Solid（s）：._FSAE_sus_leafspring.nrl_3_beam2.nrl_gra_i_13；
- Force Display：Red；
- Normal Force：Impact；
- Force Exponent：2.2；
- Damping：10；
- Coulomb Force：Coulomb；
- Coulomb Coulomb：On；
- Static Coefficient：0.3；
- Dynamic Coefficient：0.3。

(3) 其余参数保持默认，单击 Apply 按钮，完成._FSAE_sus_leafspring.CONTACT_1 接触力设置。重复上述步骤，完成所有对应接触面的接触力设置，特别强调接触面要一一对应，此模型包含 16 个接触。

9.2.2 弹簧夹

钢板弹簧弹簧夹的主要作用是保障弹簧在上下运动过程中装配（模型中为接触）的两簧片不产生分离，通过约束关系中的点面约束抽象为弹簧夹。当钢板弹簧长度较大时，在板簧接触的端部和大概中间部位约束。在大载荷冲击下，点面约束是保障整车静平衡或者板簧计算模型收敛的必要条件。

(1) 单击 Connectors > Primitives > Create an inplane Joint Primitive 命令，选择 Construction：2 Bodies-1 Location；Normal To Grid；用鼠标分别选择钢板弹簧部件._FSAE_sus_leafspring.nrl_1_beam1、._FSAE_sus_leafspring.nrl_3_beam2 及._FSAE_sus_leafspring.ground.hpl_p4 点，完成._FSAE_sus_leafspring.JPRIM_1 点面约束的创建。

(2) 在模型树上右击点面约束._FSAE_sus_leafspring.JPRIM_1，点击 Modify 或者双击点面约束._FSAE_sus_leafspring.JPRIM_1，弹出约束对话框，如图 9-15 所示。此模型建立过程中共包含 8 个点面约束。

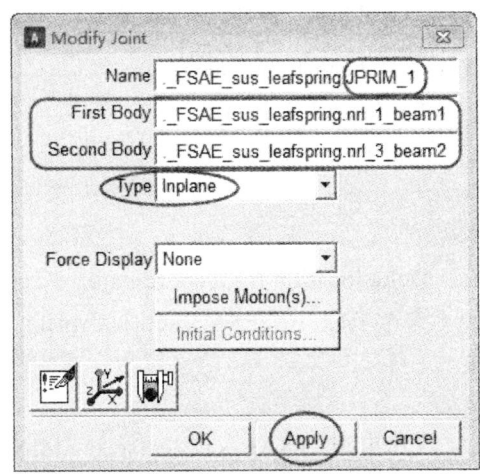

图 9-15　点面约束对话框

9.2.3 板簧约束

1. 部件 axle 与 nrl_3_beam1 之间的 fixed 约束

（1）单击 Tools > Select Mode > Switch To A/Car Template Builder 命令，切换到 ADAMS/CAR 专家界面。

（2）单击 Build > Attachments > Joint > New 命令，弹出刚性约束对话框，如图 9-16 所示。在下列对话框中输入相应的数据：

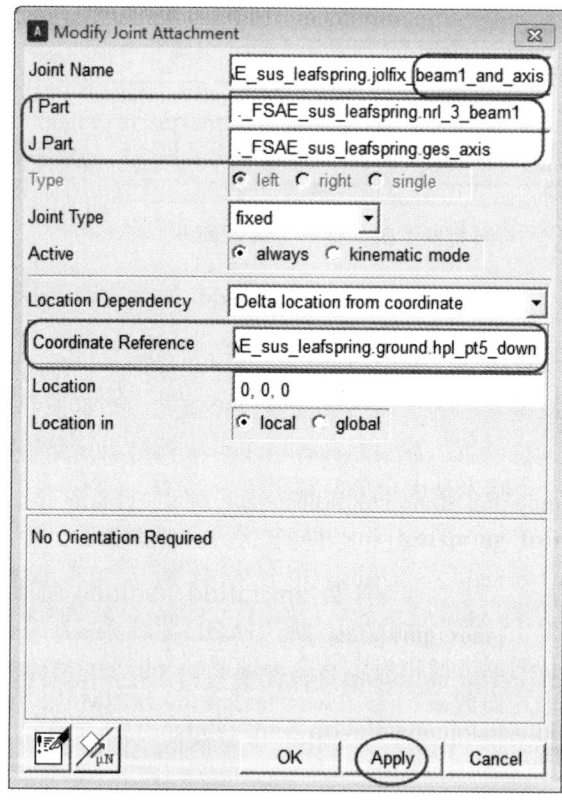

图 9-16 刚性约束对话框

- Joint Name（约束副名称）：beam1_and_axis；
- I Part：._FSAE_sus_leafspring.nrl_3_beam1；
- J Part：._FSAE_sus_leafspring.ges_axis；
- Joint Type：fixed；
- Active（激活）：always；
- Location Dependency：Delta location from coordinate；
- Coordinate Reference（参考坐标）：._FSAE_sus_leafspring.ground.hpl_p11；
- Location：0，0，0；
- Location in：local。

（3）单击 Apply 按钮，完成约束副._FSAE_sus_leafspring.jolfix_beam1_and_axis 的创建。

2. 部件 beam1 与 beam2 之间的 fixed 约束

（1）单击 Build > Attachments > Joint > New 命令，在下列对话框中输入相应的数据：
- Joint Name（约束副名称）：beam2_and_beam1；
- I Part：._FSAE_sus_leafspring.nrl_3_beam1；
- J Part：._FSAE_sus_leafspring.nrl_5_beam2；
- Joint Type：fixed；
- Active（激活）：always；
- Location Dependency：Delta location from coordinate；
- Coordinate Reference（参考坐标）：._FSAE_sus_leafspring.ground.hpl_pt5_down；
- Location：0，0，0；
- Location in：local。

（2）单击 Apply 按钮，完成约束副._FSAE_sus_leafspring.jolfix_beam2_and_beam1 的创建。

3. 部件 nrl_1_beam2 与 leafspring_to_body 之间的 bushing 约束

（1）单击 Build > Attachments > Bushing > New 命令，弹出柔性衬套约束对话框，如图 9-17 所示。在下列对话框中输入相应的数据：

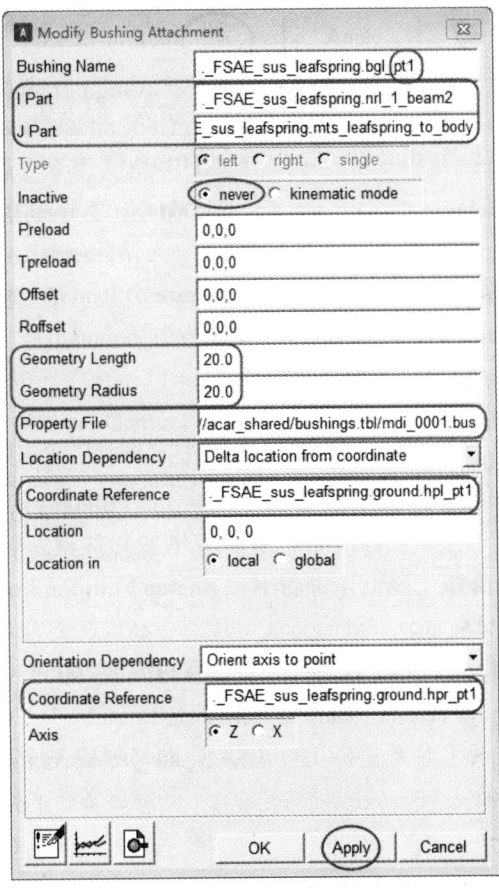

图 9-17　柔性衬套约束对话框

- Bushing Name（约束副名称）：pt1；
- I Part：._FSAE_sus_leafspring.nrl_1_beam2；
- J Part：._FSAE_sus_leafspring.mts_leafspring_to_body；
- Inactive（抑制）：never；
- Preload：0，0，0；
- Tpreload：0，0，0；
- Offset：0，0，0；
- Roffset：0，0，0；
- Geometry Length：20；
- Geometry Radius：20；
- Property File：mdids://acar_shared/bushings.tbl/mdi_0001.bus；
- Location Dependency：Delta location from coordinate；
- Coordinate Reference（参考坐标）：._FSAE_sus_leafspring.ground.hpl_pt1；
- Location：0，0，0；
- Location in：local；
- Orientation Dependency：Orient axis to point；
- Construction Frame：._FSAE_sus_leafspring.ground.hpr_pt1；
- Axis：Z。

（2）单击 Apply 按钮，完成轴套._FSAE_sus_leafspring.bgl_pt1 的创建。

4. 部件 nrl_9_beam2 与 shackle 之间的 bushing 约束

（1）单击 Build > Attachments > Bushing > New 命令，在下列对话框中输入相应的数据：

- Bushing Name（约束副名称）：pt9；
- I Part：._FSAE_sus_leafspring.nrl_9_beam2；
- J Part：._FSAE_sus_leafspring.gel_shackle；
- Inactive（抑制）：never；
- Preload：0，0，0；
- Tpreload：0，0，0；
- Offset：0，0，0；
- Roffset：0，0，0；
- Geometry Length：40；
- Geometry Radius：15；
- Property File：mdids://acar_shared/bushings.tbl/mdi_0001.bus；
- Location Dependency：Delta location from coordinate；
- Coordinate Reference（参考坐标）：._FSAE_sus_leafspring.ground.hpl_pt9；
- Location：0，0，0；
- Location in：local；
- Orientation Dependency：Orient axis to point；
- Construction Frame：._FSAE_sus_leafspring.ground.hpr_pt9；

- Axis：Z。

（2）单击 Apply 按钮，完成轴套._FSAE_sus_leafspring.bgl_pt9 的创建。

5. 部件 leafspring_to_body 与 shackle 之间的 bushing 约束

（1）单击 Build > Attachments > Bushing > New 命令，在下列对话框中输入相应的数据：
- Bushing Name（约束副名称）：pt9_up；
- I Part：._FSAE_sus_leafspring.gel_shackle；
- J Part：._FSAE_sus_leafspring.mts_leafspring_to_body；
- Inactive（抑制）：never；
- Preload：0，0，0；
- Tpreload：0，0，0；
- Offset：0，0，0；
- Roffset：0，0，0；
- Geometry Length：40；
- Geometry Radius：15；
- Property File：mdids: //acar_shared/bushings.tbl/mdi_0001.bus；
- Location Dependency：Delta location from coordinate；
- Coordinate Reference（参考坐标）：._FSAE_sus_leafspring.ground.hpl_pt9_up；
- Location：0，0，0；
- Location in：local；
- Orientation Dependency：Orient axis to point；
- Construction Frame：._FSAE_sus_leafspring.ground.hpr_pt9_up；
- Axis：Z。

（2）单击 Apply 按钮，完成轴套._FSAE_sus_leafspring.bgl_pt9_up 的创建。

9.2.4 板簧悬架约束

1. 部件 tripot 与安装件 tripot_to_differential 之间的 translational 约束

（1）单击 Build > Attachments > Joint > New 命令，在下列对话框中输入相应的数据：
- Joint Name（约束副名称）：tripot_to_differential；
- I Part：._FSAE_sus_leafspring.gel_tripot；
- J Part：._FSAE_sus_leafspring.mtl_tripot_to_differential；
- Joint Type（约束副类型）：translational；
- Active（激活）：always；
- Location Dependency：Delta location from coordinate；
- Coordinate Reference #1（参考坐标）：._FSAE_sus_leafspring.ground.hpl_drive_shaft_inr；
- Orientation Dependency：Orient to zpoint-xpoint；
- Coordinate Reference #1（参考坐标）：._FSAE_sus_leafspring.ground.hpr_drive_shaft_inr；

- Coordinate Reference #2（参考坐标）：._FSAE_sus_leafspring.ground.cfl_drive_shaft_otr；
- Axis：ZX。

（2）单击 Apply 按钮，完成约束副._FSAE_sus_leafspring.joltra_tripot_to_differential 的创建。

2. 部件 tripot 与 drive_shaft 之间的 convel 约束

（1）单击 Build > Attachments > Joint > New 命令，在下列对话框中输入相应的数据：

- Joint Name（约束副名称）：drive_sft_otr；
- I Part：._FSAE_sus_leafspring.gel_tripot；
- J Part：._FSAE_sus_leafspring.gel_drive_shaft；
- Joint Type（约束副类型）：convel（恒速副）；
- Active（激活）：always；
- Location Dependency：Delta location from coordinate；
- Coordinate Reference（参考坐标）：._FSAE_sus_leafspring.ground.hpl_drive_shaft_inr；
- Location：0，0，0；
- Location in：local；
- I-Part Axis：._FSAE_sus_leafspring.ground.hpr_drive_shaft_inr；
- J-Part Axis：._FSAE_sus_leafspring.ground.cfl_drive_shaft_otr。

（2）单击 Apply 按钮，完成约束副._FSAE_sus_leafspring.jolcon_drive_sft_int_jt 的创建。

3. 部件 spindle 与 drive_shaft 之间的 convel 约束

（1）单击 Build > Attachments > Joint > New 命令，在下列对话框中输入相应的数据：

- Joint Name（约束副名称）：drive_sft_otr；
- I Part：._FSAE_sus_leafspring.gel_drive_shaft；
- J Part：._FSAE_sus_leafspring.gel_spindle；
- Joint Type（约束副类型）：convel（恒速副）；
- Active（激活）：always；
- Location Dependency：Delta location from coordinate；
- Coordinate Reference（参考坐标）：._FSAE_sus_leafspring.ground.cfl_drive_shaft_otr；
- Location：0，0，0；
- Location in：local；
- I-Part Axis：._FSAE_sus_leafspring.ground.hpr_drive_shaft_inr；
- J-Part Axis：._FSAE_sus_leafspring.ground.hpl_wheel_center。

（2）单击 Apply 按钮，完成约束副._FSAE_sus_leafspring.jolcon_drive_sft_otr 的创建。

4. 部件 spindle 与安装件 upright 之间的 revolute 约束

（1）单击 Build > Attachments > Joint > New 命令，在下列对话框中输入相应的数据：

- Joint Name（约束副名称）：spindle_upright；
- I Part：._FSAE_sus_leafspring.gel_spindle；

- J Part：._FSAE_sus_leafspring.gel_upright；
- Joint Type（约束副类型）：revolute；
- Active（激活）：always；
- Location Dependency：Delta location from coordinate；
- Coordinate Reference（参考坐标）：._FSAE_sus_leafspring.ground.hpl_wheel_center；
- Location：0，0，0；
- Location in：local；
- Orientation Dependency：Delta orientation from coordinate；
- Construction Frame：._FSAE_sus_leafspring.ground.cfl_wheel_center；
- Orientation：0，0，0。

（2）单击 Apply 按钮，完成约束副._FSAE_sus_leafspring.jolrev_spindle_upright 的创建。

5. 部件 upright 与安装件 axis 之间的 revolute 约束

（1）单击 Build > Attachments > Joint > New 命令，在下列对话框中输入相应的数据：

- Joint Name（约束副名称）：spindle_upright；
- I Part：._FSAE_sus_leafspring.gel_upright；
- J Part：._FSAE_sus_leafspring.ges_axis；
- Joint Type（约束副类型）：revolute；
- Active（激活）：always；
- Location Dependency：Centered between coordinate；
- Coordinate Reference #1（参考坐标1）：._FSAE_sus_leafspring.ground.hpl_hub_up；
- Coordinate Reference #2（参考坐标2）：._FSAE_sus_leafspring.ground.hpl_hub_down；
- Orientation Dependency：Orient axis to point；
- Construction Frame：._FSAE_sus_leafspring.ground.hpl_hub_up；
- Axis：Z。

（2）单击 Apply 按钮，完成约束副._FSAE_sus_leafspring.jolrev_upright_to_axle 的创建。

6. 部件 tierod 与 upright 之间的 spherical 约束

（1）单击 Build > Attachments > Joint > New 命令，在下列对话框中输入相应的数据：

- Joint Name（约束副名称）：tierod_outer；
- I Part：._FSAE_sus_leafspring.gel_tierod；
- J Part：._FSAE_sus_leafspring.gel_upright；
- Joint Type（约束副类型）：spherical（转动副，约束3个自由度）；
- Active（激活）：always；
- Location Dependency：Delta location from coordinate；
- Coordinate Reference（参考坐标）：._FSAE_sus_leafspring.ground.hpl_tierod_outer；
- Location：0，0，0；
- Location in：local；
- Orientation：None。

（2）单击 Apply 按钮，完成约束副 ._FSAE_sus_leafspring.jolsph_tierod_outer 的创建。

7. 部件 tierod 与安装部件 tierod_to_steering 之间的 convel 约束

（1）单击 Build > Attachments > Joint > New 命令，在下列对话框中输入相应的数据：
- Joint Name（约束副名称）：tierod_inner；
- I Part：._FSAE_sus_leafspring.gel_tierod；
- J Part：._FSAE_sus_leafspring.mtl_tierod_to_steering；
- Joint Type（约束副类型）：convel（恒速副）；
- Active（激活）：always；
- Location Dependency：Delta location from coordinate；
- Coordinate Reference（参考坐标）：._FSAE_sus_leafspring.ground.hpl_tierod_inner；
- Location：0，0，0；
- Location in：local；
- I-Part Axis：._FSAE_sus_leafspring.ground.hpl_tierod_outer；
- J-Part Axis：._FSAE_sus_leafspring.ground.hpr_tierod_inner。

（2）单击 OK 按钮，完成约束副 ._FSAE_sus_leafspring.jolcon_tierod_inner 的创建。

9.3 板簧悬架变量参数

（1）单击 Build > Parameter Variable > New 命令，弹出参数变量对话框，如图 9-18 所示。在下列对话框中输入相应的数据：

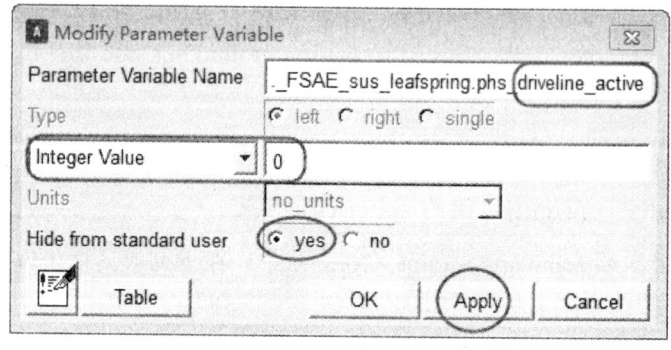

图 9-18　参数变量对话框

- Parameter Variable Name：：driveline_active；
- Integer Value（实数值）：0；
- Units：no_units；
- Hide from standard user（是否从标准界面隐藏）：yes。

（2）单击 Apply 按钮，完成变量 ._FSAE_sus_leafspring.phs_driveline_active 的创建。同理，创建其他参数变量。
- Parameter Variable Name：kinematic_flag；

- Integer Value（实数值）：0；
- Units：no_units；
- Hide from standard user（是否从标准界面隐藏）：yes。

（3）单击 OK 按钮，完成变量._FSAE_sus_leafspring.phs_kinematic_flag 的创建。

（4）单击 Build > Construction Frame > New 命令，在下列对话框中输入相应的数据：

- Construction Frame（结构框名称）：wheel_center_ref；
- Location Dependency：Delta location from coordinate；
- Coordinate Reference（参考坐标）：._FSAE_sus_leafspring.ground.hpl_wheel_center；
- Location：0，0，100；
- Location in：local；
- Orientation Dependency：Delta location from coordinate；
- Coordinate Reference（参考坐标）：._FSAE_sus_leafspring.ground.cfl_wheel_center；
- Orientation：0，0，0。

（5）单击 OK 按钮，完成._FSAE_sus_leafspring.ground.cfl_wheel_center_ref 结构框的创建。

（6）单击 Build > Suspension Parameters > Characteristics Array > Set 命令，此设置主要用于设置悬架的转向主销，如图 9-19 所示。在下列对话框中输入相应的数据：

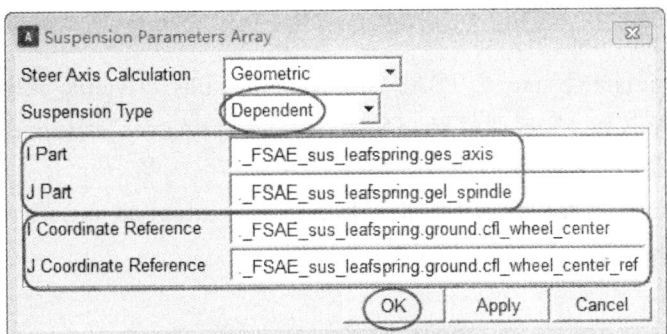

图 9-19 悬架参数变量设置

- Steer Axis Calculation：Geometric；
- Suspension Type：dependent（非独立悬架）；
- I Part：._FSAE_sus_leafspring.ges_axis；
- J Part：._FSAE_sus_leafspring.gel_spindle；
- I Coordinate Reference：._FSAE_sus_leafspring.ground.cfl_wheel_center；
- J Coordinate Reference：._FSAE_sus_leafspring.ground.cfl_wheel_center_ref。

（7）单击 OK 按钮，完成悬架参数变量设置。

9.4 板簧悬架通信器

（1）单击 Build > Communicator > Output >New 命令，弹出输出通信器对话框，如图 9-20 所示。在下列对话框中输入相应的数据：

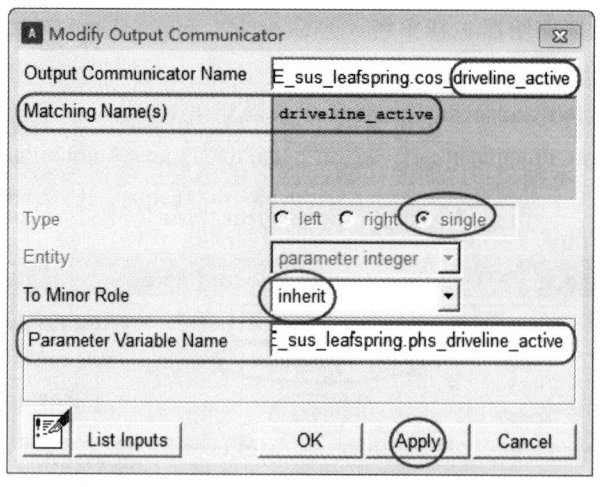

图 9-20　通信器对话框

- Output Communicator Name（输出通信器名称）：driveline_active；
- Matching Name（s）：driveline_active；
- Type：single；
- Entity：parameter integer；
- To Minor Role：inherit；
- Parameter Variable Name：._FSAE_sus_leafspring.phs_driveline_active。

（2）单击 Apply 按钮，完成通信器._FSAE_sus_leafspring.cos_driveline_active 的创建。同理，创建其他通信器。

- Output Communicator Name（输出通信器名称）：tripot_to_differential；
- Matching Name（s）：tripot_to_differential；
- Type：left；
- Entity：Location；
- To Minor Role：inherit；
- Coordinate Reference Name：._FSAE_sus_leafspring.ground.hpl_drive_shaft_inr。

（3）单击 Apply 按钮，完成通信器._FSAE_sus_leafspring.col_tripot_to_differential 的创建。

- Output Communicator Name（输出通信器名称）：suspension_mount；
- Matching Name（s）：suspension_mount；
- Type：left；
- Entity：mount；
- To Minor Role：inherit；
- Part Name：._FSAE_sus_leafspring.gel_spindle。

（4）单击 Apply 按钮，完成通信器._FSAE_sus_leafspring.col_suspension_mount 的创建。

- Output Communicator Name（输出通信器名称）：wheel_center；
- Matching Name（s）：wheel_center；
- Type：left；
- Entity：Location；

- To Minor Role：inherit；
- Coordinate Reference Name：._FSAE_sus_leafspring.ground.hpl_wheel_center。

（5）单击 Apply 按钮，完成通信器._FSAE_sus_leafspring.col_wheel_center 的创建。

- Output Communicator Name（输出通信器名称）：suspension_upright；
- Matching Name（s）：suspension_upright；
- Type：left；
- Entity：mount；
- To Minor Role：inherit；
- Part Name：._FSAE_sus_leafspring.ges_axis。

（6）单击 Apply，完成通信器._FSAE_sus_leafspring.col_suspension_upright 的创建。

- Output Communicator Name（输出通信器名称）：engine_to_subframe；
- Matching Name（s）：engine_to_subframe；
- Type：single；
- Entity：mount；
- To Minor Role：inherit；
- Part Name：._FSAE_sus_leafspring.ges_axis；

（7）单击 Apply 按钮，完成通信器._FSAE_sus_leafspring.cos_engine_to_subframe 的创建。

- Output Communicator Name（输出通信器名称）：wheel_joint；
- Matching Name（s）：wheel_joint；
- Type：left；
- Entity：joint for motion；
- To Minor Role：inherit；
- Joint Name：._FSAE_sus_leafspring.jolrev_spindle_upright。

（8）单击 Apply 按钮，完成通信器._FSAE_sus_leafspring.col_wheel_joint 的创建。

- Output Communicator Name（输出通信器名称）：rack_housing_to_suspension_subframe；
- Matching Name（s）：rack_housing_to_suspension_subframe；
- Type：single；
- Entity：mount；
- To Minor Role：inherit；
- Part Name：._FSAE_sus_leafspring.ges_axis。

（9）单击 OK 按钮，完成通信器._FSAE_sus_leafspring.cos_rack_housing_to_suspension_subframe 的创建。

9.5　板簧悬架通信器测试

（1）单击 Build > Communicator > Test 命令，弹出输出通信器测试对话框，如图 9-21 所示。设置以下信息：

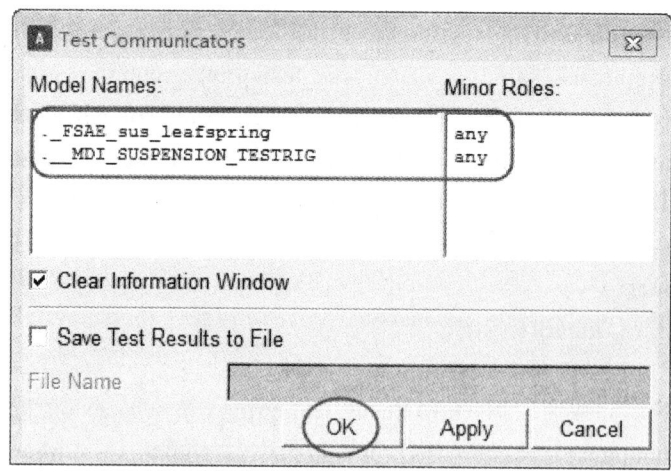

图 9-21 通信器测试对话框

- Model Names：
① ._FSAE_sus_leafspring；
② .__MDI_SUSPENSION_TESTRIGG。
- Minor Roles：并列两排输入特征 any；也可以并排输入特征 front。

（2）单击 OK 按钮，完成推杆式双叉臂悬架和悬架试验台._FSAE_sus_leafspring、.__MDI_SUSPENSION_TESTRIGG 的匹配测试；测试结果如下列信息所示。

通信器匹配信息如下：

!--------------------------- Matched communicators：---------------------------!

%以下为匹配的通信器

Communicator Matching Name：tripot_to_differential
Input Communicator Name：ci[lr]_tripot_to_differential
Located in：_FSAE_sus_leafspring
Output Communicator Name：co[lr]_tripot_to_differential
Output from：__MDI_SUSPENSION_TESTRIG

Communicator Matching Name：camber_angle
Input Communicator Name：ci[lr]_camber_angle
Located in：__MDI_SUSPENSION_TESTRIG
Output Communicator Name：co[lr]_camber_angle
Output from：_FSAE_sus_leafspring

Communicator Matching Name：toe_angle
Input Communicator Name：ci[lr]_toe_angle
Located in：__MDI_SUSPENSION_TESTRIG
Output Communicator Name：co[lr]_toe_angle

Output from: _FSAE_sus_leafspring

Communicator Matching Name: wheel_center
Input Communicator Name: ci[lr]_wheel_center
Located in: __MDI_SUSPENSION_TESTRIG
Output Communicator Name: co[lr]_wheel_center
Output from: _FSAE_sus_leafspring

Communicator Matching Name: suspension_mount
Input Communicator Name: ci[lr]_suspension_mount
Located in: __MDI_SUSPENSION_TESTRIG
Output Communicator Name: co[lr]_suspension_mount
Output from: _FSAE_sus_leafspring

Communicator Matching Name: driveline_active
Input Communicator Name: cis_driveline_active
Located in: __MDI_SUSPENSION_TESTRIG
Output Communicator Name: cos_driveline_active
Output from: _FSAE_sus_leafspring

Communicator Matching Name: suspension_parameters_array
Input Communicator Name: cis_suspension_parameters_ARRAY
Located in: __MDI_SUSPENSION_TESTRIG
Output Communicator Name: cos_suspension_parameters_ARRAY
Output from: _FSAE_sus_leafspring

Communicator Matching Name: tripot_to_differential
Input Communicator Name: ci[lr]_diff_tripot
Located in: __MDI_SUSPENSION_TESTRIG
Output Communicator Name: co[lr]_tripot_to_differential
Output from: _FSAE_sus_leafspring

Communicator Matching Name: suspension_upright
Input Communicator Name: ci[lr]_suspension_upright
Located in: __MDI_SUSPENSION_TESTRIG
Output Communicator Name: co[lr]_suspension_upright
Output from: _FSAE_sus_leafspring

!---------------------- Unmatched input communicators: ----------------------!

%以下为不匹配的输入通信器

Input Communicator Name：cis_leafspring_to_body
Class：mount
From Minor Role：any
Matching Name（s）：leafspring_to_body
In Template：_FSAE_sus_leafspring

Input Communicator Name：ci[lr]_tierod_to_steering
Class：mount
From Minor Role：any
Matching Name（s）：tierod_to_steering
In Template：_FSAE_sus_leafspring

Input Communicator Name：ci[lr]_jack_frame
Class：mount
From Minor Role：any
Matching Name（s）：jack_frame
In Template：__MDI_SUSPENSION_TESTRIG

Input Communicator Name：cis_leaf_adjustment_steps
Class：parameter_integer
From Minor Role：any
Matching Name（s）：leaf_adjustment_steps
In Template：__MDI_SUSPENSION_TESTRIG

Input Communicator Name：cis_powertrain_to_body
Class：mount
From Minor Role：any
Matching Name（s）：powertrain_to_body
In Template：__MDI_SUSPENSION_TESTRIG

Input Communicator Name：cis_steering_rack_joint
Class：joint_for_motion
From Minor Role：any
Matching Name（s）：steering_rack_joint
In Template：__MDI_SUSPENSION_TESTRIG

Input Communicator Name：cis_steering_wheel_joint

Class：joint_for_motion
From Minor Role：any
Matching Name（s）：steering_wheel_joint
In Template：__MDI_SUSPENSION_TESTRIG

!---------------------- Unmatched output communicators：----------------------!
%以下为不匹配的输出通信器

Output Communicator Name：cos_engine_to_subframe
Class：mount
To Minor Role：any
Matching Name（s）：engine_to_subframe
In Template：_FSAE_sus_leafspring

Output Communicator Name：co[lr]_wheel_joint
Class：joint_for_motion
To Minor Role：any
Matching Name（s）：wheel_joint
In Template：_FSAE_sus_leafspring

Output Communicator Name：cos_rack_housing_to_suspension_subframe
Class：mount
To Minor Role：any
Matching Name（s）：rack_housing_to_suspension_subframe
In Template：_FSAE_sus_leafspring

Output Communicator Name：cos_leaf_adjustment_multiplier
Class：array
To Minor Role：any
Matching Name（s）：leaf_adjustment_multiplier
In Template：__MDI_SUSPENSION_TESTRIG

Output Communicator Name：cos_characteristics_input_ARRAY
Class：array
To Minor Role：any
Matching Name（s）：characteristics_input_array
In Template：__MDI_SUSPENSION_TESTRIG

9.6 驱动轴显示组件

（1）在模型树栏点击 Group 菜单，在模型树栏 New Group 上右击鼠标，弹出创建组件对话框，如图 9-22 所示。设置以下信息：

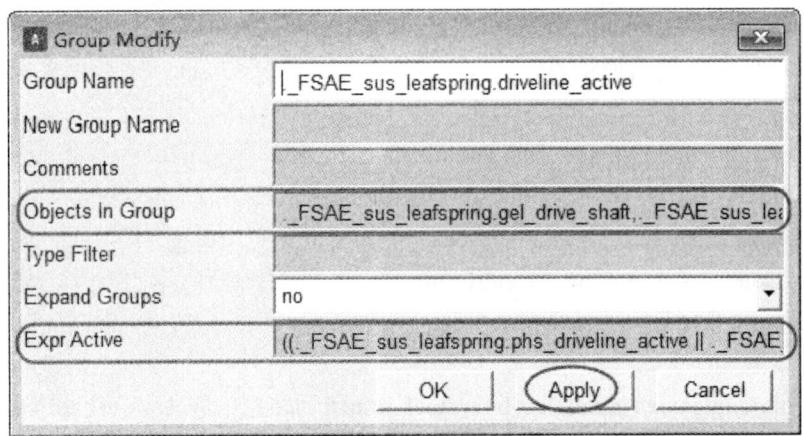

图 9-22 驱动轴显示组件对话框

- Group Name：driveline_active；
- Object In Group（显示组件包括的部件、几何体、约束等对象）：顺序输入 1~26 对象，如下信息所示。

[1] ._FSAE_sus_leafspring.gel_drive_shaft；

[2] ._FSAE_sus_leafspring.gel_tripot；

[3] ._FSAE_sus_leafspring.ger_drive_shaft；

[4] ._FSAE_sus_leafspring.ger_tripot；

[5] ._FSAE_sus_leafspring.mtl_tripot_to_differential；

[6] ._FSAE_sus_leafspring.mtr_tripot_to_differential；

[7] ._FSAE_sus_leafspring.gel_drive_shaft.gralin_drive_shaft；

[8] ._FSAE_sus_leafspring.gel_drive_shaft.graell_cv_housing；

[9] ._FSAE_sus_leafspring.gel_drive_shaft.graell_tripot_housing；

[10] ._FSAE_sus_leafspring.ger_drive_shaft.gralin_drive_shaft；

[11] ._FSAE_sus_leafspring.ger_drive_shaft.graell_cv_housing；

[12] ._FSAE_sus_leafspring.ger_drive_shaft.graell_tripot_housing；

[13] ._FSAE_sus_leafspring.gel_tripot.gracyl_tripot_housing_extention；

[14] ._FSAE_sus_leafspring.ger_tripot.gracyl_tripot_housing_extention；

[15] ._FSAE_sus_leafspring.jolcon_drive_sft_int_jt；

[16] ._FSAE_sus_leafspring.jolcon_drive_sft_otr；

[17]._FSAE_sus_leafspring.joltra_tripot_to_differential；

[18]._FSAE_sus_leafspring.jorcon_drive_sft_int_jt；

[19]._FSAE_sus_leafspring.jorcon_drive_sft_otr；

[20]._FSAE_sus_leafspring.jortra_tripot_to_differential；

[21]._FSAE_sus_leafspring.mtl_fixed_2；

[22]._FSAE_sus_leafspring.mtr_fixed_2；

[23]._FSAE_sus_leafspring.cil_tripot_to_differential；

[24]._FSAE_sus_leafspring.cir_tripot_to_differential；

[25]._FSAE_sus_leafspring.col_tripot_to_differential；

[26]._FSAE_sus_leafspring.cor_tripot_to_differential；

- Expr Active：((._FSAE_sus_leafspring.phs_driveline_active ||._FSAE_sus_leafspring.model_class == "template" ? 1：0) && DB_ACTIVE (._FSAE_sus_leafspring))。

（2）单击 Apply 按钮，完成组件 driveline_active 的创建。同理，创建相关组件。

- Group Name：driveline_inactive；
- Expr Active：((!._FSAE_sus_leafspring.phs_driveline_active ||._FSAE_sus_leafspring.model_class == "template" ? 1：0) && DB_ACTIVE (._FSAE_sus_leafspring))。

（3）单击 OK 按钮，完成组件 driveline_inactive 的创建。

（4）单击 File > Save As 命令，弹出保存模板对话框，如图 9-23 所示。在下列对话框中输入相应的数据：

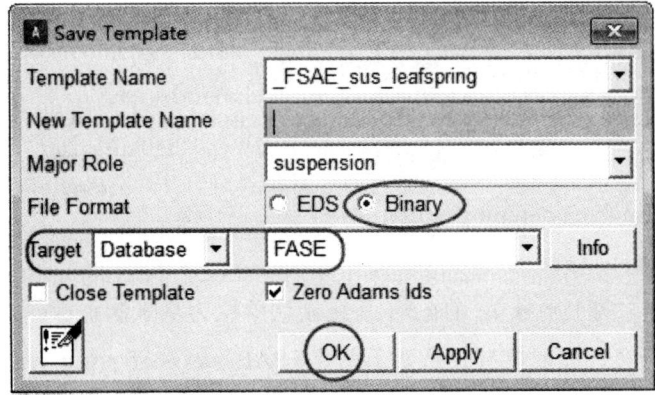

图 9-23 板簧悬架模型保存对话框

- Major Role（主特征）：suspension；
- File Format：Binary；
- Target：Database/FASE。

（5）单击 OK 按钮，完成推杆式悬架模型模板_FSAE_sus_leafspring 的保存。

（6）按 F9 键，将专家模板转换到标准模式，单击 File > New > Suspension 命令，弹出子系统对话框，如图 9-24 所示。在下列对话框中输入相应的数据：

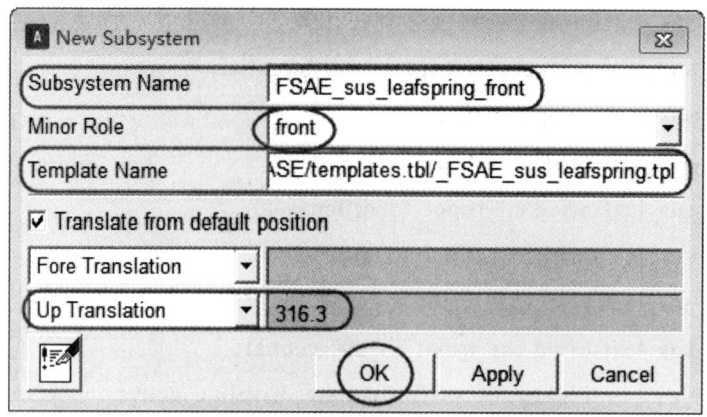

图 9-24　板簧悬架子系统（前）对话框

- Subsystem Name（系统名称）：FSAE_sus_leafspring_front；
- Minor Role（副特征）：front（指悬架为前悬架）；
- Template Name（模板路径）：mdids：//my_driveline/templates.tbl/_FSAE_sus_leafspring.tpl；
- 勾选 Translate from default position；
- Up Translate（向上平移）：316.3（系统默认单位为毫米制）。

（7）单击 Apply 按钮，完成悬架子系统 FSAE_sus_leafspring_front 的创建。同理，创建后悬架子系统。

- Subsystem Name（系统名称）：FSAE_sus_leafspring_rear；
- Minor Role（副特征）：rear（指悬架为后悬架）；
- Template Name（模板路径）：mdids：//my_driveline/templates.tbl/_FSAE_sus_leafspring.tpl；
- 勾选 Translate from default position；
- Aft Translation：1524；
- Up Translate（向上平移）：316.3（系统默认单位为毫米制）。

（8）单击 OK 按钮，完成推杆式悬架子系统 FSAE_sus_leafspring_rear 的创建。

9.7　双轮反向激振仿真

（1）单击 Simulate > Suspension Analysis > Parallel Wheel Travel 命令，弹出双轮反向激振仿真设置对话框，如图 9-25 所示。在下列对话框中输入相应的数据：

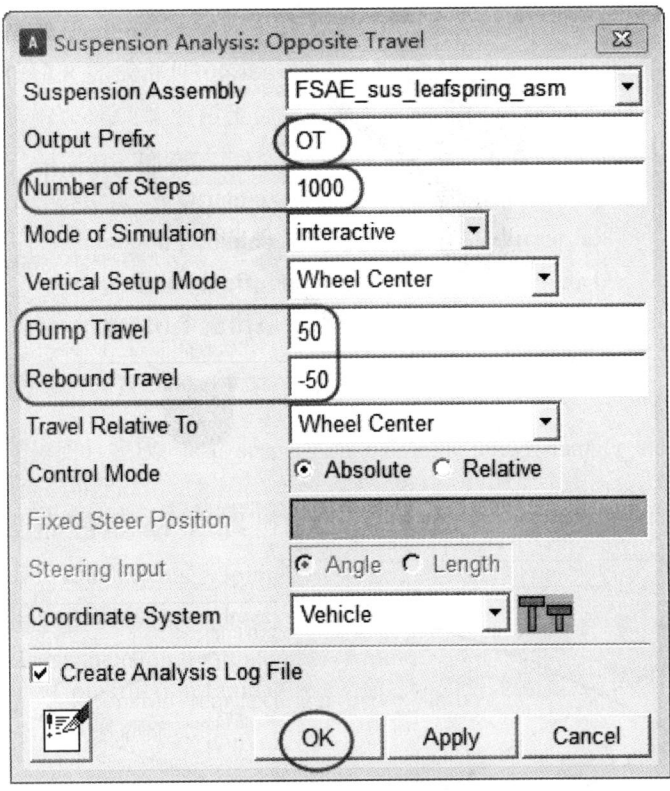

图 9-25 双轮反向激振仿真设置对话框

- Output Prefix：OT；
- Number of Steps（仿真步数）：1000；
- Mode of Simulation：interactive；
- Vertical Setup Mode：Wheel Center；
- Bump Travel：50；
- Rebound Travel：-50；
- Travel Relative To：Wheel Center；
- Control Mode：Absolute；
- Coordinate System：Vehicle。

（2）单击 OK 按钮，完成板簧悬架在 C 模式下的仿真。

（3）单击 Review > Animation Controls，开始演示动画，动画结束后悬架模型变化如图 9-26 所示。

（4）按 F8 键，界面转换到后处理模块；车轮四轮定位参数如图 9-27 ~ 图 9-30 所示。

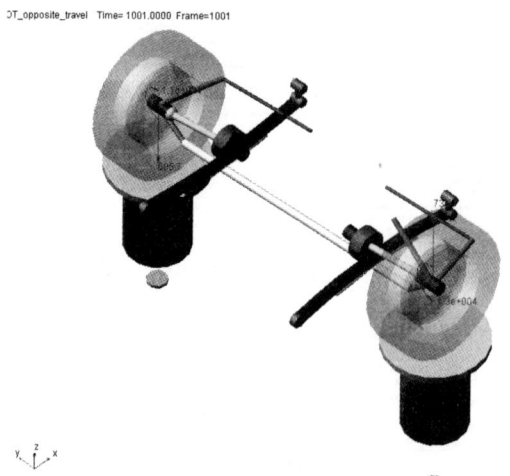

图 9-26 板簧悬架双轮反向跳动

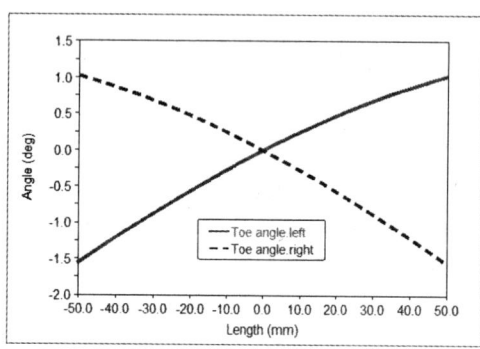

图 9-27 前束角

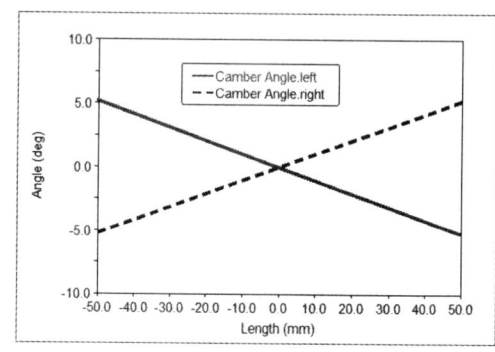

图 9-28 外倾角

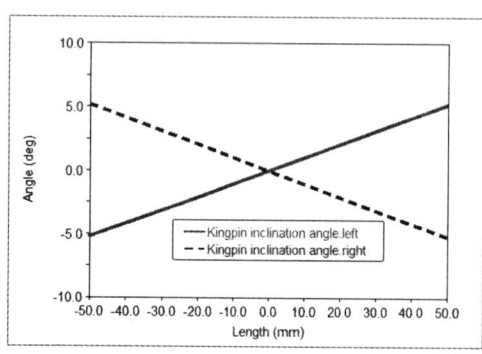

图 9-29 主销内倾角

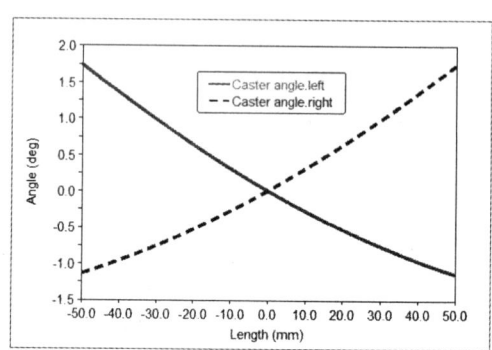

图 9-30 主销后倾角

9.8 整车模型

（1）单击 File > Assembly 命令，在 Assembly Name 中输入：mdids：//FASE/assemblies.tbl/FSAE_2020_prod.asy。

（2）单击 OK 按钮，完成推杆式悬架方程式赛车整车模型的导入。
- 前后悬架分别替换为如下子系统：

[1] mdids：//FASE/subsystems.tbl/FSAE_sus_leafspring_front.sub；

[2] mdids：//FASE/subsystems.tbl/FSAE_sus_leafspring_rear.sub。

- 硬点调试（修改对应硬点的坐标值）：

[1] hpl_tierod_outer：100.0，-400.0，228.6；

[2] hpr_tierod_outer：100.0，400.0，228.6；

[3] hpl_tierod_inner：50.8，-127.0，152.4；

[4] hpr_tierod_inner：50.8，127.0，152.4；

[5] hpl_tierod_outer：1624.0，-400.0，228.6；

[6] hpr_tierod_outer：1624.0，400.0，228.6；

[7] hpl_tierod_inner：1624.0，-150.0，228.6；

[8] hpr_tierod_inner：1624.0，150.0，228.6。

（3）替换完成后整车模型另存为：FSAE_2020_leafspring，整车模型如图 9-31 所示。

图 9-31　整车模型（前后为建立的板簧悬架模型）

9.9　Fish-Hook 仿真

（1）单击 Simulate > Full-Vehicle Analysis > Open-loop Steering Events > Fish Hook 命令，弹出蛇形绕桩仿真对话框，如图 9-32 所示。在下列对话框中输入相应的数据：

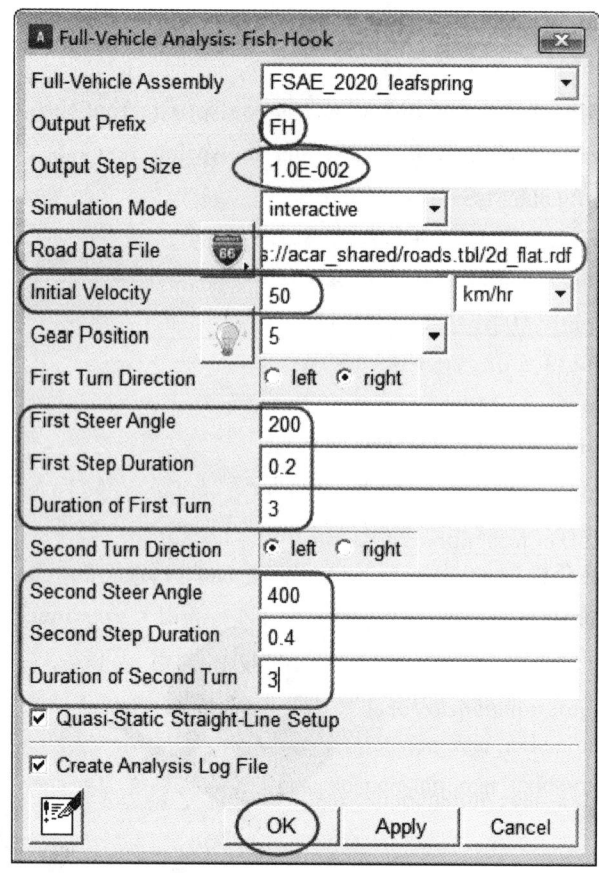

图 9-32　Fish-Hook 仿真设置对话框

- Output Prefix：FH；
- Output Step Size（仿真步数）：0.01；
- Mode of Simulation：interactive；
- Road Date File：mdids：//FASE/roads.tbl/2d_flat.rdf，在仿真过程中，由于路面场地过小，可能导致整车运行时驶出场地范围，但不影响仿真的正常运行，可以打开路面属性文件，对路面的长度与宽度参数进行修改，本路面尺寸为 500 000×400 000，单位为 mm；
- Initial Velocity：50；
- First Turn Direction：right；
- First Steer Angle：200；
- Duration of First Turn：3；
- Second Turn Direction：left；
- Second Steer Angle：400；
- Duration of First Turn：3；
- 勾选 Quasi-Static Straight-Line Setup。

（2）单击 OK 按钮，完成蛇形绕桩仿真设置并提交运算。

仿真结束后，整车的运行轨迹如图 9-33 所示，在运行过程中，计算参数如图 9-34～图 9-45 所示。

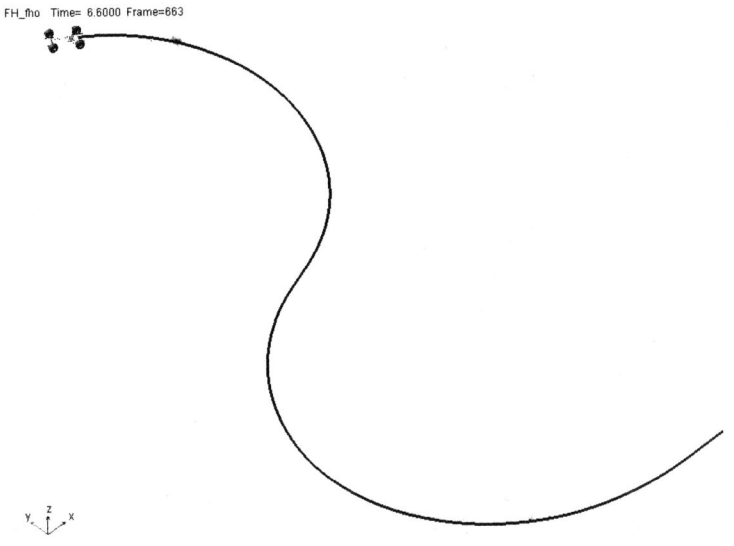

图 9-33　整车运行轨迹

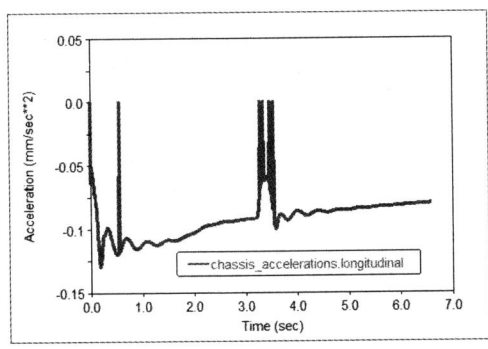

图 9-34　纵向加速度

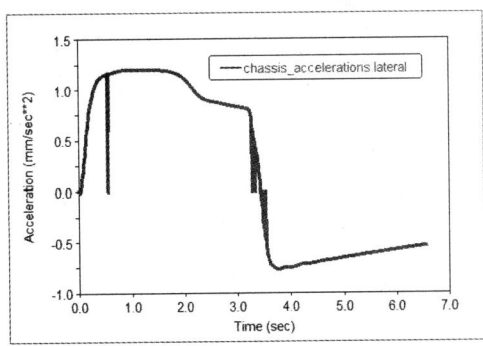

图 9-35　侧向加速度

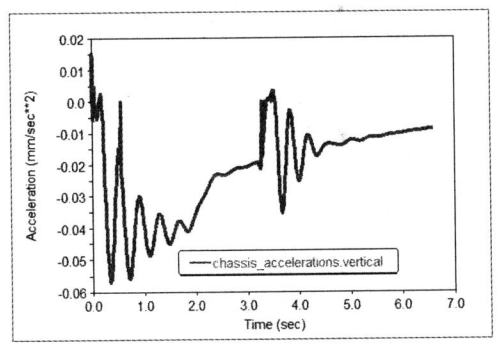

图 9-36　垂向加速度

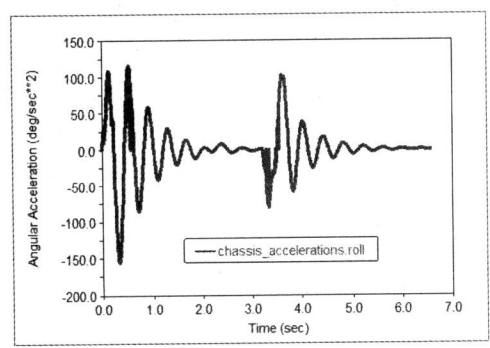

图 9-37　俯仰角加速度

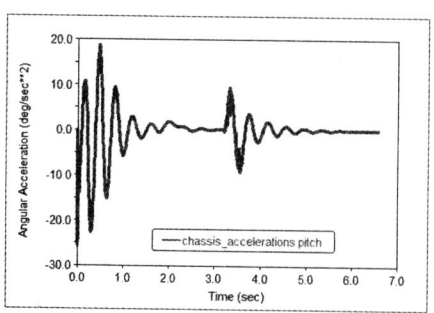

图 9-38　侧倾角加速度

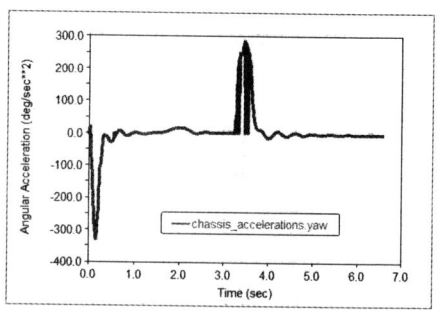

图 9-39　横摆角加速度

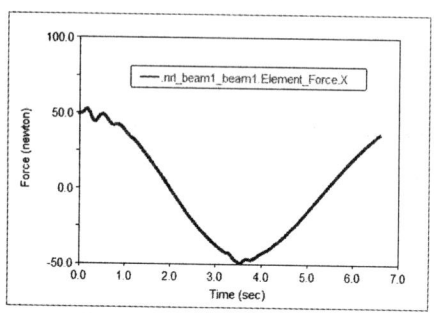

图 9-40　beam1 梁 X 方向受力

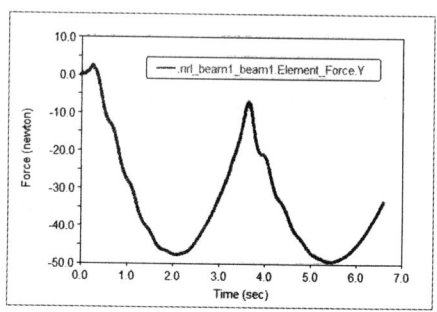

图 9-41　beam1 梁 Y 方向受力

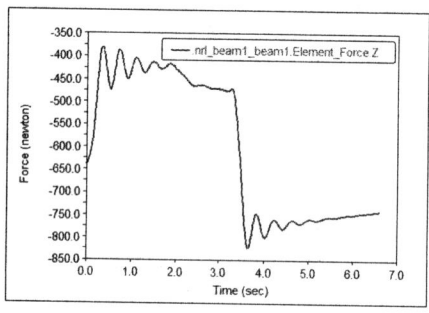

图 9-42　beam1 梁 Z 方向受力

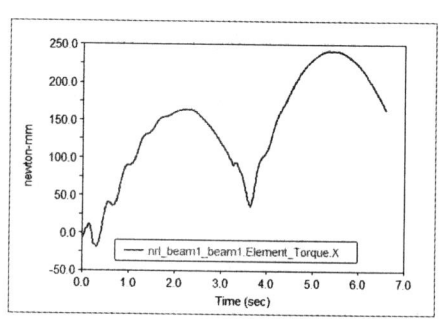

图 9-43　beam1 梁 X 方向力矩

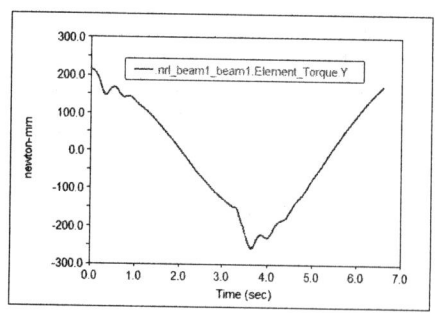

图 9-44　beam1 梁 Y 方向力矩

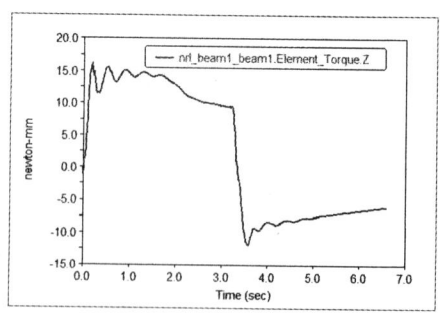

图 9-45　beam1 梁 Z 方向力矩

第 10 章　避震器纵置式后推杆悬架

推力杆式悬架较少见，多用于方程式赛车及超级跑车，在快速过弯过程中，可以减少车身的侧倾特性，提升整车的操控性。另一方面，采用推力杆式悬架的赛车，有利于整车动力总成及传动系统的布置。建立好的后推力杆式双横臂悬挂如图 10-1 所示。

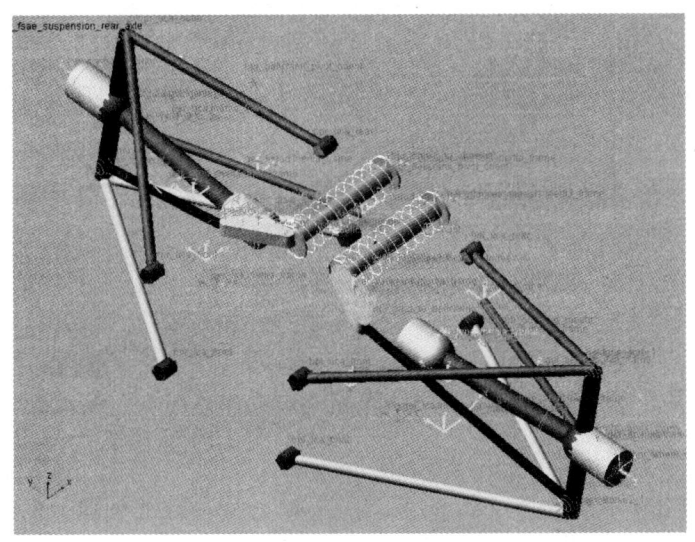

图 10-1　后推力杆式双横臂悬架

学习目标

（1）了解推杆悬架部件。
（2）会推杆悬架刚性约束。
（3）会推杆悬架柔性约束。
（4）了解驱动轴。
（5）了解通信器。
（6）会车轮跳动仿真。
（7）会硬点调试。

10.1 后推杆悬架

（1）启动 ADAMS/CAR，选择专家模块进入建模界面。

（2）单击 File > New 命令，弹出建模对话框，如图 10-2 所示；在模板名称里输入：fsae_suspension_rear_axle，主特征选择 suspension，单击 OK 按钮。

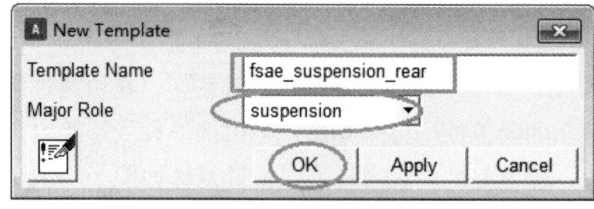

图 10-2 模板对话框

（3）单击 Build > Hardpoind > New 命令，弹出创建硬点对话框，如图 10-3 所示；在硬点名称里输入：drive_shaft_inr，类型选择 left；在位置文本框输入：1500，-200.0，225.0。

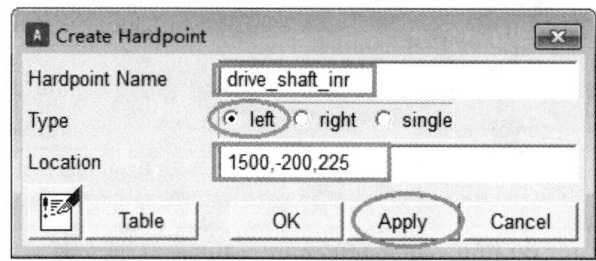

图 10-3 硬点对话框

（4）单击 Apply 按钮，完成 drive_shaft_inr 硬点的创建。此时，在屏幕上显示出左右对称的两个硬点；以此类推，重复上述步骤完成图 10-4 中硬点的创建，创建完成后单击 OK 按钮。

	loc_x	loc_y	loc_z
hpl_arb_bushing_mount	1651.0	-127.0	101.6
hpl_arblink_to_bellcrank	1447.8	-50.8	317.5
hpl_bellcrank_pivot	1447.8	-114.3	304.8
hpl_bellcrank_pivot_orient	1473.2	-146.05	547.3
hpl_drive_shaft_inr	1500.0	-200.0	225.0
hpl_lca_front	1270.0	-127.0	127.0
hpl_lca_outer	1498.6	-482.6	101.6
hpl_lca_rear	1651.0	-127.0	127.0
hpl_prod_outer	1498.6	-482.6	127.0
hpl_prod_to_bellcrank	1409.7	-139.7	304.8
hpl_shock_to_bellcrank	1460.5	-50.8	304.8
hpl_shock_to_chassis	1651.0	-50.8	304.8
hpl_tierod_inner	1676.4	-127.0	152.4
hpl_tierod_outer	1574.8	-457.2	152.4
hpl_uca_front	1270.0	-152.4	304.8
hpl_uca_outer	1549.4	-482.6	355.6
hpl_uca_rear	1625.6	-152.4	304.8
hpl_wheel_center	1524.0	-558.8	228.6
hps_global	1524.0	0.0	0.0

图 10-4 后推力杆式双横臂悬架硬点数据

10.1.1 上控制臂部件（UCA）

（1）单击 Build > Part > General Part > New 命令，弹出创建部件对话框，如图 10-5 所示；图 10-5 为已经建立好的推杆式双横臂悬架模型，通过右击 ._fsae_suspension_rear_axle.gel_uca 部件，在弹出的快捷菜单中设置。

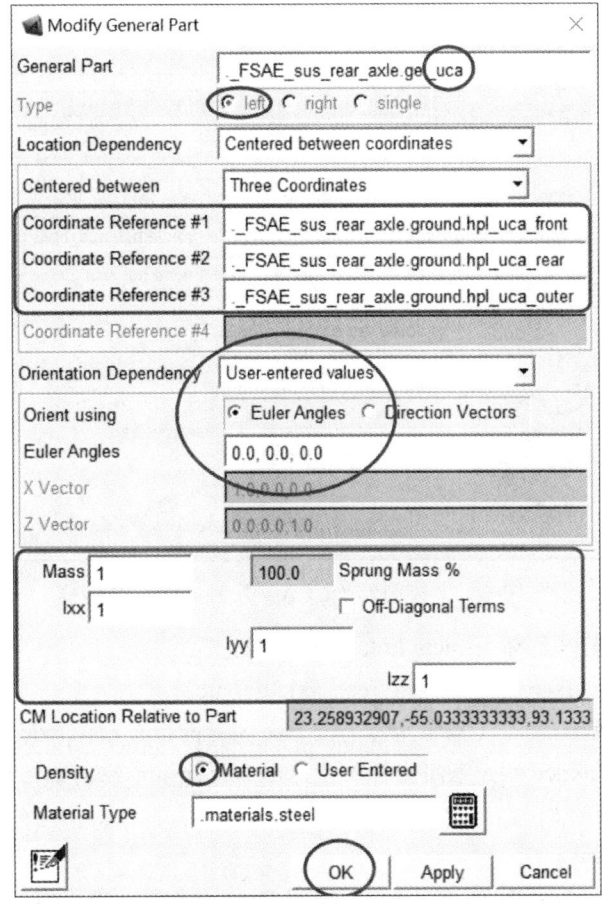

图 10-5 上控制臂（UCA）部件创建对话框

- General Part：uca；
- Location Dependency：Centered between coordinates；
- Centered between：Three Coordinates（上控制臂部件位于三点坐标的中心位置）；
- Coordinate Reference #1（参考坐标）：._fsae_suspension_rear_axle.ground.hpl_uca_front；
- Coordinate Reference #2（参考坐标）：._fsae_suspension_rear_axle.ground.hpl_uca_rear；
- Coordinate Reference #3（参考坐标）：._fsae_suspension_rear_axle.ground.hpl_uca_outer；
- Orient using：Euler Angles（部件定向采用欧拉角模式）；
- Euler Angles：0，0，0；
- Mass：1；
- Ixx：1；

- Iyy：1；
- Izz：1；
- Density：Material；
- Material Type：.materials.steel。

（2）单击 OK 按钮，完成部件._fsae_suspension_rear_axle.gel_uca 的创建。

（3）单击 Build > Geometry > Link > New 命令，弹出创建部件对话框，如图 10-6 所示。在下列对话框中输入相应的数据：

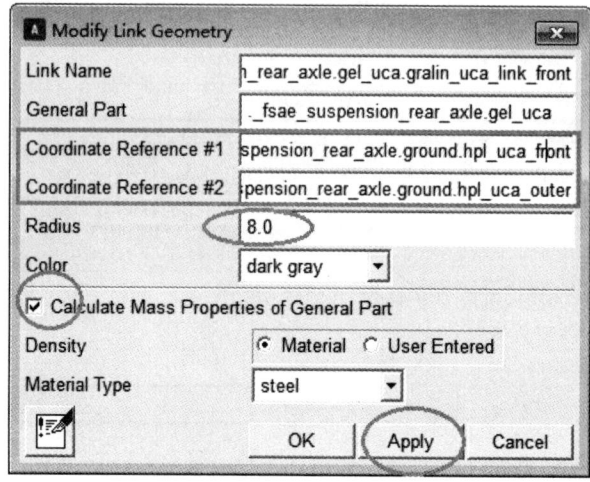

图 10-6　上控制臂（UCA）几何体创建对话框

- Link Name（连杆名称）：uca_link_front；
- General Part：._fsae_suspension_rear_axle.gel_uca；
- Coordinate Reference #1（参考坐标）：._fsae_suspension_rear_axle.ground.hpl_uca_front；
- Coordinate Reference #2（参考坐标）：._fsae_suspension_rear_axle.ground.hpl_uca_outer；
- Radius（半径）：8；
- Color：dark gray；
- 选择 Calculate Mass Properties of General Part 复选框，当几何体建立好之后，会更新对应部件的质量和惯量参数；
- Density：Material；
- Material Type：steel。

（4）单击 Apply 按钮，完成 uca_link_front 几何体的创建。同时，创建其他几何体。

- Link Name（连杆名称）：uca_link_rear；
- General Part：._fsae_suspension_rear_axle.gel_uca；
- Coordinate Reference #1（参考坐标）：._fsae_suspension_rear_axle.ground.hpl_uca_rear；
- Coordinate Reference #2（参考坐标）：._fsae_suspension_rear_axle.ground.hpl_uca_outer；
- Radius（半径）：8；
- Color：dark gray；
- 勾选 Calculate Mass Properties of General Part 复选框；

- Density：Material；
- Material Type：steel。

（5）单击 OK 按钮，完成 uca_link_rear 几何体的创建。

10.1.2 下控制臂部件（LCA）

（1）单击 Build > Part > General Part > New 命令，弹出创建部件对话框，如图 10-5 所示。在下列对话框中输入相应的数据：

- General Part：lca；
- Location Dependency：Centered between coordinates；
- Centered between：Three Coordinates；
- Coordinate Reference #1（参考坐标）：._fsae_suspension_rear_axle.ground.hpl_lca_front；
- Coordinate Reference #2（参考坐标）：._fsae_suspension_rear_axle.ground.hpl_lca_rear；
- Coordinate Reference #3（参考坐标）：._fsae_suspension_rear_axle.ground.hpl_lca_outer；
- Orient using：Euler Angles；
- Euler Angles：0，0，0；
- Mass：1；
- Ixx：1；
- Iyy：1；
- Izz：1；
- Density：Material；
- Material Type：.materials.steel。

（2）单击 OK 按钮，完成部件._fsae_suspension_rear_axle.gel_lca 的创建。

（3）单击 Build > Geometry > Link > New 命令，创建连杆几何体，如图 10-6 所示。在下列对话框中输入相应的数据：

- Link Name（连杆名称）：lca_link_front；
- General Part：._fsae_suspension_rear_axle.gel_lca；
- Coordinate Reference #1（参考坐标）：._fsae_suspension_rear_axle.ground.hpl_lca_front；
- Coordinate Reference #2（参考坐标）：._fsae_suspension_rear_axle.ground.hpl_lca_outer；
- Radius（半径）：8；
- Color：yellow；
- 选择 Calculate Mass Properties of General Part 复选框，当几何体建立好之后，会更新对应部件的质量和惯量参数；
- Density：Material；
- Material Type：steel。

（4）单击 Apply 按钮，完成 lca_link_front 几何体的创建。同理，创建其他几何体。

- Link Name（连杆名称）：lca_link_rear；
- General Part：._fsae_suspension_rear_axle.gel_lca；
- Coordinate Reference #1（参考坐标）：._fsae_suspension_rear_axle.ground.hpl_lca_rear；

- Coordinate Reference #2（参考坐标）：._fsae_suspension_rear_axle.ground.hpl_lca_outer；
- Radius（半径）：8；
- Color：yellow；
- 勾选 Calculate Mass Properties of General Part 复选框；
- Density：Material；
- Material Type：steel。

（5）单击 OK 按钮，完成 lca_link_rear 几何体的创建。

10.1.3 转向节 upright 部件

（1）单击 Build > Part > General Part > New 命令，弹出创建部件对话框，如图 10-5 所示。在下列对话框中输入相应的数据：

- General Part：upright；
- Location Dependency：Centered between coordinates；
- Centered between：Two Coordinates；
- Coordinate Reference #1（参考坐标）：._fsae_suspension_rear_axle.ground.hpl_uca_outer；
- Coordinate Reference #2（参考坐标）：._fsae_suspension_rear_axle.ground.hpl_lca_outer；
- Orient using：Euler Angles；
- Euler Angles：0，0，0；
- Mass：1；
- Ixx：1；
- Iyy：1；
- Izz：1；
- Density：Material；
- Material Type：.materials.steel。

（2）单击 OK 按钮，完成部件._fsae_suspension_rear_axle.gel_upright 的创建。

（3）单击 Build > Geometry > Link > New 命令，在下列对话框中输入相应的数据：

- Link Name（连杆名称）：upright；
- General Part：._fsae_suspension_rear_axle.gel_upright；
- Coordinate Reference #1（参考坐标）：._fsae_suspension_rear_axle.ground.hpl_uca_outer；
- Coordinate Reference #2（参考坐标）：._fsae_suspension_rear_axle.ground.hpl_lca_outer；
- Radius（半径）：13；
- Color（杆件几何体颜色）：blue；
- 选择 Calculate Mass Properties of General Part 复选框，当几何体建立好之后，会更新对应部件的质量和惯量参数；
- Density：Material；
- Material Type：steel。

（4）单击 OK 按钮，完成 upright 几何体的创建。

10.1.4 推力杆 prod 部件

（1）单击 Build > Part > General Part > New 命令，弹出创建部件对话框，如图 10-5 所示。在下列对话框中输入相应的数据：

- General Part：prod；
- Location Dependency：Centered between coordinates；
- Centered between：Two Coordinates；
- Coordinate Reference #1（参考坐标）：._fsae_suspension_rear_axle.ground.hpl_prod_outer；
- Coordinate Reference #2（参考坐标）：._fsae_suspension_rear_axle.ground.hpl_prod_to_bellcrank；
- Orientation Dependency（部件坐标轴方向）：Orient axis along line，结构框的轴沿两参考方向，一般指 Z 轴；
- Coordinate Reference #1（参考坐标）：._fsae_suspension_rear_axle.ground.hpl_prod_outer；
- Coordinate Reference #2（参考坐标）：._fsae_suspension_rear_axle.ground.hpl_prod_to_bellcrank；
- Axis：Z；
- Mass：1；
- Ixx：1；
- Iyy：1；
- Izz：1；
- Density：Material；
- Material Type：.materials.steel。

（2）单击 OK 按钮，完成部件._fsae_suspension_rear_axle.gel_prod 的创建。

（3）单击 Build > Geometry > Link > New 命令，在下列对话框中输入相应的数据：

- Link Name（连杆名称）：prod；
- General Part：._fsae_suspension_rear_axle.gel_prod；
- Coordinate Reference #1（参考坐标）：._fsae_suspension_rear_axle.ground.hpl_prod_outer；
- Coordinate Reference #2（参考坐标）：._fsae_suspension_rear_axle.ground.hpl_prod_to_bellcrank；
- Radius（半径）：7；
- Color（杆件几何体颜色）：blue；
- 选择 Calculate Mass Properties of General Part 复选框，当几何体建立好之后，会更新对应部件的质量和惯量参数；
- Density：Material；
- Material Type：steel。

（4）单击 OK 按钮，完成 prod 几何体的创建。

10.1.5 曲柄 bellcrank 部件

（1）单击 Build > Part > General Part > New 命令，弹出创建部件对话框，如图 10-5 所示。在下列对话框中输入相应的数据：
- General Part：bellcrank；
- Location Dependency：Centered between coordinates；
- Centered between：T Coordinates；
- Coordinate Reference #1（参考坐标）：._fsae_suspension_rear_axle.ground.hpl_prod_to_bellcrank；
- Coordinate Reference #2（参考坐标）：._fsae_suspension_rear_axle.ground.hpl_shock_to_bellcrank；
- Coordinate Reference #3（参考坐标）：._fsae_suspension_rear_axle.ground.hpl_bellcrank_pivot；
- Coordinate Reference #4（参考坐标）：._fsae_suspension_rear_axle.ground.hpl_arblink_to_bellcrank；
- Orientation Dependency：User-entered values；
- Orient using：Euler Angles；
- Euler Angles：0，0，0；
- Mass：1；
- Ixx：1；
- Iyy：1；
- Izz：1；
- Density：Material；
- Material Type：.materials.steel。

（2）单击 OK 按钮，完成部件._fsae_suspension_rear_axle.gel_bellcrank 的创建。

（3）单击 Build > Geometry > Arm > New 命令，在下列对话框中输入相应的数据：
- Link Name（连杆名称）：bellcrank；
- General Part：._fsae_suspension_rear_axle.gel_bellcrank；
- Coordinate Reference #1（参考坐标）：._fsae_suspension_rear_axle.ground.hpl_prod_to_bellcrank；
- Coordinate Reference #2（参考坐标）：._fsae_suspension_rear_axle.ground.hpl_bellcrank_pivot；
- Coordinate Reference #3（参考坐标）：._fsae_suspension_rear_axle.ground.hpl_arblink_to_bellcrank；
- Thinkness：20；
- Color（杆件几何体颜色）：yellow；
- 选择 Calculate Mass Properties of General Part 复选框，当几何体建立好之后，会更新对应部件的质量和惯量参数；
- Density：Material；
- Material Type：steel。

（4）单击 OK 按钮，完成._fsae_suspension_rear_axle.gel_bellcrank.graarm_bellcrank 几何体的创建。

10.1.6 转向横拉杆 tierod 部件

（1）单击 Build > Part > General Part > New 命令，弹出创建部件对话框，如图 10-5 所示，在下列对话框中输入相应的数据：

- General Part：tierod；
- Location Dependency：Centered between coordinates；
- Centered between：Two Coordinates；
- Coordinate Reference #1（参考坐标）：._fsae_suspension_rear_axle.ground.hpl_tierod_inner；
- Coordinate Reference #2（参考坐标）：._fsae_suspension_rear_axle.ground.hpl_tierod_outer；
- Orientation Dependency：User-entered values；
- Orient using：Euler Angles；
- Euler Angles：0，0，0；
- Mass：1；
- Ixx：1；
- Iyy：1；
- Izz：1；
- Density：Material；
- Material Type：.materials.steel。

（2）单击 OK 按钮，完成部件._fsae_suspension_rear_axle.gel_tierod 的创建。

（3）单击 Build > Geometry > Link > New 命令，在下列对话框中输入相应的数据：

- Link Name（连杆名称）：tierod；
- General Part：._fsae_suspension_rear_axle.gel_tierod；
- Coordinate Reference #1（参考坐标）：._fsae_suspension_rear_axle.ground.hpl_tierod_innerr；
- Coordinate Reference #2（参考坐标）：._fsae_suspension_rear_axle.ground.hpl_tierod_outer；
- Radius（半径）：7；
- Color（杆件几何体颜色）：magenta；
- 选择 Calculate Mass Properties of General Part 复选框，当几何体建立好之后，会更新对应部件的质量和惯量参数；
- Density：Material；
- Material Type：steel。

（4）单击 OK 按钮，完成横拉杆._fsae_suspension_rear_axle.gel_tierod.gralin_tierod 几何体的创建。

10.1.7 阻尼器坐垫 damper_chassis 部件

（1）单击 Build > Part > General Part > New 命令，弹出创建部件对话框，如图 10-5 所示。在下列对话框中输出相应的数据：

- General Part：damper_chassis；
- Location Dependency：Delta location from coordinate；
- Coordinate Reference（参考坐标）：._fsae_suspension_rear_axle.ground.hpl_shock_to_chassis；
- Location：0，0，0；
- Location in：local；
- Orientation Dependency：User-entered values；
- Orient using：Euler Angles；
- Euler Angles：0，0，0；
- Mass：1；
- Ixx：1；
- Iyy：1；
- Izz：1；
- Density：Material；
- Material Type：.materials.steel。

（2）单击 OK 按钮，完成部件._fsae_suspension_rear_axle.gel_damper_chassis 的创建。

10.1.8 阻尼器曲柄 damper_bellcrank 部件

（1）单击 Build > Part > General Part > New 命令，弹出创建部件对话框，如图 10-5 所示。在下列对话框中输出相应的数据：

- General Part：damper_bellcrank；
- Location Dependency：Delta location from coordinate；
- Coordinate Reference（参考坐标）：._fsae_suspension_rear_axle.ground.hpl_shock_to_bellcrank；
- Location：0，0，0；
- Location in：local；
- Orientation Dependency：User-entered values；
- Orient using：Euler Angles；
- Euler Angles：0，0，0；
- Mass：1；
- Ixx：1；
- Iyy：1；
- Izz：1；
- Density：Material；

- Material Type：.materials.steel。

（2）单击 OK 按钮，完成部件._fsae_suspension_rear_axle.gel_damper_bellcrank 的创建。

10.1.9 轮毂 spindle 部件

（1）单击 Build > Suspension Parameters > Toe/Camber Values > Set 命令，弹出悬架参数对话框，如图 10-7 所示。前束角输入：0；外倾角输入：-1.5；单击 OK 按钮，完成参数的创建。与此同时，系统自动建立两个输出通信器：col[r]_toe_angle、col[r]_camber_angle。

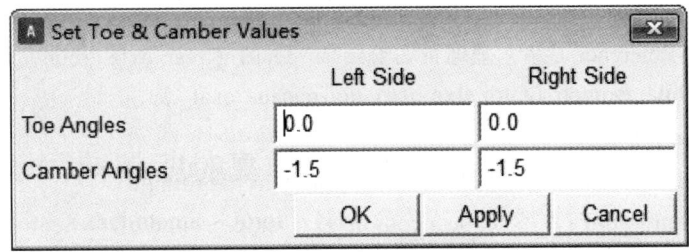

图 10-7 悬架参数对话框

（2）单击 Build > Construction Frame > New 命令，弹出创建结构框，如图 10-8 所示。在下列对话框中输入相应的数据。

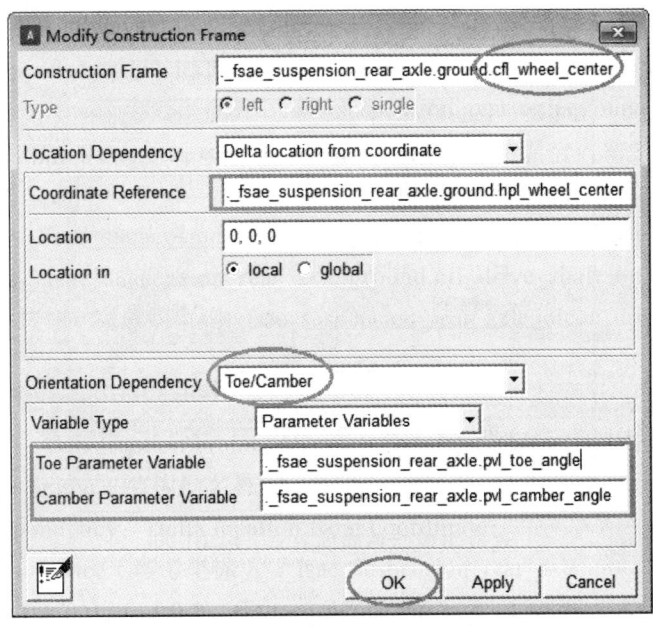

图 10-8 wheel_center 结构框

- Construction Frame（结构框名称）：wheel_center；
- Coordinate Reference（参考坐标）：_fsae_suspension_rear_axle.ground.hpl_wheel_center；
- Location：0，0，0；
- Location in：local；

- Orientation Dependency：Toe/Camber；
- Variable Type（变量类型）：Parameter Variable（参数变量）；
- Toe Parameter Values（前束变量值）：._fsae_suspension_rear_axle.pvl_toe_angle；
- Camber Parameter Values（外倾变量值）：._fsae_suspension_rear_axle.pvl_camber_angle。

（3）单击 OK 按钮，完成._fsae_suspension_rear_axle.ground.cfl_wheel_center 结构框的创建。

（4）单击 Build > Part > General Part > New 命令，弹出创建部件对话框，如图 10-5 所示，在下列对话框中输入相应的数据：

- General Part：spindle；
- Location Dependency：Delta location from coordinate；
- Coordinate Reference（参考坐标）：._fsae_suspension_rear_axle.ground.cfl_wheel_center；
- Location：0，0，0；
- Location in：local；
- Orientation Dependency：Delta orientation from coordinate；
- Construction Frame：._fsae_suspension_rear_axle.ground.cfl_wheel_center；
- Orientation：0，0，0；
- Mass：1；
- Ixx：1；
- Iyy：1；
- Izz：1；
- Density：Material；
- Material Type：.materials.steel。

（5）单击 OK 按钮，完成部件._fsae_suspension_rear_axle.gel_spindle 的创建。

（6）单击 Build > Geometry > Cylinder（圆柱体）> New 命令，弹出创建部件对话框，如图 10-9 所示。在下列对话框输入相应的数据：

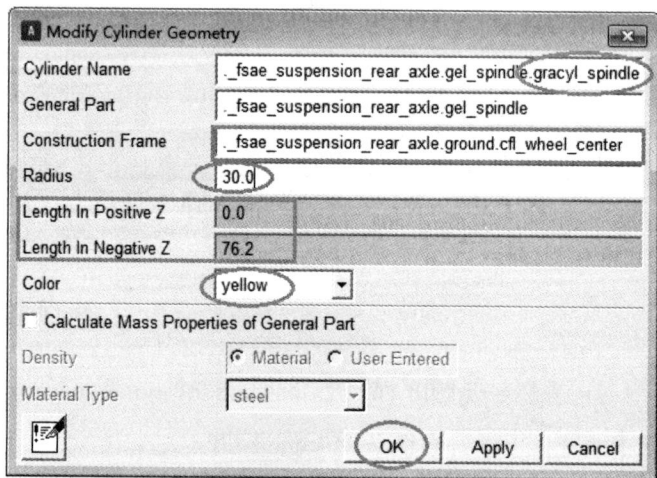

图 10-9 轮毂几何体创建对话框

- Cylinder Name（连杆名称）：spindle；

- 170 -

- General Part：._fsae_suspension_rear_axle.gel_spindle；
- Radius（半径）：30；
- Length In Positive Z（Z 轴正方向长度）：0；
- Length In Negative Z（Z 轴负方向长度）：76.2；
- Color（圆柱体几何体颜色）：yellow；
- 选择 Calculate Mass Properties of General Part 复选框。

（7）单击 OK 按钮，完成轮毂圆柱体._fsae_suspension_rear_axle.gel_spindle.gracyl_spindle 几何体的创建。

10.1.10 驱动轴 drive_shaft 部件

（1）单击 Build > Parameter Variable > New 命令，弹出参数变量对话框，如图 10-10 所示。在下列对话框中输入相应的数据：

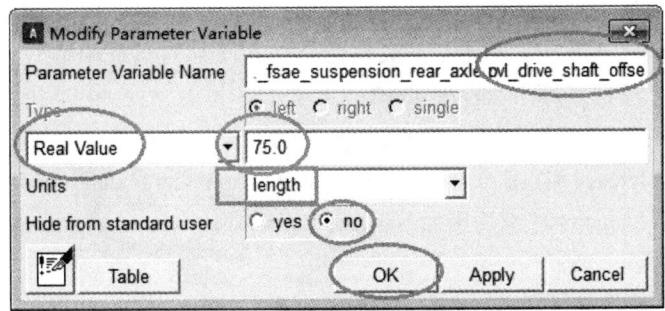

图 10-10 drive_shaft_offset 变量对话框

- Parameter Variable Name：drive_shaft_offset；
- 参数类型：Real Value（实数值），数值为 75；
- Units：length；
- Hide from standard user（是否从标准界面隐藏）：no。

（2）单击 OK 按钮，完成变量._fsae_suspension_rear_axle.pvl_drive_shaft_offset 的创建。
（3）单击 Build > Construction Frame > New 命令，在下列对话框中输入相应的数据：
- Construction Frame（结构框名称）：drive_shaft_otr；
- Location Dependency：Delta location from coordinate；
- Coordinate Reference（参考坐标）：._fsae_suspension_rear_axle.ground.cfl_wheel_center；
- Location：0.0，0.0，(-1.0 * ._fsae_suspension_rear_axle.pvl_drive_shaft_offset)；
- Location in：local；
- Orientation Dependency：Orient axis to point；
- Coordinate Reference（参考坐标）：._fsae_suspension_rear_axle.ground.hpl_wheel_center；
- Axis：Z。

（4）单击 OK 按钮，完成._fsae_suspension_rear_axle.ground.cfl_drive_shaft_otr 结构框的创建。

（5）单击 Build > Part > General Part > New 命令，在下列对话框中输入相应的数据：
- General Part：drive_shaft；
- Location Dependency：Delta location from coordinate；
- Coordinate Reference #1（参考坐标）：._fsae_suspension_rear_axle.ground.cfl_drive_shaft_otr；
- Coordinate Reference #2（参考坐标）：._fsae_suspension_rear_axle.ground.hpl_drive_shaft_inr；
- Coordinate Reference#3（参考坐标）：._fsae_suspension_rear_axle.ground.hpl_wheel_center；
- Axis：ZX；
- Mass：1；
- Ixx：1；
- Iyy：1；
- Izz：1；
- Density：Material；
- Material Type：.materials.steel。

（6）单击 OK 按钮，完成部件._fsae_suspension_rear_axle.gel_drive_shaft 的创建。

（7）单击 Build > Geometry > Link > New 命令，在下列对话框中输入相应的数据：
- Link Name（连杆名称）：drive_shaft；
- General Part：._fsae_suspension_rear_axle.gel_drive_shaft；
- Coordinate Reference #1（参考坐标）：._fsae_suspension_rear_axle.ground.hpl_drive_shaft_inr；
- Coordinate Reference #2（参考坐标）：._fsae_suspension_rear_axle.ground.cfl_drive_shaft_otr；
- Radius（半径）：15；
- Color（杆件几何体颜色）：red；
- 选择 Calculate Mass Properties of General Part 复选框，当几何体建立好之后，会更新对应部件的质量和惯量参数；
- Density：Material；
- Material Type：steel。

（8）单击 OK 按钮，完成._fsae_suspension_rear_axle.gel_drive_shaft.gralin_drive_shaft 几何体的创建。

（9）单击 Build > Geometry > Ellipsoid > New 命令，在下列对话框中输入相应的数据：
- Ellipsoid Name（连杆名称）：otr_cv_housing；
- Coordinate Reference（参考坐标）：._fsae_suspension_rear_axle.ground.cfl_drive_shaft_otr；
- Link：._fsae_suspension_rear_axle.gel_drive_shaft.gralin_drive_shaft；
- X Scale：2；
- Y Scale：2；
- Z Scale：2；
- Color（杆件几何体颜色）：red；
- 选择 Calculate Mass Properties of General Part 复选框，当几何体建立好之后，会更新对应部件的质量和惯量参数；
- Density：Material；
- Material Type：steel。

（10）单击 Apply 按钮，完成._fsae_suspension_rear_axle.gel_drive_shaft.graell_otr_cv_housing 几何体的创建。

（11）单击 Build > Geometry > Ellipsoid > New 命令，在下列对话框中输入相应的数据：
- Ellipsoid Name（连杆名称）：tripot_housing；
- Coordinate Reference（参考坐标）：._fsae_suspension_rear_axle.ground.hpl_drive_shaft_inr；
- Link：._fsae_suspension_rear_axle.gel_drive_shaft.gralin_drive_shaft；
- X　Scale：2；
- Y　Scale：2；
- Z　Scale：2；
- Color（杆件几何体颜色）：yellow；
- 选择 Calculate Mass Properties of General Part 复选框；
- Density：Material；
- Material Type：steel。

（12）单击 OK 按钮，完成._fsae_suspension_rear_axle.gel_drive_shaft.graell_tripot_housing 几何体的创建。

10.1.11　等速万向节 tripot 部件

（1）单击 Build > Construction Frame > New 命令，在下列对话框中输入相应的数据：
- Construction Frame（结构框名称）：drive_shaft_inr；
- Location Dependency：Delta location from coordinate；
- Coordinate Reference（参考坐标）：._fsae_suspension_rear_axle.ground.hpl_drive_shaft_inr；
- Location：0，0，0；
- Location in：local；
- Orientation Dependency：Orient in plane；
- Coordinate Reference #1（参考坐标）：._fsae_suspension_rear_axle.ground.hpl_drive_shaft_inr；
- Coordinate Reference #2（参考坐标）：._fsae_suspension_rear_axle.ground.hpr_drive_shaft_inr；
- Coordinate Reference#3（参考坐标）：._fsae_suspension_rear_axle.ground.cfl_drive_shaft_otr；
- Axis：ZX。

（2）单击 OK 按钮，完成._fsae_suspension_rear_axle.ground.cfl_drive_shaft_inr 结构框的创建。

（3）单击 Build > Part > General Part > New 命令，弹出创建部件对话框，如图 10-5 所示。在下列对话框中输入相应的数据：
- General Part：tripot；
- Location Dependency：Delta location from coordinate；
- Coordinate Reference（参考坐标）：._fsae_suspension_rear_axle.ground.hpl_drive_shaft_inr；
- Location：0，0，0；
- Location in：local；
- Orientation Dependency：Delta orientation from coordinate；

- Construction Frame：._fsae_suspension_rear_axle.ground.cfl_drive_shaft_inr；
- Orientation：0，0，0；
- Mass：1；
- Ixx：1；
- Iyy：1；
- Izz：1；
- Density：Material；
- Material Type：.materials.steel。

（4）单击 OK 按钮，完成部件._fsae_suspension_rear_axle.gel_tripot 的创建。

（5）单击 Build > Geometry > Cylinder（圆柱体）> New 命令，弹出创建部件对话框，如图 10-9 所示。在下列对话框中输入相应的数据：
- Cylinder Name（连杆名称）：tripot_housing_extention；
- General Part：._fsae_suspension_rear_axle.gel_tripot；
- Radius（半径）：30；
- Length In Positive Z（Z 轴正方向长度）：50；
- Length In Negative Z（Z 轴负方向长度）：0；
- Color（圆柱体几何体颜色）：yellow；
- 选择 Calculate Mass Properties of General Part 复选框。

（6）单击 OK 按钮，完成轮毂圆柱体._fsae_suspension_rear_axle.gel_tripot.gracyl_tripot_housing_extention 几何体的创建。

10.1.12 弹簧与避震器

（1）单击 Build > Force > Spring > New 命令，弹出创建部件对话框，如图 10-11 所示。在下列对话框中输入相应的数据：

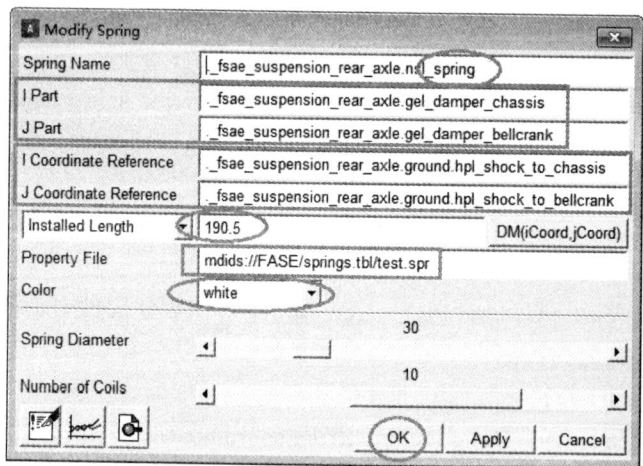

图 10-11 spring 弹簧创建对话框

- Spring Name（减震器名称）：spring；
- I Part：._fsae_suspension_rear_axle.gel_damper_chassis；

- J Part：._fsae_suspension_rear_axle.gel_damper_bellcrank；
- I Coordinate Reference（参考坐标）：._fsae_suspension_rear_axle.ground.hpl_shock_to_chassis；
- J Coordinate Reference（参考坐标）：._fsae_suspension_rear_axle.ground.hpl_shock_to_bellcrank；
- Installed Length（安装长度）：单击 DM（iCoord，jCoord）自动计算弹簧的安装长度并将数值填入方框中，此模型的安装长度为 190.5；
- Property File（属性文件）：mdids：//FASE/springs.tbl/test.spr，弹簧属性文件用记事本文件打开后如下列信息所示，可以根据实验情况测出弹簧的参数（力与位移之间的关系，即刚度）后修改如下信息，可修改部分为下划线部分；

```
弹簧属性文件信息：
$--------------------------------------------------------------MDI_HEADER
[MDI_HEADER]
 FILE_TYPE      = 'spr'
 FILE_VERSION   = 4.0
 FILE_FORMAT    = 'ASCII'
$--------------------------------------------------------------------UNITS
[UNITS]
 LENGTH  =  'mm'
 ANGLE   =  'degrees'
 FORCE   =  'newton'
 MASS    =  'kg'
 TIME    =  'second'
$--------------------------------------------------------------SPRING_DATA
[SPRING_DATA]
 FREE_LENGTH  =  205.7
$--------------------------------------------------------------------CURVE
[CURVE]
{  disp         force}
 -100.0        -100000.0
 -50.0         -50000.0
  0.0           0.0
  50.0          50000.0
  100.0         100000.0
```

- Spring Diameter（弹簧直径）：拖动滑块选择 30 mm；
- Number of Coils（弹簧圈数）：拖动滑块选择 10。

（2）单击 OK 按钮，完成弹簧._fsae_suspension_rear_axle.nsl_spring 的创建。

（3）单击 Build > Force > Damper > New 命令，弹出避震器创建对话框如图 10-12 所示。在下列对话框中输入相应的数据：

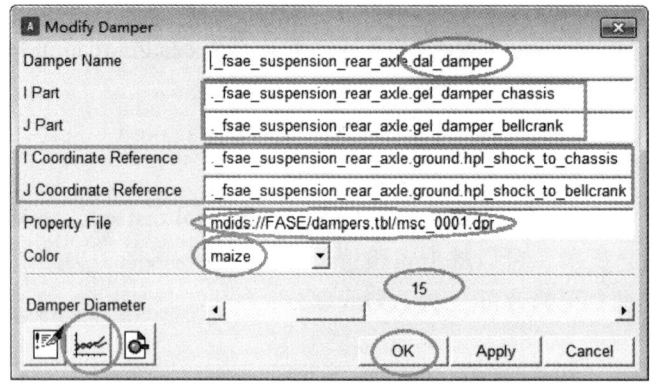

图 10-12　damper 避震器创建对话框

- Damper Name（减震器名称）：damper；
- I Part：._fsae_suspension_rear_axle.gel_damper_chassis；
- J Part：._fsae_suspension_rear_axle.gel_damper_bellcrank；
- I Coordinate Reference（参考坐标）：._fsae_suspension_rear_axle.ground.hpl_shock_to_chassis；
- J Coordinate Reference（参考坐标）：._fsae_suspension_rear_axle.ground.hpl_shock_to_bellcrank；
- Property File（属性文件）：mdids://FASE/dampers.tbl/MSC_default.dpr，避震器系数曲线如图 10-13 所示，具体数据如下列避震器信息；

```
避震器属性文件信息：
$--------------------------------------------------MDI_HEADER
[MDI_HEADER]
 FILE_TYPE      = 'dpr'
 FILE_VERSION   = 4.0
 FILE_FORMAT    = 'ASCII'
$--------------------------------------------------UNITS
[UNITS]
 LENGTH  = 'mm'
 ANGLE   = 'degrees'
 FORCE   = 'newton'
 MASS    = 'kg'
 TIME    = 'second'
$-------------------------------------------------- CURVE
[CURVE]
{    vel                     force}
```

−4916.935	−8.889
−1000.0	−3.0
−500.0	−1.5
−250.0	−0.75
−100.0	−0.3
0.0	**0.0**
100.0	0.3
250.0	0.75
500.0	1.5
1000.0	3.0
4914.298	9.0416

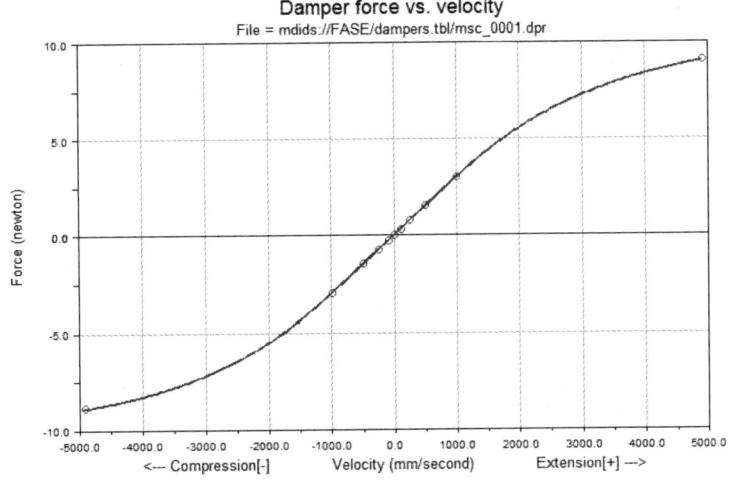

图 10-13　避震器系数曲线

- Damper Diameter（避震器直径）：拖动滑块选择 15 mm；
- Color：maize。

（4）单击 OK 按钮，完成避震器._fsae_suspension_rear_axle.dal_damper 的创建。

10.1.13　安装部件

（1）单击 Build > Part > Mount > New 命令，弹出创建部件对话框，如图 10-14 所示。在下列对话框中输入相应的数据：

- Mount name（安装件名称）：suspension_to_chassis；
- Coordinate Reference（参考坐标）：._fsae_suspension_rear_axle.ground.hps_global；
- From Minor Role：inherit（继承特性）。

（2）单击 Apply 按钮，完成._fsae_suspension_rear_axle.mts_suspension_to_chassis 安装部件的创建。同理，创建其他安装部件。

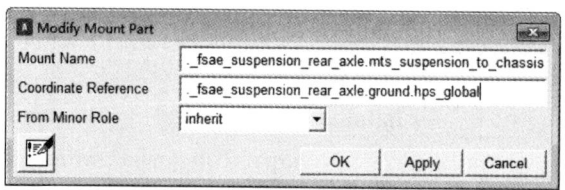

图 10-14　安装部件对话框

- Mount name（安装件名称）：tierod_to_steering；
- Coordinate Reference（参考坐标）：._fsae_suspension_rear_axle.ground.hpl_tierod_inner；
- From Minor Role：inherit（继承特性）。

（3）单击 Apply 按钮，完成 ._fsae_suspension_rear_axle.mtl_tierod_to_steering 安装部件的创建。

- Mount name（安装件名称）：tripot_to_differential；
- Coordinate Reference（参考坐标）：._fsae_suspension_rear_axle.ground.hpl_drive_shaft_inr；
- From Minor Role：inherit（继承特性）。

（4）单击 OK 按钮，完成 ._fsae_suspension_rear_axle.mtl_tripot_to_differential 安装部件的创建。

10.1.14　刚性约束

1. 部件 uca 与安装件 suspension_to_chassis 之间的 revolute 约束

（1）单击 Build > Attachments > Joint > New 命令，弹出创约束件对话框，如图 10-15 所示。在下列对话框中输入相应的数据：

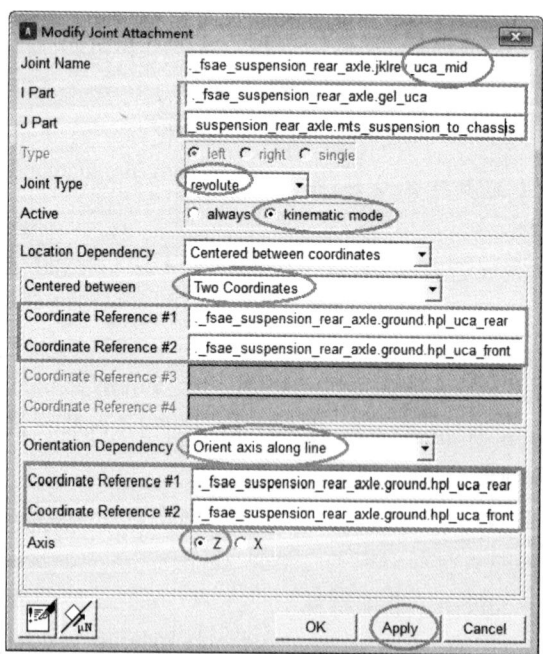

图 10-15　刚性约束对话框-revolute

- Joint Name（约束副名称）：uca_mid；
- I Part：._fsae_suspension_rear_axle.gel_uca；
- J Part：._fsae_suspension_rear_axle.mts_suspension_to_chassis；
- Joint Type（约束副类型）：revolute，转动副，约束 5 个自由度；
- Active（激活）：kinematic mode（运动学模式）；
- Location Dependency：Centered between coordinates；
- Centered between：Two Coordinates；
- Coordinate Reference #1（参考坐标）：._fsae_suspension_rear_axle.ground.hpl_uca_rear；
- Coordinate Reference #2（参考坐标）：._fsae_suspension_rear_axle.ground.hpl_uca_front；
- Orientation Dependency：Orient axis along line；
- Coordinate Reference #1（参考坐标）：._fsae_suspension_rear_axle.ground.hpl_uca_rear；
- Coordinate Reference #2（参考坐标）：._fsae_suspension_rear_axle.ground.hpl_uca_front。

（2）单击 Apply 按钮，完成._fsae_suspension_rear_axle.jklrev_uca_mid 转动副的创建。

2. 部件 uca 与 upright 之间的 spherical 约束

（1）单击 Build > Attachments > Joint > New 命令，在下列对话框中输入相应的数据：
- Joint Name（约束副名称）：uca_outer；
- I Part：._fsae_suspension_rear_axle.gel_uca；
- J Part：._fsae_suspension_rear_axle.gel_upright；
- Joint Type（约束副类型）：spherical，转动副，约束 3 个自由度；
- Active（激活）：always；
- Location Dependency：Delta location from coordinate；
- Coordinate Reference（参考坐标）：._fsae_suspension_rear_axle.ground.hpl_uca_outer；
- Location：0，0，0；
- Location in：local；
- Orientation：None。

（2）单击 Apply 按钮，完成约束副._fsae_suspension_rear_axle.jolsph_uca_outer 的创建。

3. 部件 spindle 与 upright 之间的 revolute 约束

（1）单击 Build > Attachments > Joint > New 命令，在下列对话框中输入相应的数据：
- Joint Name（约束副名称）：spindle；
- I Part：._fsae_suspension_rear_axle.gel_spindle；
- J Part：._fsae_suspension_rear_axle.gel_upright；
- Joint Type（约束副类型）：revolute；
- Active（激活）：always；
- Location Dependency：Delta location from coordinate；
- Coordinate Reference（参考坐标）：._fsae_suspension_rear_axle.ground.hpl_wheel_center；
- Location：0，0，0；

- Location in：local；
- Orientation Dependency：Delta orientation from coordinate；
- Construction Frame：._fsae_suspension_rear_axle.ground.cfl_wheel_center。

（2）单击 Apply 按钮，完成约束副._fsae_suspension_rear_axle.jolrev_spindle 的创建。

4. 部件 damper_chassis 与安装件 suspension_to_chassis 之间的 hook 约束

（1）单击 Build > Construction Frame > New 命令，在下列对话框中输入相应的数据：
- Construction Frame（结构框名称）：damper_chassis_orient；
- Location Dependency：Delta location from coordinate；
- Coordinate Reference（参考坐标）：._fsae_suspension_rear_axle.ground.hpl_shock_to_chassis；
- Location：20，0，0；
- Location in：local；
- Orientation Dependency：User entered values；
- Orient using：Euler Angles；
- Euler Angles：0，0，0。

（2）单击 OK 按钮，完成._fsae_suspension_rear_axle.ground.cfl_damper_chassis_orient 结构框的创建。

（3）单击 Build > Attachments > Joint > New 命令，在下列对话框中输入相应的数据：
- Joint Name（约束副名称）：damper_to_chassis；
- I Part：._fsae_suspension_rear_axle.gel_damper_chassis；
- J Part：._fsae_suspension_rear_axle.mts_suspension_to_chassis；
- Joint Type（约束副类型）：hook；
- Active（激活）：always；
- Location Dependency：Delta location from coordinate；
- Coordinate Reference（参考坐标）：._fsae_suspension_rear_axle.ground.hpl_shock_to_chassis；
- Location：0，0，0；
- Location in：local；
- I-Part Axis：._fsae_suspension_rear_axle.ground.hpl_shock_to_bellcrank；
- J-Part Axis：._fsae_suspension_rear_axle.ground.cfl_damper_chassis_orient。

（4）单击 Apply 按钮，完成约束副._fsae_suspension_rear_axle.jolhoo_damper_to_chassis 的创建。

5. 部件 damper_bellcrank 与安装件 bellcrank 之间的 hook 约束

（1）单击 Build > Construction Frame > New 命令，在下列对话框中输入相应的数据：
- Construction Frame（结构框名称）：damper_bellcrank_orient；
- Location Dependency：Delta location from coordinate；
- Coordinate Reference（参考坐标）：._fsae_suspension_rear_axle.hpl_shock_to_

bellcrank；
- Location：-20，0，0；
- Location in：local；
- Orientation Dependency：User entered values；
- Orient using：Euler Angles；
- Euler Angles：0，0，0。

（2）单击 OK 按钮，完成 ._fsae_suspension_rear_axle.ground.cfl_damper_bellcrank_orient 结构框的创建。

（3）单击 Build > Attachments > Joint > New 命令，在下列对话框中输入相应的数据：
- Joint Name（约束副名称）：damper_to_bellcrank；
- I Part：._fsae_suspension_rear_axle.gel_damper_bellcrank；
- J Part：._fsae_suspension_rear_axle.gel_bellcrank；
- Joint Type（约束副类型）：hook；
- Active（激活）：always；
- Location Dependency：Delta location from coordinate；
- Coordinate Reference（参考坐标）：._fsae_suspension_rear_axle.ground.hpl_shock_to_bellcrank；
- Location：0，0，0；
- Location in：local；
- I-Part Axis：._fsae_suspension_rear_axle.ground.hpl_shock_to_chassis；
- J-Part Axis：._fsae_suspension_rear_axle.ground.cfl_damper_bellcrank_orient。

（4）单击 Apply 按钮，完成约束副 ._fsae_suspension_rear_axle.jolhoo_damper_to_bellcrank 的创建。

6. 部件 lca 与安装件 suspension_to_chassis 之间的 revolute 约束

（1）单击 Build > Attachments > Joint > New 命令，在下列对话框中输入相应的数据：
- Joint Name（约束副名称）：lca_inner_mid；
- I Part：._fsae_suspension_rear_axle.gel_lca；
- J Part：._fsae_suspension_rear_axle.mts_suspension_to_chassis；
- Joint Type（约束副类型）：revolute，转动副，约束 5 个自由度；
- Active（激活）：kinematic mode（运动学模式）；
- Location Dependency：Centered between coordinates；
- Centered between：Two Coordinates；
- Coordinate Reference #1（参考坐标）：._fsae_suspension_rear_axle.ground.hpl_lca_rear；
- Coordinate Reference #2（参考坐标）：._fsae_suspension_rear_axle.ground.hpl_lca_front；
- Orientation Dependency：Orient axis along line；
- Coordinate Reference #1（参考坐标）：._fsae_suspension_rear_axle.ground.hpl_lca_rear；
- Coordinate Reference #2（参考坐标）：._fsae_suspension_rear_axle.ground.hpl_lca_front；
- Axis：Z。

（2）单击 Apply 按钮，完成 ._fsae_suspension_rear_axle.jklrev_lca_inner_mid 铰接副的创建。

7. 部件 prod 与 bellcrank 之间的 hook 约束

（1）单击 Build > Attachments > Joint > New 命令，在下列对话框中输入相应的数据：
- Joint Name（约束副名称）：prod_to_bellcrank；
- I Part：._fsae_suspension_rear_axle.gel_prod；
- J Part：._fsae_suspension_rear_axle.gel_bellcrank；
- Joint Type（约束副类型）：hook；
- Active（激活）：always；
- Location Dependency：Delta location from coordinate；
- Coordinate Reference（参考坐标）：._fsae_suspension_rear_axle.ground.hpl_prod_to_bellcrank；
- Location：0，0，0；
- Location in：local；
- I-Part Axis：._fsae_suspension_rear_axle.ground.hpl_prod_outer；
- J-Part Axis：._fsae_suspension_rear_axle.ground.hpl_bellcrank_pivot；

（2）单击 Apply 按钮，完成约束副 ._fsae_suspension_rear_axle.jolhoo_prod_to_bellcrank 的创建。

8. 部件 bellcrank 与安装件 suspension_to_chassis 之间的 revolute 约束

（1）单击 Build > Construction Frame > New 命令，在下列对话框中输入相应的数据：
- Construction Frame（结构框名称）：bellcrank_pivot；
- Location Dependency：Delta location from coordinate；
- Coordinate Reference（参考坐标）：._fsae_suspension_rear_axle.ground.hpl_bellcrank_pivot；
- Location：0，0，0；
- Location in：local；
- Orientation Dependency：Orient axie to point；
- Coordinate Reference（参考坐标）：._fsae_suspension_rear_axle.ground.hpl_bellcrank_pivot_orient；
- Axis：Z。

（2）单击 OK 按钮，完成 ._fsae_suspension_rear_axle.ground.cfl_bellcrank_pivot 结构框的创建。

（3）单击 Build > Attachments > Joint > New 命令，在下列对话框中输入相应的数据：
- Joint Name（约束副名称）：bellcrank_pivot；
- I Part：._fsae_suspension_rear_axle.gel_bellcrank；
- J Part：._fsae_suspension_rear_axle.mts_suspension_to_chassis；
- Joint Type（约束副类型）：revolute；
- Active（激活）：always；

- Location Dependency: Delta location from coordinate;
- Coordinate Reference（参考坐标）: ._fsae_suspension_rear_axle.ground.hpl_bellcrank_pivot;
- Location: 0, 0, 0。
- Location in: local;
- Orientation Dependency: Delta orientation from coordinate;
- Construction Frame: ._fsae_suspension_rear_axle.ground.cfl_bellcrank_pivot;
- Orientation: 0, 0, 0。

（4）单击 Apply 按钮，完成约束副 ._fsae_suspension_rear_axle.jolrev_bellcrank_pivot 的创建。

9. 部件 lca 与 prod 之间的 spherical 约束

（1）单击 Build > Attachments > Joint > New 命令，在下列对话框中输入相应的数据：

- Joint Name（约束副名称）: prod_outer;
- I Part: ._fsae_suspension_rear_axle.gel_lca;
- J Part: ._fsae_suspension_rear_axle.gel_prod;
- Joint Type（约束副类型）: spherical, 转动副, 约束 3 个自由度;
- Active（激活）: always;
- Location Dependency: Delta location from coordinate;;
- Coordinate Reference（参考坐标）: ._fsae_suspension_rear_axle.ground.hpl_prod_outer;
- Location: 0, 0, 0;
- Location in: local;
- Orientation: None。

（2）单击 Apply 按钮，完成约束副 ._fsae_suspension_rear_axle.jolsph_prod_outer 的创建。

10. 部件 tierod 与安装件 tierod_to_steering 之间的 convel 约束

（1）单击 Build > Attachments > Joint > New 命令，在下列对话框中输入相应的数据：

- Joint Name（约束副名称）: tierod_inner;
- I Part: ._fsae_suspension_rear_axle.gel_tierod;
- J Part: ._fsae_suspension_rear_axle.mtl_tierod_to_steering;
- Joint Type（约束副类型）: convel, 恒速副;
- Active（激活）: always;
- Location Dependency: Delta location from coordinate;
- Coordinate Reference（参考坐标）: ._fsae_suspension_rear_axle.ground.hpl_tierod_inner;
- Location: 0, 0, 0;
- Location in: local;
- I-Part Axis: ._fsae_suspension_rear_axle.ground.hpl_tierod_outer;
- J-Part Axis: ._fsae_suspension_rear_axle.ground.hpr_tierod_inner。

（2）单击 Apply 按钮，完成约束副 ._fsae_suspension_rear_axle.jolcon_tierod_inner 的创建。

11. 部件 tierod 与 upright 之间的 spherical 约束

（1）单击 Build > Attachments > Joint > New 命令，在下列对话框中输入相应的数据：

- Joint Name（约束副名称）：tierod_outer；
- I Part：._fsae_suspension_rear_axle.gel_tierod；
- J Part：._fsae_suspension_rear_axle.gel_upright；
- Joint Type（约束副类型）：spherical，约束 3 个自由度；
- Active（激活）：always；
- Location Dependency：Delta location from coordinate；；
- Coordinate Reference（参考坐标）：._fsae_suspension_rear_axle.ground.hpl_tierod_outer；
- Location：0，0，0；
- Location in：local；
- Orientation：None。

（2）单击 Apply 按钮，完成约束副._fsae_suspension_rear_axle.jolsph_tierod_outer 的创建。

12. 部件 lca 与 upright 之间的 spherical 约束

（1）单击 Build > Attachments > Joint > New 命令，在下列对话框中输入相应的数据：

- Joint Name（约束副名称）：lca_outer；
- I Part：._fsae_suspension_rear_axle.gel_lca；
- J Part：._fsae_suspension_rear_axle.gel_upright；
- Joint Type（约束副类型）：spherical，约束 3 个自由度；
- Active（激活）：always；
- Location Dependency：Delta location from coordinate；；
- Coordinate Reference（参考坐标）：._fsae_suspension_rear_axle.ground.hpl_lca_outer；
- Location：0，0，0；
- Location in：local；
- Orientation：None。

（2）单击 Apply 按钮，完成约束副._fsae_suspension_rear_axle.jolsph_lca_outer 的创建。

13. 部件 damper_chassis 与 damper_bellcrank 之间的 cylindrical 约束

（1）单击 Build > Attachments > Joint > New 命令，在下列对话框中输入相应的数据：

- Joint Name（约束副名称）：damper_slide；
- I Part：._fsae_suspension_rear_axle.gel_damper_chassis；
- J Part：._fsae_suspension_rear_axle.gel_damper_bellcrank；
- Joint Type（约束副类型）：cylindrical；
- Active（激活）：always；
- Location Dependency：Centered between coordinates；
- Centered between：Two Coordinates；

- Coordinate Reference #1（参考坐标）：._fsae_suspension_rear_axle.ground.hpl_shock_to_chassis；
- Coordinate Reference #2（参考坐标）：._fsae_suspension_rear_axle.ground.hpl_shock_to_bellcrank；
- Orientation Dependency：Orient axis along line；
- Coordinate Reference #1（参考坐标）：._fsae_suspension_rear_axle.ground.hpl_shock_to_chassis；
- Coordinate Reference #2（参考坐标）：._fsae_suspension_rear_axle.ground.hpl_shock_to_bellcrank；
- Axis：Z。

（2）单击 Apply 按钮，完成 ._fsae_suspension_rear_axle.jolcyl_damper_slide 转动副的创建。

14. 部件 tripot 与 drive_shaft 之间的 convel 约束

（1）单击 Build > Attachments > Joint > New 命令，在下列对话框中输入相应的数据：
- Joint Name（约束副名称）：drive_sft_int_jt；
- I Part：._fsae_suspension_rear_axle.gel_tripot；
- J Part：._fsae_suspension_rear_axle.gel_drive_shaft；
- Joint Type（约束副类型）：convel，恒速副；
- Active（激活）：always；
- Location Dependency：Delta location from coordinate；
- Coordinate Reference（参考坐标）：._fsae_suspension_rear_axle.ground.hpl_drive_shaft_inr；
- Location：0，0，0；
- Location in：local；
- I-Part Axis：._fsae_suspension_rear_axle.ground.hpr_drive_shaft_inr；
- J-Part Axis：._fsae_suspension_rear_axle.ground.cfl_drive_shaft_otr。

（2）单击 Apply 按钮，完成约束副 ._fsae_suspension_rear_axle.jolcon_drive_sft_int_jt 的创建。

15. 部件 spindle 与 drive_shaft 之间的 convel 约束

（1）单击 Build > Construction Frame > New 命令，在下列对话框中输入相应的数据：
- Construction Frame（结构框名称）：drive_shaft_otr；
- Location Dependency：Delta location from coordinate；
- Coordinate Reference（参考坐标）：._fsae_suspension_rear_axle.ground.cfl_wheel_center；
- Location：0.0，0.0，（-1.0 * ._fsae_suspension_rear_axle.pvl_drive_shaft_offset）；
- Location in：local；
- Orientation Dependency：Orient axie to point；
- Coordinate Reference（参考坐标）：._fsae_suspension_rear_axle.ground.hpl_wheel_center；
- Axis：Z。

（2）单击 OK 按钮，完成 ._fsae_suspension_rear_axle.ground.cfl_drive_shaft_otr 结构框的创建。

（3）单击 Build > Attachments > Joint > New 命令，在下列对话框中输入相应的数据：
- Joint Name（约束副名称）：drive_sft_otr；
- I Part：._fsae_suspension_rear_axle.gel_drive_shaft；
- J Part：._fsae_suspension_rear_axle.gel_spindle；
- Joint Type（约束副类型）：convel，恒速副；
- Active（激活）：always；
- Location Dependency：Delta location from coordinate；
- Coordinate Reference（参考坐标）：._fsae_suspension_rear_axle.ground.cfl_drive_shaft_otr；
- Location：0，0，0；
- Location in：local；
- I-Part Axis：._fsae_suspension_rear_axle.ground.hpl_drive_shaft_inr；
- J-Part Axis：._fsae_suspension_rear_axle.ground.hpl_wheel_center。

（4）单击 Apply 按钮，完成约束副_fsae_suspension_rear_axle.jolcon_drive_sft_otr 的创建。

16. 部件 tripot 与安装件 tripot_to_differential 之间的 translational 约束

（1）单击 Build > Attachments > Joint > New 命令，在下列对话框中输入相应的数据：
- Joint Name（约束副名称）：tripot_to_differential；
- I Part：._fsae_suspension_rear_axle.gel_tripot；
- J Part：._fsae_suspension_rear_axle.mtl_tripot_to_differential；
- Joint Type（约束副类型）：translational；
- Active（激活）：always；
- Location Dependency：Delta location from coordinate；
- Coordinate Reference #1（参考坐标）：._fsae_suspension_rear_axle.ground.hpl_drive_shaft_inr；
- Orientation Dependency：Orient to zpoint-xpoint；
- Coordinate Reference #1（参考坐标）：._fsae_suspension_rear_axle.ground.hpr_drive_shaft_inr；
- Coordinate Reference #2（参考坐标）：._fsae_suspension_rear_axle.ground.cfl_drive_shaft_otr；
- Axis：ZX。

（2）单击 OK 按钮，完成约束副._fsae_suspension_rear_axle.joltra_tripot_to_differential 的创建。

10.1.15 柔性约束

1. 部件 uca 与 suspension_to_chassis 之间的 bushing 约束

（1）单击 Build > Attachments > Bushing > New 命令，弹出衬套约束对话框，如图 10-16 所示。在下列对话框中输入相应的数据：

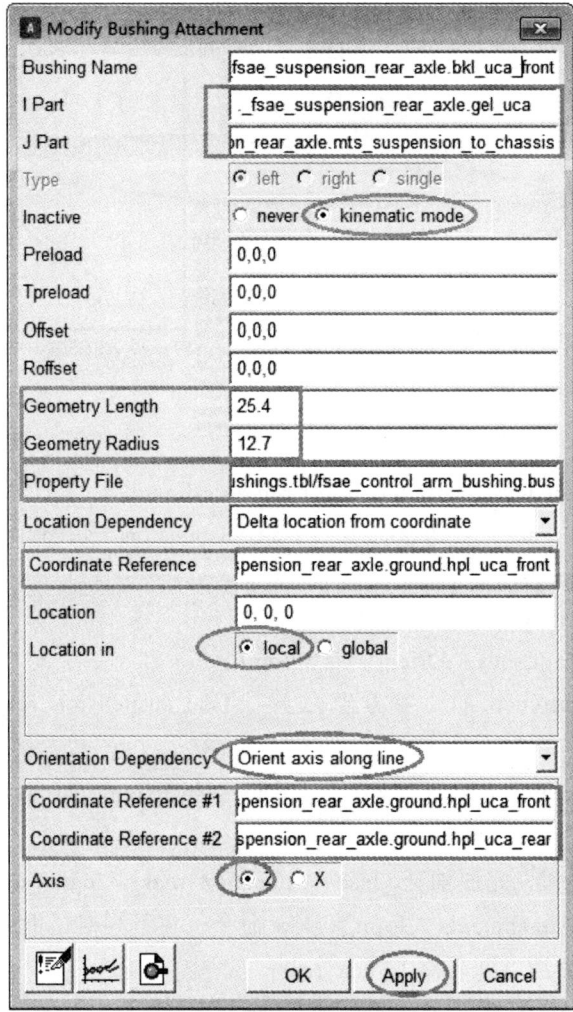

图 10-16 衬套约束对话框-bushing

- Bushing Name（约束副名称）：uca_front；
- I Part：._fsae_suspension_rear_axle.gel_uca；
- J Part：._fsae_suspension_rear_axle.mts_suspension_to_chassis；
- Inactive（抑制）：kinematic mode（运动学模式）；
- Preload：0，0，0；
- Tpreload：0，0，0；
- Offset：0，0，0；
- Roffset：0，0，0；
- Geometry Length：25.4；
- Geometry Radius：12.7；
- Property File：mdids：//FASE/bushings.tbl/fsae_control_arm_bushing.bus，用记事本文件打开衬套属性文件，用 MATLAB 软件绘制在 X、Y、Z 方向的垂向刚度及扭转刚度如图 10-17 和图 10-18 所示。

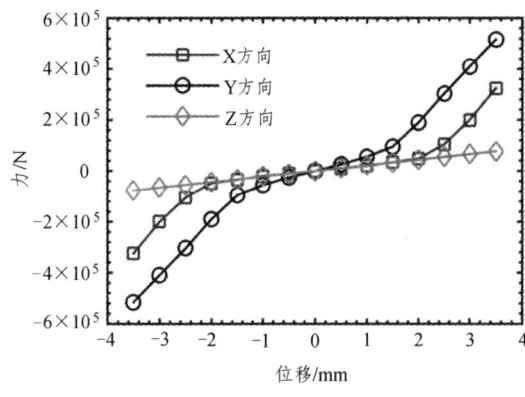

图 10-17 衬套垂向刚度　　　　图 10-18 衬套扭转刚度

- Location Dependency：Delta location from coordinate；
- Coordinate Reference（参考坐标）：._fsae_suspension_rear_axle.ground.hpl_uca_front；
- Location：0，0，0；
- Location in：local；
- Orientation Dependency：Orient axis along line；
- Coordinate Reference #1（参考坐标）：._fsae_suspension_rear_axle.ground.hpl_uca_front；
- Coordinate Reference #2（参考坐标）：._fsae_suspension_rear_axle.ground.hpl_uca_rear；
- Axis：Z。

（2）单击 Apply 按钮，完成轴套._fsae_suspension_rear_axle.bkl_uca_front 的创建。

（3）单击 Build > Attachments > Joint > New 命令，在下列对话框中输入相应的数据：

- Bushing Name（约束副名称）：uca_rear；
- I Part：._fsae_suspension_rear_axle.gel_uca；
- J Part：._fsae_suspension_rear_axle.mts_suspension_to_chassis；
- Inactive（抑制）：kinematic mode（运动学模式）；
- Preload：0，0，0；
- Tpreload：0，0，0；
- Offset：0，0，0；
- Roffset：0，0，0；
- Geometry Length：25.4；
- Geometry Radius：12.7；
- Property File：mdids：//FASE/bushings.tbl/fsae_control_arm_bushing.bus；
- Location Dependency：Delta location from coordinate；
- Coordinate Reference（参考坐标）：._fsae_suspension_rear_axle.ground.hpl_uca_rear；
- Location：0，0，0；
- Location in：local；

- Orientation Dependency：Orient axis along line；
- Coordinate Reference #1（参考坐标）：._fsae_suspension_rear_axle.ground.hpl_uca_front；
- Coordinate Reference #2（参考坐标）：._fsae_suspension_rear_axle.ground.hpl_uca_rear；
- Axis：Z。

（4）单击 Apply 按钮，完成轴套._fsae_suspension_rear_axle.bkl_uca_rear 的创建。

2. 部件 lca 与 suspension_to_chassis 之间的 bushing 约束

（1）单击 Build > Attachments > Joint > New 命令，在下列对话框中输入相应的数据：
- Bushing Name（约束副名称）：lca_front；
- I Part：._fsae_suspension_rear_axle.gel_lca；
- J Part：._fsae_suspension_rear_axle.mts_suspension_to_chassis；
- Inactive（抑制）：kinematic mode（运动学模式）；
- Preload：0，0，0；
- Tpreload：0，0，0；
- Offset：0，0，0；
- Roffset：0，0，0；
- Geometry Length：25.4；
- Geometry Radius：12.7；
- Property File：mdids：//FASE/bushings.tbl/fsae_control_arm_bushing.bus；
- Location Dependency：Delta location from coordinate；
- Coordinate Reference（参考坐标）：._fsae_suspension_rear_axle.ground.hpl_lca_front；
- Location：0，0，0；
- Location in：local；
- Orientation Dependency：Orient axis along line；
- Coordinate Reference #1（参考坐标）：._fsae_suspension_rear_axle.ground.hpl_lca_front；
- Coordinate Reference #2（参考坐标）：._fsae_suspension_rear_axle.ground.hpl_lca_rear；
- Axis：Z。

（2）单击 Apply 按钮，完成轴套._fsae_suspension_rear_axle.bkl_lca_front 的创建。

（3）单击 Build > Attachments > Joint > New 命令，在下列对话框中输入相应的数据：
- Bushing Name（约束副名称）：lca_rear；
- I Part：._fsae_suspension_rear_axle.gel_lca；
- J Part：._fsae_suspension_rear_axle.mts_suspension_to_chassis；
- Inactive（抑制）：kinematic mode（运动学模式）；
- Preload：0，0，0；
- Tpreload：0，0，0；
- Offset：0，0，0；
- Roffset：0，0，0；
- Geometry Length：25.4；

- Geometry Radius：12.7；
- Property File：mdids：//FASE/bushings.tbl/fsae_control_arm_bushing.bus；
- Location Dependency：Delta location from coordinate；
- Coordinate Reference（参考坐标）：._fsae_suspension_rear_axle.ground.hpl_lca_rear；
- Location：0，0，0；
- Location in：local；
- Orientation Dependency：Orient axis along line；
- Coordinate Reference #1（参考坐标）：._fsae_suspension_rear_axle.ground.hpl_lca_front；
- Coordinate Reference #2（参考坐标）：._fsae_suspension_rear_axle.ground.hpl_lca_rear；
- Axis：Z。

（4）单击 OK 按钮，完成轴套._fsae_suspension_rear_axle.bkl_lca_rear 的创建。

10.1.16 推杆式悬架变量参数

（1）单击 Build > Parameter Variable > New 命令，在下列对话框中输入相应的数据：
- Parameter Variable Name：：driveline_active；
- Integer Value（实数值）：1；
- Units：length；
- Hide from standard user（是否从标准界面隐藏）：yes。

（2）单击 Apply 按钮，完成变量._fsae_suspension_rear_axle.phs_driveline_active 的创建。

（3）单击 Build > Parameter Variable > New 命令，在下列对话框中输入相应的数据：
- Parameter Variable Name：kinematic_flag；
- Integer Value（实数值）：1；
- Units：length；
- Hide from standard user（是否从标准界面隐藏）：yes。

（4）单击 OK 按钮，完成变量._fsae_suspension_rear_axle.phs_kinematic_flag 的创建。

（5）单击 Build > Suspension Parameters > Characteristics Array > Set 命令，设置悬架的转向主销参数，如图 10-19 所示。在下列对话框中输入相应的数据：

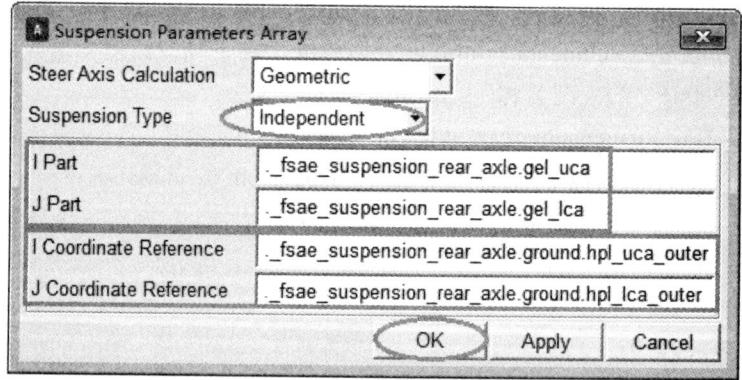

图 10-19 参数变量设置对话框

- Steer Axis Calculation：Geometric；
- Suspension Type：Independent，非独立悬架；
- I Part：._fsae_suspension_rear_axle.gel_uca；
- J Part：._fsae_suspension_rear_axle.gel_lca；
- I Coordinate Reference：._fsae_suspension_rear_axle.ground.hpl_uca_outer；
- J Coordinate Reference：._fsae_suspension_rear_axle.ground.hpl_lca_outer。

（6）单击 OK 按钮，完成悬架参数变量设置。

10.1.17 推杆式悬架通信器的建立

（1）单击 Build > Communicator > Output > New 命令，在下列对话框中输入相应的数据：
- Output Communicator Name（输出通信器名称）：driveline_active；
- Matching Name（s）：driveline_active；
- Type：single；
- Entity：parameter integer；
- To Minor Role：inherit；
- Parameter Variable Name：._fsae_suspension_rear_axle.phs_driveline_active。

（2）单击 Apply 按钮，完成通信器._fsae_suspension_rear_axle.cos_driveline_active 的创建。同理，创建其他通信器。
- Output Communicator Name（输出通信器名称）：tripot_to_differential；
- Matching Name（s）：tripot_to_differential；
- Type：left；
- Entity：Location；
- To Minor Role：inherit；
- Coordinate Reference Name：._fsae_suspension_rear_axle.ground.hpl_drive_shaft_inr。

（3）单击 Apply 按钮，完成通信器._fsae_suspension_rear_axle.col_tripot_to_differential 的创建。
- Output Communicator Name（输出通信器名称）：suspension_mount；
- Matching Name（s）：suspension_mount；
- Type：left；
- Entity：mount；
- To Minor Role：inherit；
- Part Name：._fsae_suspension_rear_axle.gel_spindle。

（4）单击 Apply 按钮，完成通信器._fsae_suspension_rear_axle.col_suspension_mount 的创建。
- Output Communicator Name（输出通信器名称）：wheel_center；
- Matching Name（s）：wheel_center；
- Type：left；
- Entity：Location；

- To Minor Role：inherit；
- Coordinate Reference Name：._fsae_suspension_rear_axle.ground.hpl_wheel_center。

（5）单击 Apply 按钮，完成通信器._fsae_suspension_rear_axle.col_wheel_center 的创建。
- Output Communicator Name（输出通信器名称）：suspension_upright；
- Matching Name（s）：suspension_upright；
- Type：left；
- Entity：mount；
- To Minor Role：inherit；
- Part Name：._fsae_suspension_rear_axle.gel_upright。

（6）单击 OK 按钮，完成通信器._fsae_suspension_rear_axle.col_suspension_upright 的创建。

10.1.18 推杆式悬架通信器测试

（1）单击 Build > Communicator > Test 命令，弹出输出通信器测试对话框，如图 10-20 所示。在下列对话框中输入相应的数据：

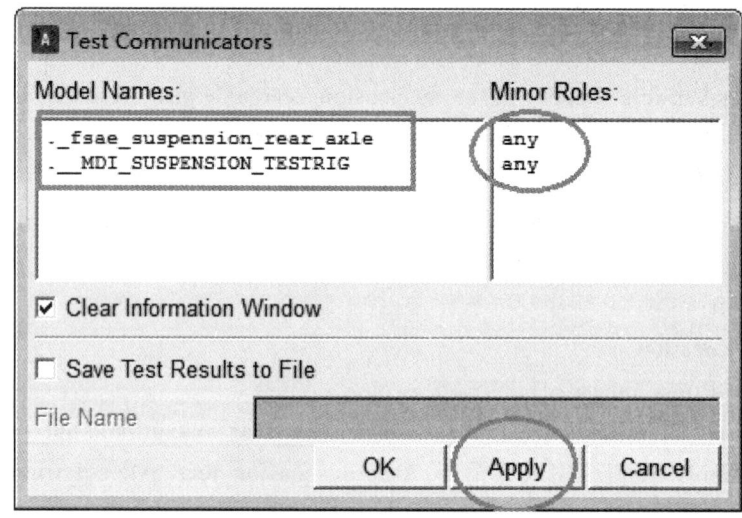

图 10-20 通信器测试对话框

- Model Names：._fsae_suspension_rear_axle、.__MDI_SUSPENSION_TESTRIG；
- Minor Roles：并列两排输入特征 any，也可以并排输入特征 front。

（2）单击 OK 按钮，完成推杆式双叉臂悬架和悬架试验台._fsae_suspension_rear_axle、.__MDI_SUSPENSION_TESTRIG 的匹配测试，测试结果如下列信息所示。

通信器匹配信息如下：

!------ Matched communicators：--------!   %以下为匹配的通信器

Communicator Matching Name：tripot_to_differential
Input Communicator Name：ci[lr]_tripot_to_differential

Located in：_fsae_suspension_rear_axle
Output Communicator Name：co[lr]_tripot_to_differential
Output from：__MDI_SUSPENSION_TESTRIG

Communicator Matching Name：camber_angle
Input Communicator Name：ci[lr]_camber_angle
Located in：__MDI_SUSPENSION_TESTRIG
Output Communicator Name：co[lr]_camber_angle
Output from：_fsae_suspension_rear_axle

Communicator Matching Name：toe_angle
Input Communicator Name：ci[lr]_toe_angle
Located in：__MDI_SUSPENSION_TESTRIG
Output Communicator Name：co[lr]_toe_angle
Output from：_fsae_suspension_rear_axle

Communicator Matching Name：wheel_center
Input Communicator Name：ci[lr]_wheel_center
Located in：__MDI_SUSPENSION_TESTRIG
Output Communicator Name：co[lr]_wheel_center
Output from：_fsae_suspension_rear_axle

Communicator Matching Name：suspension_mount
Input Communicator Name：ci[lr]_suspension_mount
Located in：__MDI_SUSPENSION_TESTRIG
Output Communicator Name：co[lr]_suspension_mount
Output from：_fsae_suspension_rear_axle

Communicator Matching Name：driveline_active
Input Communicator Name：cis_driveline_active
Located in：__MDI_SUSPENSION_TESTRIG
Output Communicator Name：cos_driveline_active
Output from：_fsae_suspension_rear_axle

Communicator Matching Name：suspension_parameters_array
Input Communicator Name：cis_suspension_parameters_ARRAY
Located in：__MDI_SUSPENSION_TESTRIG
Output Communicator Name：cos_suspension_parameters_ARRAY

Output from: _fsae_suspension_rear_axle

Communicator Matching Name: tripot_to_differential
Input Communicator Name: ci[lr]_diff_tripot
Located in: __MDI_SUSPENSION_TESTRIG
Output Communicator Name: co[lr]_tripot_to_differential
Output from: _fsae_suspension_rear_axle

Communicator Matching Name: suspension_upright
Input Communicator Name: ci[lr]_suspension_upright
Located in: __MDI_SUSPENSION_TESTRIG
Output Communicator Name: co[lr]_suspension_upright
Output from: _fsae_suspension_rear_axle

!----------Unmatched input communicators: -----------! %以下为不匹配的输入通信器
Input Communicator Name: cis_suspension_to_chassis
Class: mount
From Minor Role: any
Matching Name（s）: suspension_to_chassis
In Template: _fsae_suspension_rear_axle

Input Communicator Name: ci[lr]_tierod_to_steering
Class: mount
From Minor Role: any
Matching Name（s）: tierod_to_steering
In Template: _fsae_suspension_rear_axle

Input Communicator Name: ci[lr]_jack_frame
Class: mount
From Minor Role: any
Matching Name（s）: jack_frame
In Template: __MDI_SUSPENSION_TESTRIG

Input Communicator Name: cis_leaf_adjustment_steps
Class: parameter_integer
From Minor Role: any
Matching Name（s）: leaf_adjustment_steps
In Template: __MDI_SUSPENSION_TESTRIG

Input Communicator Name：cis_powertrain_to_body
Class：mount
From Minor Role：any
Matching Name（s）：powertrain_to_body
In Template：__MDI_SUSPENSION_TESTRIG

Input Communicator Name：cis_steering_rack_joint
Class：joint_for_motion
From Minor Role：any
Matching Name（s）：steering_rack_joint
In Template：__MDI_SUSPENSION_TESTRIG

Input Communicator Name：cis_steering_wheel_joint
Class：joint_for_motion
From Minor Role：any
Matching Name（s）：steering_wheel_joint
In Template：__MDI_SUSPENSION_TESTRIG

!-----------Unmatched output communicators：------------!　％以下为不匹配的输出通信器
Output Communicator Name：cos_leaf_adjustment_multiplier
Class：array
To Minor Role：any
Matching Name（s）：leaf_adjustment_multiplier
In Template：__MDI_SUSPENSION_TESTRIG

Output Communicator Name：cos_characteristics_input_ARRAY
Class：array
To Minor Role：any
Matching Name（s）：characteristics_input_array
In Template：__MDI_SUSPENSION_TESTRIG

10.1.19　驱动轴显示组件

1．创建组件

（1）在模型树栏点击 Group 菜单，右击鼠标 New Group，弹出创建组件对话框，如图 10-21 所示。在下列对话框中输入相应的数据：

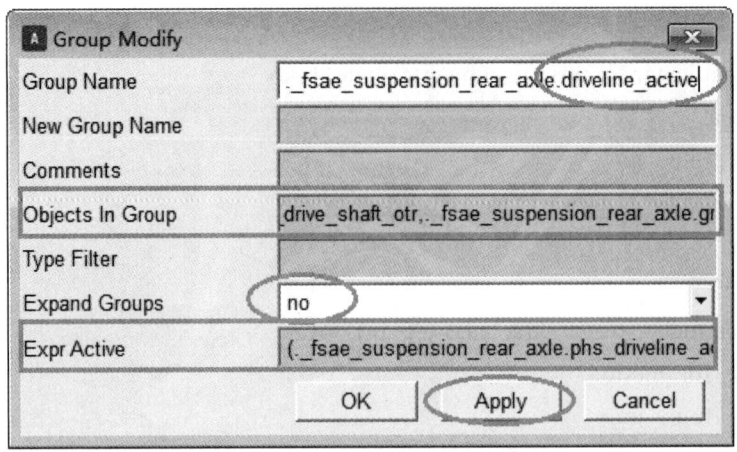

图 10-21 驱动轴显示组件对话框

- Group Name：driveline_active。
- Object In Group（显示组件包括的部件、几何体、约束等对象）：顺序输入 1~26 对象，如下信息所示。

① ._fsae_suspension_rear_axle.gel_drive_shaft；

② ._fsae_suspension_rear_axle.ger_drive_shaft；

③ ._fsae_suspension_rear_axle.gel_tripot；

④ ._fsae_suspension_rear_axle.ger_tripot；

⑤ ._fsae_suspension_rear_axle.ground.cfl_drive_shaft_otr；

⑥ ._fsae_suspension_rear_axle.ground.cfr_drive_shaft_otr；

⑦ ._fsae_suspension_rear_axle.ground.cfl_drive_shaft_inr；

⑧ ._fsae_suspension_rear_axle.ground.cfr_drive_shaft_inr；

⑨ ._fsae_suspension_rear_axle.mtl_tripot_to_differential；

⑩ ._fsae_suspension_rear_axle.mtr_tripot_to_differential；

⑪ ._fsae_suspension_rear_axle.jolcon_drive_sft_int_jt；

⑫ ._fsae_suspension_rear_axle.jorcon_drive_sft_int_jt；

⑬ ._fsae_suspension_rear_axle.jolcon_drive_sft_otr；

⑭ ._fsae_suspension_rear_axle.jorcon_drive_sft_otr；

⑮ ._fsae_suspension_rear_axle.joltra_tripot_to_differential；

⑯ ._fsae_suspension_rear_axle.jortra_tripot_to_differential；

⑰ ._fsae_suspension_rear_axle.gel_drive_shaft.gralin_drive_shaft；

⑱ ._fsae_suspension_rear_axle.gel_drive_shaft.graell_otr_cv_housing；

⑲ ._fsae_suspension_rear_axle.gel_drive_shaft.graell_tripot_housing；

⑳ ._fsae_suspension_rear_axle.gel_tripot.gracyl_tripot_housing_extention；

㉑ ._fsae_suspension_rear_axle.ger_drive_shaft.gralin_drive_shaft；

㉒ ._fsae_suspension_rear_axle.ger_drive_shaft.graell_otr_cv_housing；

㉓._fsae_suspension_rear_axle.ger_drive_shaft.graell_tripot_housing；

㉔._fsae_suspension_rear_axle.ger_tripot.gracyl_tripot_housing_extention；

㉕._fsae_suspension_rear_axle.mtl_fixed_2；

㉖._fsae_suspension_rear_axle.mtr_fixed_2。

- Expr Active：（._fsae_suspension_rear_axle.phs_driveline_active ||._fsae_suspension_rear_axle.model_class == "template" ? 1：0）。

（2）单击 Apply 按钮，完成组件 driveline_active 的创建。同理，创建其他组件。

- Group Name：driveline_inactive；
- Expr Active：（!._fsae_suspension_rear_axle.phs_driveline_active ||._fsae_suspension_rear_axle.model_class == "template" ? 1：0）。

（3）单击 OK 按钮，完成组件 driveline_inactive 的创建。

（4）单击 File > Save As 命令，弹出保存模板对话框，如图 10-22 所示。在下列对话框中输入相应的数据：

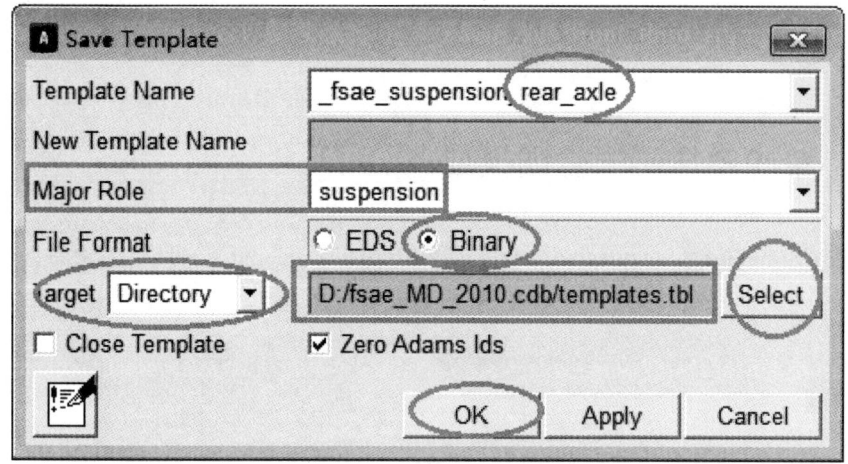

图 10-22　推杆式悬架模型保存

- Major Role（主特征）：suspension；
- File Format：Binary；
- Target：Directory。

（5）单击 Select 按钮，选择存储路径为：D：/fsae_MD_2010.cdb/templates.tbl。

（6）单击 OK 按钮，完成推杆式悬架模型模板._fsae_suspension_rear_axle 的保存。

2. 推杆式悬架子系统

（1）按 F9 键，将专家模板转换到标准模式，单击 File > New > Suspension 命令，弹出子系统对话框，如图 10-23 所示。在下列对话框中输入相应的数据：

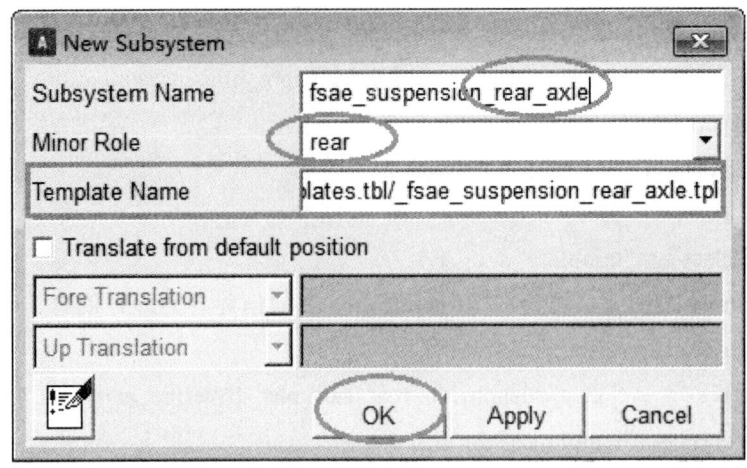

图 10-23 推杆式悬架子系统创建对话框

- Subsystem Name（系统名称）：fsae_suspension_rear_axle；
- Minor Role（副特征）：rear（悬架为后悬架）；
- Template Name（模板路径）：mdids：//FASE/templates.tbl/_fsae_suspension_rear_axle.tpl。

（2）单击 OK 按钮，完成推杆式悬架子系统 fsae_suspension_rear_axle 的创建。

10.2 单轮激振测试验证模型

（1）单击 Simulate > Suspension Analysis > Single Travel 命令，弹出单轮激振对话框，如图 10-24 所示。在下列对话框中输入相应的数据：

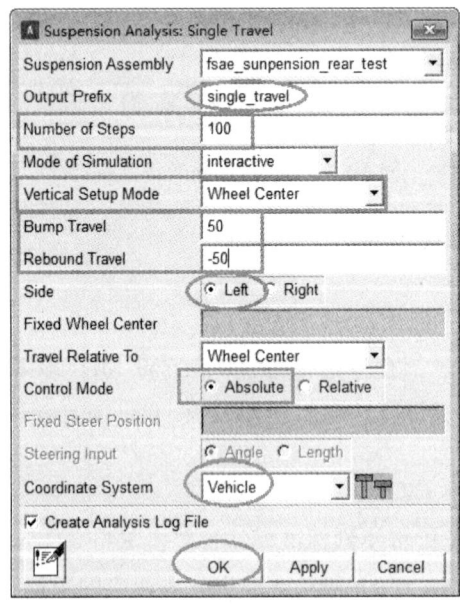

图 10-24 左单轮激振仿真设置

- Output Prefix：single travel；
- Number of Steps（仿真步数）：100；
- Mode of Simulation：interactive；
- Vertical Setup Mode：Wheel Center；
- Bump Travel：50；
- Rebound Travel：-50；
- Side：Left；
- Travel Relative To：Wheel Center；
- Control Mode：Absolute；
- Coordinate System：Vehicle。

（2）单击 Apply 按钮，完成推杆式双横臂悬架在 C 模式下的仿真。

（3）单击 Review > Animation Controls，开始观看动画，动画结束后悬架模型变化如图 10-25 所示。

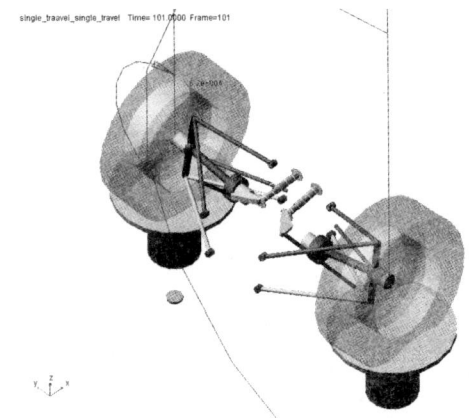

图 10-25　左单轮激振仿真动画

（4）按 F8 键，界面转换到后处理模块；点击 Independent Axis / date，弹出坐标轴参数索引对话框，如图 10-26 所示。在下列对话框中选择相应的数据：

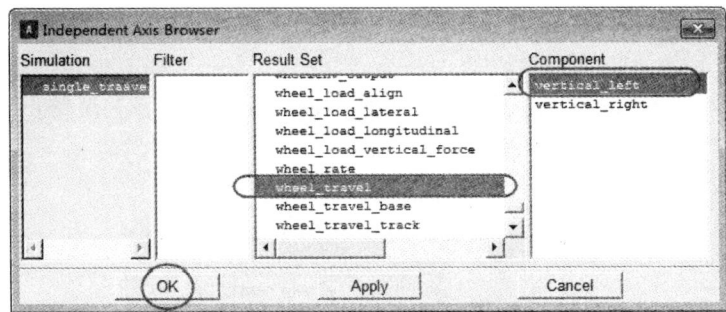

图 10-26　坐标轴参数索引对话框

- Result Set：wheel travel；
- Component：vertical left（左侧车轮垂向跳动的距离，即仿真设置的上下跳动 50 mm）。

(5)单击 OK 按钮,返回到后处理主界面。在下列对话框中输入相应的数据:
- Filter:user defined;
- Request:选择左前轮 toe_angle;
- Component:left/right。

(6)绘制其他参数曲线,如图 10-27~图 10-30 所示。

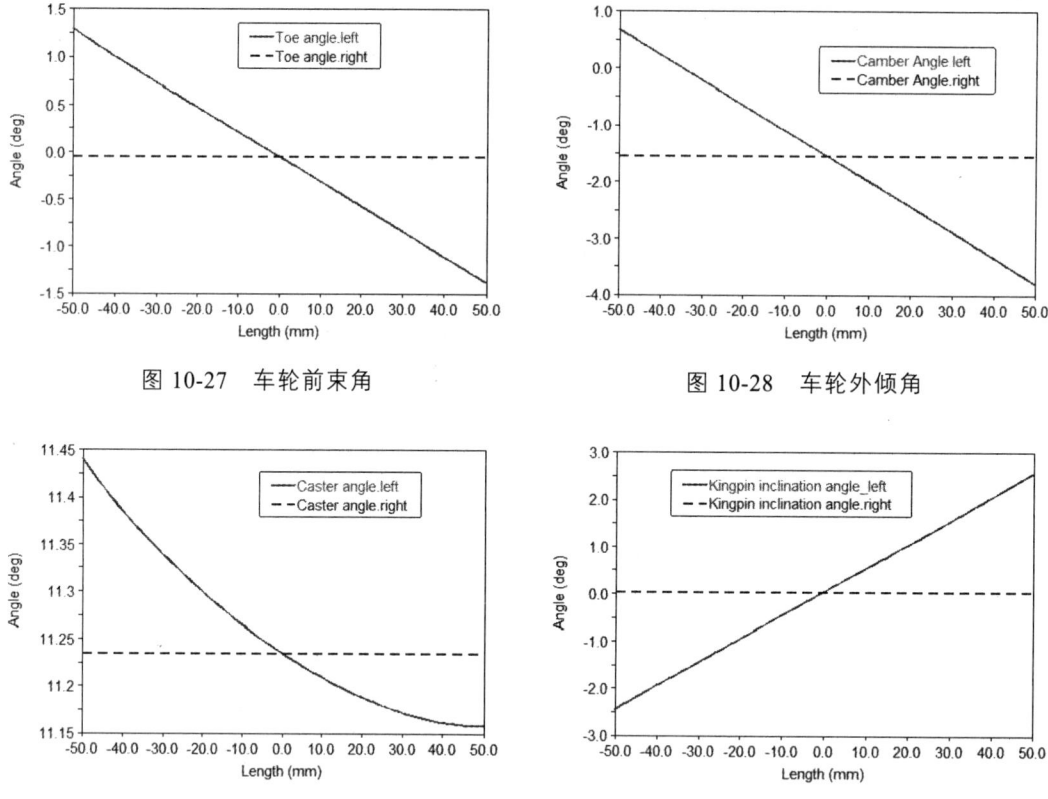

图 10-27　车轮前束角　　　　　　　　图 10-28　车轮外倾角

图 10-29　主销后倾角　　　　　　　　图 10-30　主销内倾角

10.3　悬架模型调试

(1)选择部件 bellcrank,右击 Modify,在部件 bellcrank 下选择三角臂:graarm_bellcrank,如图 10-31 所示;在 Thickness 中输入:10。

(2)单击 OK 按钮,完成三角臂厚度的修改。

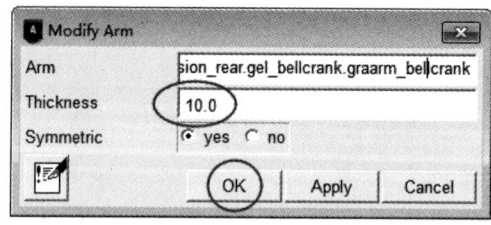

图 10-31　三角臂厚度修改对话框

（3）选择点 arblink_to_bellcrank，右击 Modify，弹出硬点修改对话框，如图 10-32 所示；Location 更改为：1447.8，-50.8，304.8。

（4）单击 OK 按钮，完成硬点修改，此时三角臂平面与 XY 平面保持平行。

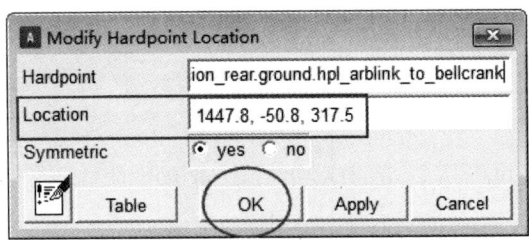

图 10-32　硬点修改对话框

10.4　车轮同向跳动仿真

（1）单击 Simulate > Suspension Analysis > Parallel Wheel Travel 命令，弹出车轮同向激振对话框，如图 10-33 所示。在下列对话框中输入相应的数据：

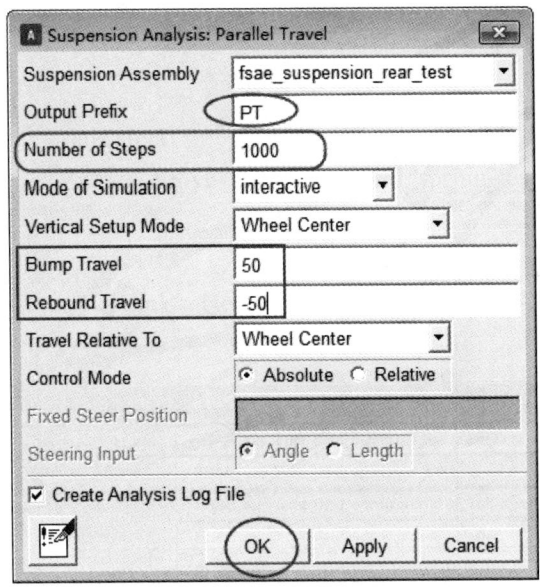

图 10-33　车轮同向跳动激振仿真

- Output Prefix：PT；
- Number of Steps（仿真步数）：1000；
- Mode of Simulation：interactive；
- Vertical Setup Mode：Wheel Center；
- Bump Travel：50；
- Rebound Travel：-50；
- Travel Relative To：Wheel Center；

- Control Mode：Absolute。

（2）单击 OK 按钮，完成推杆悬架设置并提交运算。

（3）计算完成后，按 F8 键切换到后处理模块；绘制其他参数曲线，如图 10-34～图 10-38 所示。

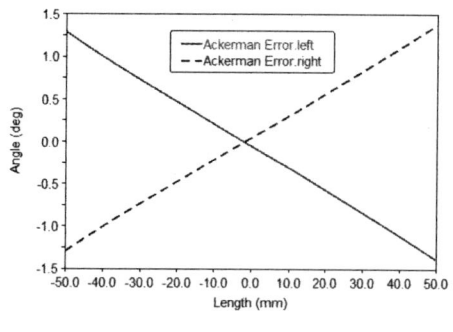

图 10-34　阿克曼角误差

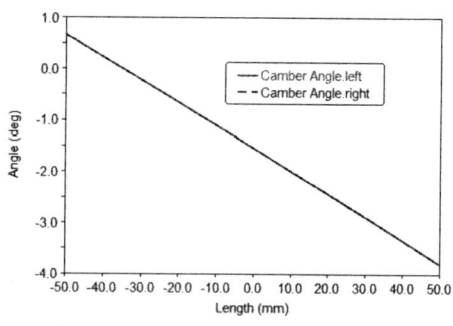

图 10-35　车轮外倾角

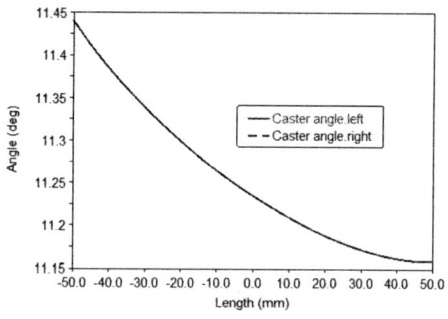

图 10-36　主销后倾角

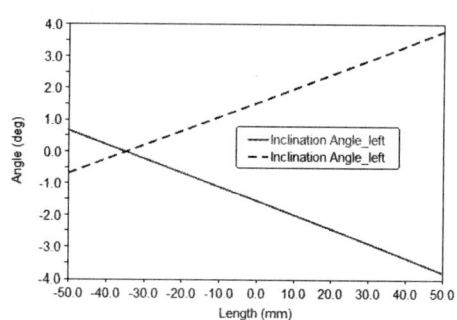

图 10-37　主销内倾角

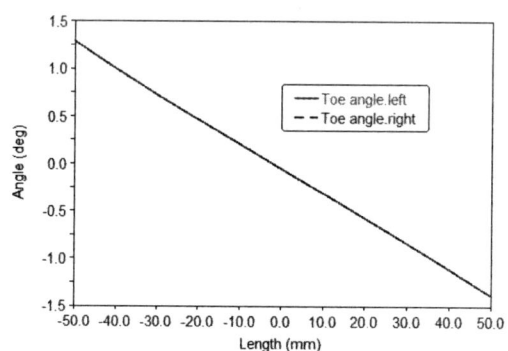

图 10-38　车轮前束角

第 11 章 避震器纵置式前推杆悬架

FSAE 赛车前推力杆式双横臂悬架模型与后推力杆式悬架模型构架相似，区别在于前悬架硬点参数不同，前束角与外倾角不同，相对后悬架设置参数较大，以保证转向后 FSAE 整车方向快速回正，同时便于提升轮胎与地面的侧向接触力，这也是较多赛车改装车轮为"外八字"形的原因。

本章不再重复悬架的建模过程与步骤，在此仅提供悬架的硬点参数；部件特性、约束关系、通信器、结构框参数与后推力杆式悬架一样，读者可参考上述后推力杆悬架建模过程自行建立，或者修改后推杆悬架模型的硬点信息完成前推杆悬架模型的创建。已经建立好的前推力杆式双横臂悬架_FSAE_sus_front.tpl 存放在章节文件夹中，读者可自行查阅。建立好的纵置式前推杆悬架模型如图 11-1 所示。

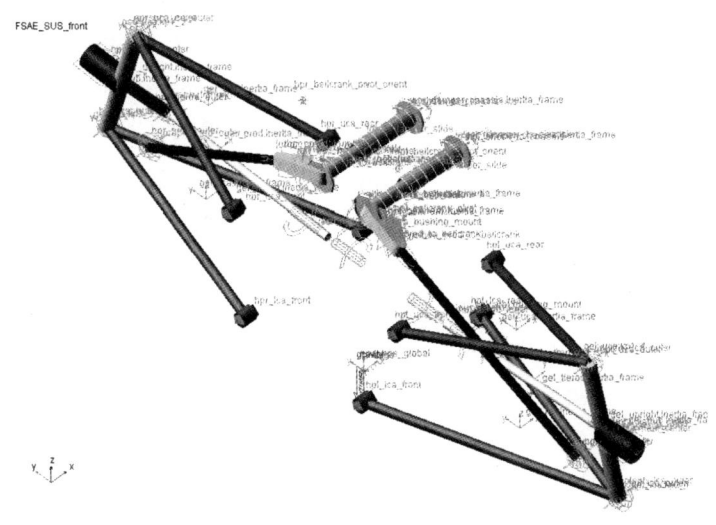

图 11-1 前推力杆式双横臂悬架模型

学习目标

（1）会硬点信息调试。
（2）会 FSAE 整车装配。
（3）会匀速直线仿真。
（4）进行稳定性参数对比。

11.1 硬点信息

纵置式前推杆悬架模板特征及硬点信息如下:

```
Info for subsystem:  FSAE_SUS_front

  File Name      : <FASE>/subsystems.tbl/FSAE_SUS_front.sub
  Template       : mdids://FASE/templates.tbl/_FSAE_sus_front.tpl
  Comments       : *no comments found*
  Major Role     : suspension
  Minor Role     : front

HARDPOINTS:      %前推理杆式双横臂悬架硬点参数
```

hardpoint name	symmetry	x_value	y_value	z_value
hps_global	single	0.0	0.0	0.0
arblink_to_bellcrank	left/right	-38.1	-50.8	393.7
arb_bushing_mount	left/right	127.0	-127.0	101.6
bellcrank_pivot	left/right	-25.4	-88.9	381.0
bellcrank_pivot_orient	left/right	-25.4	-101.6	508.0
lca_front	left/right	-127.0	-127.0	114.3
lca_outer	left/right	0.0	-546.1	120.65
lca_rear	left/right	127.0	-127.0	114.3
prod_outer	left/right	0.0	-457.2	139.7
prod_to_bellcrank	left/right	-50.8	-127.0	381.0
shock_to_bellcrank	left/right	-38.1	-50.8	393.7
shock_to_chassis	left/right	152.4	-50.8	381.0
tierod_inner	left/right	50.8	-127.0	152.4
tierod_outer	left/right	63.5	-482.6	152.4
uca_front	left/right	-101.6	-177.8	279.4
uca_outer	left/right	0.0	-482.6	355.6
uca_rear	left/right	101.6	-177.8	279.4
wheel_center	left/right	0.0	-558.8	241.3

11.2 FSAE 整车模型装配

（1）按 F9 键切换到标准模板，单击 File > Full-Vehicle Assembly 命令，弹出整车装配对话框，如图 11-2 所示。在下列对话框中输入相应的数据：

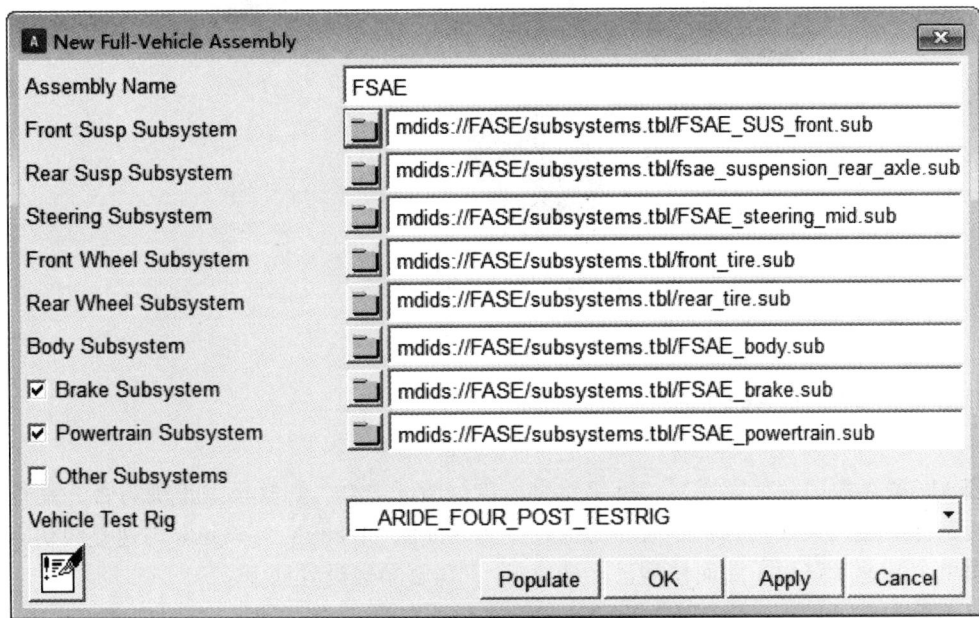

图 11-2　FSAE 整车装配对话框

- Assembly Name：FSAE；
- Front Susp Subsystem：mdids：//FASE/subsystems.tbl/FSAE_SUS_front.sub；
- Rear Susp Subsystem：mdids：//FASE/subsystems.tbl/fsae_suspension_rear_axle.sub；
- Steering Subsystem：mdids：//FASE/subsystems.tbl/FSAE_steering_mid.sub；
- Front Wheel Subsystem：mdids：//FASE/subsystems.tbl/front_tire.sub；
- Rear Wheel Subsystem：mdids：//FASE/subsystems.tbl/rear_tire.sub；
- Body Subsystem：mdids：//FASE/subsystems.tbl/FSAE_body.sub；
- 勾选 Brake Subsystem：mdids：//FASE/subsystems.tbl/FSAE_brake.sub；
- 勾选 Powertrain Subsystem：mdids：//FASE/subsystems.tbl/FSAE_powertrain.sub；
- Vehicle Test Rig（整车实验台架）：_MDI_SDI_TESTRIG。

（2）单击 OK 按钮，完成整车模型装配，装配好的模型如图 11-3 所示。

图 11-3　FSAE 整车模型

11.3　匀速直线仿真

FSAE 赛车模型装配完成后，可以对整车进行验证仿真，观察模型正确与否，是否能够根据实验标准达到预期状态；分别对模型进行 20 km/h、30 km/h、40 km/h 匀速直线仿真，路面采用单个减速带路面（路面模型已经建立好），仿真参数设置如下：

（1）单击 Simulate > Full-Vehicle Analysis > Straight-line Events > Maintain 命令，弹出直线仿真对话框，如图 11-4 所示。在下列对话框中输入相应的数据：

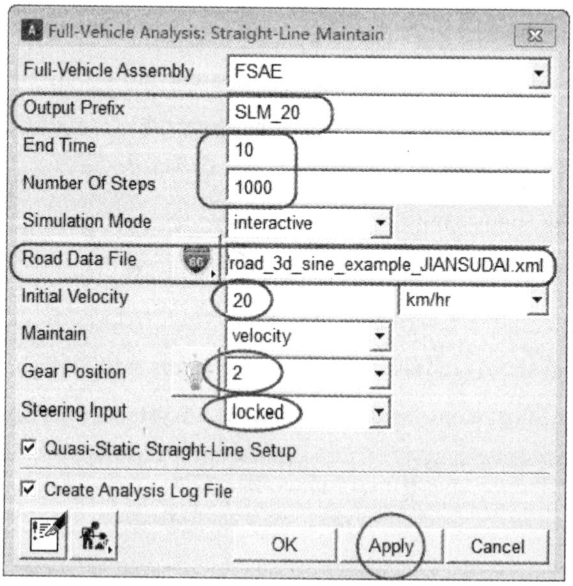

图 11-4　匀速直线仿真

- Output Prefix：SLM_20；
- End Time：10；
- Number of steps：1000；
- Mode of Simulation：interactive；
- Road Date File：mdids：//FASE/roads.tbl/road_3d_sine_example_JIANSUDAI.xml，单条减速带路面如图 11-5 所示；

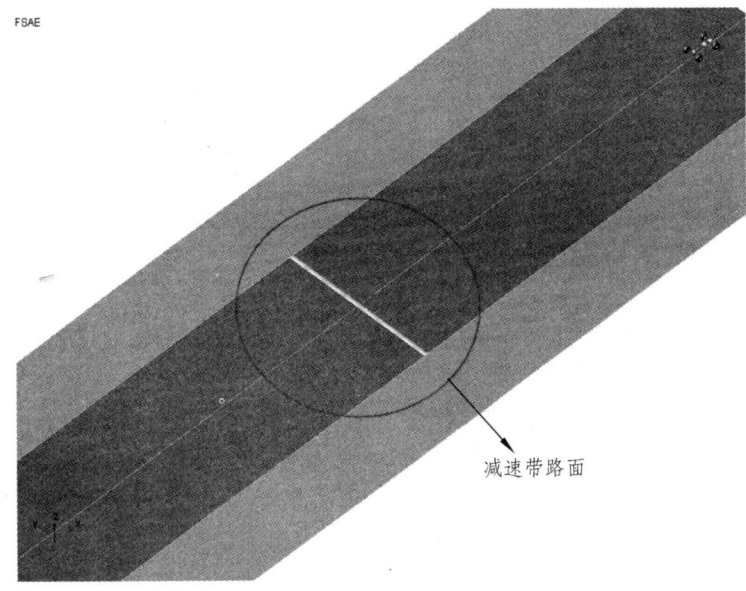

图 11-5　减速带路面

- Initial Velocity：20，单位 km/h；
- First Turn Direction：right；
- Maintain：velocity；
- Gear Position：2；
- Steering Input：locked；
- 勾选 Quasi-Static Straight-Line Setup。

（2）单击 Apply 按钮，完成 20 km/h 的蛇形绕桩仿真设置并提交运算。同理，进行其他仿真。

- Output Prefix：SLM_30；
- End Time：10；
- Number of steps：1000；
- Mode of Simulation：interactive；
- Road Date File：mdids：//FASE/roads.tbl/road_3d_sine_example_JIANSUDAI.xml；
- Initial Velocity：30，单位 km/h；
- First Turn Direction：right；
- Maintain：velocity；

- Gear Position：3；
- Steering Input：locked；
- 勾选 Quasi-Static Straight-Line Setup。

（3）单击 Apply 按钮，完成 30 km/h 的蛇形绕桩仿真设置并提交运算。

- Output Prefix：SLM_40；
- End Time：10；
- Number of steps：1000；
- Mode of Simulation：interactive；
- Road Date File：mdids：//FASE/roads.tbl/road_3d_sine_example_JIANSUDAI.xml；
- Initial Velocity：40，单位 km/h；
- First Turn Direction：right；
- Maintain：velocity；
- Gear Position：4；
- Steering Input：locked；
- 勾选 Quasi-Static Straight-Line Setup。

（4）单击 OK 按钮，完成 40 km/h 的蛇形绕桩仿真设置并提交运算。仿真结束后，整车稳定性的相关参数如图 11-6~图 11-9 所示。

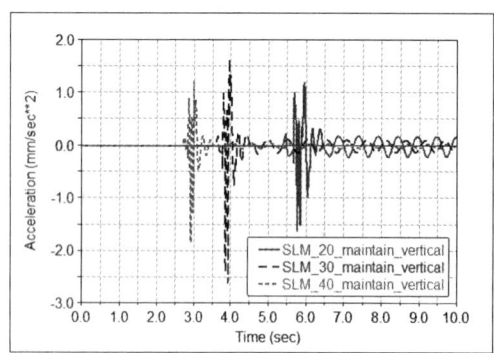

图 11-6　车身垂向加速度

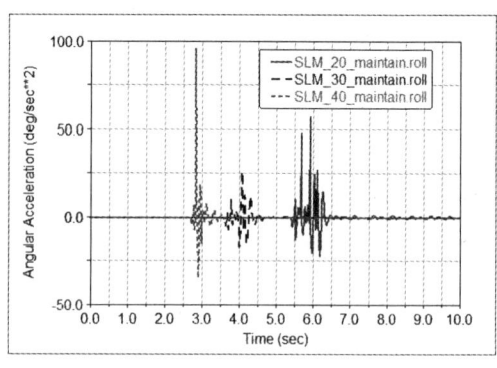

图 11-7　车身侧倾角加速度

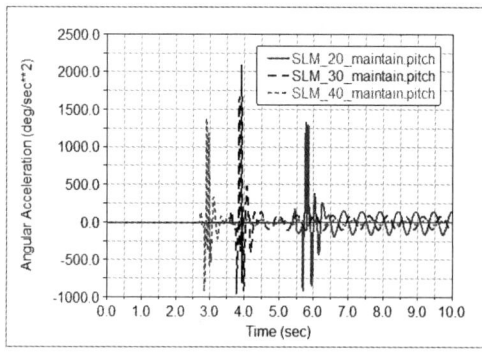

图 11-8　车身俯仰角加速度

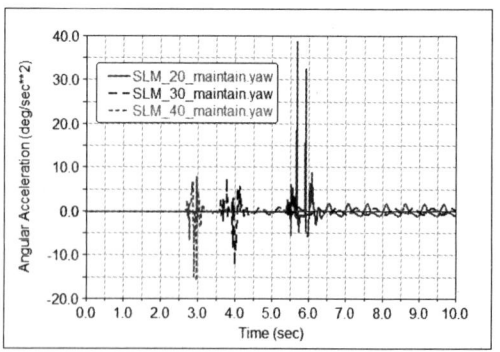

图 11-9　车身横摆角加速度

第 12 章 避震器纵置式推杆悬架

推杆悬架具有多种布置形式，不同的布置形式需要不同的参数对应匹配。推杆悬架将避震器及附属部件放置在车身上，减轻了非簧载质量，相对于家用轿车采用的悬架，其可调节参数较多，相对技术含量较高。采用推杆悬架后，如果连接推杆与弹簧之间的旋转支架摆臂较短，则需要采用较大刚度的弹簧与其匹配；反之如果旋转支架摆臂长，则匹配的弹簧刚度小，但此时摆臂震动空间范围大，针对此问题，在避震器纵置式推杆悬架的基础上通过结构改进，增加辅助弹簧与减震器，改善摆臂震动特性，同时提升赛车在起步加速时或制动时的"抬头"与"点头"现象。本章直接在模型_FSAE_sus_front.tpl 上添加改进辅助弹簧与避震器机构，完成后的前推力杆式双横臂悬架：_FSAE_sus_front_third_spring.tpl 存放在章节文件夹中，读者可自行查阅。纵置式推杆悬架模型如图 12-1 所示。

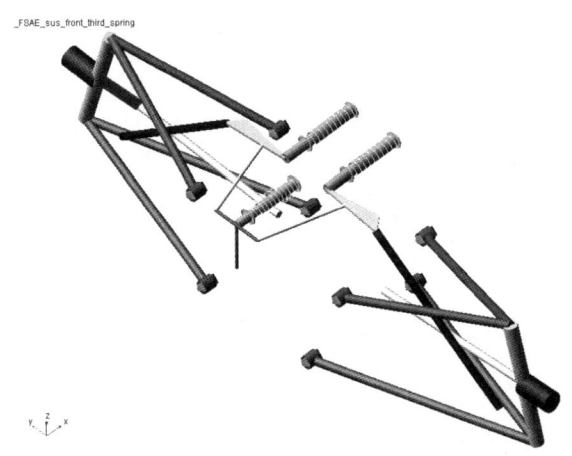

图 12-1 纵置式推杆悬架

学习目标

（1）了解纵置式推杆悬架模型。
（2）熟悉辅助弹簧与避震器结构。
（3）会漂移仿真。
（4）会稳定性能指标对比。

12.1 纵置式前推杆悬架模型

1. 模型导入

(1) 启动 ADAMS/CAR，选择专家模块进入建模界面。

(2) 单击 File > Open 命令，弹出打开模板对话框，如图 12-2 所示。

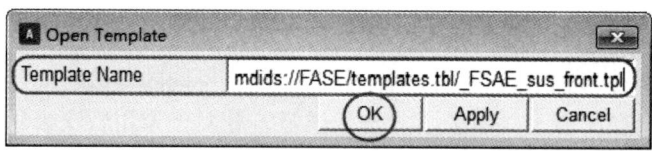

图 12-2 模板对话框

(3) 在 Template Name 中输入：mdids：//FASE/templates.tbl/_FSAE_sus_front.tpl，单击 OK 按钮，完成模型导入。

2. 调试、添加硬点

(1) 单击 Build > Hardpoind > Modify 命令，在下列对话框中输入相应的数据：
- Hardpoint：prod_to_bellcrank；
- Location：-20.8，-180.0，381.0。

(2) 单击 Apply 按钮，完成 prod_to_bellcrank 硬点的修改。

重复上述硬点修改步骤，完成图 12-3 中标注线内的硬点参数修改及添加。

	loc x	loc y	loc z
hpl_arb_bushing_mount	127.0	-127.0	101.6
hpl_arblink_to_bellcrank	-10.0	-50.8	381.0
hpl_bellcrank_pivot	5.0	-160.0	381.0
hpl_bellcrank_pivot_orient	-25.4	-101.6	508.0
hpl_lca_front	-127.0	-127.0	114.3
hpl_lca_outer	0.0	-546.1	120.65
hpl_lca_rear	127.0	-127.0	114.3
hpl_link	-180.0	-50.8	380.0
hpl_prod_outer	0.0	-457.2	139.7
hpl_prod_to_bellcrank	-20.8	-180.0	381.0
hpl_shock_to_bellcrank	-10.0	-50.8	381.0
hpl_shock_to_chassis	152.4	-50.8	381.0
hpl_tierod_inner	50.8	-127.0	152.4
hpl_tierod_outer	63.5	-482.6	152.4
hpl_uca_front	-101.6	-177.8	279.4
hpl_uca_outer	0.0	-482.6	355.6
hpl_uca_rear	101.6	-177.8	279.4
hpl_wheel_center	0.0	-558.8	241.3
hps_hps_global	0.0	0.0	0.0

图 12-3 硬 点

3. 约束调试

（1）单击 Build > Attachments > Joint >Modify 命令，弹出创建约束修改对话框，如图 12-4 所示，在下列对话框中输入相应的数据：

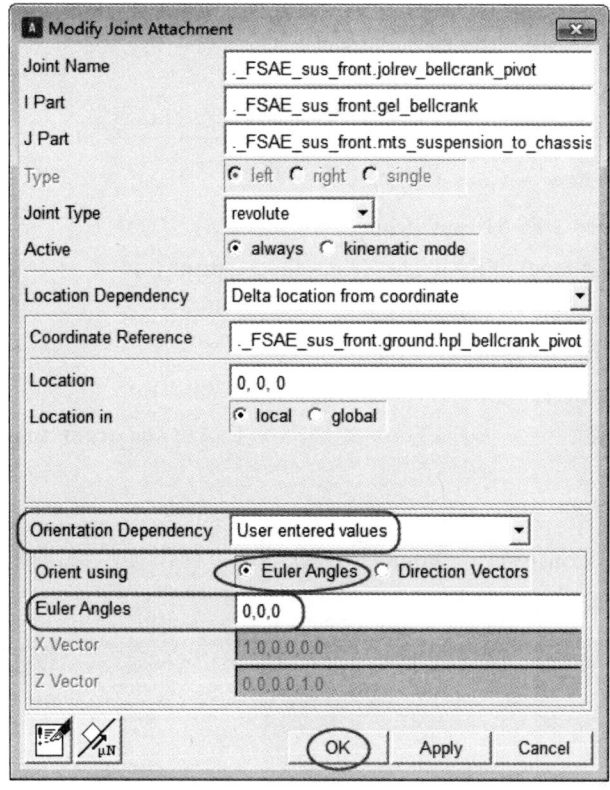

图 12-4　约束修改对话框

- Joint Name（约束副名称）：._FSAE_sus_front.jolrev_bellcrank_pivot；
- I Part：._FSAE_sus_front.gel_bellcrank；
- J Part：._FSAE_sus_front.mts_suspension_to_chassis；
- Joint Type（约束副类型）：revolute；
- Active（激活）：always；
- Location Dependency：Delta location from coordinate；
- Coordinate Reference（参考坐标）：._FSAE_sus_front.ground.hpl_bellcrank_pivot；
- Location：0，0，0；
- Location in：local；
- Orientation Dependency：User entered values；
- Orient using：Euler Angles；
- Euler Angles：0，0，0。

（2）单击 OK 按钮，完成约束副._FSAE_sus_front.jolrev_bellcrank_pivot 的修改。

（3）单击 File > Save As 命令，弹出保存模板对话框，如图 12-5 所示，在下列对话框中输入相应的数据：

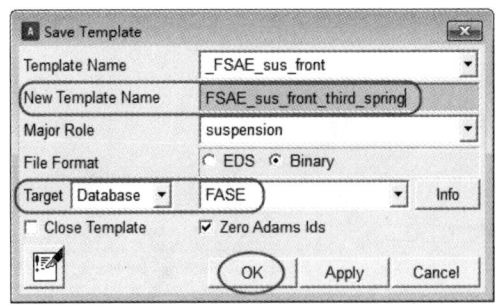

图 12-5　模型保存对话框

- Template Name：_FSAE_sus_front；
- New Template Name：FSAE_sus_front_third_spring；
- Major Role（主特征）：suspension；
- File Format：Binary；
- Target：Datebase/FASE。

（4）单击 OK 按钮，完成推杆式悬架模型模板 FSAE_sus_front_third_spring 的保存。

4. 结构框

（1）单击 Build > Construction Frame > New 命令，弹出创建结构框对话框，如图 12-6 所示，在下列对话框中输入相应的数据：

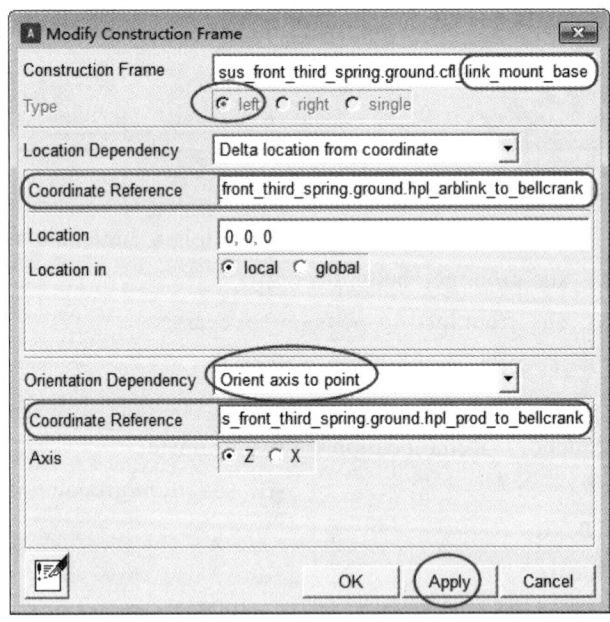

图 12-6　结构框对话框

- Construction Frame（结构框名称）：link_mount_base；
- Location Dependency：Delta location from coordinate；
- Coordinate Reference（参考坐标）：._FSAE_sus_front_third_spring.ground.hpl_arblink_to_bellcrank；

- Location：0，0，0；
- Location in：local；
- Orientation Dependency：Orient axis to point；
- Coordinate Reference（参考坐标）：._FSAE_sus_front_third_spring.ground.hpl_prod_to_bellcrank；
- Axis：Z。

（2）单击 Apply 按钮，完成._FSAE_sus_front_third_spring.ground.cfl_link_mount_base 结构框的创建。用同样的方法创建其他结构框。

- Construction Frame（结构框名称）：link_mount；
- Location Dependency：Delta location from coordinate；
- Coordinate Reference（参考坐标）：._FSAE_sus_front_third_spring.ground.cfl_link_mount_base；
- Location：0，0，0；
- Location in：local；
- Orientation Dependency：Orient axis to point；
- Coordinate Reference（参考坐标）：._FSAE_sus_front_third_spring.ground.hpl_prod_to_bellcrank；
- Axis：Z。

（3）单击 Apply 按钮，完成._FSAE_sus_front_third_spring.ground.cfl_link_mount 结构框的创建。

- Construction Frame（结构框名称）：link_center；
- Location Dependency：Centered between coordinates；
- Centered between：Two Coordinates；
- Coordinate Reference #1（参考坐标）：._FSAE_sus_front_third_spring.ground.hpl_link；
- Coordinate Reference #2（参考坐标）：._FSAE_sus_front_third_spring.ground.hpr_link；
- Orientation Dependency：User entered values；
- Orient using：Euler Angles；
- Euler Angles：0，0，0。

（4）单击 Apply 按钮，完成._FSAE_sus_front_third_spring.ground.cfs_link_center 结构框的创建。

- Construction Frame（结构框名称）：link_down；
- Location Dependency：Delta location from coordinate；
- Coordinate Reference（参考坐标）：._FSAE_sus_front_third_spring.ground.cfs_link_center；
- Location：0，0，-100；
- Location in：local；
- Orientation Dependency：User entered values；
- Orient using：Euler Angles；
- Euler Angles：0，0，0。

（5）单击 Apply 按钮，完成._FSAE_sus_front_third_spring.ground.cfs_link_down 结构框的创建。

- Construction Frame（结构框名称）：spring_mount；
- Location Dependency：Delta location from coordinate；
- Coordinate Reference（参考坐标）：._FSAE_sus_front_third_spring.ground.cfs_link_center；
- Location：30，0，0；
- Location in：local；
- Orientation Dependency：User entered values；
- Orient using：Euler Angles；
- Euler Angles：0，0，0。

（6）单击 OK 按钮，完成._FSAE_sus_front_third_spring.ground.cfs_spring_mount 结构框的创建。

5. 部件 link

（1）单击 Build > Part > General Part > New 命令，弹出创建部件对话框，如图 12-7 所示，在下列对话框中输入相应的数据：

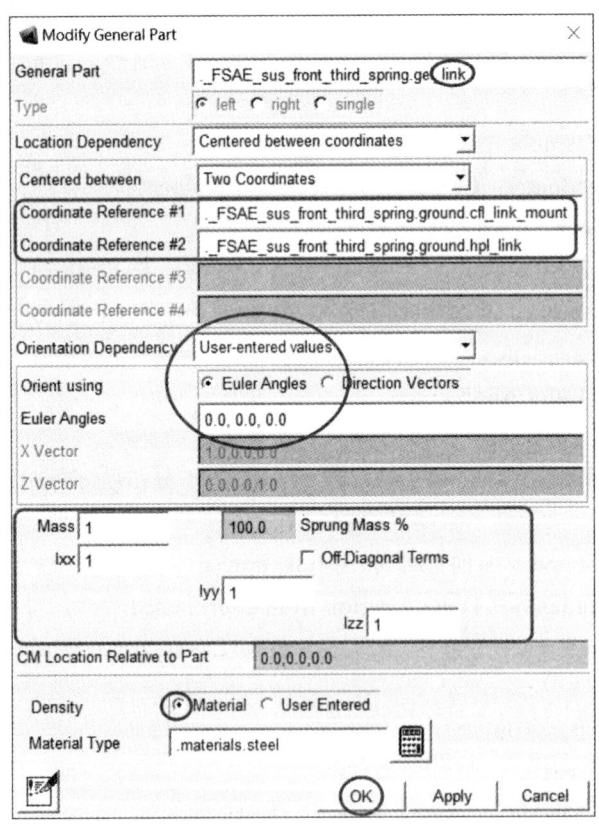

图 12-7 部件_link 对话框

- General Part：link；
- Type：left；
- Location Dependency：Centered between coordinates；
- Centered between：Two Coordinates；
- Coordinate Reference #1（参考坐标）：._FSAE_sus_front_third_spring.ground.cfl_link_mount；
- Coordinate Reference #2（参考坐标）：._FSAE_sus_front_third_spring.ground.hpl_link；
- Orientation Dependency：User-entered values；
- Orient using：Euler Angles；
- Euler Angles：0，0，0；
- Mass：1；
- Ixx：1；
- Iyy：1；
- Izz：1；
- Density：Material；
- Material Type：.materials.steel；

（2）单击 Apply 按钮，完成部件._FSAE_sus_front_third_spring.gel_link 的创建。

6. 部件 link_front

（1）单击 Build > Part > General Part > New 命令，在下列对话框中输入相应的数据：
- General Part：link_front；
- Type：single；
- Location Dependency：Delta location from coordinate；
- Coordinate Reference（参考坐标）：._FSAE_sus_front_third_spring.ground.cfs_link_center；
- Location：0，0，0；
- Location in：local；
- Orientation Dependency：User-entered values；
- Orient using：Euler Angles；
- Euler Angles：0，0，0；
- Mass：1；
- Ixx：1；
- Iyy：1；
- Izz：1；
- Density：Material；
- Material Type：.materials.steel。

（2）单击 Apply 按钮，完成部件._FSAE_sus_front_third_spring.ges_link_front 的创建。

7. 部件 link_down

（1）单击 Build > Part > General Part > New 命令，在下列对话框中输入相应的数据：
- General Part：link_down；
- Type：single；
- Location Dependency：Centered between coordinates；
- Centered between：Two Coordinates；
- Coordinate Reference #1（参考坐标）：._FSAE_sus_front_third_spring.ground.cfs_link_center；
- Coordinate Reference #2（参考坐标）：._FSAE_sus_front_third_spring.ground.cfs_link_down；
- Orientation Dependency：User-entered values；
- Orient using：Euler Angles；
- Euler Angles：0，0，0；
- Mass：1；
- Ixx：1；
- Iyy：1；
- Izz：1；
- Density：Material；
- Material Type：.materials.steel。

（2）单击 Apply 按钮，完成部件._FSAE_sus_front_third_spring.ges_link_down 的创建。

8. 部件 damper_up

（1）单击 Build > Part > General Part > New 命令，在下列对话框中输入相应的数据：
- General Part：damper_up；
- Type：single；
- Location Dependency：Delta location from coordinate；
- Coordinate Reference（参考坐标）：._FSAE_sus_front_third_spring.ground.cfs_damper_up；
- Location：0，0，0；
- Location in：local；
- Orientation Dependency：User-entered values；
- Orient using：Euler Angles；
- Euler Angles：0，0，0；
- Mass：1；
- Ixx：1；
- Iyy：1；
- Izz：1；
- Density：Material；
- Material Type：.materials.steel。

（2）单击 Apply 按钮，完成部件 ._FSAE_sus_front_third_spring.ges_damper_up 的创建。

9. 部件 damper_down

（1）单击 Build > Part > General Part > New 命令，在下列对话框中输入相应的数据：
- General Part：damper_down；
- Type：single；
- Location Dependency：Delta location from coordinate；
- Coordinate Reference（参考坐标）：._FSAE_sus_front_third_spring.ground.cfs_link_center；
- Location：0，0，0；
- Location in：local；
- Orientation Dependency：User-entered values；
- Orient using：Euler Angles；
- Euler Angles：0，0，0；
- Mass：1；
- Ixx：1；
- Iyy：1；
- Izz：1；
- Density：Material；
- Material Type：.materials.steel。

（2）单击 OK 按钮，完成部件 ._FSAE_sus_front_third_spring.ges_damper_down 的创建。

10. 几何体 link

（1）单击 Build > Geometry > Link > New 命令，弹出创建几何体对话框，如图 12-8 所示，在下列对话框中输入相应的数据：

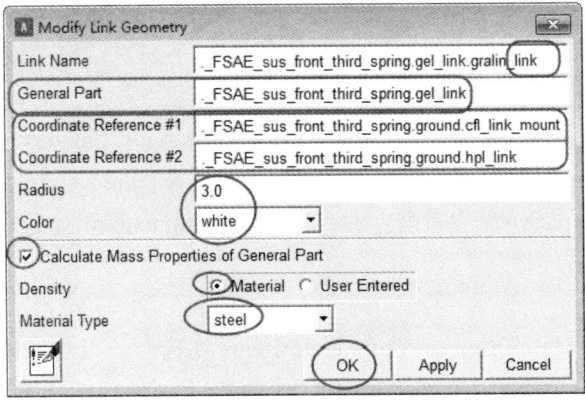

图 12-8 连杆几何体_link

- Link Name（连杆名称）：link；
- General Part 输入：._FSAE_sus_front_third_spring.gel_link；

- Coordinate Reference #1（参考坐标）：._FSAE_sus_front_third_spring.ground.cfl_link_mount；
- Coordinate Reference #2（参考坐标）：._FSAE_sus_front_third_spring.ground.hpl_link；
- Radius（半径）：3；
- Color：white；
- 勾选 Calculate Mass Properties of General Part 复选框，当几何体建立好之后，会更新对应部件的质量和惯量参数；
- Density：Material；
- Material Type：steel。

（2）单击 Apply 按钮，完成._FSAE_sus_front_third_spring.gel_link.gralin_link 几何体的创建。

11. 几何体 link_front

（1）单击 Build > Geometry > Link > New 命令，在下列对话框中输入相应的数据：
- Link Name（连杆名称）：link_front；
- General Part 输入：._FSAE_sus_front_third_spring.ges_link_front；
- Coordinate Reference #1（参考坐标）：._FSAE_sus_front_third_spring.ground.hpl_link；
- Coordinate Reference #2（参考坐标）：._FSAE_sus_front_third_spring.ground.hpr_link；
- Radius（半径）：3；
- Color：red；
- 勾选 Calculate Mass Properties of General Part 复选框，当几何体建立好之后，会更新对应部件的质量和惯量参数；
- Density：Material；
- Material Type：steel。

（2）单击 Apply 按钮，完成._FSAE_sus_front_third_spring.ges_link_front.gralin_link_front 几何体的创建。

12. 几何体 link_down

（1）单击 Build > Geometry > Link > New 命令，在下列对话框中输入相应的数据：
- Link Name（连杆名称）：link_down；
- General Part 输入：._FSAE_sus_front_third_spring.ges_link_down；
- Coordinate Reference #1（参考坐标）：._FSAE_sus_front_third_spring.ground.cfs_link_center；
- Coordinate Reference #2（参考坐标）：._FSAE_sus_front_third_spring.ground.cfs_link_down；
- Radius（半径）：3；
- Color：green；
- 勾选 Calculate Mass Properties of General Part 复选框，当几何体建立好之后，会更新对应部件的质量和惯量参数；
- Density：Material；
- Material Type：steel。

（2）单击 OK 按钮，完成._FSAE_sus_front_third_spring.ges_link_down.gralin_link_down 几何体的创建。

13. 弹　簧

（1）单击 Build > Force > Spring > New 命令，弹出创建部件对话框，如图 12-9 所示。在下列对话框中输入相应的数据：

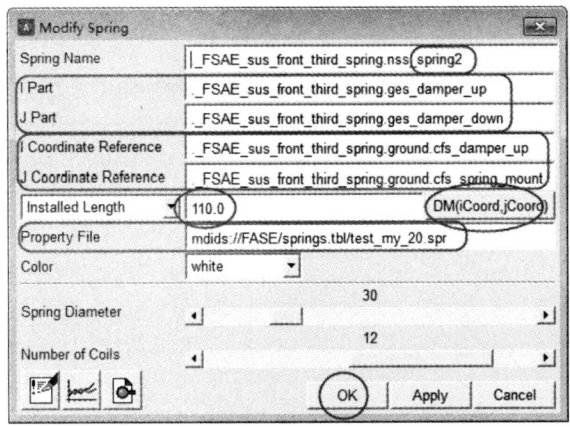

图 12-9　spring2 弹簧创建对话框

- Spring Name（减震器名称）：spring 2；
- I Part：._FSAE_sus_front_third_spring.ges_damper_up；
- J Part：._FSAE_sus_front_third_spring.ges_damper_down；
- I Coordinate Reference（参考坐标）：._FSAE_sus_front_third_spring.ground.cfs_damper_up；
- J Coordinate Reference（参考坐标）：._FSAE_sus_front_third_spring.ground.cfs_spring_mount；
- Installed Length（安装长度）：单击 DM（iCoord，jCoord），自动计算弹簧的安装长度并填入到方框中，此模型的安装长度为 110，弹簧刚度曲线如图 12-10 所示；

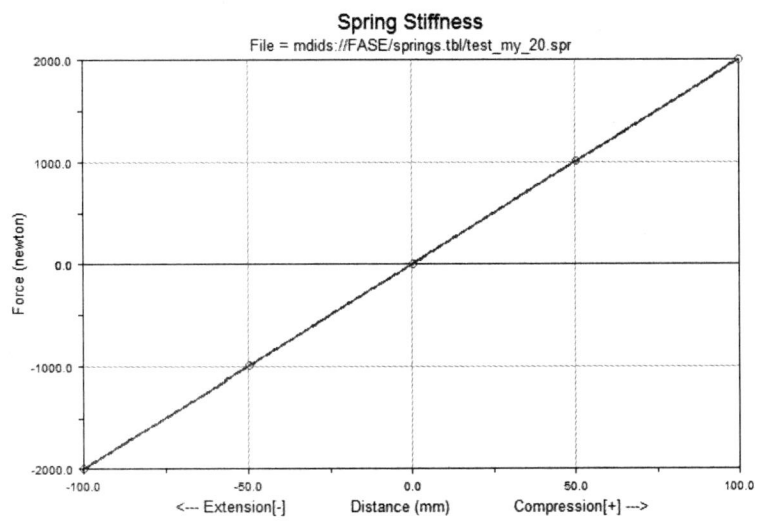

图 12-10　弹簧刚度曲线

- Property File（属性文件）：mdids：//FASE/springs.tbl/test_my_20.spr，弹簧属性文件用记事本文件打开后如下列信息所示，可以根据实验情况测出弹簧的参数（力与位移之间的关系，即刚度）并填写到如下信息列表中，可修改部分为下划线部分；
  - Spring Diameter（弹簧直径）：拖动滑块选择 30 mm；
  - Spring of Coils（弹簧圈数）：拖动滑块选择 12。

（2）单击 OK 按钮，完成弹簧._FSAE_sus_front_third_spring.nss_spring2 的创建。

```
弹簧刚度信息：
$--------------------------------------------------------------MDI_HEADER
[MDI_HEADER]
 FILE_TYPE       =  'spr'
 FILE_VERSION    =   4.0
 FILE_FORMAT     =  'ASCII'
$--------------------------------------------------------------UNITS
[UNITS]
 LENGTH   =  'mm'
 ANGLE    =  'degrees'
 FORCE    =  'newton'
 MASS     =  'kg'
 TIME     =  'second'
$--------------------------------------------------------------SPRING_DATA
[SPRING_DATA]
 FREE_LENGTH  =  205.7
$--------------------------------------------------------------CURVE
[CURVE]
{   disp        force}
 -100.0       -2000.0
  -50.0       -1000.0
    0.0           0.0
   50.0        1000.0
  100.0        2000.0
```

14. 避震器

（1）单击 Build > Force > Damper > New 命令，弹出避震器创建对话框，如图 12-11 所示，在下列对话框中输入相应的数据：

- Damper Name（减震器名称）：damper2；
- I Part：._FSAE_sus_front_third_spring.ges_damper_up；
- J Part：._FSAE_sus_front_third_spring.ges_damper_down；
- I Coordinate Reference（参考坐标）：._FSAE_sus_front_third_spring.ground.cfs_damper_up；

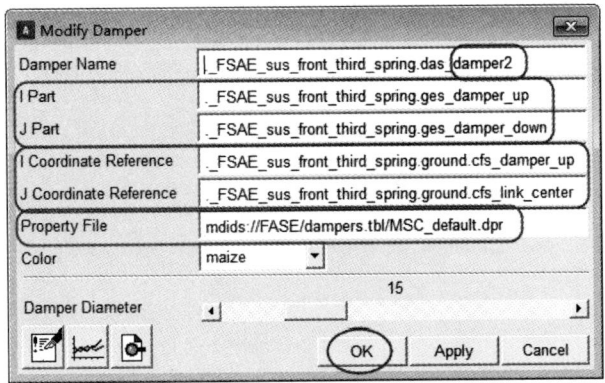

图 12-11 damper 避震器创建对话框

- J Coordinate Reference（参考坐标）：._FSAE_sus_front_third_spring.ground.cfs_link_center；
- Property File（属性文件）：mdids：//FASE/dampers.tbl/MDI_default.dpr，避震器系数曲线如图 12-12 所示，具体数据如下列避震器信息；

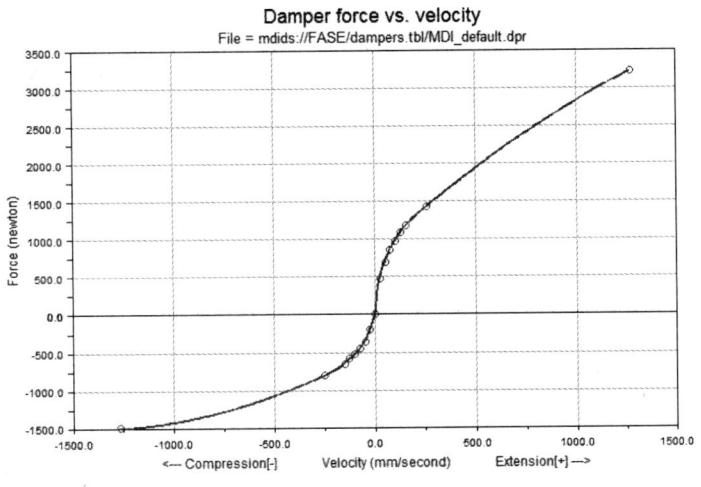

图 12-12 避震器系数曲线

```
$------------------------------------------------MDI_HEADER
[MDI_HEADER]
 FILE_TYPE    = 'dpr'
 FILE_VERSION = 4.0
 FILE_FORMAT  = 'ASCII'
$------------------------------------------------UNITS
[UNITS]
 LENGTH = 'mm'
 ANGLE  = 'degrees'
 FORCE  = 'newton'
 MASS   = 'kg'
```

```
    TIME   =  'second'
$------------------------------------------------------CURVE
    [CURVE]
    {vel           force}
    -1270.0        -1495.5
    -254.0         -809.5
    -152.4         -654.8
    -127.0         -587.1
    -101.6         -533.8
    -76.2          -455.5
    -50.8          -370.1
    -25.4          -206.4
    0.0            0.0
    25.4           462.6
    50.8           695.4
    76.2           854.0
    101.6          966.4
    127.0          1085.1
    152.4          1171.4
    254.0          1423.4
    1270.0         3218.1
```

- Damper Diameter（避震器直径）：拖动滑块选择 15 mm；
- Color：maize。

（2）单击 OK 按钮，完成避震器 ._FSAE_sus_front_third_spring.das_damper2 的创建。

15. 部件 link_down 与安装件 suspension_to_chassis 之间的 revolute 约束

（1）单击 Build > Attachments > Joint > New 命令，弹出创建约束件对话框，如图 12-13 所示，在下列对话框中输入相应的数据：

- Joint Name（约束副名称）：link_down；
- I Part：._FSAE_sus_front_third_spring.ges_link_down；
- J Part：._FSAE_sus_front_third_spring.mts_suspension_to_chassis；
- Type：single；
- Joint Type（约束副类型）：revolute；
- Active（激活）：always；
- Location Dependency：Delta location from coordinate；
- Coordinate Reference（参考坐标）：._FSAE_sus_front_third_spring.ground.cfs_link_down；

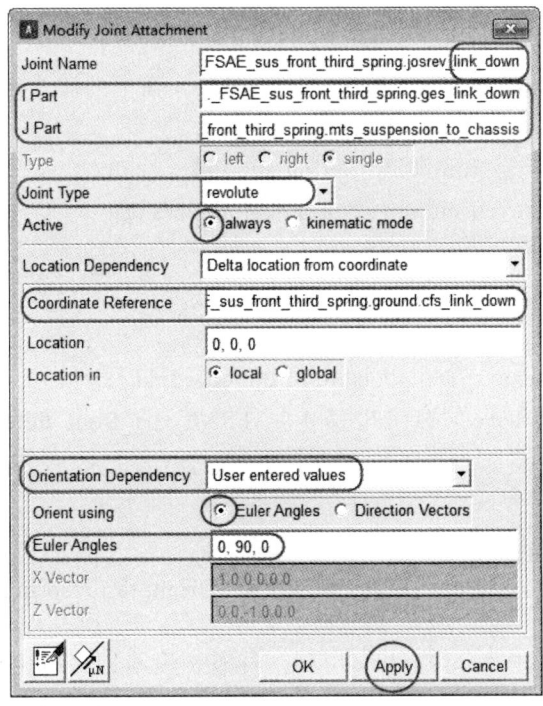

图 12-13 刚性约束对话框-revolute

- Location：0，0，0；
- Location in：local；
- Orientation Dependency：User entered values；
- Orient using：Euler Angles；
- Euler Angles：0，90，0。

（2）单击 Apply 按钮，完成约束副 ._FSAE_sus_front_third_spring.josrev_link_down 的创建。

16. 部件 link_front 与 link_down 之间的 fixed 约束

（1）单击 Build > Attachments > Joint > New 命令，在下列对话框中输入相应的数据：

- Joint Name（约束副名称）：link_center；
- I Part：._FSAE_sus_front_third_spring.ges_link_front；
- J Part：._FSAE_sus_front_third_spring.ges_link_down；
- Type：single；
- Joint Type（约束副类型）：fixed；
- Active（激活）：always；
- Location Dependency：Delta location from coordinate；
- Coordinate Reference（参考坐标）：._FSAE_sus_front_third_spring.ground.cfs_link_center；
- Location：0，0，0；
- Location in：local。

（2）单击 Apply 按钮，完成约束副 ._FSAE_sus_front_third_spring.josfix_link_center 的创建。

17. 部件 link 与 bellcrank 之间的 spherical 约束

（1）单击 Build > Attachments > Joint > New 命令，在下列对话框中输入相应的数据：
- Joint Name（约束副名称）：link1；
- I Part：._FSAE_sus_front_third_spring.gel_link；
- J Part：._FSAE_sus_front_third_spring.gel_bellcrank；
- Type：left；
- Joint Type（约束副类型）：spherical；
- Active（激活）：always；
- Location Dependency：Delta location from coordinate；
- Coordinate Reference（参考坐标）：._FSAE_sus_front_third_spring.ground.cfl_link_mount；
- Location：0，0，0；
- Location in：local。

（2）单击 Apply 按钮，完成约束副._FSAE_sus_front_third_spring.jolsph_link1 的创建。

18. 部件 damper_up 与 suspension_to_chassis 之间的 convel 约束

（1）单击 Build > Attachments > Joint > New 命令，在下列对话框中输入相应的数据：
- Joint Name（约束副名称）：damper_up；
- I Part：._FSAE_sus_front_third_spring.ges_damper_up；
- J Part：._FSAE_sus_front_third_spring.mts_suspension_to_chassis；
- Joint Type（约束副类型）：convel；
- Active（激活）：always；
- Location Dependency：Delta location from coordinate；
- Coordinate Reference（参考坐标）：._FSAE_sus_front_third_spring.ground.cfs_damper_up；
- Location：0，0，0；
- Location in：local；
- I-Part Axis：._FSAE_sus_front_third_spring.ground.cfs_damper_up_ref；
- J-Part Axis：._FSAE_sus_front_third_spring.ground.cfs_link_center。

（2）单击 Apply 按钮，完成约束副._FSAE_sus_front_third_spring.joscon_damper_up 的创建。

19. 部件 damper_down 与 link_front 之间的 spherical 约束

（1）单击 Build > Attachments > Joint > New 命令，在下列对话框中输入相应的数据：
- Joint Name（约束副名称）：damper_down；
- I Part：._FSAE_sus_front_third_spring.ges_damper_down；
- J Part：._FSAE_sus_front_third_spring.ges_link_front；
- Type：left；
- Joint Type（约束副类型）：spherical；
- Active（激活）：always；

- Location Dependency：Delta location from coordinate；
- Coordinate Reference（参考坐标）：._FSAE_sus_front_third_spring.ground.cfs_link_center；
- Location：0，0，0；
- Location in：local。

（2）单击 Apply 按钮，完成约束副._FSAE_sus_front_third_spring.jossph_damper_down 的创建。

20. 部件 damper_down 与 damper_up 之间的 translational 约束

（1）单击 Build > Attachments > Joint > New 命令，在下列对话框中输入相应的数据：
- Joint Name（约束副名称）：damper；
- I Part：._FSAE_sus_front_third_spring.ges_damper_up；
- J Part：._FSAE_sus_front_third_spring.ges_damper_down；
- Joint Type（约束副类型）：translational；
- Active（激活）：always；
- Location Dependency：Centered between coordinates；
- Centered between：Two Coordinates；
- Coordinate Reference #1（参考坐标）：._FSAE_sus_front_third_spring.ground.cfs_damper_up；
- Coordinate Reference #2（参考坐标）：._FSAE_sus_front_third_spring.ground.cfs_link_center；
- Location Dependency：Orient axis to point；
- Coordinate Reference（参考坐标）：._FSAE_sus_front_third_spring.ground.cfs_damper_up；
- Axis：Z。

（2）单击 Apply 按钮，完成约束副._FSAE_sus_front_third_spring.jossph_damper_down 的创建。

21. 部件 link 与 link_front 之间的 convel 约束

（1）单击 Build > Attachments > Joint > New 命令，在下列对话框中输入相应的数据：
- Joint Name（约束副名称）：link2；
- I Part：._FSAE_sus_front_third_spring.ger_link；
- J Part：._FSAE_sus_front_third_spring.ges_link_front；
- Joint Type（约束副类型）：convel；
- Active（激活）：always；
- Location Dependency：Delta location from coordinate；
- Coordinate Reference（参考坐标）：._FSAE_sus_front_third_spring.ground.hpr_link；
- Location：0，0，0；
- Location in：local；
- I-Part Axis：._FSAE_sus_front_third_spring.ground.hpr_shock_to_bellcrank；
- J-Part Axis：._FSAE_sus_front_third_spring.ground.hpl_link。

（2）单击 OK 按钮，完成约束副._FSAE_sus_front_third_spring.jorcon_link2 的创建。

至此，避震器纵置式推杆悬架模型建立完成，建立好的模型如图 12-1 所示，保存模型参考图 12-5，参数保持默认。

22. 纵置式推杆悬架子系统

（1）按 F9 键，将专家模板转换到标准模式，单击 File > New > Suspension 命令，弹出子系统对话框，如图 12-14 所示，在下列对话框中输入相应的数据：

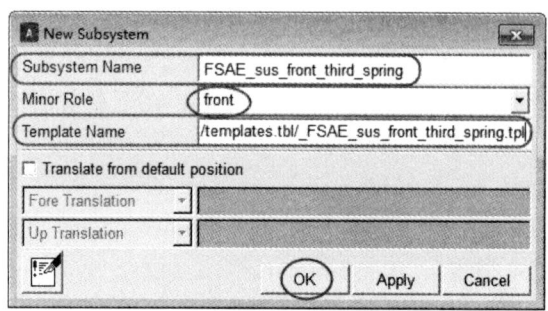

图 12-14　纵置式推杆悬架子系统对话框

- Subsystem Name（系统名称）：FSAE_sus_front_third_spring；
- Minor Role（副特征）：front；
- Template Name（模板路径）：mdids：//FASE/templates.tbl/_FSAE_sus_front_third_spring.tpl。

（2）单击 OK 按钮，完成推杆式悬架子系统 FSAE_sus_front_third_spring 的创建。

12.2　纵置式后推杆悬架模型

纵置式后推杆悬架模型辅助机构及弹簧建模与前推杆悬架建模类似，不同点在于对应的硬点位置不同。硬点变量参数信息如下，读者可参考上述建模过程完成模型建立。建立好的纵置式后推杆悬架模型如图 12-15 所示。

```
Info for subsystem:  FSAE_sus_rear_thirdspring


File Name      :   <FASE>/subsystems.tbl/FSAE_sus_rear_thirdspring.sub
Template       :   mdids://FASE/templates.tbl/_FSAE_sus_rear_thirdspring.tpl
Comments       :   *no comments found*
Major Role     :   suspension
Minor Role     :   rear


HARDPOINTS:
```

hardpoint name	symmetry	x_value	y_value	z_value
global	single	1524.0	0.0	0.0
arblink_to_bellcrank	left/right	1547.8	-50.8	305.0
arb_bushing_mount	left/right	1651.0	-127.0	101.6
bellcrank_pivot	left/right	1547.8	-170.0	305.0
bellcrank_pivot_orient	left/right	1473.2	-146.05	547.3
drive_shaft_inr	left/right	1550.0	-200.0	225.0
lca_front	left/right	1270.0	-127.0	127.0
lca_outer	left/right	1498.6	-482.6	101.6
lca_rear	left/right	1651.0	-127.0	127.0
link	left/right	1350.5	-60.8	304.8
prod_outer	left/right	1498.6	-407.2	127.0
prod_to_bellcrank	left/right	1509.7	-200.0	304.8
shock_to_bellcrank	left/right	1560.5	-50.8	304.8
shock_to_chassis	left/right	1700.0	-50.8	304.8
tierod_inner	left/right	1676.4	-127.0	152.4
tierod_outer	left/right	1574.8	-457.2	152.4
uca_front	left/right	1270.0	-152.4	304.8
uca_outer	left/right	1549.4	-482.6	355.6
uca_rear	left/right	1625.6	-152.4	304.8
wheel_center	left/right	1524.0	-558.8	228.6

PARAMETERS:

parameter name	symmetry	type	value
driveline_active	single	integer	1
kinematic_flag	single	integer	0
camber_angle	left/right	real	-1.5
drive_shaft_offset	left/right	real	75.0
toe_angle	left/right	real	0.0

图 12-15　纵置式后推杆悬架系统（包含驱动轴）

12.3　漂移仿真

1. 整车模型装配

（1）按 F9 键切换到标准模板。

（2）单击 File > Open > Assembly 命令，弹出整车装配对话框，如图 12-16 所示，在 Assembly Name 中输入：mdids：//FASE/assemblies.tbl/FSAE.asy。

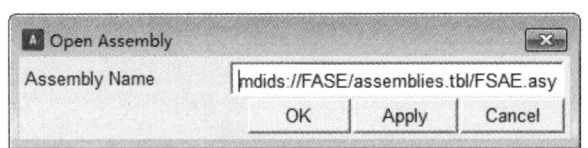

图 12-16　FSAE 整车模型装配对话框

（3）单击 OK 按钮，打开 FSAE 整车模型；此时整车模型如图 12-17 所示。

图 12-17　FSAE 整车模型

（4）单击 File > Manage Assemblies > Replace Subsystem（s）命令，弹出替换子系统对话框，如图 12-18 所示，在下列对话框中输入相应的数据：

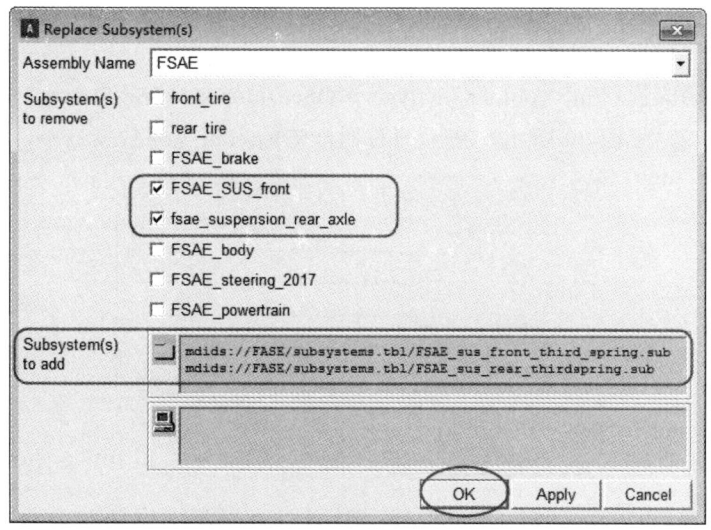

图 12-18　替换子系统对话框

- 勾选 FSAE_SUS_front；
- 勾选 fsae_suspension_rear_axle；
- Subsystem（s）To add：
① mdids：//FASE/subsystems.tbl/FSAE_sus_front_third_spring.sub；
② mdids：//FASE/subsystems.tbl/FSAE_sus_rear_thirdspring.sub。

（5）单击 OK 按钮，完成前后悬架子系统的替换，此时整车模型如图 12-19 所示。

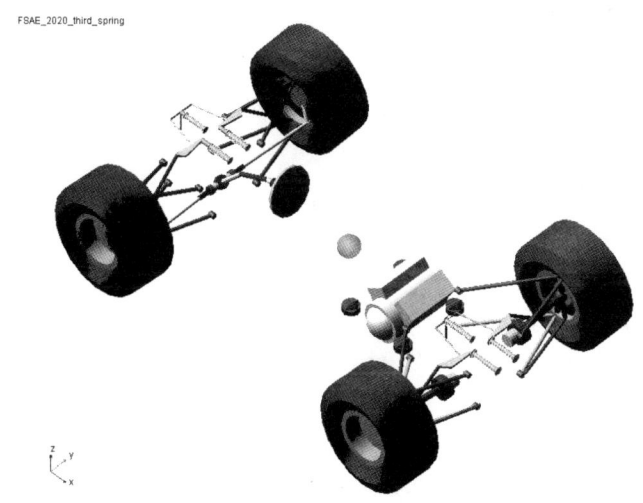

图 12-19　FSAE 整车模型

（6）单击 File > Save as > Assembly 命令，在下列对话框中输入相应的数据：
- Assembly Name：FSAE；
- New Assembly Name：FSAE_2020_third_spring；

- Target：Database/FSAE。

（7）单击 OK 按钮，完成基于避震器横置式整车 FSAE_2020_third_spring 的存储。

2. 漂移仿真

（1）单击 Simulate > Full-Vehicle Analysis > Open-loop steering Events > Drift 命令，弹出阶跃仿真对话框，如图 12-20 所示，在下列对话框中输入相应的数据：

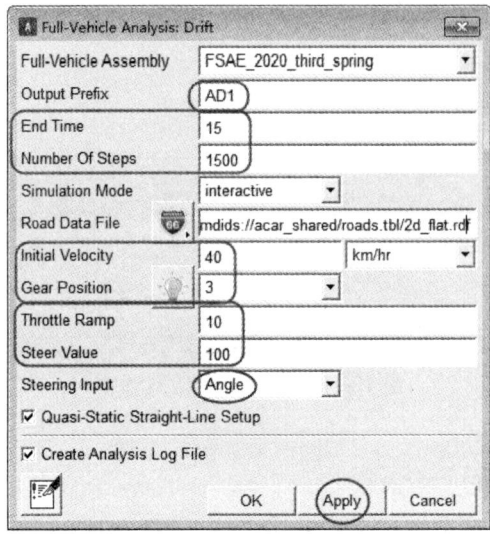

图 12-20　漂移仿真参数设置

- Full-Vehicle Assembly：FSAE_2020_third_spring；
- Output Prefix：AD1；
- End Time：15；
- Number of steps：1500；
- Road Date File：mdids：//acar_shared/roads.tbl/2d_flat.rdf；
- Simulation Mode：interactive；
- Initial Velocity（单位：km/h）：40；
- Gear Position：3；
- Throttle Value：10；
- Steer Value（单位：度）：100；
- 勾选 Quasi-Static Straight-Line Setup。

（2）单击 Apply 按钮，完成 FSAE_2020_third_spring 赛车漂移仿真设置并提交运算，运算完成后，整车运行轨迹如图 12-21 所示，从图中可以看出，整车具有不足转向的特性，符合车轮设计要求。同理，完成其他仿真设备。

- Full-Vehicle Assembly：FSAE；
- Output Prefix：AD2；
- End Time：15；
- Number of steps：1500；

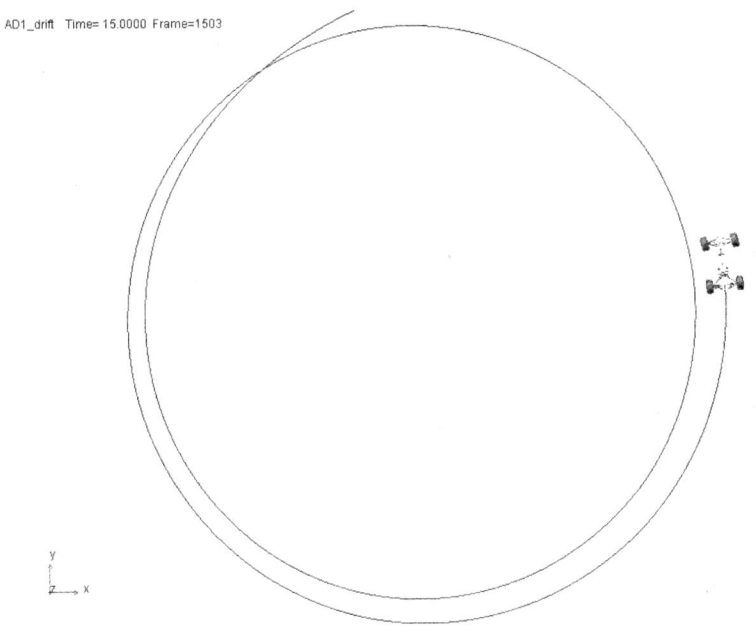

图 12-21 车辆运行轨迹

- Road Date File：mdids：//acar_shared/roads.tbl/2d_flat.rdf；
- Initial Velocity（单位：km/h）：40；
- Gear Position：3；
- Throttle Value：10；
- Steer Value（单位：度）：100；
- 勾选 Quasi-Static Straight-Line Setup。

单击 Apply 按钮，完成 FSAE 赛车漂移仿真设置并提交运算。

两种不同底盘赛车的稳定性参数如图 12-22 ~ 图 12-27 所示，从图中可以看出，除了车身侧向加速度和横摆角加速度基本保持一致外，其余参数指标性能均有提升；证明了通过增加辅助机构与弹簧减震器，可以改善车辆的稳定性能。

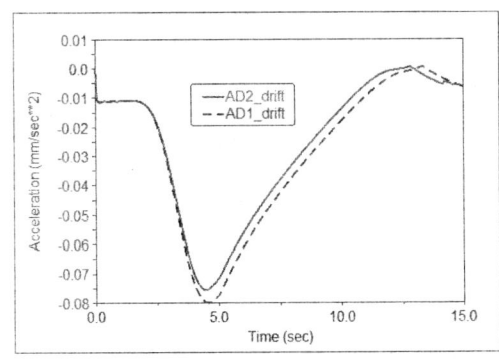

图 12-22 车身纵向加速度

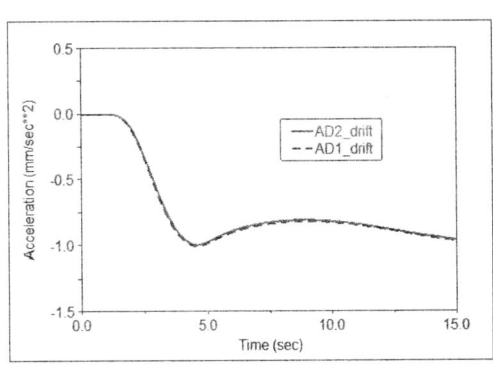

图 12-23 车身侧向加速度

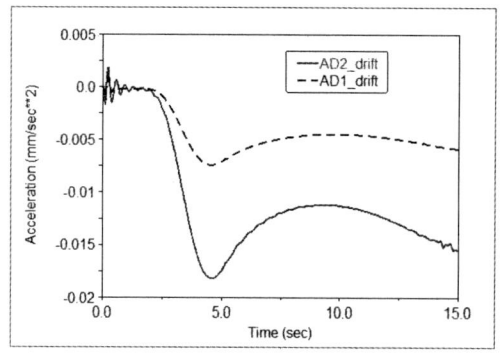

图 12-24　车身垂向加速度

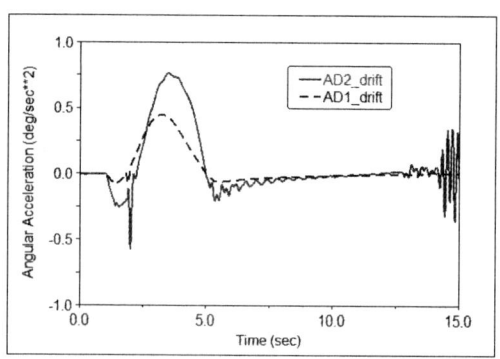

图 12-25　车身俯仰角加速度

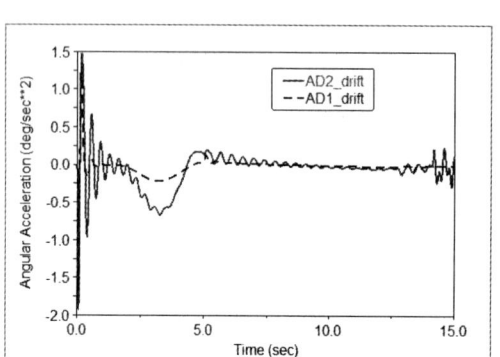

图 12-26　车身侧倾角加速度

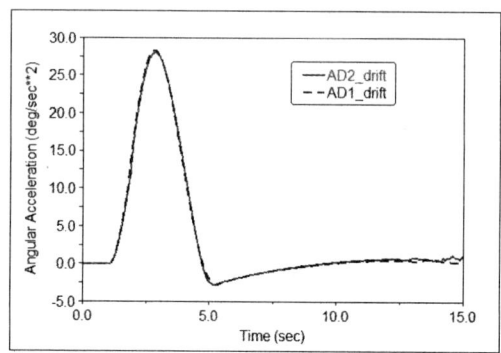

图 12-27　车身横摆角加速度

第 13 章 避震器横置式推杆悬架

FSAE 赛车悬架有多种形式，推杆与避震器的布置结构也多有不同，如弹簧避震器纵置式、横置式、斜置式等，不同的布置形式取决于多种因素的耦合，如空间限制、风阻影响等。与家用轿车多采用的垂直式（大多接近垂向）相比，横置或纵置弹簧的刚度对推杆位置、旋转支撑架、避震器弹簧安装角度极为敏感，角度的稍微变化可能需要匹配的刚度变化较大，因此需要对关键硬点位置多次优化才能得到优良的综合特性。本章不再重复悬架的建模过程与步骤，通过在避震器纵置式后推杆悬架模型_fsae_suspension_rear_axle.tpl 中修改部分部件、约束、力等建立而成，读者可自行查阅。建立好的横置式后推杆悬架模型如图 13-1 所示。

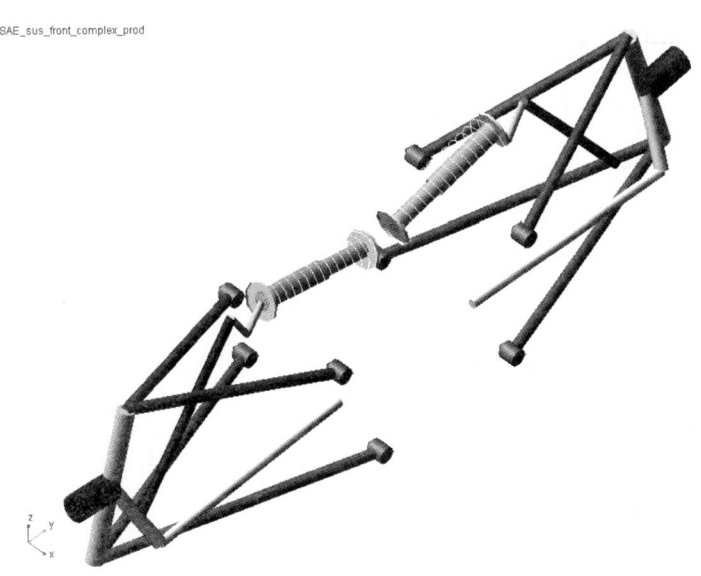

图 13-1 避震器横置式推杆前悬架模型

学习目标

（1）会进行刚度匹配。
（2）会模型修改调试。
（3）了解前横置推杆悬架信息。
（4）会阶跃转向仿真。

13.1 刚度匹配

FSAE_sus_front_third_spring 推杆悬架的传力模型如图 13-2 所示，OB 为下控制臂到车轮中心的长度，AC 为推杆部件，CDE 为旋转支撑架部件，EG 为弹簧减震器安装位置；推杆与旋转支撑架在 C 点通过胡克副连接，旋转支撑架与车架在 D 点通过旋转副连接，旋转支撑架与减震器在 E 点通过胡克副连接，弹簧减震器与车架在 G 点通过胡克副连接，推杆与下控制臂在 A 点通过球副连接。当车轮处于静止状态时，减震器不工作，此时悬架的传力公式如下所示：

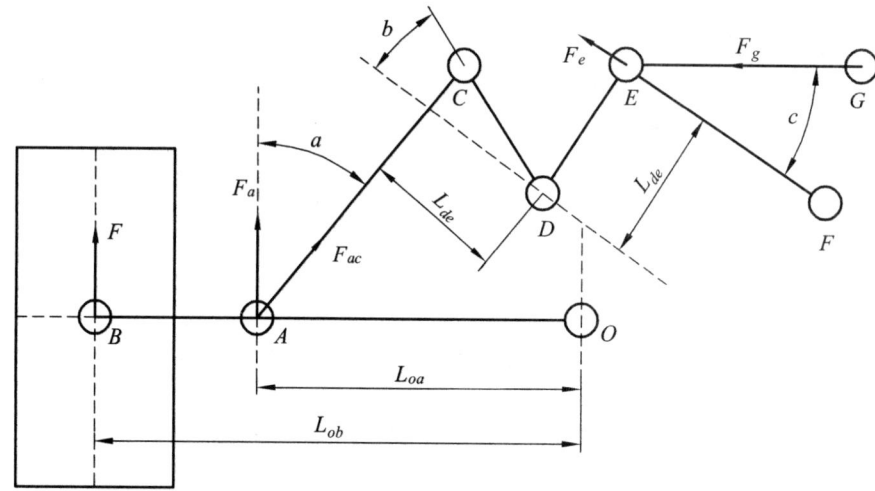

图 13-2 推杆悬架传力模型

$$FL_{ob} = -F_a L_{oa} \tag{13-1}$$

$$F_{ac}\cos a = F_a \tag{13-2}$$

$$F_{ac} L_{cd} = F_e L_{de} \tag{13-3}$$

$$F_e = F_g \cos c \tag{13-4}$$

$$F_g = Kx \tag{13-5}$$

$$|\overline{CD}|\cos b = L_{cd} \tag{13-6}$$

将式（13-1）~（13-4）、（13-6）代入式（13-5）并整理得：

$$K = \frac{|\overline{CD}|\cos b L_{ob} F}{x L_{de} L_{oa} \cos c} \tag{13-7}$$

式中，L_{ob} 为下控制臂与车身铰接中心到车轮中心的水平距离；L_{oa} 为下控制臂与车身铰接中心到推杆与下控制臂铰接中心的距离；F 为地面对车轮的支撑力；F_{ac} 为推杆 AC 之间的传递力；F_a 为 F_{ac} 在垂直方向上的分力；L_{cd} 为 D 点到推杆 AC 之间的垂直距离，即力臂；L_{de} 为 DE 之间的距离；F_g 为弹簧力；K 为弹簧刚度；x 为弹簧压缩或拉伸最大行程；F_e 为弹簧力 F_g 在力臂 L_{de} 上的分力；a 为推杆 AC 与垂直方向的夹角；b 为线段 CD 与力臂 L_{cd} 之间的夹角；c 为力 F_e 与 F_g 之间的夹角，即弹簧与水平面之间的安装角度。

通过分析式（13-7），弹簧刚度与角 a、b、c 有关，其中角 a、b 可调节的范围不大，角 c 可调的范围较大，当角 c 与水平面的夹角越小时，弹簧刚度越小；因此在推杆式悬架硬件匹配完成后，可以通过角度 c 的大小来调节整车稳定特性。

13.2 避震器横置式后推杆悬架

1. 模型导入

（1）启动 ADAMS/CAR，选择专家模块进入建模界面。

（2）单击 File > Open 命令，弹出打开模板对话框，如图 13-3 所示，在 Template Name 中输入：mdids: //FASE/templates.tbl/_fsae_suspension_rear_axle.tpl。

（3）单击 OK 按钮，完成模型导入。

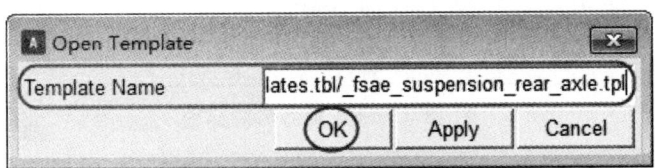

图 13-3 打开模板对话框

2. 删除部件、约束、硬点

（1）删除以下部件，在删除部件的同时，与部件有关的约束也会同时删除，硬点保留：

① ._fsae_suspension_rear_axle.gel_damper_bellcrank；
② ._fsae_suspension_rear_axle.ger_damper_bellcrank；
③ ._fsae_suspension_rear_axle.gel_damper_chassis；
④ ._fsae_suspension_rear_axle.ger_damper_chassis；
⑤ ._fsae_suspension_rear_axle.gel_bellcrank；
⑥ ._fsae_suspension_rear_axle.ger_bellcrank。

（2）删除以下硬点，以硬点为基础对应的结构框也会自动删除：

① ._fsae_suspension_rear_axle.ground.hpl_shock_to_bellcrank；
② ._fsae_suspension_rear_axle.ground.hpr_shock_to_bellcrank；
③ ._fsae_suspension_rear_axle.ground.hpl_bellcrank_pivot；
④ ._fsae_suspension_rear_axle.ground.hpr_bellcrank_pivot；
⑤ ._fsae_suspension_rear_axle.ground.hpl_arblink_to_bellcrank；
⑥ ._fsae_suspension_rear_axle.ground.hpr_arblink_to_bellcrank；
⑦ ._fsae_suspension_rear_axle.ground.hpl_shock_to_chassis；

⑧ ._fsae_suspension_rear_axle.ground.hpr_shock_to_chassis；
⑨ ._fsae_suspension_rear_axle.ground.hpl_bellcrank_pivot_orient；
⑩ ._fsae_suspension_rear_axle.ground.hpr_bellcrank_pivot_orient。

（3）删除弹簧、避震器：

① ._fsae_suspension_rear_axle.nsl_spring；
② ._fsae_suspension_rear_axle.nsr_spring；
③ ._fsae_suspension_rear_axle.dal_damper；
④ ._fsae_suspension_rear_axle.dar_damper。

（4）在模型树中选择硬点：._fsae_suspension_rear_axle.ground.hpl_prod_to_bellcrank；右击选择 Rename，在 New Name 中输入：prod_to_zhijia。

（5）单击 OK 按钮，完成硬点重命名，如图 13-4 所示。

（6）单击 File > Save As 命令，弹出保存模板对话框，如图 13-5 所示，在下列对话框中输入相应的数据：

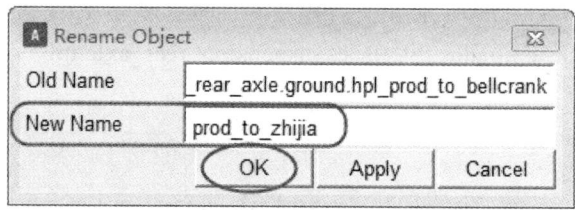

图 13-4　硬点重命名对话框

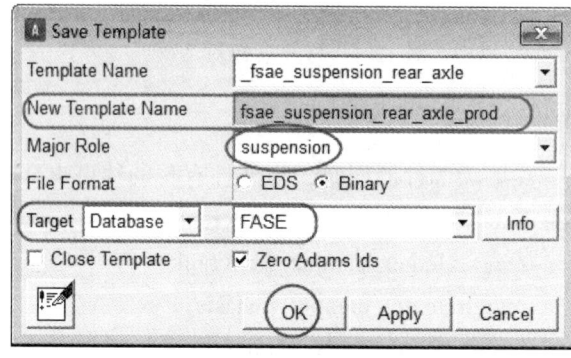

图 13-5　保存模板对话框

- Template Name：_fsae_suspension_rear_axle；
- New Template Name：fsae_suspension_rear_axle_prod；
- Major Role（主特征）：suspension；
- File Format：Binary；
- Target：Directory/FASE。

（7）单击 OK 按钮，完成推杆式悬架模型模板._fsae_suspension_rear_axle 的保存。

（8）单击 Build > Hardpoind > New 命令，弹出创建硬点对话框，如图 13-6 所示，在下列对话框中输入相应的数据：

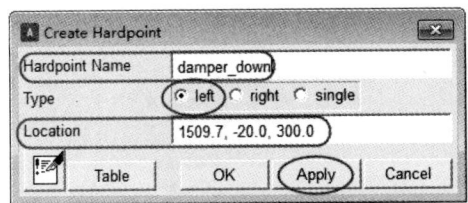

图 13-6　创建硬点对话框

- Hardpoint name：damper_down；
- Type：left；
- Location：1509.7，-20.0，300.0。

（9）单击 Apply 按钮，完成 drive_shaft_inr 硬点的创建。

（10）重复上述创建硬点步骤，完成图 13-7 中标注线内的硬点创建，同时对应图 13-7 完成其余硬点数值大小的调试。

	loc x	loc y	loc z
hpl_arb_bushing_mour	1651.0	-127.0	101.6
hpl_damper_down	1509.7	-20.0	300.0
hpl_drive_shaft_inr	1550.0	-200.0	225.0
hpl_lca_front	1270.0	-127.0	127.0
hpl_lca_outer	1498.6	-482.6	101.6
hpl_lca_rear	1651.0	-127.0	127.0
hpl_prod_outer	1498.6	-482.6	127.0
hpl_prod_to_zhijia	1509.7	-250.0	381.0
hpl_tierod_inner	1676.4	-127.0	152.4
hpl_tierod_outer	1574.8	-457.2	152.4
hpl_uca_front	1270.0	-152.4	304.8
hpl_uca_outer	1549.4	-482.6	355.6
hpl_uca_rear	1625.6	-152.4	304.8
hpl_wheel_center	1524.0	-558.8	228.6
hpl_zhijia_to_body	1509.7	-200.0	330.0
hpl_zhijie_out	1509.7	-150.0	380.0
hps_global	1524.0	0.0	0.0

图 13-7　硬点信息

3. 部件 damper_down

（1）单击 Build > Part > General Part > New 命令，弹出创建部件对话框，如图 13-8 所示，在下列对话框中输入相应的数据：

- General Part：damper_down；
- Location Dependency：Delta location from coordinate；
- Coordinate Reference（参考坐标）：._fsae_suspension_rear_axle_prod.ground.hpl_damper_down；
- Location：0，0，0；
- Location in：local；
- Orientation Dependency：User-entered values；
- Orient using：Euler Angles；

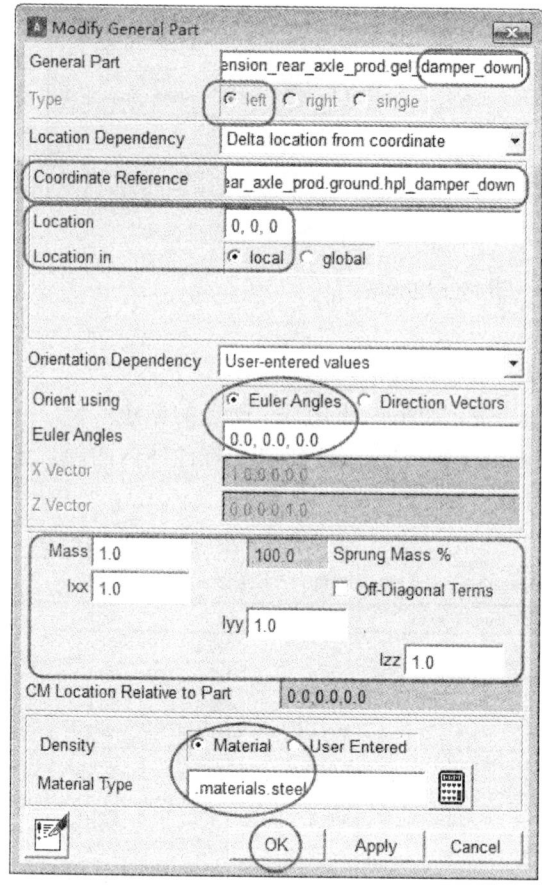

图 13-8　部件 damper_down

- Euler Angles：0，0，0；
- Mass：1；
- Ixx：1；
- Iyy：1；
- Izz：1；
- Density：Material；
- Material Type：.materials.steel。

（2）单击 Apply 按钮，完成部件 ._fsae_suspension_rear_axle_prod.gel_damper_down 的创建。

4. 部件 damper_up

（1）单位 Build > Part > General Part > New 命令，在下列对话框中输入相应的数据：

- General Part：damper_up；
- Location Dependency：Delta location from coordinate；
- Coordinate Reference（参考坐标）：._fsae_suspension_rear_axle_prod.ground.hpl_zhijie_out；
- Location：0，0，0；
- Location in：local；

- Orientation Dependency：User-entered values；
- Orient using：Euler Angles；
- Euler Angles：0，0，0；
- Mass：1；
- Ixx：1；
- Iyy：1；
- Izz：1；
- Density：Material；
- Material Type：.materials.steel。

（2）单击 Apply 按钮，完成部件._fsae_suspension_rear_axle_prod.gel_damper_up 的创建。

5. 部件 zhijie

（1）单位 Build > Part > General Part > New 命令，在下列对话框中输入相应的数据：
- General Part：zhijie；
- Location Dependency：Centered between coordinates；
- Centered between：Three Coordinates；
- Coordinate Reference #1（参考坐标）：._fsae_suspension_rear_axle_prod.ground.hpl_prod_to_zhijia；
- Coordinate Reference #2（参考坐标）：._fsae_suspension_rear_axle_prod.ground.hpl_zhijia_to_body；
- Coordinate Reference #3（参考坐标）：._fsae_suspension_rear_axle_prod.ground.hpl_zhijie_out；
- Orient using：Euler Angles；
- Euler Angles：0，0，0；
- Mass：1；
- Ixx：1；
- Iyy：1；
- Izz：1；
- Density：Material；
- Material Type：.materials.steel。

（2）单击 OK 按钮，完成部件._fsae_suspension_rear_axle_prod.gel_zhijie 的创建。

6. 支架部件几何体

（1）单击 Build > Geometry > Link > New 命令，弹出创建部件对话框，如图 13-9 所示，在下列对话框中输入相应的数据：
- Link Name（连杆名称）：zhijia1；
- General Part 输入：._fsae_suspension_rear_axle_prod.gel_zhijia；
- Coordinate Reference #1（参考坐标）：._fsae_suspension_rear_axle_prod.ground.hpl_zhijia_to_body；

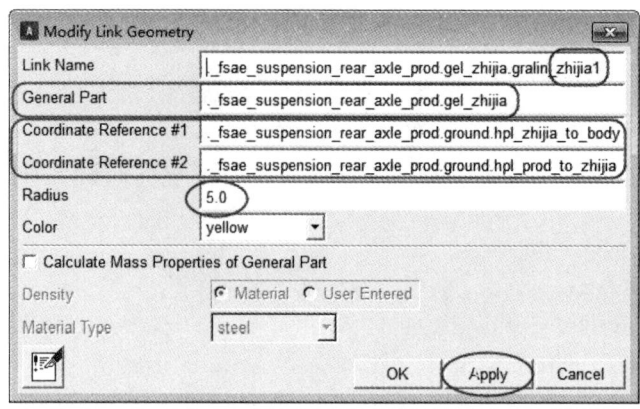

图 13-9　连杆几何体

- Coordinate Reference #2（参考坐标）：._fsae_suspension_rear_axle_prod.ground.hpl_prod_to_zhijia；
- Radius（半径）：5；
- Color：yellow；
- 选择 Calculate Mass Properties of General Part 复选框，当几何体建立好之后，会更新对应部件的质量和惯量参数；
- Density：Material；
- Material Type：steel。

（2）单击 Apply 按钮，完成._fsae_suspension_rear_axle_prod.gel_zhijia.gralin_zhijia1 几何体的创建。同理，创建其他几何体。

- Link Name（连杆名称）：zhijia2；
- General Part 输入：._fsae_suspension_rear_axle_prod.gel_zhijia；
- Coordinate Reference #1（参考坐标）：._fsae_suspension_rear_axle_prod.ground.hpl_zhijia_to_body；
- Coordinate Reference #2（参考坐标）：._fsae_suspension_rear_axle_prod.ground.hpl_zhijie_out；
- Radius（半径）：5；
- Color：yellow；
- 选择 Calculate Mass Properties of General Part 复选框，当几何体建立好之后，会更新对应部件的质量和惯量参数；
- Density：Material；
- Material Type：steel。

（3）单击 OK 按钮，完成._fsae_suspension_rear_axle_prod.gel_zhijia.gralin_zhijia2 几何体的创建。

7. 弹　簧

（1）单击 Build > Force > Spring > New 命令，弹出创建弹簧对话框，如图 13-10 所示，在下列对话框中输入相应的数据：

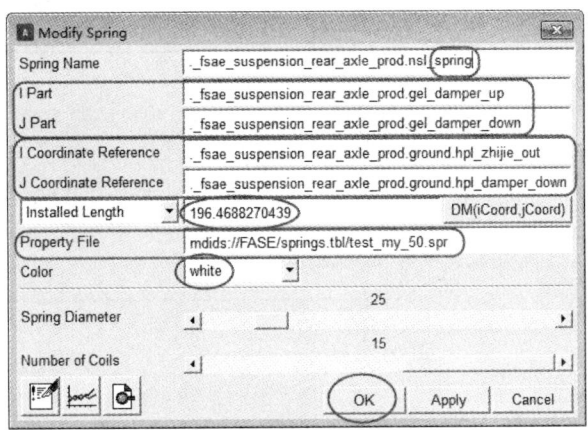

图 13-10 spring 弹簧创建对话框

- Spring Name（减震器名称）：spring；
- I Part：._fsae_suspension_rear_axle_prod.gel_damper_up；
- J Part：._fsae_suspension_rear_axle_prod.gel_damper_down；
- I Coordinate Reference（参考坐标）：._fsae_suspension_rear_axle_prod.ground.hpl_zhijie_out；
- J Coordinate Reference（参考坐标）：._fsae_suspension_rear_axle_prod.ground.hpl_damper_down；
- Installed Length（安装长度）：单击 DM（iCoord，jCoord），自动计算弹簧的安装长度并填入到方框中，此模型的安装长度为 196.468 827 043 9，弹簧刚度曲线如图 13-11 所示；

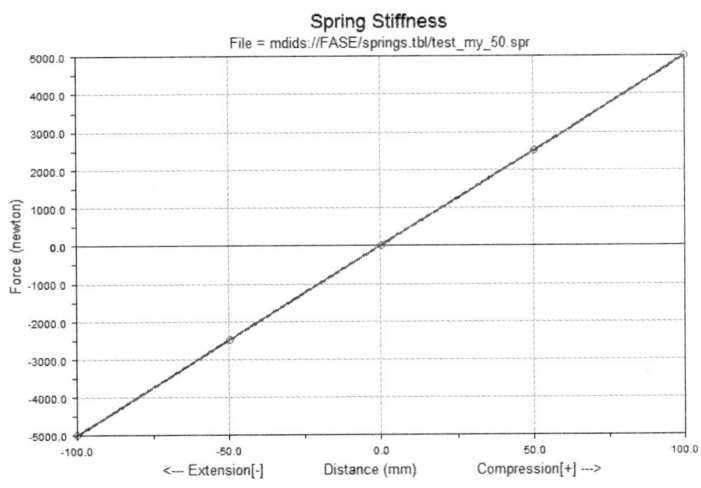

图 13-11 弹簧刚度曲线

- Property File（属性文件）：mdids：//FASE/springs.tbl/test_my_50.spr，弹簧属性文件用记事本文件打开后如下列信息所示，可以根据实验情况测出弹簧的参数（力与位移之间的关系，即刚度）并填写到如下信息列表中，可修改部分为下划线部分；
- Spring Diameter（弹簧直径）：拖动滑块选择 25 mm；
- Spring of Coils（弹簧圈数）：拖动滑块选择 15。

（2）单击 OK 按钮，完成弹簧 ._fsae_suspension_rear_axle_prod.nsl_spring 的创建。

```
弹簧刚度信息：
$------------------------------------------------------------MDI_HEADER
[MDI_HEADER]
 FILE_TYPE     = 'spr'
 FILE_VERSION  = 4.0
 FILE_FORMAT   = 'ASCII'
$------------------------------------------------------------UNITS
[UNITS]
 LENGTH = 'mm'
 ANGLE  = 'degrees'
 FORCE  = 'newton'
 MASS   = 'kg'
 TIME   = 'second'
$------------------------------------------------------------SPRING_DATA
[SPRING_DATA]
 FREE_LENGTH = 205.7
$------------------------------------------------------------CURVE
[CURVE]
{disp           force}
-100.0         -5000.0
 -50.0         -2500.0
   0.0             0.0
  50.0          2500.0
 100.0          5000.0
```

8. 避震器

（1）单击 Build > Force > Damper > New 命令，弹出避震器创建对话框，如图 13-12 所示，在下列对话框中输入相应的数据：

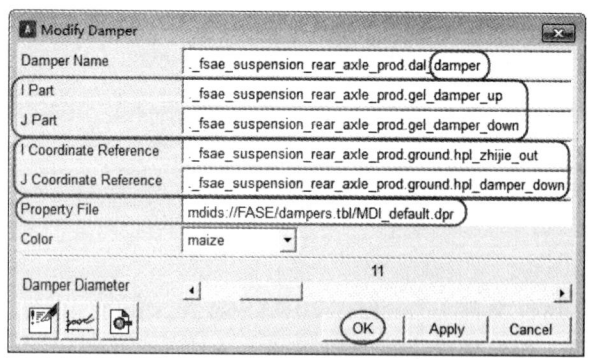

图 13-12 damper 避震器创建对话框

- Damper Name（减震器名称）：damper；
- I Part：._fsae_suspension_rear_axle_prod.gel_damper_up；

- J Part：._fsae_suspension_rear_axle_prod.gel_damper_down；
- I Coordinate Reference（参考坐标）：._fsae_suspension_rear_axle_prod.ground.hpl_zhijie_out；
- J Coordinate Reference（参考坐标）：._fsae_suspension_rear_axle_prod.ground.hpl_damper_down；
- Property File（属性文件）：mdids：//FASE/dampers.tbl/MDI_default.dpr，避震器系数曲线如图 13-13 所示，具体数据如下列避震器信息；

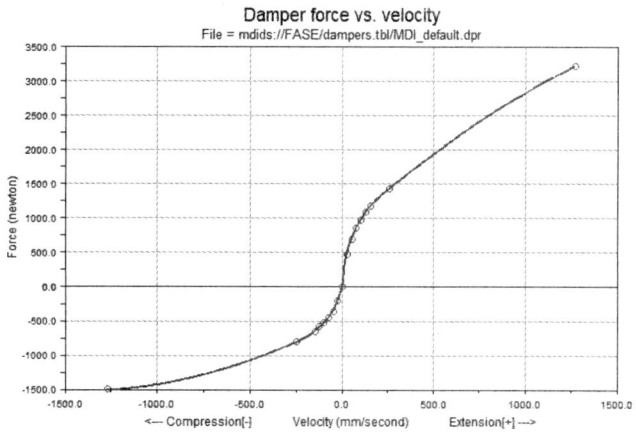

图 13-13　避震器系数曲线图

- Damper Diameter（避震器直径）：拖动滑块选择 11 mm；
- Color：maize。

（2）单击 OK 按钮，完成避震器._fsae_suspension_rear_axle_prod.dal_damper 的创建。

```
$------------------------------------------------MDI_HEADER
[MDI_HEADER]
 FILE_TYPE     = 'dpr'
 FILE_VERSION  = 4.0
 FILE_FORMAT   = 'ASCII'
$------------------------------------------------UNITS
[UNITS]
 LENGTH  = 'mm'
 ANGLE   = 'degrees'
 FORCE   = 'newton'
 MASS    = 'kg'
 TIME    = 'second'
$------------------------------------------------CURVE
[CURVE]
{vel          force}
-1270.0       -1495.5
-254.0        -809.5
-152.4        -654.8
```

−127.0	−587.1
−101.6	−533.8
−76.2	−455.5
−50.8	−370.1
−25.4	−206.4
0.0	0.0
25.4	462.6
50.8	695.4
76.2	854.0
101.6	966.4
127.0	1085.1
152.4	1171.4
254.0	1423.4
1270.0	3218.1

9. 部件 zhijia 与安装件 suspension_to_chassis 之间的 revolute 约束

（1）单击 Build > Attachments > Joint > New 命令，弹出创建约束件对话框，如图 13-14 所示，在下列对话框中输入相应的数据：

- Joint Name（约束副名称）：zhijia_to_body；
- I Part：._fsae_suspension_rear_axle_prod.gel_zhijia；
- J Part：._fsae_suspension_rear_axle_prod.mts_suspension_to_chassis；

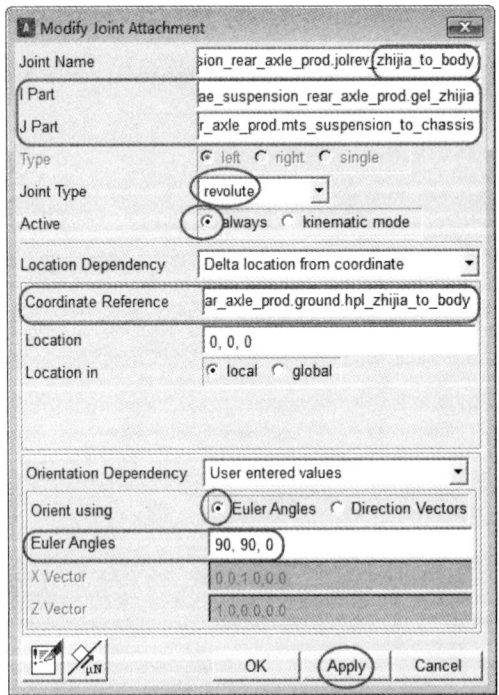

图 13-14　刚性约束对话框-revolute

- Joint Type（约束副类型）：revolute；
- Active（激活）：always；
- Location Dependency：Delta location from coordinate；
- Coordinate Reference（参考坐标）：._fsae_suspension_rear_axle_prod.ground.hpl_zhijia_to_body；
- Location：0，0，0；
- Location in：local；
- Orientation Dependency：User entered values；
- Orient using：Euler Angles；
- Euler Angles：90，90，0；

（2）单击 Apply 按钮，完成约束副._fsae_suspension_rear_axle_prod.jolrev_zhijia_to_body 的创建。

10. 部件 prod 与 zhijia 之间的 convel 约束

（1）单击 Build > Attachments > Joint > New 命令，在下列对话框中输入相应的数据。
- Joint Name（约束副名称）：prod_to_zhijia；
- I Part：._fsae_suspension_rear_axle_prod.gel_prod；
- J Part：._fsae_suspension_rear_axle_prod.gel_zhijia；
- Joint Type（约束副类型）：convel；
- Active（激活）：always；
- Location Dependency：Delta location from coordinate；
- Coordinate Reference（参考坐标）：._fsae_suspension_rear_axle_prod.ground.hpl_prod_to_zhijia；
- Location：0，0，0；
- Location in：local；
- I-Part Axis：._fsae_suspension_rear_axle_prod.ground.hpl_prod_outer；
- J-Part Axis：._fsae_suspension_rear_axle_prod.ground.hpl_zhijia_to_body；

（2）单击 Apply 按钮，完成约束副._fsae_suspension_rear_axle_prod.jolcon_prod_to_zhijia 的创建。

11. 部件 damper_up 与 zhijia 之间的 convel 约束

（1）单击 Build > Attachments > Joint > New 命令，在下列对话框中输入相应的数据。
- Joint Name（约束副名称）：zhijie_to_damper；
- I Part：._fsae_suspension_rear_axle_prod.gel_zhijia；
- J Part：._fsae_suspension_rear_axle_prod.gel_damper_up；
- Joint Type（约束副类型）：convel；
- Active（激活）：always；
- Location Dependency：Delta location from coordinate；
- Coordinate Reference（参考坐标）：._fsae_suspension_rear_axle_prod.ground.hpl_

zhijie_out；
- Location：0，0，0；
- Location in：local；
- I-Part Axis：._fsae_suspension_rear_axle_prod.ground.hpl_zhijia_to_body；
- J-Part Axis：._fsae_suspension_rear_axle_prod.ground.hpl_damper_down。

（2）单击 Apply 按钮，完成约束副 ._fsae_suspension_rear_axle_prod.jolcon_zhijie_to_damper 的创建。

12. 部件 damper_up 与 damper_down 之间的 cylindrical 约束

（1）单击 Build > Attachments > Joint > New 命令，在下列对话框中输入相应的数据。
- Joint Name（约束副名称）：damper；
- I Part：._fsae_suspension_rear_axle_prod.gel_damper_up；
- J Part：._fsae_suspension_rear_axle_prod.gel_damper_down；
- Joint Type（约束副类型）：cylindrical；
- Active（激活）：always；
- Location Dependency：Centered between coordinates；
- Centered between：Two Coordinates；
- Coordinate Reference #1（参考坐标）：._fsae_suspension_rear_axle_prod.ground.hpl_zhijie_out；
- Coordinate Reference #2（参考坐标）：._fsae_suspension_rear_axle_prod.ground.hpl_damper_down；
- Orientation Dependency：Orient axis to point；
- Coordinate Reference（参考坐标）：._fsae_suspension_rear_axle_prod.ground.hpl_damper_down；
- Axis：Z。

（2）单击 Apply 按钮，完成 ._fsae_suspension_rear_axle_prod.jolcyl_damper 圆柱副的创建。

13. 部件 damper_down 与 suspension_to_chassis 之间的 spherical 约束

（1）单击 Build > Attachments > Joint > New 命令，在下列对话框中输入相应的数据。
- Joint Name（约束副名称）：damper_down；
- I Part：._fsae_suspension_rear_axle_prod.gel_damper_down；
- J Part：._fsae_suspension_rear_axle_prod.mts_suspension_to_chassis；
- Joint Type（约束副类型）：spherical；
- Active（激活）：always；
- Location Dependency：Delta location from coordinate；
- Coordinate Reference（参考坐标）：._fsae_suspension_rear_axle_prod.ground.hpl_damper_down；

- Location：0，0，0；
- Location in：local。

（2）单击 Apply 按钮，完成约束副._fsae_suspension_rear_axle_prod.jolsph_damper_down 的创建。

14. 部件 arb_bushing_mount 与 suspension_to_chassis 之间的 fixed 约束

（1）单击 Build > Attachments > Joint > New 命令，在下列对话框中输入相应的数据。
- Joint Name（约束副名称）：arb_bushing_to_ground；
- I Part：._fsae_suspension_rear_axle_prod.gel_arb_bushing_mount；
- J Part：._fsae_suspension_rear_axle_prod.mts_suspension_to_chassis；
- Joint Type（约束副类型）：fixed；
- Active（激活）：always；
- Location Dependency：Delta location from coordinate；
- Coordinate Reference（参考坐标）：._fsae_suspension_rear_axle_prod.ground.hpl_arb_bushing_mount；
- Location：0，0，0；
- Location in：local。

（2）单击 OK 按钮，完成约束副._fsae_suspension_rear_axle_prod.jolfix_arb_bushing_to_ground 的创建。

13.3 避震器横置式推杆悬架通信器

需要添加以下通信器，作用是保证横向稳定杆与悬架的装配：

（1）单击 Build > Communicator > Output >New 命令，弹出输出通信器对话框，如图 13-15 所示。在下列对话框中输入相应的数据：

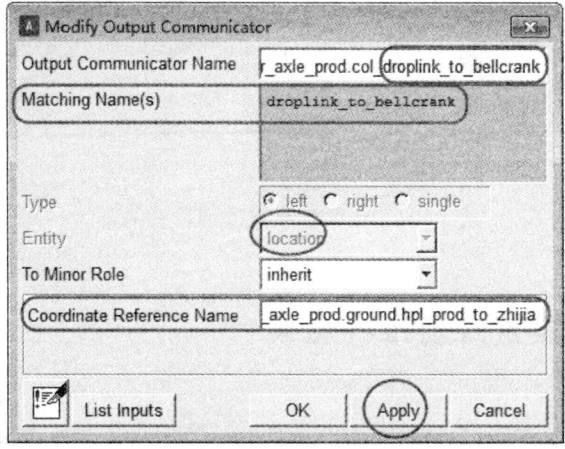

图 13-15　输出通信器对话框

- Output Communicator Name（输出通信器名称）：droplink_to_bellcrank；
- Matching Name（s）：droplink_to_bellcrank；
- Type：left；
- Entity：location；
- To Minor Role：inherit；
- Coordinate Reference Name：._fsae_suspension_rear_axle_prod.ground.hpl_prod_to_zhijia。

（2）单击 Apply 按钮，完成通信器._fsae_suspension_rear_axle_prod.col_droplink_to_bellcrank 的创建。同理，创建其他通信器。

- Output Communicator Name（输出通信器名称）：arb_bushing_mount；
- Matching Name（s）：arb_bushing_mount；
- Type：left；
- Entity：mount；
- To Minor Role：inherit；
- Part Name：._fsae_suspension_rear_axle_prod.gel_arb_bushing_mount。

（3）单击 OK 按钮，完成通信器._fsae_suspension_rear_axle_prod.col_arb_bushing_mount 的创建。

（4）删除输入通信器：droplink_to_bellcrank，此通信器与 FSAE 赛车车身里包含的通信器重复；至此避震器横置式后推杆悬架模型建立完成，如图 13-16 所示，保存模板。

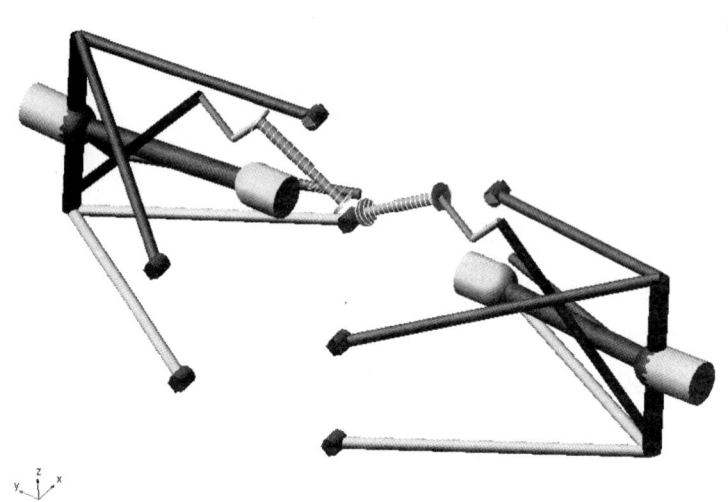

图 13-16　避震器横置式后推杆悬架

13.4　避震器横置式前推杆悬架

避震器横置式前推杆悬架模型的建立参考后推杆式悬架模型，删除对应的部件、约束、硬点、结构框等；然后在模型上添加其他部件、约束等，完成建模后的前推杆悬架如图 13-1

所示。前推杆式悬架的硬点信息如下：

模板及硬点信息

```
Info for subsystem:    FSAE_sus_front_complex_prod
    File Name       :  <FASE>/subsystems.tbl/FSAE_sus_front_complex_prod.sub
    Template        :  mdids://FASE/templates.tbl/_FSAE_sus_front_complex_prod.tpl
    Comments        :  *no comments found*
    Major Role      :  suspension
    Minor Role      :  front


    HARDPOINTS:
    hardpoint name            symmetry         x_value      y_value      z_value
    -----------------         ---------        -------      -------      -------
    hps_global                single             0.0          0.0          0.0
    arb_bushing_mount         left/right       127.0        -127.0        101.6
    damper_down               left/right         0.0         -30.0        300.0
    lca_front                 left/right      -157.0        -127.0        114.3
    lca_outer                 left/right         0.0        -546.1        120.65
    lca_rear                  left/right       127.0        -127.0        114.3
    link                      left/right         0.0         -80.0        230.0
    prod_outer                left/right         0.0        -457.2        139.7
    prod_to_zhijia            left/right         0.0        -230.0        371.0
    tierod_inner              left/right        50.8        -127.0        152.4
    tierod_outer              left/right        63.5        -482.6        152.4
    uca_front                 left/right      -101.6        -177.8        279.4
    uca_outer                 left/right         0.0        -482.6        355.6
    uca_rear                  left/right       101.6        -177.8        279.4
    wheel_center              left/right         0.0        -558.8        241.3
    zhijia_rear               left/right        50.0        -120.0        330.0
    zhijia_to_body            left/right         0.0        -200.0        330.0
    zhijie_out                left/right         0.0        -170.0        370.0
```

13.5 阶跃转向仿真

1. 整车模型装配

（1）按 F9 键切换到标准模板。

（2）单击 File > Open > Assembly 命令，弹出整车装配对话框，如图 13-17 所示，在 Assembly Name 中输入：mdids：//FASE/assemblies.tbl/FSAE.asy。

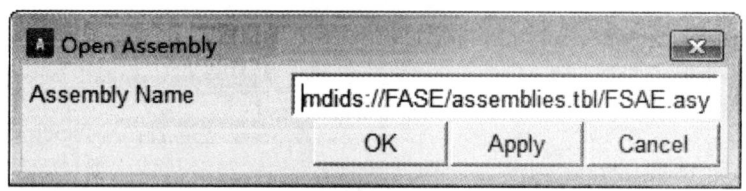

图 13-17　FSAE 整车模型装配对话框

（3）单击 OK 按钮，打开 FSAE 整车模型。

（4）单击 File > Manage Assemblies > Replace Subsystem（s）命令，弹出替换子系统对话框，如图 13-18 所示，在下列对话框中输入相应的数据：

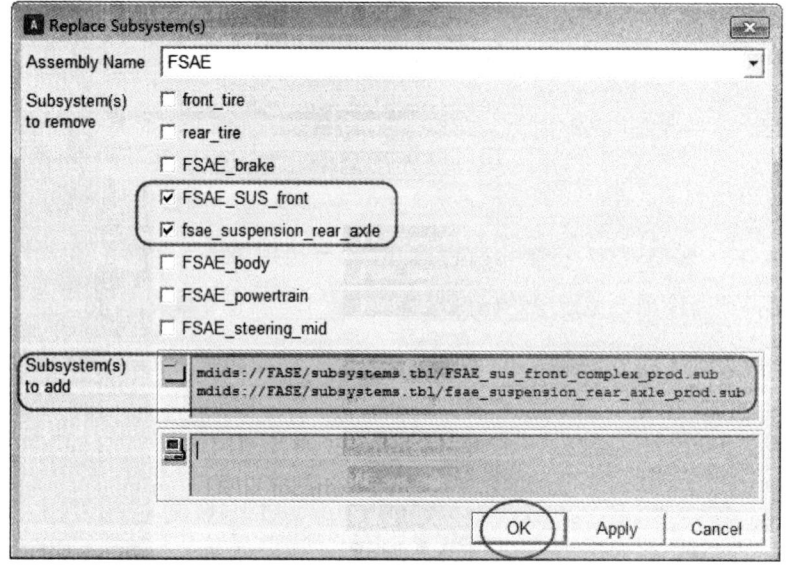

图 13-18　替换子系统对话框

- 勾选 FSAE_SUS_front；
- 勾选 fsae_suspension_rear_axle；
- Subsystem（s）To add：
① mdids：//FASE/subsystems.tbl/FSAE_sus_leafspring_front.sub；
② mdids：//FASE/subsystems.tbl/FSAE_sus_leafspring_rear.sub。

（5）单击 OK 按钮，完成前后悬架子系统的替换，此时整车模型如图 13-19 所示。

图 13-19　FSAE 整车模型

（6）单击 File > Save as > Assembly 命令，弹出保存整车模型对话框，如图 13-20 所示，在下列对话框中输入相应的数据：

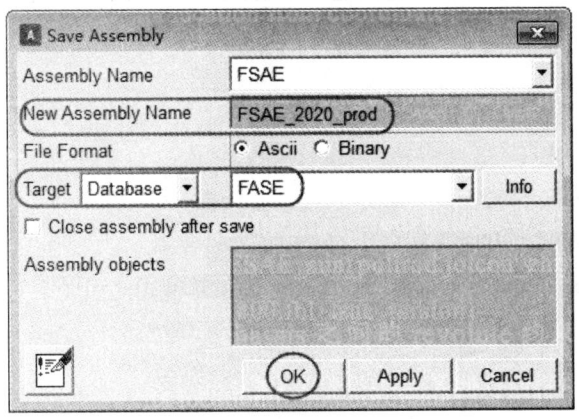

图 13-20　FSAE 整车模型保存对话框

- Assembly Name：FSAE；
- New Assembly Name：FSAE_2020_prod；
- Target：Database / FSAE。

（7）单击 OK 按钮，完成基于避震器横置式整车 FSAE_2020_prod 的存储。

2. 仿　真

FSAE 赛车模型替换完成后，可以对整车进行验证仿真，观察模型正确与否，是否能够根据实验标准达到预期状态，阶跃仿真参数设置如下：

（1）单击 Simulate > Full-Vehicle Analysis > Open-loop steering Events > Step steer 命令，弹出阶跃仿真对话框，如图 13-21 所示，在下列对话框中输入相应的数据：

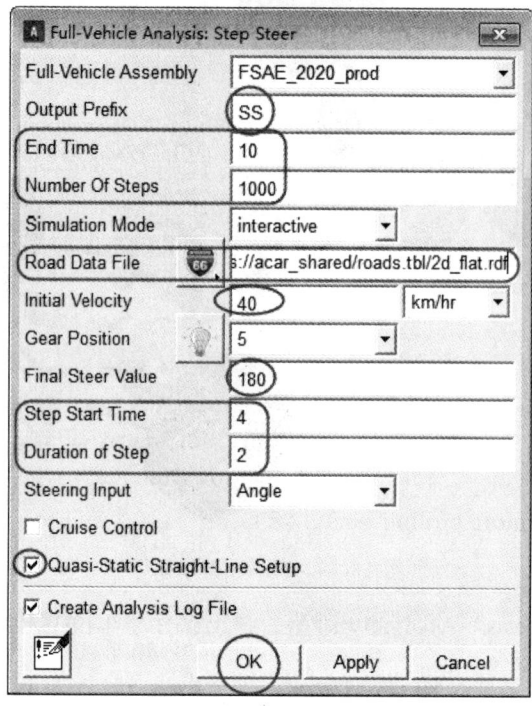

图 13-21　阶跃转向参数设置

- Output Prefix：SS；
- End Time：10；
- Number of steps：1000；
- Simulation Mode：interactive；
- Road Date File：mdids：//acar_shared/roads.tbl/2d_flat.rdf；
- Initial Velocity（单位：km/h）：40；
- Gear Position：5；
- First Steer Value（单位：度）：180；
- Step Start Time：4；
- Duration of Step：2；
- 勾选 Quasi-Static Straight-Line Setup。

（2）单击 OK 按钮，完成阶跃转向仿真设置并提交运算。

仿真完成后整车运行轨迹如图 13-22 所示，从轨迹中可以看出，在终了时刻 FSAE 赛车已经打滑，整车失去稳定性；切换到后处理模式，绘制整车稳定性参数，如图 13-23～图 13-26 所示，从图 13-25 和图 13-26 中也可以看出，角加速度急剧增加，整车失去稳定性。

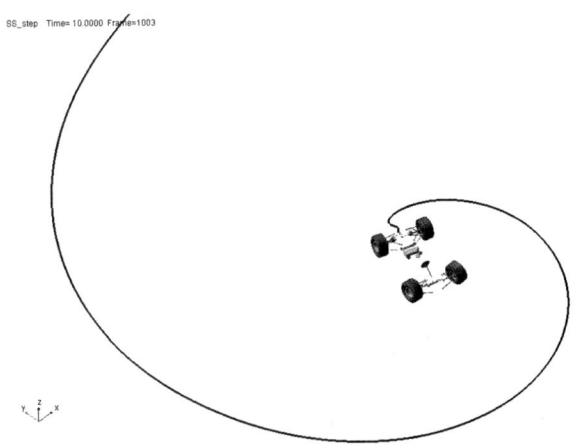

图 13-22 阶跃转向运行轨迹

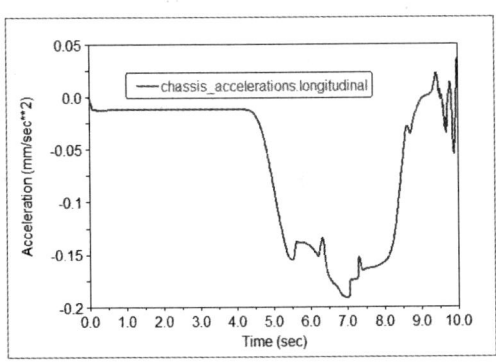

图 13-23 车身纵向加速度

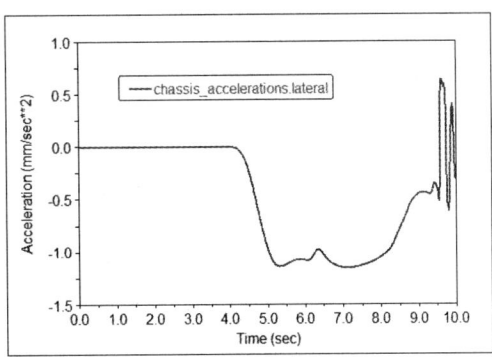

图 13-24 车身侧向加速度

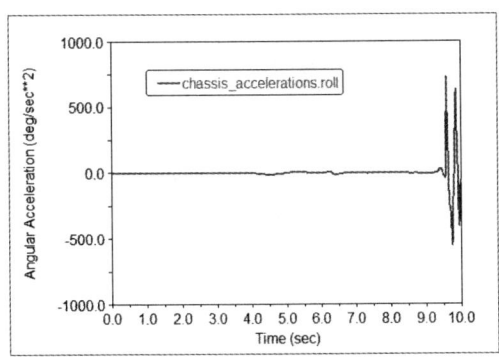

图 13-25 车身侧倾角加速度

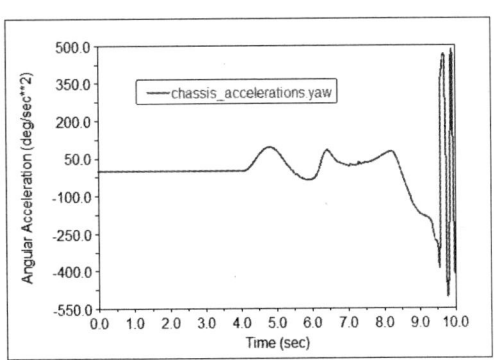

图 13-26 车身横摆角加速度

第 14 章　拉杆式悬架模型 I

拉杆式悬架近些年极为少见，多在 20 世纪 F1 赛车上采用。拉杆式悬架最大的优势是可以降低车身高度，提升轮胎的抓地性能，进而改善整车的稳定性能。2009 年，红牛 F1 车队的 RB5 赛车在后悬架上采用的是基于扭杆弹簧式的拉杆悬架，其模型如图 14-1 所示。悬架系统的优劣（底盘优劣）与采用什么类型的悬架无关，关键在于悬架与整车的匹配与调试。本章首先将建立基于螺旋弹簧式的拉杆悬架，通过整车仿真发现当车速高于 50 km/h 时，存在不稳定现象；通过调试改为扭杆式弹簧，整车性能提升较大；对扭杆弹簧刚度计算及扭杆弹簧的细节处理在后续章节详细讨论。

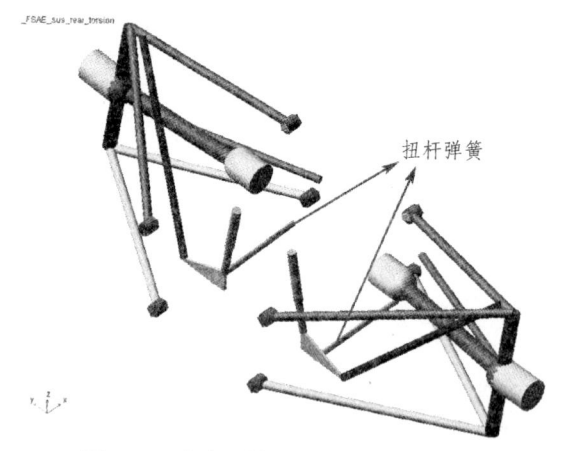

图 14-1　扭杆弹簧式拉杆悬架（后）

学习目标

（1）了解螺旋弹簧式拉杆悬架。
（2）会单线移动超车仿真。
（3）了解扭杆弹簧 MNF。
（4）了解扭杆弹簧式拉杆悬架。
（5）会扭杆弹簧调试。
（6）会加速仿真。
（7）了解横向避震器。

14.1　前螺旋弹簧式拉杆悬架

1. 模型导入

（1）启动 ADAMS/CAR，选择专家模块进入建模界面。

（2）单击 File > Open 命令，弹出打开模板对话框，如图 14-2 所示，在 Template Name 中输入：mdids: //FASE/templates.tbl/_FSAE_sus_front.tpl。

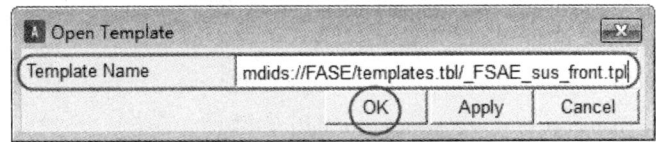

图 14-2　打开模板对话框

（3）单击 OK 按钮，导入模型。

2. 删除部件、弹簧、避震器、约束、结构框

通过删除部件，与部件相关联的约束、弹簧、避震器等都会全部删除，最后删除多余的硬点和结构框，此种方法删除速度较快，同时不容易出现错误。需要删除的信息如下：

① ._FSAE_sus_front.gel_bellcrank；
② ._FSAE_sus_front.ger_bellcrank；
③ ._FSAE_sus_front.gel_prod；
④ ._FSAE_sus_front.ger_prod；
⑤ ._FSAE_sus_front.gel_damper_chassis；
⑥ ._FSAE_sus_front.ger_damper_chassis；
⑦ ._FSAE_sus_front.gel_damper_bellcrank；
⑧ ._FSAE_sus_front.ger_damper_bellcrank；
⑨ ._FSAE_sus_front.ground.hpl_prod_outer；
⑩ ._FSAE_sus_front.ground.hpr_prod_outer；
⑪ ._FSAE_sus_front.ground.hpl_prod_to_bellcrank；
⑫ ._FSAE_sus_front.ground.hpr_prod_to_bellcrank；
⑬ ._FSAE_sus_front.ground.hpl_shock_to_bellcrank；
⑭ ._FSAE_sus_front.ground.hpr_shock_to_bellcrank；
⑮ ._FSAE_sus_front.ground.hpl_bellcrank_pivot；
⑯ ._FSAE_sus_front.ground.hpr_bellcrank_pivot；
⑰ ._FSAE_sus_front.ground.hpl_arblink_to_bellcrank；
⑱ ._FSAE_sus_front.ground.hpr_arblink_to_bellcrank；
⑲ ._FSAE_sus_front.ground.hpl_shock_to_chassis；
⑳ ._FSAE_sus_front.ground.hpr_shock_to_chassis；

㉑ ._FSAE_sus_front.ground.hpl_bellcrank_pivot_orient；
㉒ ._FSAE_sus_front.ground.hpr_bellcrank_pivot_orient；
㉓ ._FSAE_sus_front.ground.cfl_damper_bellcrank_orient；
㉔ ._FSAE_sus_front.ground.cfr_damper_bellcrank_orient；
㉕ ._FSAE_sus_front.ground.cfl_bellcrank_pivot；
㉖ ._FSAE_sus_front.ground.cfr_bellcrank_pivot；
㉗ ._FSAE_sus_front.ground.cfl_damper_chassis_orient；
㉘ ._FSAE_sus_front.ground.cfr_damper_chassis_orient；
㉙ ._FSAE_sus_front.ground.cfl_shock_to_bellcrank；
㉚ ._FSAE_sus_front.ground.cfr_shock_to_bellcrank；
㉛ ._FSAE_sus_front.ground.cfl_spring_to_bellcrank；
㉜ ._FSAE_sus_front.ground.cfr_spring_to_bellcrank。

3．保存模型

（1）单击 File > Save As 命令，弹出保存模板对话框，如图 14-3 所示，在下列对话框中输入相应的数据：

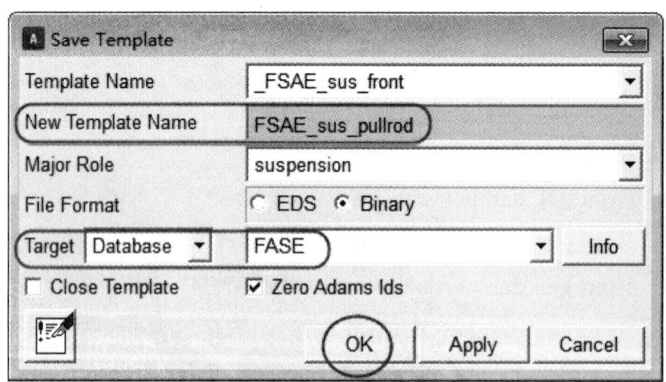

图 14-3　模型保存对话框

- Template Name：_FSAE_sus_front；
- New Template Name：FSAE_sus_pullrod；
- Major Role（主特征）：suspension；
- File Format：Binary；
- Target：Datebase/FASE。

（2）单击 OK 按钮，完成推杆式悬架模型模板 FSAE_sus_pullrod 的保存。

4．添加硬点

（1）单击 Build > Hardpoind > New 命令，弹出创建硬点对话框，如图 14-4 所示，在下列对话框中输入相应的数据：

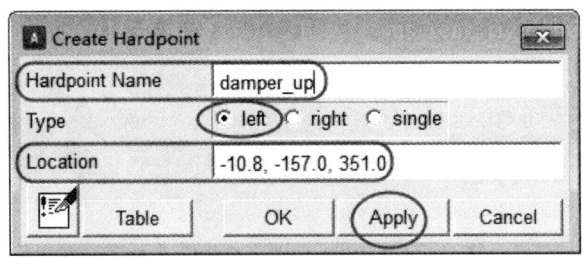

图 14-4 硬点创建

- Hardpoint：damper_up；
- Typer：left；
- Location：-10.8，-157.0，351.0。

（2）单击 Apply 按钮，完成._FSAE_sus_pullrod.ground.hpl_damper_up 硬点的创建。

（3）重复上述创建硬点步骤，完成图 14-5 中标注线内的硬点创建。

	loc x	loc y	loc z
hpl_arb_bushing_mour	127.0	-127.0	101.6
hpl_damper_up	-10.8	-157.0	351.0
hpl_damper_up_ref	-10.8	-157.0	401.0
hpl_lca_front	-127.0	-127.0	114.3
hpl_lca_outer	0.0	-546.1	120.65
hpl_lca_rear	127.0	-127.0	114.3
hpl_prod_outer	0.0	-452.6	355.6
hpl_prod_to_bellcrank	-10.8	-207.0	131.0
hpl_tierod_inner	50.8	-127.0	152.4
hpl_tierod_outer	63.5	-482.6	152.4
hpl_uca_front	-101.6	-177.8	279.4
hpl_uca_outer	0.0	-482.6	355.6
hpl_uca_rear	101.6	-177.8	279.4
hpl_wheel_center	0.0	-558.8	241.3
hpl_zhijia_to_chassis	-10.8	-107.0	131.0
hpl_zhijia_to_damper	-10.8	-187.0	151.0
hps_hps_global	0.0	0.0	0.0

图 14-5 硬点信息

5. 部件 prod

（1）单击 Build > Part > General Part > New 命令，弹出创建部件对话框，如图 14-6 所示，在下列对话框中输入相应的数据：

- General Part：prod；
- Type：left；
- Location Dependency：Centered between coordinates；
- Centered between：Two Coordinates；

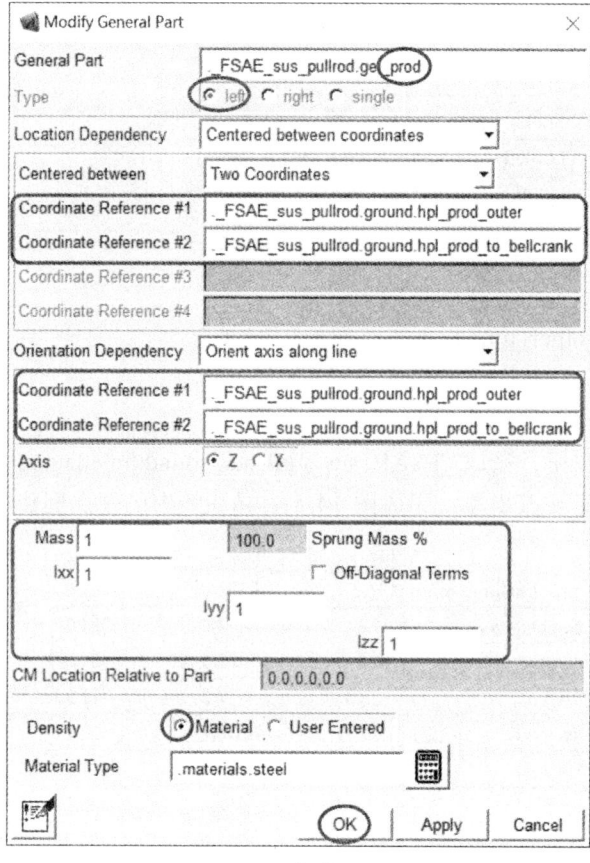

图 14-6　部件_prod

- Coordinate Reference #1（参考坐标）：._FSAE_sus_pullrod.ground.hpl_prod_outer；
- Coordinate Reference #2（参考坐标）：._FSAE_sus_pullrod.ground.hpl_prod_to_bellcrank；
- Orientation Dependency：Orient axis along line；
- Coordinate Reference #1（参考坐标）：._FSAE_sus_pullrod.ground.hpl_prod_outer；
- Coordinate Reference #2（参考坐标）：._FSAE_sus_pullrod.ground.hpl_prod_to_bellcrank；
- Axis：Z；
- Mass：1；
- Ixx：1；
- Iyy：1；
- Izz：1；
- Density：Material；
- Material Type：.materials.steel。

（2）单击 OK 按钮，完成部件._FSAE_sus_pullrod.gel_prod 的创建。

6. 几何体 prod

（1）单击 Build > Geometry > Link > New 命令，弹出创建几何体对话框，如图 14-7 所示，在下列对话框中输入相应的数据：

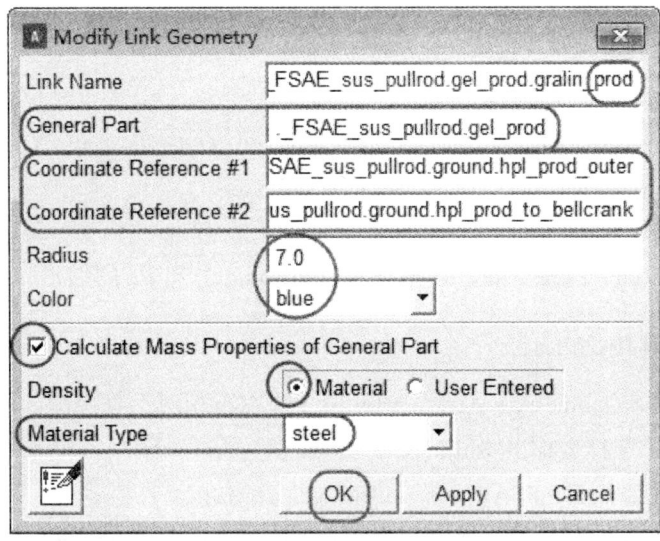

图 14-7　几何体_prod

- Link Name（连杆名称）：prod；
- General Part：._FSAE_sus_pullrod.gel_prod；
- Coordinate Reference #1（参考坐标）：._FSAE_sus_pullrod.ground.hpl_prod_outer；
- Coordinate Reference #2（参考坐标）：._FSAE_sus_pullrod.ground.hpl_prod_to_bellcrank；
- Radius（半径）：7；
- Color：blue；
- 勾选 Calculate Mass Properties of General Part 复选框，当几何体建立好之后，会更新对应部件的质量和惯量参数；
- Density：Material；
- Material Type：steel。

（2）单击 OK 按钮，完成._FSAE_sus_pullrod.gel_prod.gralin_prod 几何体的创建。

7. 部件 zhijia

（1）单击 Build > Part > General Part > New 命令，在下列对话框中输入相应的数据：

- General Part：zhijia；
- Type：left；
- Location Dependency：Centered between coordinates；
- Centered between：Three Coordinates；
- Coordinate Reference #1（参考坐标）：._FSAE_sus_pullrod.ground.hpl_prod_to_bellcrank；
- Coordinate Reference #2（参考坐标）：._FSAE_sus_pullrod.ground.hpl_zhijia_to_chassis；
- Coordinate Reference #2（参考坐标）：._FSAE_sus_pullrod.ground.hpl_zhijia_to_damper；
- Orientation Dependency：User-entered values；
- Orient using：Euler Angles；
- Euler Angles：0，0，0；

- Mass：1；
- Ixx：1；
- Iyy：1；
- Izz：1；
- Density：Material；
- Material Type：.materials.steel。

（2）单击 OK 按钮，完成部件._FSAE_sus_pullrod.gel_zhijia 的创建。

8. 三角臂几何体 zhijia

（1）单击 Build > Geometry > Arm > New 命令，在下列对话框中输入相应的数据：

- Link Name（连杆名称）：zhijia；
- General Part 输入：._FSAE_sus_pullrod.gel_zhijia；
- Coordinate Reference #1（参考坐标）：._FSAE_sus_pullrod.ground.hpl_prod_to_bellcrank；
- Coordinate Reference #2（参考坐标）：._FSAE_sus_pullrod.ground.hpl_zhijia_to_chassis；
- Coordinate Reference #2（参考坐标）：._FSAE_sus_pullrod.ground.hpl_zhijia_to_damper；
- Radius（半径）：5；
- Color：yellow；
- 勾选 Calculate Mass Properties of General Part 复选框，当几何体建立好之后，会更新对应部件的质量和惯量参数；
- Density：Material；
- Material Type：steel。

（2）单击 OK 按钮，完成._FSAE_sus_pullrod.gel_zhijia.graarm_zhijia 几何体的创建。

9. 部件 damper_up

（1）单击 Build > Part > General Part > New 命令，在下列对话框中输入相应的数据：

- General Part：damper_up；
- Type：single；
- Location Dependency：Delta location from coordinate；
- Coordinate Reference（参考坐标）：._FSAE_sus_pullrod.ground.hpl_damper_up；
- Location：0，0，0；
- Location in：local；
- Orientation Dependency：User-entered values；
- Orient using：Euler Angles；
- Euler Angles：0，0，0；
- Mass：1；
- Ixx：1；

- Iyy：1；
- Izz：1；
- Density：Material；
- Material Type：.materials.steel。

（2）单击 Apply 按钮，完成部件._FSAE_sus_pullrod.gel_damper_up 的创建。

10. 部件 damper_down

（1）单击 Build > Part > General Part > New 命令，在下列对话框中输入相应的数据：
- General Part：damper_down；
- Type：single；
- Location Dependency：Delta location from coordinate；
- Coordinate Reference（参考坐标）：._FSAE_sus_pullrod.ground.hpl_zhijia_to_damper；
- Location：0，0，0；
- Location in：local；
- Orientation Dependency：User-entered values；
- Orient using：Euler Angles；
- Euler Angles：0，0，0；
- Mass：1；
- Ixx：1；
- Iyy：1；
- Izz：1；
- Density：Material；
- Material Type：.materials.steel。

（2）单击 OK 按钮，完成部件._FSAE_sus_pullrod.gel_damper_down 的创建。

11. 结构框

（1）单击 Build > Construction Frame > New 命令，弹出创建结构框对话框，如图 14-8 所示，在下列对话框中输入相应的数据：
- Construction Frame（结构框名称）：spring_mount_base；
- Location Dependency：Delta location from coordinate；
- Coordinate Reference（参考坐标）：._FSAE_sus_pullrod.ground.hpl_zhijia_to_damper；
- Location：0，0，0；
- Location in：local；
- Orientation Dependency：Orient axis to point；
- Coordinate Reference（参考坐标）：._FSAE_sus_pullrod.ground.hpl_damper_up；
- Axis：Z。

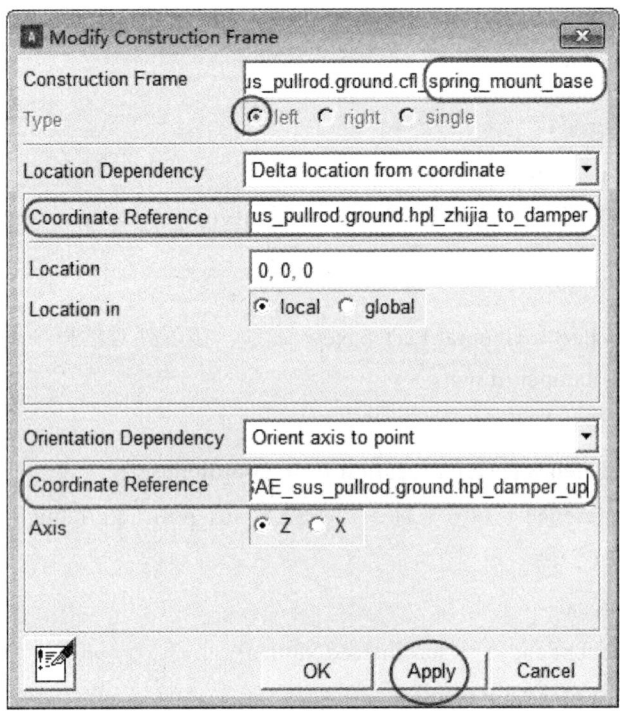

图 14-8　模结构框对话框

（2）单击 Apply 按钮，完成 ._FSAE_sus_pullrod.ground.cfl_spring_mount_base 结构框的创建。同理，创建其他结构框。

- Construction Frame（结构框名称）：spring_mount；
- Location Dependency：Delta location from coordinate；
- Coordinate Reference（参考坐标）：._FSAE_sus_pullrod.ground.cfl_spring_mount_base；
- Location：0，0，50；
- Location in：local；
- Orientation Dependency：Orient axis to point；
- Coordinate Reference（参考坐标）：._FSAE_sus_pullrod.ground.hpl_damper_up；
- Axis：Z。

（3）单击 OK 按钮，完成 ._FSAE_sus_pullrod.ground.cfl_spring_mount 结构框的创建。

12. 弹　簧

（1）单击 Build > Force > Spring > New 命令，弹出创建弹簧对话框，如图 14-9 所示，在下列结构框中输入相应的数据：

- Spring Name（减震器名称）：spring；
- I Part：._FSAE_sus_pullrod.gel_damper_up；
- J Part：._FSAE_sus_pullrod.gel_damper_down；
- I Coordinate Reference（参考坐标）：._FSAE_sus_pullrod.ground.hpl_damper_up；
- J Coordinate Reference（参考坐标）：._FSAE_sus_pullrod.ground.cfl_spring_mount；

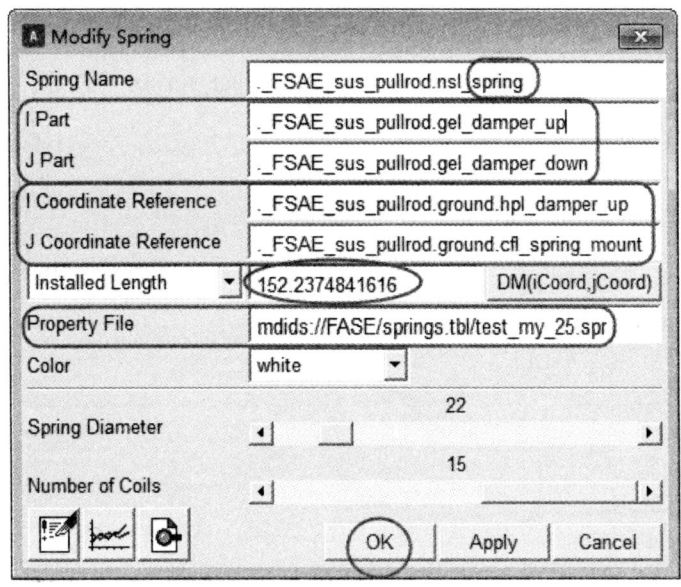

图 14-9 spring 弹簧创建对话框

- Installed Length（安装长度）：单击 DM（iCoord，jCoord），自动计算弹簧的安装长度并填入到方框中，此模型的安装长度为 152.237 484 161 6，弹簧刚度曲线如图 14-10 所示；

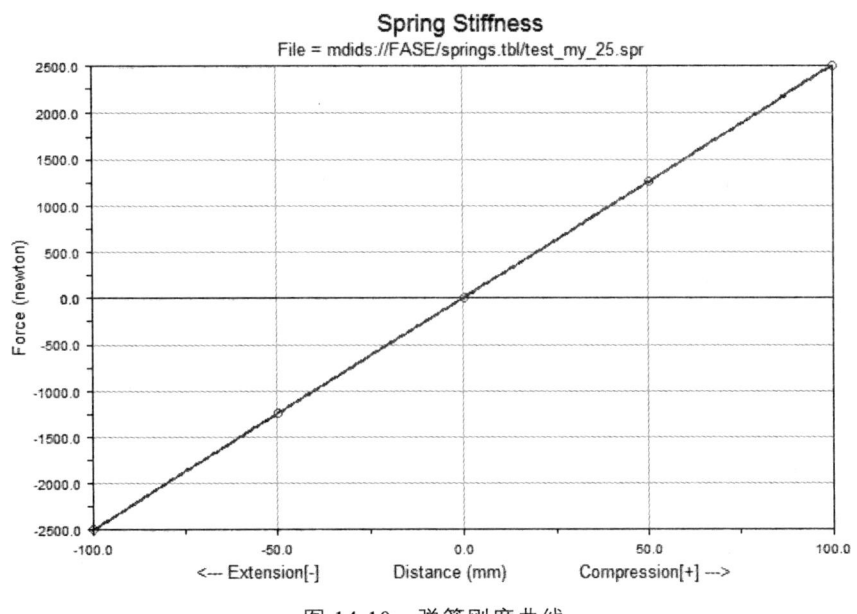

图 14-10 弹簧刚度曲线

- Property File（属性文件）：mdids：//FASE/springs.tbl/test_my_25.spr，弹簧属性文件用记事本文件打开后如下列信息所示，可以根据实验情况测出弹簧的参数（力与位移之间的关系，即刚度）填写到如下信息列表中，可修改部分为下划线部分；
- Spring Diameter（弹簧直径）：拖动滑块选择 22 mm；

- Spring of Coils（弹簧圈数）：拖动滑块选择 15。

（2）单击 OK 按钮，完成弹簧._FSAE_sus_pullrod.nsl_spring 的创建。

```
弹簧刚度信息：
$----------------------------------------------------------------MDI_HEADER
[MDI_HEADER]
 FILE_TYPE      =  'spr'
 FILE_VERSION   =  4.0
 FILE_FORMAT    =  'ASCII'
$--------------------------------------------------------------------UNITS
[UNITS]
 LENGTH  =  'mm'
 ANGLE   =  'degrees'
 FORCE   =  'newton'
 MASS    =  'kg'
 TIME    =  'second'
$--------------------------------------------------------------SPRING_DATA
[SPRING_DATA]
 FREE_LENGTH  =  205.7
$--------------------------------------------------------------------CURVE
[CURVE]
{  disp         force}
 -100.0        -2500.0
  -50.0        -1250.0
    0.0            0.0
   50.0         1250.0
  100.0         2500.0
```

13. 避震器

（1）单击 Build > Force > Damper > New 命令，弹出避震器创建对话框，如图 14-11 所示，在下列对话框中输入相应的数据：

- Damper Name（减震器名称）：damper；
- I Part：._FSAE_sus_pullrod.gel_damper_up；
- J Part：._FSAE_sus_pullrod.gel_damper_down；
- I Coordinate Reference（参考坐标）：._FSAE_sus_pullrod.ground.hpl_damper_up；
- J Coordinate Reference（参考坐标）：._FSAE_sus_pullrod.ground.hpl_zhijia_to_damper；

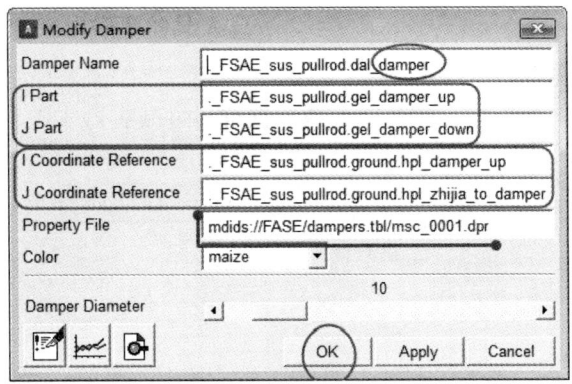

图 14-11 damper 避震器创建对话框

- Property File（属性文件）：mdids：//FASE/dampers.tbl/msc_0001.dpr，避震器系数曲线如图 14-12 所示，具体数据如下列避震器信息；

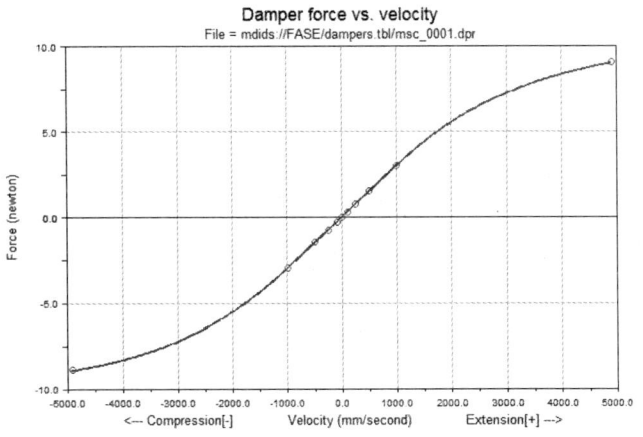

图 14-12 避震器系数曲线

- Damper Diameter（避震器直径）：拖动滑块选择 10 mm；
- Color：maize。

（2）单击 OK 按钮，完成避震器._FSAE_sus_pullrod.dal_damper 的创建。

```
$--------------------------------------------MDI_HEADER
[MDI_HEADER]
 FILE_TYPE    = 'dpr'
 FILE_VERSION = 4.0
 FILE_FORMAT  = 'ASCII'
$--------------------------------------------UNITS
[UNITS]
 LENGTH = 'mm'
 ANGLE  = 'degrees'
 FORCE  = 'newton'
```

```
   MASS  =  'kg'
   TIME  =  'second'
$-------------------------CURVE      %数据可通过减震器实验获取
[CURVE]
 {vel          force}
 -4916.935    -8.889
 -1000.0      -3.0
 -500.0       -1.5
 -250.0       -0.75
 -100.0       -0.3
  0.0          0.0
  100.0        0.3
  250.0        0.75
  500.0        1.5
  1000.0       3.0
  4914.298     9.0416
```

14. 部件 uca 与 prod 之间的 spherical 约束

（1）单击 Build > Attachments > Joint > New 命令，弹出约束件对话框，如图 14-13 所示，在下列对话框中输入相应的数据：

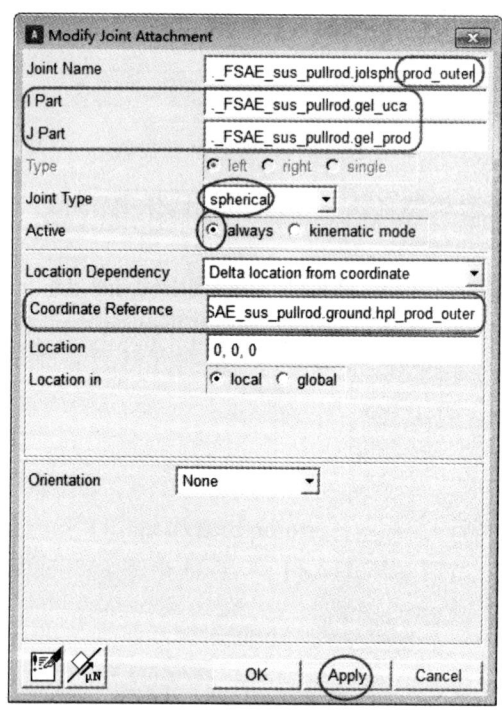

图 14-13　刚性约束-spherical

- Joint Name（约束副名称）：prod_outer；
- Type：left；
- I Part：._FSAE_sus_pullrod.gel_uca；
- J Part：._FSAE_sus_pullrod.gel_prod；
- Joint Type（约束副类型）：spherical；
- Active（激活）：always；
- Location Dependency：Delta location from coordinate；
- Coordinate Reference（参考坐标）：._FSAE_sus_pullrod.ground.hpl_prod_outer；
- Location：0，0，0；
- Location in：local。

（2）单击 Apply 按钮，完成约束副._FSAE_sus_pullrod.jolsph_prod_outer 的创建。

15. 部件 zhijia 与 prod 之间的 convel 约束

（1）单击 Build > Attachments > Joint > New 命令，在下列对话框中输入相应的数据：
- Joint Name（约束副名称）：zhijia_to_pullrod；
- I Part：._FSAE_sus_pullrod.gel_zhijia；
- J Part：._FSAE_sus_pullrod.gel_prod；
- Joint Type（约束副类型）：convel；
- Active（激活）：always；
- Location Dependency：Delta location from coordinate；
- Coordinate Reference（参考坐标）：._FSAE_sus_pullrod.ground.hpl_prod_to_bellcrank；
- Location：0，0，0；
- Location in：local；
- I-Part Axis：._FSAE_sus_pullrod.ground.hpl_zhijia_to_chassis；
- J-Part Axis：._FSAE_sus_pullrod.ground.hpl_prod_outer。

（2）单击 Apply 按钮，完成约束副._FSAE_sus_pullrod.jolcon_zhijia_to_pullrod 的创建。

16. 部件 zhijia 与 suspension_to_chassis 之间的 revolute 约束

（1）单击 Build > Attachments > Joint > New 命令，在下列对话框中输入相应的数据：
- Joint Name（约束副名称）：zhijia_to_chassis；
- Type：left；
- I Part：._FSAE_sus_pullrod.gel_zhijia；
- J Part：._FSAE_sus_pullrod.mts_suspension_to_chassis；
- Joint Type（约束副类型）：revolute；
- Active（激活）：always；
- Location Dependency：Delta location from coordinate；
- Coordinate Reference（参考坐标）：._FSAE_sus_pullrod.ground.hpl_zhijia_to_chassis；
- Location：0，0，0；
- Location in：local；

- Orientation Dependency：User entered values；
- Orient using：Euler Angles；
- Euler Angles：90，90，0。

（2）单击 Apply 按钮，完成约束副 ._FSAE_sus_pullrod.jolrev_zhijia_to_chassis 的创建。

17. 部件 zhijia 与 damper_down 之间的 hook 约束

（1）单击 Build > Attachments > Joint > New 命令，在下列对话框中输入相应的数据：
- Joint Name（约束副名称）：damper_down；
- Type：left；
- I Part：._FSAE_sus_pullrod.gel_zhijia；
- J Part：._FSAE_sus_pullrod.gel_damper_down；
- Joint Type（约束副类型）：hook；
- Active（激活）：always；
- Location Dependency：Delta location from coordinate；
- Coordinate Reference（参考坐标）：._FSAE_sus_pullrod.ground.hpl_zhijia_to_damper；
- Location：0，0，0；
- Location in：local；
- I-Part Axis：._FSAE_sus_pullrod.ground.hpl_prod_to_bellcrank；
- J-Part Axis：._FSAE_sus_pullrod.ground.hpl_damper_up。

（2）单击 Apply 按钮，完成约束副 ._FSAE_sus_pullrod.jolhoo_damper_down 的创建。

18. 部件 damper_up 与 suspension_to_chassis 之间的 hook 约束

（1）单击 Build > Attachments > Joint > New 命令，在下列对话框中输入相应的数据：
- Joint Name（约束副名称）：damper_up；
- Type：left；
- I Part：._FSAE_sus_pullrod.gel_damper_up；
- J Part：._FSAE_sus_pullrod.mts_suspension_to_chassis；
- Joint Type（约束副类型）：hook；
- Active（激活）：always；
- Location Dependency：Delta location from coordinate；
- Coordinate Reference（参考坐标）：._FSAE_sus_pullrod.ground.hpl_damper_up；
- Location：0，0，0；
- Location in：local；
- I-Part Axis：._FSAE_sus_pullrod.ground.hpl_damper_up_ref；
- J-Part Axis：._FSAE_sus_pullrod.ground.hpl_zhijia_to_damper。

（2）单击 Apply 按钮，完成约束副 ._FSAE_sus_pullrod.jolhoo_damper_up 的创建。

19. 部件 damper_up 与 damper_down 之间的 cylindrical 约束

（1）单击 Build > Attachments > Joint > New 命令，在下列对话框中输入相应的数据：
- Joint Name（约束副名称）：damper；
- Type：left；
- I Part：._FSAE_sus_pullrod.gel_damper_up；
- J Part：._FSAE_sus_pullrod.gel_damper_down；
- Joint Type（约束副类型）：cylindrical；
- Active（激活）：always；
- Location Dependency：Centered between coordinates；
- Centered between：Two Coordinates；
- Coordinate Reference #1（参考坐标）：._FSAE_sus_pullrod.ground.hpl_damper_up；
- Coordinate Reference #2（参考坐标）：._FSAE_sus_pullrod.ground.hpl_zhijia_to_ damper；
- Location Dependency：Orient axis to point；
- Coordinate Reference（参考坐标）：._FSAE_sus_pullrod.ground.hpl_damper_up。

（2）单击 OK 按钮，完成约束副._FSAE_sus_pullrod.jolcyl_damper 的创建。

至此，拉杆悬架模型建立完成，如图 14-14 所示，保存模型参考图 14-3，参数保持默认。

图 14-14　螺旋弹簧式拉杆悬架

20. 螺旋弹簧式拉杆悬架子系统

- 按 F9 键，将专家模板转换到标准模式，单击 File > New > Suspension 命令，弹出子系统对话框，如图 14-15 所示，在下列对话框中输入相应的数据：

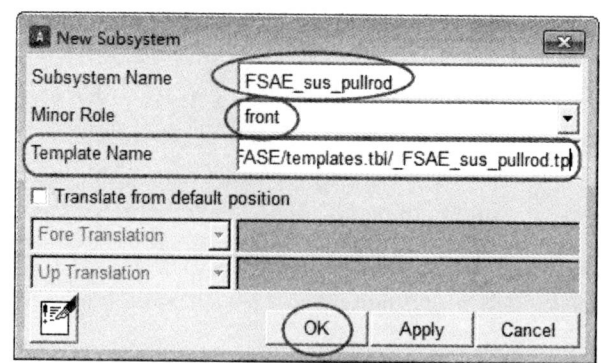

图 14-15 纵置式推杆悬架子系统

- Subsystem Name（系统名称）：FSAE_sus_pullrod；
- Minor Role（副特征）：front；
- Template Name（模板路径）：mdids：//FASE/templates.tbl/_FSAE_sus_pullrod.tpl。

（2）单击 OK 按钮，完成拉杆式悬架子系统 FSAE_sus_pullrod 的创建。

14.2 后螺旋弹簧式拉杆悬架

后螺旋弹簧式拉杆悬架模型硬点及变量参数信息如下，建立好的模型如图 14-1 所示。

```
Info for subsystem:   FSAE_sus_rear_pullrod

File Name       :  <FASE>/subsystems.tbl/FSAE_sus_rear_pullrod.sub
Template        :  mdids://FASE/templates.tbl/_FSAE_sus_rear_pullrod.tpl
Comments        :  *no comments found*
Major Role      :  suspension
Minor Role      :  rear

HARDPOINTS:

hardpoint name          symmetry        x_value     y_value     z_value
--------------          --------        -------     -------     -------
global                  single          1524.0      0.0         0.0
arb_bushing_mount       left/right      1651.0      -127.0      101.6
damper_up               left/right      1459.7      -140.0      354.8
damper_up_ref           left/right      1459.7      -100.0      454.8
drive_shaft_inr         left/right      1550.0      -200.0      225.0
```

lca_front	left/right	1270.0	-127.0	127.0
lca_outer	left/right	1498.6	-482.6	101.6
lca_rear	left/right	1651.0	-127.0	127.0
prod_outer	left/right	1549.4	-452.6	355.6
prod_to_bellcrank	left/right	1459.7	-200.0	144.8
tierod_inner	left/right	1676.4	-127.0	152.4
tierod_outer	left/right	1574.8	-457.2	152.4
uca_front	left/right	1270.0	-152.4	304.8
uca_outer	left/right	1549.4	-482.6	355.6
uca_rear	left/right	1625.6	-152.4	304.8
wheel_center	left/right	1524.0	-558.8	228.6
zhijia_to_chassis	left/right	1459.7	-100.0	144.8
zhijia_to_damper	left/right	1459.7	-170.0	164.8

PARAMETERS:

parameter name	symmetry	type	value
driveline_active	single	integer	1
kinematic_flag	single	integer	0
camber_angle	left/right	real	-1.5
drive_shaft_offset	left/right	real	75.0
toe_angle	left/right	real	0.0

14.3 单线移动超车仿真

1. 整车模型装配

（1）按 F9 键切换到标准模板。

（2）单击 File > Open > Assembly 命令，弹出整车装配对话框，如图 14-16 所示，在 Assembly Name 中输入：mdids://FASE/assemblies.tbl/FSAE.asy。

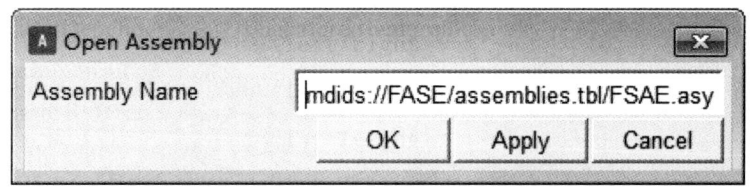

图 14-16　FSAE 整车模型打开

（3）单击 OK 按钮，打开 FSAE 整车模型。

（4）单击 File > Manage Assemblies > Replace Subsystem（s）命令，弹出替换子系统对话框，如图 14-17 所示，在下列对话框中输入相应的数据：

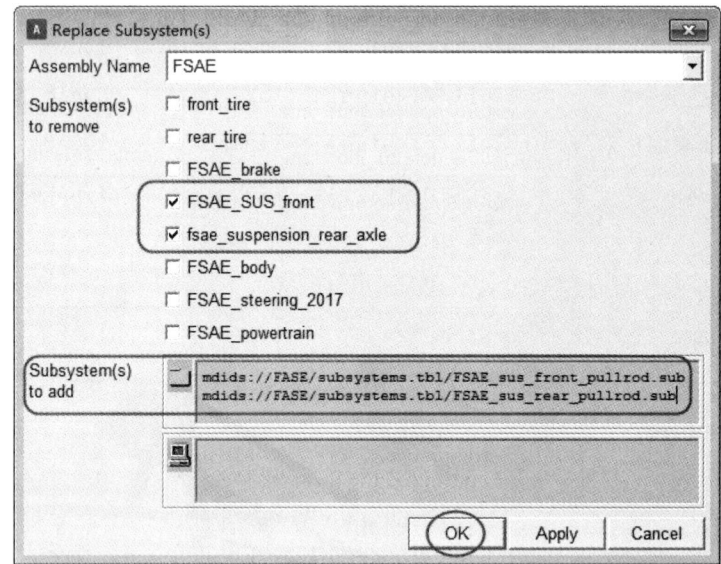

图 14-17　替换子系统对话框

- 勾选 FSAE_SUS_front；
- 勾选 fsae_suspension_rear_axle；
- Subsystem（s）To add：
① mdids：//FASE/subsystems.tbl/FSAE_sus_front_pullrod.sub；
② mdids：//FASE/subsystems.tbl/FSAE_sus_rear_pullrod.sub。

（5）单击 OK 按钮，完成前后悬架子系统的替换，此时整车模型如图 14-18 所示。

（6）单击 File > Save as > Assembly 命令，在下列对话框中输入相应的数据：

- Assembly Name：FSAE；
- New Assembly Name：FSAE_2020_pullrod；
- Target：Database/FSAE。

（7）单击 OK 按钮，完成基于螺旋弹簧式拉杆悬架整车 FSAE_2020_pullrod 的存储。

图 14-18　拉杆式悬架整车模型（发动机系统隐藏）

2．单线移动超车仿真

（1）单击 Simulate > Full-Vehicle Analysis > Open-loop steering Events > Single Lane Change 命令，弹出仿真对话框，如图 14-19 所示，在下列对话框中输入相应的数据：

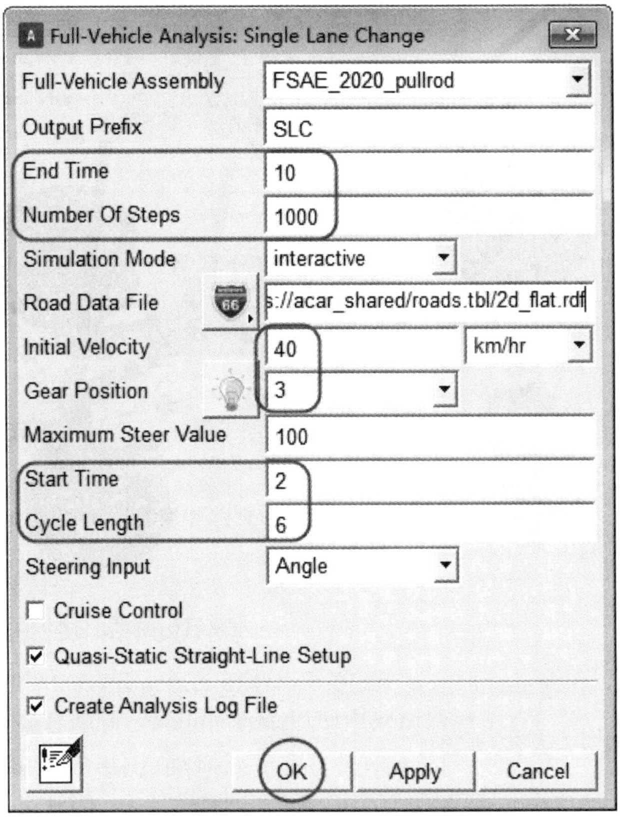

图 14-19　单线移动超车仿真参数

- Full-Vehicle Assembly：FSAE_2020_pullrod；
- Output Prefix：SLC；
- End Time：10；
- Number of steps：1000；
- Road Date File：mdids：//acar_shared/roads.tbl/2d_flat.rdf；
- Initial Velocity（单位：km/h）：40；
- Gear Position：3；
- Maximum Steer Value：100；
- Start Time：10；
- Cycle Length（指方向盘从转动开始到回正完毕所用的时间，即一个运行周期，单位为秒）：6；
- 勾选 Quasi-Static Straight-Line Setup。

（2）单击 Apply 按钮，完成 FSAE_2020_pullrod 赛车超车仿真设置并提交运算。

整车仿真完成后，从后处理动画可以看出整车各项振动特性均较大。如果车速超过 40 km/h，前轴转向横拉杆由于车身跳动过大而失效，前轴发生"塌陷"，从后处理车身曲线图可以验证判定的准确性。50 km/h 运行时前轴失效如图 14-20 所示。40 km/h 运行时整车俯仰、侧倾、横摆角加速度如图 14-21 ~ 图 14-23 所示，从图中可以看出，振动参数均过大，同时伴有高频振荡现象。

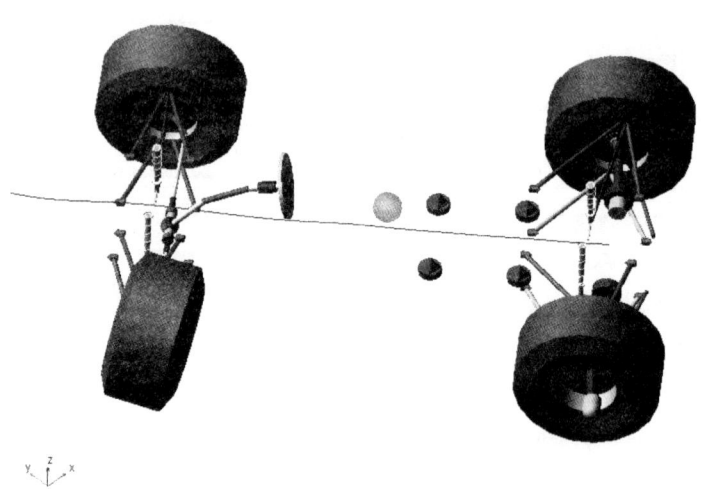

图 14-20　前轮"塌陷"（转向拉杆失效导致）

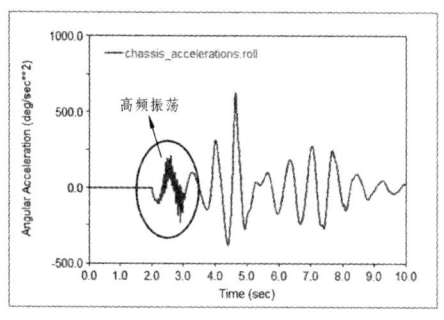
图 14-21 俯仰角加速度

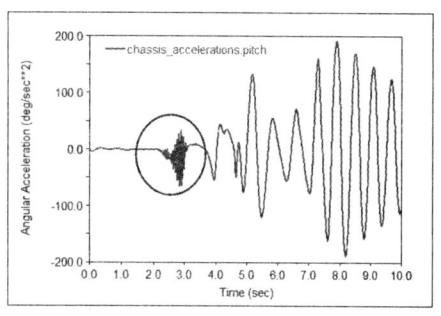
图 14-22 侧倾角加速度

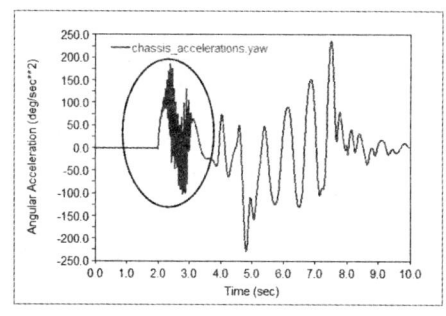
图 14-23 横摆角加速度

当车速过大时，整车发生失稳，失稳的原因是转向横拉杆因车身垂向振动过大，而导致车身振荡过大的原因可能是弹簧与减震器匹配不合理。此悬架弹簧刚度为 20 N/mm，根据经验，该刚度足以支撑车身的重量（FSAE 赛车车身质量一般为 200 kg 左右），因此悬架刚度与阻尼不存在问题（首先考虑弹簧刚度，阻尼一般不存在问题，阻尼即使匹配不精准，也会减小车身的振动参数）。通过观看动画，发现支架上下摆动过大，如图 14-24 所示，因此问题转化为如何减少支架的上下摆动。

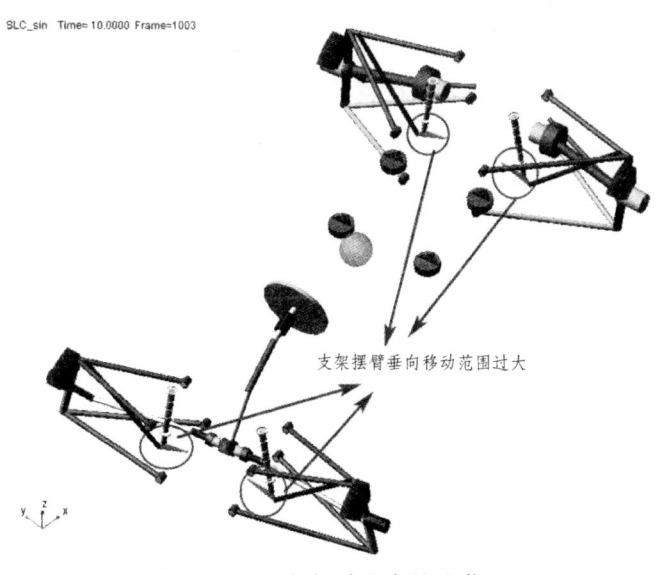
图 14-24 支架摆臂（车轮隐藏）

14.4 螺旋弹簧式拉杆悬架调试

减小支架的摆臂长度可以减小摆臂上下摆动的振幅，继而减小车身的振动，摆臂长度在Y方向减小50 mm，具体修改硬点如下：

① .FSAE_2020_pullrod.FSAE_sus_front_pullrod.ground.hpl_zhijia_to_damper：-10.8，-157.0，151.0；

② .FSAE_2020_pullrod.FSAE_sus_front_pullrod.ground.hpr_zhijia_to_damper：-10.8，157.0，151.0；

③ .FSAE_2020_pullrod.FSAE_sus_front_pullrod.ground.hpl_damper_up：-10.8，-107.0，351.0；

④ .FSAE_2020_pullrod.FSAE_sus_front_pullrod.ground.hpr_damper_up：-10.8，107.0，351.0；

⑤ .FSAE_2020_pullrod.FSAE_sus_front_pullrod.ground.hpl_prod_outer：0.0，-452.6，255.6；

⑥ .FSAE_2020_pullrod.FSAE_sus_front_pullrod.ground.hpr_prod_outer：0.0，452.6，255.6；

⑦ .FSAE_2020_pullrod.FSAE_sus_rear_pullrod.ground.hpl_zhijia_to_damper：1459.7，-140.0，164.8；

⑧ .FSAE_2020_pullrod.FSAE_sus_rear_pullrod.ground.hpr_zhijia_to_damper：1459.7，140.0，164.8；

⑨ .FSAE_2020_pullrod.FSAE_sus_rear_pullrod.ground.hpl_damper_up：1459.7，-110.0，354.8；

⑩ .FSAE_2020_pullrod.FSAE_sus_rear_pullrod.ground.hpr_damper_up：1459.7，110.0，354.8；

⑪ .FSAE_2020_pullrod.FSAE_sus_rear_pullrod.ground.hpl_prod_outer：1549.4，-452.6，255.6；

⑫ .FSAE_2020_pullrod.FSAE_sus_rear_pullrod.ground.hpr_prod_outer：1549.4，452.6，255.6。

重复单线移动超车仿真，设置相同的参数，车身稳定参数如图14-25~图14-27所示，发现性能有所改善，整体幅值变化明显，但依然伴有高频振荡现象。

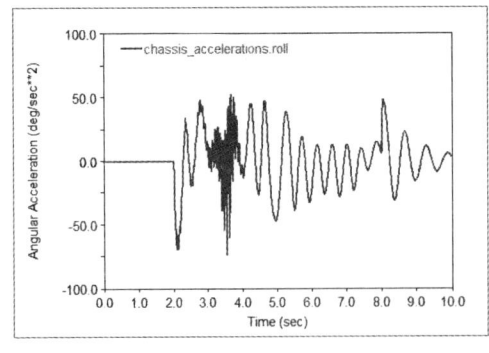

图14-25 俯仰角加速度

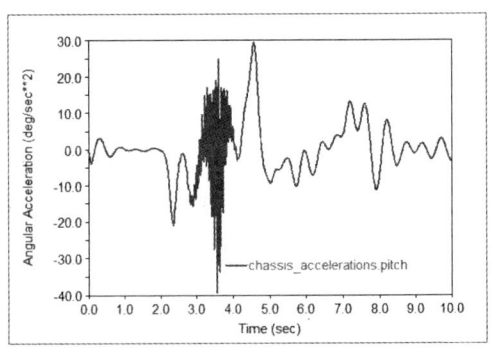

图14-26 侧倾角加速度

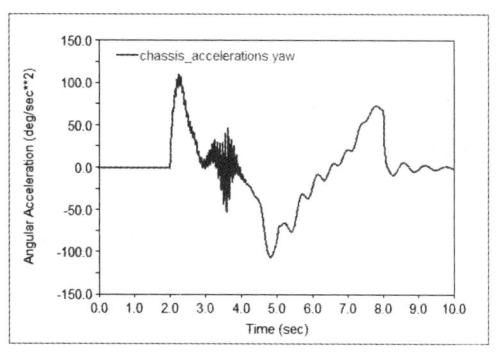

图 14-27　横摆角加速度

14.5　扭杆弹簧 MNF

扭杆弹簧悬架在轿车、赛车、工程车中均少见，2009 年，红牛 F1 车队 RB5 车型的后悬架上采用了扭杆弹簧悬架。扭杆弹簧悬架的优势是振幅小，弹簧的韧度比较大，高速过弯能提供较好的轮胎抓地力（转弯时保证轮胎与地面的贴合，提供较大的侧向力）；如果扭杆弹簧与拉杆悬架匹配，能进一步降低车身质心，提升赛车高速行驶时的下压力，最大限度地改善整车的性能。扭杆截面半径为 8 mm，长度为 200 mm；扭杆弹簧的模态中性文件 MNF 制作不再重复，本章扭杆弹簧 MNF 文件存储在章节文件中。

扭杆弹簧制作时，提取前 20 阶约束模态频率，前 6 阶模态变形如图 14-28～图 14-33 所示；20 阶频率数值如图 14-34 所示。

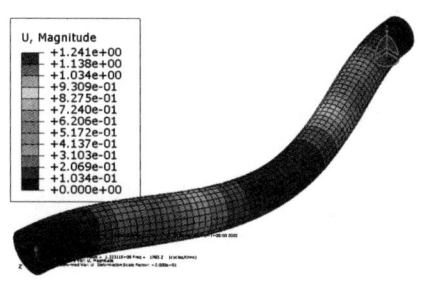

图 14-28　一阶模态

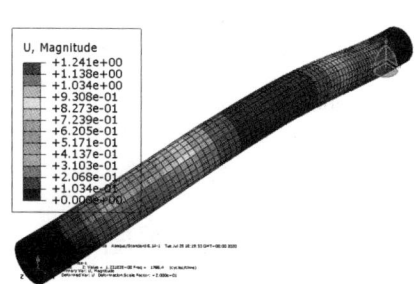

图 14-29　二阶模态

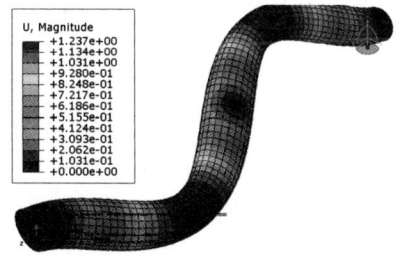

图 14-30　三阶模态

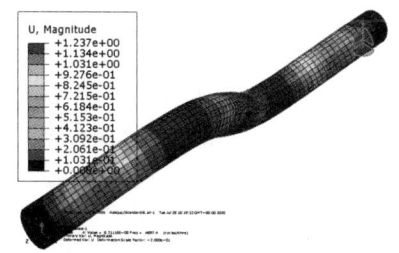

图 14-31　四阶模态

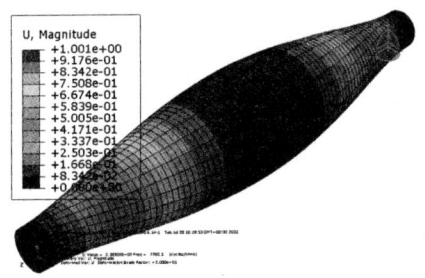
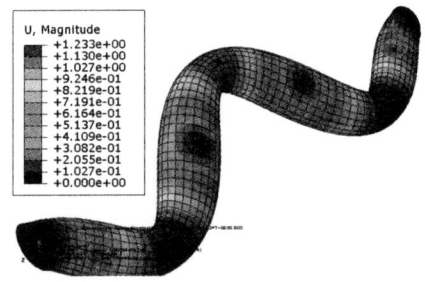

图 14-32　五阶模态　　　　　　　　　图 14-33　六阶模态

```
 1    Mode    1: Value = 1.22311E+08  Freq =  1760.2   (cycles/time)
 2    Mode    2: Value = 1.23183E+08  Freq =  1766.4   (cycles/time)
 3    Mode    3: Value = 8.65255E+08  Freq =  4681.6   (cycles/time)
 4    Mode    4: Value = 8.71118E+08  Freq =  4697.4   (cycles/time)
 5    Mode    5: Value = 2.38988E+09  Freq =  7780.5   (cycles/time)
 6    Mode    6: Value = 3.04739E+09  Freq =  8785.9   (cycles/time)
 7    Mode    7: Value = 3.06701E+09  Freq =  8814.1   (cycles/time)
 8    Mode    8: Value = 6.61087E+09  Freq =  12940.   (cycles/time)
 9    Mode    9: Value = 7.54063E+09  Freq =  13821.   (cycles/time)
10    Mode   10: Value = 7.58682E+09  Freq =  13863.   (cycles/time)
11    Mode   11: Value = 9.55630E+09  Freq =  15558.   (cycles/time)
12    Mode   12: Value = 1.51324E+10  Freq =  19578.   (cycles/time)
13    Mode   13: Value = 1.52209E+10  Freq =  19635.   (cycles/time)
14    Mode   14: Value = 2.14896E+10  Freq =  23331.   (cycles/time)
15    Mode   15: Value = 2.63814E+10  Freq =  25851.   (cycles/time)
16    Mode   16: Value = 2.64577E+10  Freq =  25888.   (cycles/time)
17    Mode   17: Value = 2.66064E+10  Freq =  25961.   (cycles/time)
18    Mode   18: Value = 3.81736E+10  Freq =  31096.   (cycles/time)
19    Mode   19: Value = 4.19970E+10  Freq =  32616.   (cycles/time)
20    Mode   20: Value = 4.22250E+10  Freq =  32704.   (cycles/time)
```

图 14-34　前 20 阶约束模态频率

14.6　刚度虚拟实验

（1）启动 ADAMS/View，单击 Bodies > Flexible Bodies > Create a Flexible Body，弹出导入 MNF 文件对话框，如图 14-35 所示，在下列对话框中输入相应的数据：
- Flexible Body Name：.MODEL_1.FSAE_torsion；
- MNF：D：\ADAMS_MNF\fsae_torsion.mnf。

（2）单击 OK 按钮，完成模态中性文件的导入。
- 设置角度单位：弧度，rad；
- 在扭杆弹簧一端施加固定约束，约束部件为扭杆弹簧与大地，如图 14-36 所示；
- 在扭杆弹簧另一端施加扭矩，如图 14-37 所示；
- 约束施加完成，扭杆弹簧刚度测试前处理完成，如图 14-28 所示。

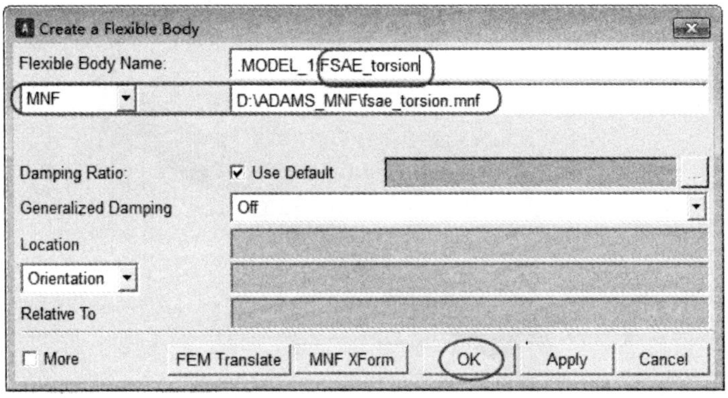

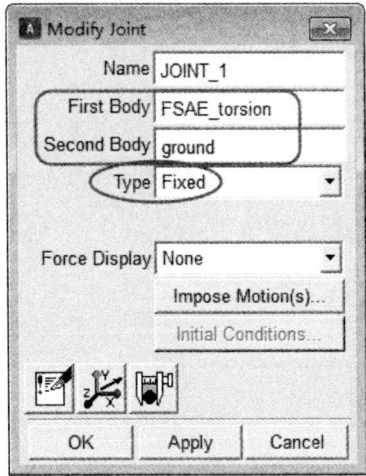

图 14-35　导入 MNF 文件

图 14-36　固定约束

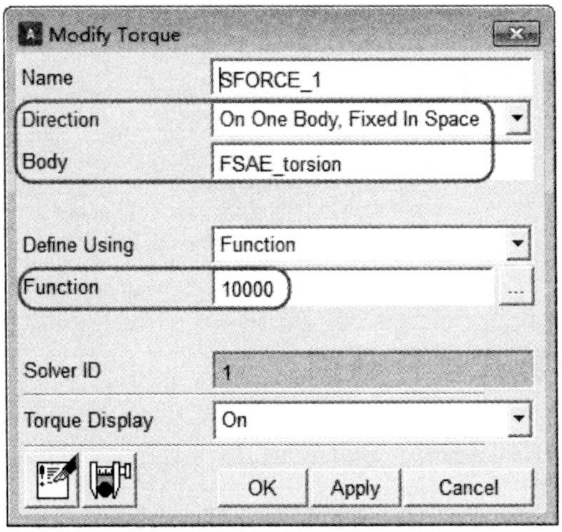

图 14-37　扭矩（单位：N·mm）

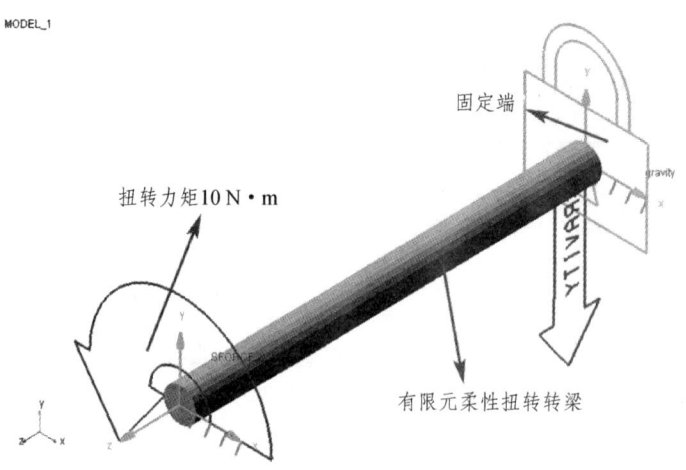

图 14-38　扭杆弹簧刚度测试模型

（3）创建测量函数：AZ（MARKER_3，MARKER_2），此函数测量扭杆弹簧在 10 N·m 的扭矩作用下扭杆弹簧前端相对于后端转动的角度。

设置仿真时间为 1 s，仿真步数为 1 000，仿真完成后，测量角如图 14-39 所示，从图中可以看出，扭杆弹簧最大的转动角度为 0.004 1 rad，弧度与度之间的关系变量为 57.3。根据刚度计算公式：$\phi = \dfrac{Tl}{GI_p}$，计算出扭转刚度为 487.80 N·m/rad。

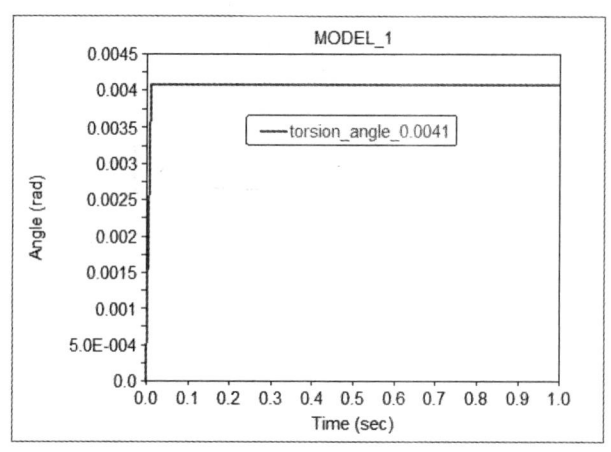

图 14-39　扭杆弹簧转动弧度

14.7　前扭杆弹簧式拉杆悬架

1. 模型导入

（1）启动 ADAMS/CAR，选择专家模块进入建模界面。

（2）单击 File > Open 命令，弹出打开模板对话框，如图 14-40 所示，在 Template Name 中输入：mdids：//FASE/templates.tbl/_FSAE_sus_pullrod.tpl。

（3）单击 OK 按钮，完成模型的导入。

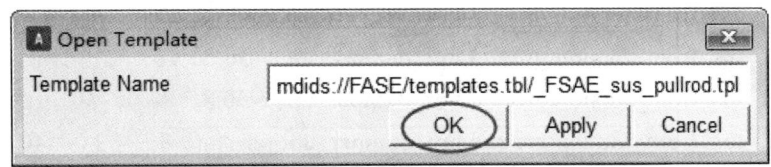

图 14-40　模板对话框

2. 删除螺旋弹簧、约束

（1）删除弹簧：

① ._FSAE_sus_pullrod.nsl_spring；

② ._FSAE_sus_pullrod.nsr_spring。

（2）删除约束：

① ._FSAE_sus_pullrod.jolrev_zhijia_to_chassis；

② ._FSAE_sus_pullrod.jorrev_zhijia_to_chassis。

（3）单击 File > Save As 命令，弹出保存模板对话框，如图 14-41 所示，在下列对话框中输入相应的数据：

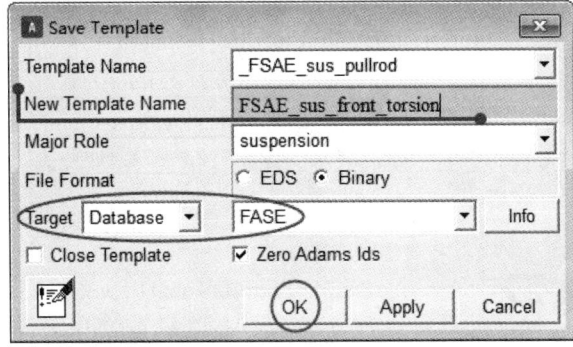

图 14-41　模型保存_FSAE_sus_front_torsion

- Template Name：_FSAE_sus_pullrod；
- New Template Name：FSAE_sus_front_torsion；
- Major Role（主特征）：suspension；；
- File Format：Binary；
- Target：Datebase/FASE。

（4）单击 OK 按钮，完成推杆式悬架模型模板 FSAE_sus_front_torsion 的保存。

3. 调试硬点

① ._FSAE_sus_front_torsion.ground.hpl_zhijia_to_damper：-50.8，-137.0，121.0；

② ._FSAE_sus_front_torsion.ground.hpr_zhijia_to_damper：-50.8，137.0，121.0；

③ ._FSAE_sus_front_torsion.ground.hpl_prod_to_bellcrank：-50.8，-237.0，81.0；

④ ._FSAE_sus_front_torsion.ground.hpr_prod_to_bellcrank：-50.8，237.0，81.0；

⑤ ._FSAE_sus_front_torsion.ground.hpl_zhijia_to_chassis：-50.8，-137.0，81.0；
⑥ ._FSAE_sus_front_torsion.ground.hpr_zhijia_to_chassis：-50.8，137.0，81.0；
⑦ ._FSAE_sus_front_torsion.ground.hpl_damper_up：-50.8，-87.0，301.0；
⑧ ._FSAE_sus_front_torsion.ground.hpr_damper_up：-50.8，87.0，301.0；
⑨ ._FSAE_sus_front_torsion.ground.hpl_damper_up_ref：-50.8，-87.0，401.0；
⑩ ._FSAE_sus_front_torsion.ground.hpr_damper_up_ref：-50.8，87.0，401.0；

4. 结构框

（1）单击 Build > Construction Frame > New 命令，弹出创建结构框对话框，如图 14-42 所示，在下列对话框中输入相应的数据：

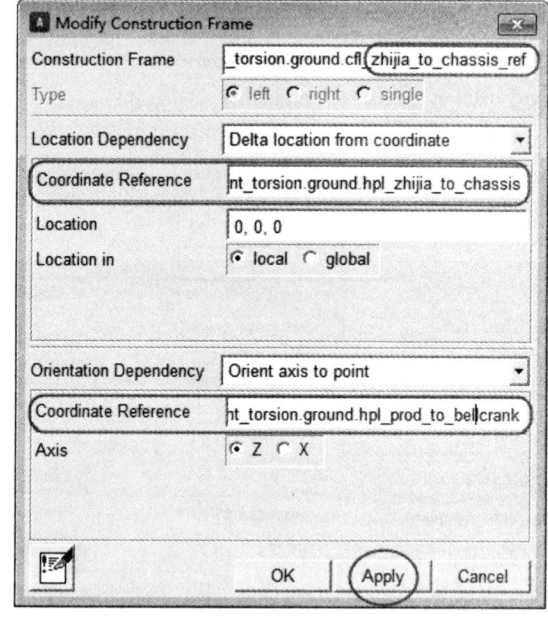

图 14-42　模结构框_zhijia_to_chassis_ref

- Construction Frame（结构框名称）：zhijia_to_chassis_ref；
- Location Dependency：Delta location from coordinate；
- Coordinate Reference（参考坐标）：._FSAE_sus_front_torsion.ground.hpl_zhijia_to_chassis；
- Location：0，0，0；
- Location in：local；
- Orientation Dependency：Orient axis to point；
- Coordinate Reference（参考坐标）：._FSAE_sus_front_torsion.ground.hpl_prod_to_bellcrank；
- Axis：Z。

（2）单击 Apply 按钮，完成._FSAE_sus_front_torsion.ground.cfl_zhijia_to_chassis_ref 结构框的创建。同理，创建其他结构框。

- Construction Frame（结构框名称）：zhijia_to_chassis；
- Location Dependency：Delta location from coordinate；
- Coordinate Reference（参考坐标）：._FSAE_sus_front_torsion.ground.cfl_zhijia_to_chassis_ref；
- Location：0，0，30；
- Location in：local；
- Orientation Dependency：Orient axis to point；
- Coordinate Reference（参考坐标）：._FSAE_sus_front_torsion.ground.hpl_prod_to_bellcrank；
- Axis：Z。

（3）单击 Apply 按钮，完成._FSAE_sus_front_torsion.ground.cfl_zhijia_to_chassis 结构框的创建。

- Construction Frame（结构框名称）：torsion_mount；
- Location Dependency：Delta location from coordinate；
- Coordinate Reference（参考坐标）：._FSAE_sus_front_torsion.ground.cfl_zhijia_to_chassis；
- Location：200，0，0；
- Location in：local；
- Orientation Dependency：User entered values；
- Orient using：Euler Angles；
- Euler Angles：0，0，0。

（4）单击 OK 按钮，完成._FSAE_sus_front_torsion.ground.cfl_torsion_mount 结构框的创建。

5. 导入扭杆弹簧

（1）单击 Build > Part > Flexible Body > New 命令，弹出创建部件对话框，如图 14-43 所示，在下列对话框中输入相应的数据：

- Flexible Body Name：fsae_torsion；
- Type：left；
- Location Dependency：Delta location from coordinate；
- Coordinate Reference（参考坐标）：._FSAE_sus_front_torsion.ground.cfl_zhijia_to_chassis；
- Location：0，0，0；
- Location in：local；
- Orientation Dependency：User-entered values；
- Orient using：Euler Angles；
- Euler Angles：90，90，0；
- Left MNF File：file：//D：/ADAMS_MNF/fsae_torsion.mnf；
- Right MNF File：file：//D：/ADAMS_MNF/fsae_torsion.mnf；
- Color：white。

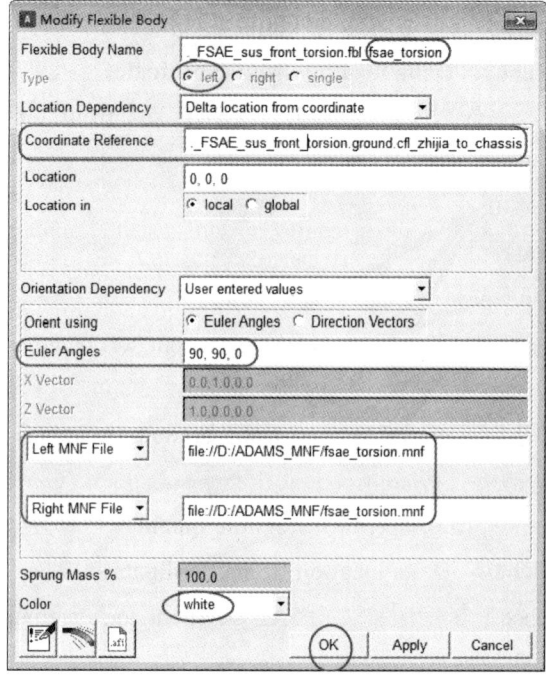

图 14-43 扭杆弹簧部件

（2）单击 OK 按钮，完成部件._FSAE_sus_front_torsion.fbl_fsae_torsion 的创建。

6. 部件 fsae_torsion 与安装件 suspension_to_chassis 之间的 revolute 约束

（1）单击 Build > Attachments > Joint > New 命令，弹出创建约束件对话框，如图 14-44 所示，在下列对话框中输入相应的数据：

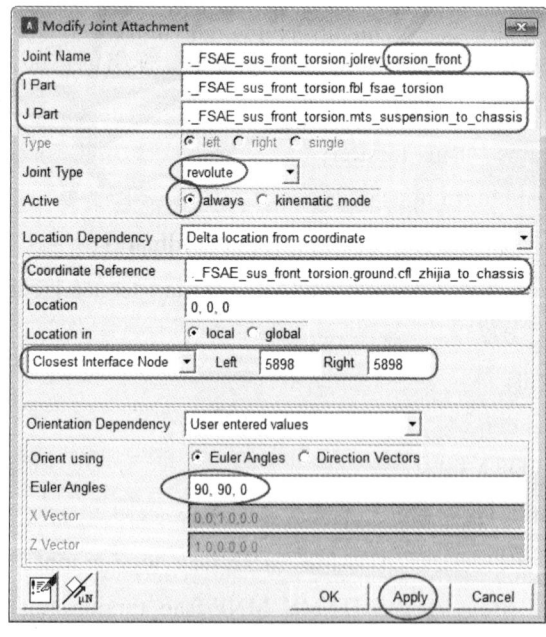

图 14-44 刚性约束对话框_torsion_front

- Joint Name（约束副名称）：torsion_front；
- Type：Left；
- I Part：._FSAE_sus_front_torsion.fbl_fsae_torsion；
- J Part：._FSAE_sus_front_torsion.mts_suspension_to_chassis；
- Joint Type（约束副类型）：revolute；
- Active（激活）：always；
- Location Dependency：Delta location from coordinate；
- Coordinate Reference（参考坐标）：._FSAE_sus_front_torsion.ground.cfl_zhijia_to_ chassis；
- Location：0，0，0；
- Location in：local；
- Closest Interface Node（指参考点：._FSAE_sus_front_torsion.ground.cfl_zhijia_to_chassis 附近的节点，注意节点指的是有限元节点或者是在有限元软件中创建的接口点）：left/5898，right/5898（5898 指有限元软件中创建的接口点）；
- Orientation Dependency：User entered values；
- Orient using：Euler Angles；
- Euler Angles：0，90，0。

（2）单击 Apply 按钮，完成约束副._FSAE_sus_front_torsion.jolrev_torsion_front 的创建。

7. 部件 fsae_torsion 与安装件 suspension_to_chassis 之间的 fixed 约束

（1）单击 Build > Attachments > Joint > New 命令，在下列对话框中输入相应的数据：

- Joint Name（约束副名称）：torsion_rear；
- Type：left；
- I Part：._FSAE_sus_front_torsion.fbl_fsae_torsion；
- J Part：._FSAE_sus_front_torsion.mts_suspension_to_chassis；
- Joint Type（约束副类型）：fixed；
- Active（激活）：always；
- Location Dependency：Delta location from coordinate；
- Coordinate Reference（参考坐标）：._FSAE_sus_front_torsion.ground.cfl_torsion_mount；
- Location：0，0，0；
- Location in：local；
- Closest Interface Node：left/5897，right/5897（5897 指有限元软件中创建的接口点）。

（2）单击 Apply 按钮，完成约束副._FSAE_sus_front_torsion.jolfix_torsion_rear 的创建。

8. 部件 fsae_torsion 与 zhijia 之间的 fixed 约束

（1）单击 Build > Attachments > Joint > New 命令，在下列对话框中输入相应的数据：

- Joint Name（约束副名称）：torsion_to_zhijia；
- Type：left；
- I Part：._FSAE_sus_front_torsion.fbl_fsae_torsion；
- J Part：._FSAE_sus_front_torsion.gel_zhijia；

- Joint Type（约束副类型）：fixed；
- Active（激活）：always；
- Location Dependency：Delta location from coordinate；
- Coordinate Reference（参考坐标）：._FSAE_sus_front_torsion.ground.cfl_zhijia_to_chassis；
- Location：0，0，0；
- Location in：local；
- Closest Interface Node：left/5898，right/5898（5898 指有限元软件中创建的接口点）。

（2）单击 OK 按钮，完成约束副._FSAE_sus_front_torsion.jolfix_torsion_to_zhijia 的创建；至此，纵置扭杆弹簧式拉杆悬架创建完成，如图 14-45 所示。

图 14-45　前纵置扭杆弹簧式拉杆悬架

9. 纵置扭杆弹簧式拉杆悬架子系统

（1）按 F9 键，将专家模板转换到标准模式，单击 File > New > Suspension 命令，弹出子系统对话框，如图 14-46 所示，在下列对话框中输入相应的数据：

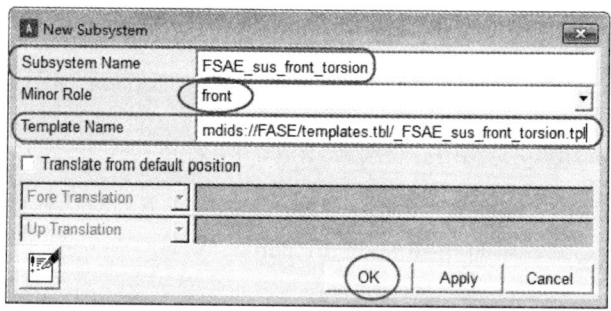

图 14-46　纵置式推杆悬架子系统

- Subsystem Name（系统名称）：FSAE_sus_front_torsion；
- Minor Role（副特征）：front；
- Template Name（模板路径）：mdids://FASE/templates.tbl/_FSAE_sus_front_torsion.tpl。

（2）单击 OK 按钮，完成推杆式悬架子系统 FSAE_sus_front_torsion 的创建。

14.8 后扭杆弹簧式拉杆悬架

后扭杆弹簧式拉杆悬架模型建模过程不再重复，请参考前扭杆弹簧式拉杆悬架模型建模过程，后扭杆弹簧式拉杆悬架的硬点与变量参数信息如下，建立好的后扭杆弹簧式拉杆悬架模型子系统如图 14-47 所示。

```
Info for subsystem:   FSAE_sus_rear_torsion

File Name       :   <FASE>/subsystems.tbl/FSAE_sus_rear_torsion.sub
Template        :   mdids://FASE/templates.tbl/_FSAE_sus_rear_torsion.tpl
Comments        :   *no comments found*
Major Role      :   suspension
Minor Role      :   rear

HARDPOINTS:

hardpoint name              symmetry          x_value      y_value      z_value
--------------              --------          -------      -------      -------

global                      single            1524.0       0.0          0.0
arb_bushing_mount           left/right        1651.0       -127.0       101.6
damper_up                   left/right        1409.7       -70.0        304.8
damper_up_ref               left/right        1409.7       -100.0       404.8
drive_shaft_inr             left/right        1550.0       -200.0       225.0
lca_front                   left/right        1270.0       -127.0       127.0
lca_outer                   left/right        1498.6       -482.6       101.6
lca_rear                    left/right        1651.0       -127.0       127.0
prod_outer                  left/right        1549.4       -452.6       355.6
prod_to_bellcrank           left/right        1409.7       -200.0       94.8
tierod_inner                left/right        1676.4       -127.0       152.4
tierod_outer                left/right        1574.8       -457.2       152.4
```

uca_front	left/right	1270.0	-152.4	304.8
uca_outer	left/right	1549.4	-482.6	355.6
uca_rear	left/right	1625.6	-152.4	304.8
wheel_center	left/right	1524.0	-558.8	228.6
zhijia_to_chassis	left/right	1409.7	-100.0	94.8
zhijia_to_damper	left/right	1409.7	-100.0	134.8

PARAMETERS:

parameter name	symmetry	type	value
driveline_active	single	integer	1
kinematic_flag	single	integer	0
camber_angle	left/right	real	-1.5
drive_shaft_offset	left/right	real	75.0
toe_angle	left/right	real	0.0

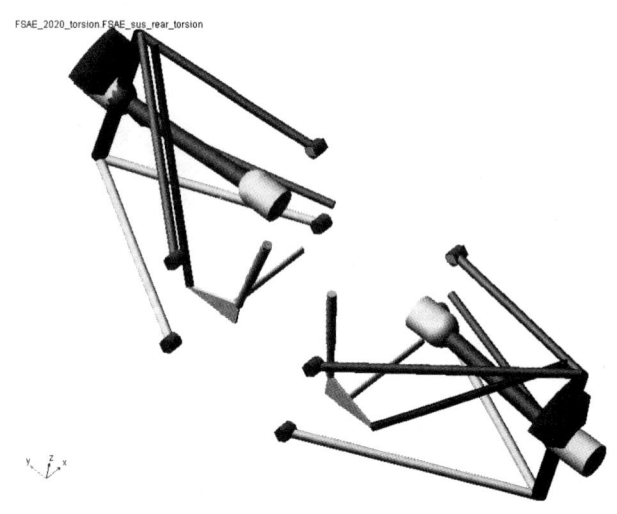

图 14-47 后纵置扭杆弹簧式拉杆悬架

14.9 转向盘角脉冲仿真

整车装配信息如下，读者可根据装配信息重新装配整车或者通过上述章节的整车模型替换前后悬架完成整车 FSAE_2020_torsion 的建立；装配完成后，采用前后扭转弹簧式拉杆悬架的整车模型如图 14-48 所示。

图 14-48 整车 FSAE_2020_torsion

```
Assembly Name   :   FSAE_2020_torsion
Assembly Class  :   full_vehicle
File Name       :   <FASE>/assemblies.tbl/FSAE_2020_torsion.asy


SUBSYSTEM NAME                  MAJOR ROLE              MINOR ROLE


front_tire                      wheel                   front
rear_tire                       wheel                   rear
FSAE_brake                      brake_system            any
FSAE_body                       body                    any
FSAE_steering_2017              steering                front
FSAE_powertrain                 powertrain              rear
FSAE_sus_front_torsion          suspension              front
FSAE_sus_rear_torsion           suspension              rear


Info for subsystem:   front_tire


File Name       :   <FASE>/subsystems.tbl/front_tire.sub
Template        :   mdids://FASE/templates.tbl/_handling_tire.tpl
Comments        :   *no comments found*
```

```
Major Role      :   wheel
Minor Role      :   front


Info for subsystem:   rear_tire

File Name       :   <FASE>/subsystems.tbl/rear_tire.sub
Template        :   mdids://FASE/templates.tbl/_handling_tire.tpl
Comments        :   *no comments found*
Major Role      :   wheel
Minor Role      :   rear


Info for subsystem:   FSAE_brake

File Name       :   <FASE>/subsystems.tbl/FSAE_brake.sub
Template        :   mdids://FASE/templates.tbl/_brake_system_4Wdisk.tpl
Comments        :
Template     :  4 Wheel Disk Brake System
Subsystem    :  *no subsystem comments found*
Major Role      :   brake_system
Minor Role      :   any


Info for subsystem:   FSAE_body

File Name       :   <FASE>/subsystems.tbl/FSAE_body.sub
Template        :   mdids://FASE/templates.tbl/_fsae_chassis.tpl
Comments        :
Template     :  Simple One Part Rigid Chassis
Subsystem    :  *no subsystem comments found*
Major Role      :   body
Minor Role      :   any


Info for subsystem:   FSAE_steering_2017
```

```
File Name       : <FASE>/subsystems.tbl/FSAE_steering_2017.sub
Template        : mdids://FASE/templates.tbl/_FSAE_steering_mid.tpl
Comments        : *no comments found*
Major Role      : steering
Minor Role      : front


Info for subsystem:  FSAE_powertrain

File Name       : <FASE>/subsystems.tbl/FSAE_powertrain.sub
Template        : mdids://FASE/templates.tbl/_powertrain.tpl
Comments        :
Template    :   Example of a non-spinning powertrain
Subsystem   :   *no subsystem comments found*
Major Role      : powertrain
Minor Role      : rear


Info for subsystem:  FSAE_sus_front_torsion

File Name       : <FASE>/subsystems.tbl/FSAE_sus_front_torsion.sub
Template        : mdids://FASE/templates.tbl/_FSAE_sus_front_torsion.tpl
Comments        : *no comments found*
Major Role      : suspension
Minor Role      : front


Info for subsystem:  FSAE_sus_rear_torsion

File Name       : <FASE>/subsystems.tbl/FSAE_sus_rear_torsion.sub
Template        : mdids://FASE/templates.tbl/_FSAE_sus_rear_torsion.tpl
Comments        : *no comments found*
Major Role      : suspension
Minor Role      : rear
```

（1）单击 Simulate > Full-Vehicle Analysis > Open-loop steering Events > Impulse Steer 命令，弹出阶跃仿真对话框，如图 14-49 所示，在下列对话框中输入相应的数据：

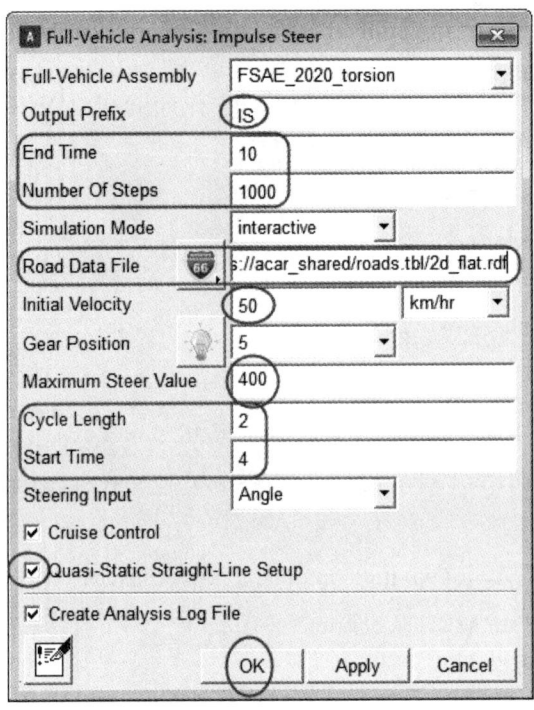

图 14-49 角脉冲仿真

- Full-Vehicle Assembly：FSAE_2020_torsion；
- Output Prefix：IS；
- End Time：10；
- Number of steps：1000；
- Simulation Mode：interactive；
- Road Date File：mdids：//acar_shared/roads.tbl/2d_flat.rdf；
- Initial Velocity（单位：km/h）：50；
- Gear Position：5；
- Maximum Steer Value（单位：度）：400；
- Cycle Length（单位：s）：2；
- Start Time：4；
- 勾选 Quasi-Static Straight-Line Setup。

（2）单击 OK 按钮，完成 FSAE_2020_torsion 赛车脉冲转向仿真设置并提交运算，运算完成后，整车运行轨迹如图 14-50 所示。

整车运行参数如图 14-51～图 14-60 所示，从图中可以看出，曲线急剧变化均从 4 s 开始，轮胎侧向力在 4 s 开始急剧增大，前后轮胎的滑移率亦是急剧变大。从滑移率及横摆角加速度图中可以判定，此时整车瞬间失去稳定性。

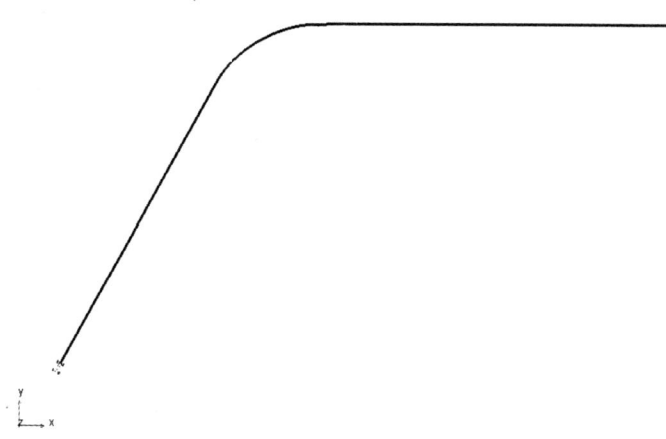

图 14-50 车辆运行轨迹

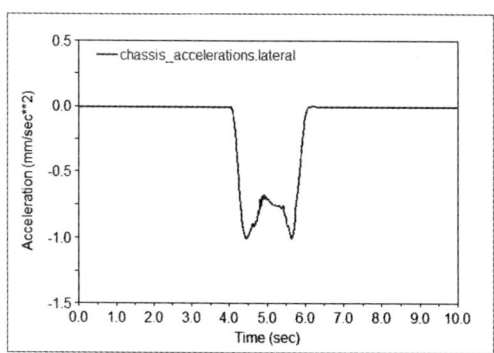

图 14-51 侧向加速度

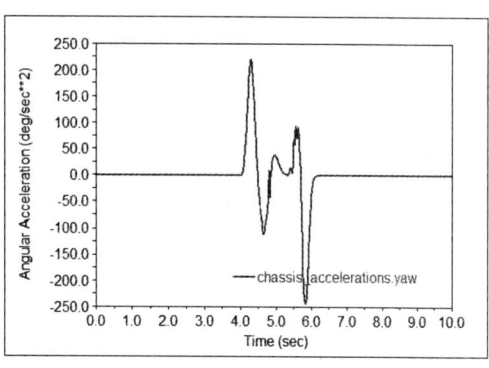

图 14-52 横摆角加速度

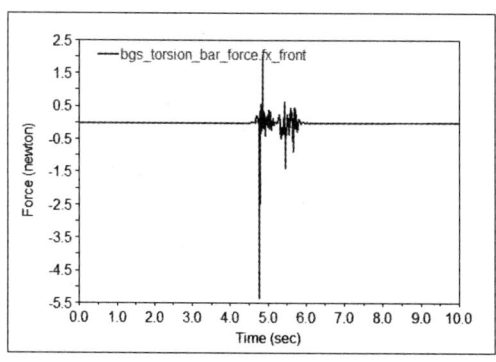

图 14-53 前扭杆弹簧 X 方向受力

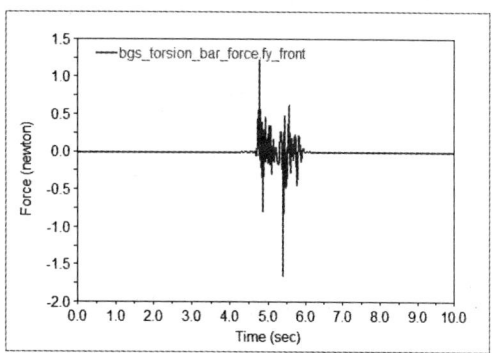

图 14-54 前扭杆弹簧 Y 方向受力

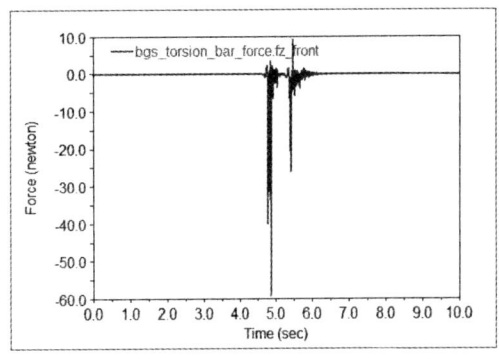

图 14-55　前扭杆弹簧 Z 方向受力

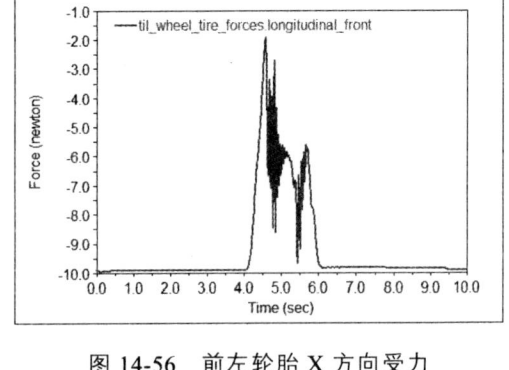

图 14-56　前左轮胎 X 方向受力

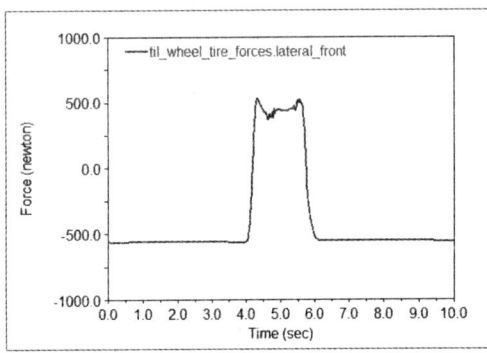

图 14-57　前左轮胎 Y 方向受力

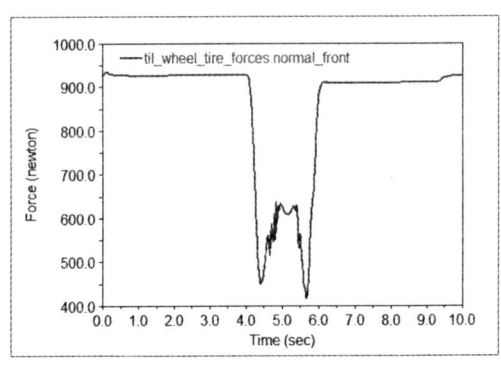

图 14-58　前左轮胎 Z 方向受力

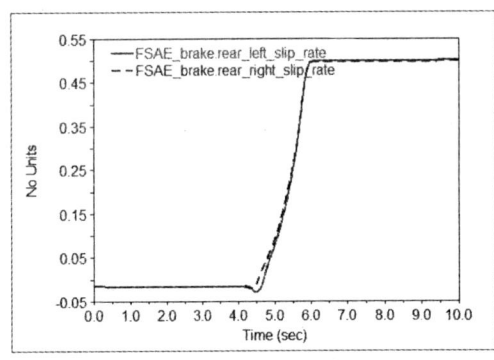

图 14-59　后轮胎滑移率

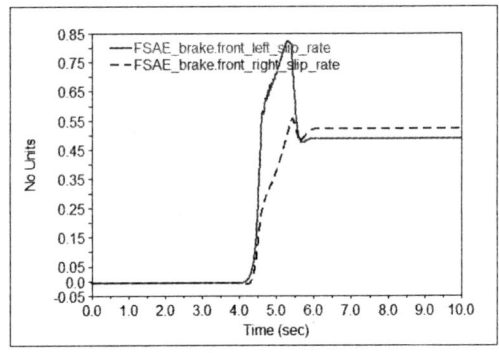

图 14-60　前轮胎滑移率

14.10　扭杆弹簧调试

上述扭杆弹簧在有限元模型中左接口时较为粗糙，RP 点与对应的同心圆周边采用梁单元连接，这种连接方法的缺点是圆周边节点受力较大，与真实扭杆弹簧连接不符；同时上述扭杆弹簧的刚度过大，针对此问题，重新制作扭杆弹簧模态中性文件 MNF。扭杆弹簧的截面半径为 5 mm，长度为 400 mm；此时 RP 点与同心圆所在的两个厚度单元内所有的节点采用梁单元连接（并非最外圈的节点），此时 RP 点位为对外连接的接口，外力通过 RP 点把力均匀地传递到两个厚度内所有节点上，跟实际状态符合，如图 14-61 所示；整根扭杆弹簧约束模型如图 14-62 所示，与上述扭杆弹簧相比，增加 RP-3，扭转弹簧与支撑架的固定及车架旋转的接口分开；此时 RP-1 与车架旋转约束，RP-2 与车架固定约束，RP-3 与支架固定约束。

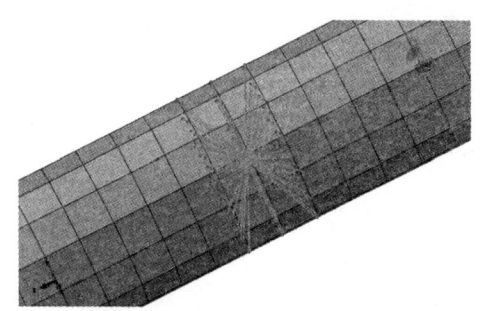

图 14-61　MPC 多点约束

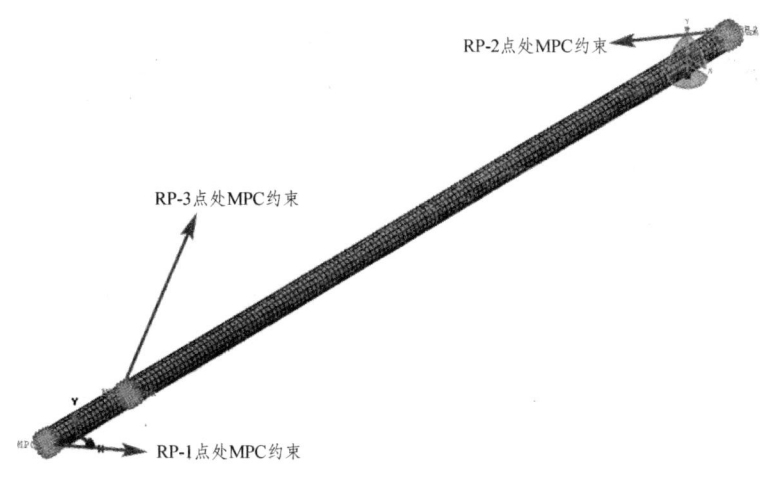

图 14-62　扭杆弹簧约束模型

前处理完成后计算扭杆弹簧的前 20 阶模态，前六阶模态如图 14-63 ~ 图 14-68 所示，模态中性文件制作参考横向稳定杆Ⅱ；测量扭杆弹簧的扭转角度最大值为 0.055 rad，如图 14-69 所示，施加扭转力矩为 10 N·m；根据刚度计算公式：$\phi = \dfrac{Tl}{GI_p}$，计算出刚度为 72.73 N·m/rad。此时扭转弹簧刚度变小。

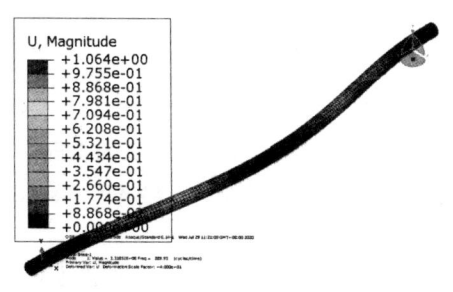

图 14-63　一阶变形图

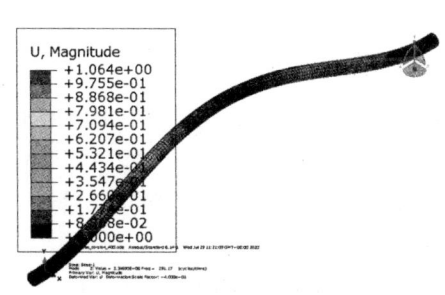

图 14-64　二阶变形图

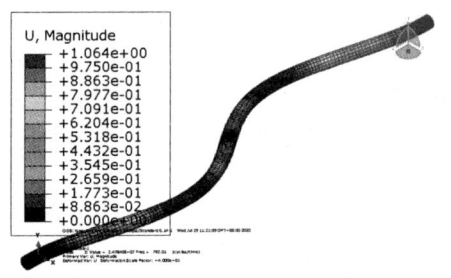
图 14-65 三阶变形图

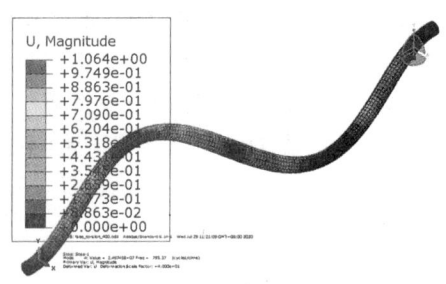
图 14-66 四阶变形图

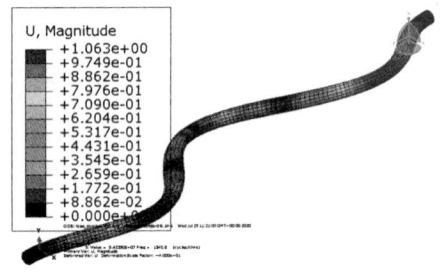
图 14-67 五阶变形图

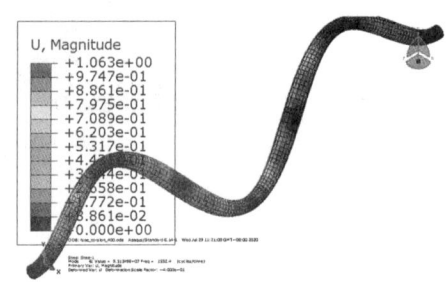
图 14-68 六阶变形图

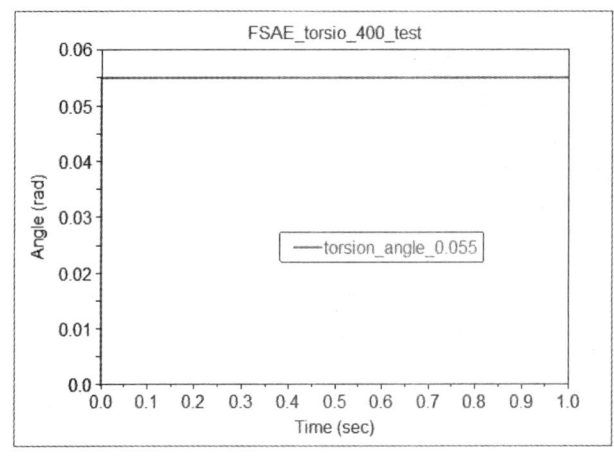
图 14-69 扭转角度

1. 结构框调试

扭杆弹簧制作完成后，因为长度增加 200 mm，即一倍，此时对应的结构框需要调整，同时还需新建 RP-1 点对应的结构框，RP-1 与 RP-3 之间的距离为 50 mm，RP-2 与 RP-3 之间的距离为 350 mm；调试结构框前，将模型 FSAE_sus_front_torsion 另存为：FSAE_sus_front_torsion_400。

（1）单击 Build > Construction Frame > New 命令，弹出创建结构对话框，如图 14-70 所示，在下列对话框中输入相应的数据：

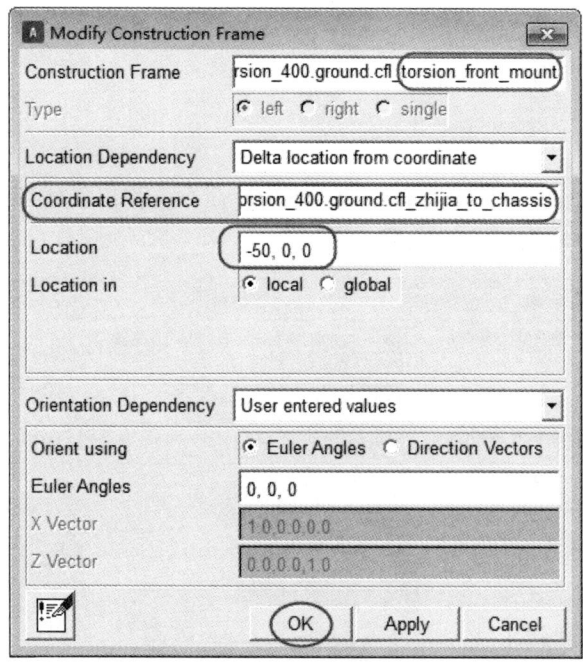

图 14-70 结构框_torsion_front_mount

- Construction Frame（结构框名称）：torsion_front_mount；
- Location Dependency：Delta location from coordinate；
- Coordinate Reference（参考坐标）：._FSAE_sus_front_torsion_400.ground.cfl_zhijia_to_chassis；
- Location：-50，0，0，此位置与扭杆弹簧中的 RP-1 位置对应；
- Location in：local；
- Orientation Dependency：User-entered values；
- Orient using：Euler Angles；
- Euler Angles：0，0，0。

（2）单击 OK 按钮，完成._FSAE_sus_front_torsion_400.ground.cfl_torsion_front_mount 结构框的创建。

（3）单击 Build > Construction Frame > Modify 命令，在下列对话框中输入相应的数据：

- Construction Frame（结构框名称）：torsion_rear_mount；
- Location Dependency：Delta location from coordinate；
- Coordinate Reference（参考坐标）：._FSAE_sus_front_torsion_400.ground.cfl_zhijia_to_chassis；
- Location：350，0，0，此位置与扭杆弹簧中的 RP-2 位置对应；
- Location in：local；
- Orientation Dependency：User-entered values；
- Orient using：Euler Angles；
- Euler Angles：0，0，0。

（4）单击 OK 按钮，完成 ._FSAE_sus_front_torsion_400.ground.cfl_torsion_rear_mount 结构框的修改。

2．替换扭杆弹簧

（1）单击 Build > Part > Flexible Body > Modify 命令，弹出创建部件对话框，如图 14-71 所示，在下列对话框中输入相应的数据：

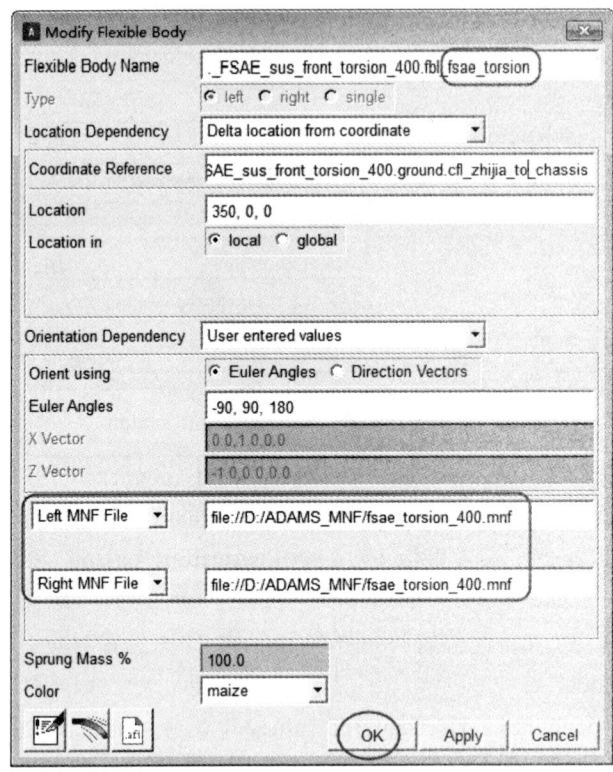

图 14-71　柔性体修改（替换）

- Flexible Body Name：fsae_torsion；
- Type：left；
- Location Dependency：Delta location from coordinate；
- Coordinate Reference（参考坐标）：._FSAE_sus_front_torsion_400.ground.cfl_zhijia_to_chassis；
- Location：350，0，0；
- Location in：local；
- Orientation Dependency：User-entered values；
- Orient using：Euler Angles；
- Euler Angles：-90，90，180；
- Left MNF File：file：//D：/ADAMS_MNF/fsae_torsion_400.mnf；
- Right MNF File：file：//D：/ADAMS_MNF/fsae_torsion_400.mnf；
- Color：maize。

（2）单击 OK 按钮，完成部件._FSAE_sus_front_torsion_400.fbl_fsae_torsion 的替换；替换完成后，RP-2 与 RP-3 对应的约束不用修改；需要把 RP-3 处扭杆弹簧与副车架的旋转约束移动到 RP-1 处。

3. 调试约束

（1）单击 Build > Attachments > Joint > Modify 命令，弹出修改约束对话框，如图 14-72 所示，在下列对话框中输入相应的数据：

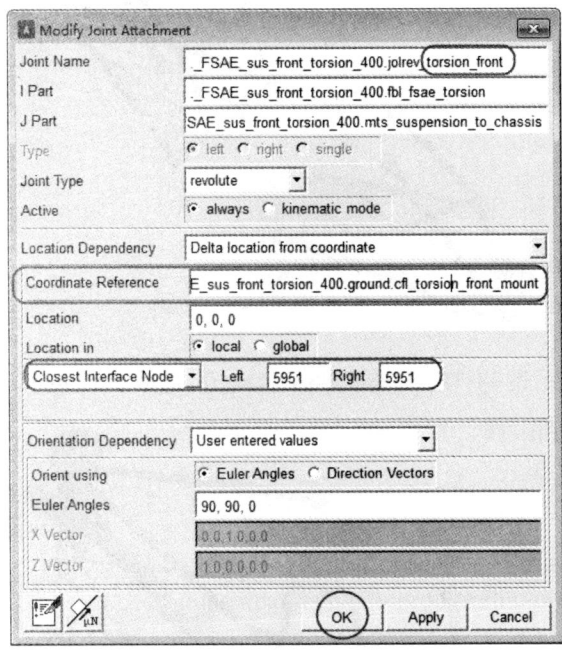

图 14-72　约束_torsion_front

- Joint Name（约束副名称）：torsion_front；
- Type：single；
- I Part：._FSAE_sus_front_torsion_400.fbl_fsae_torsion；
- J Part：._FSAE_sus_front_torsion_400.mts_suspension_to_chassis；
- Joint Type（约束副类型）：revolute；
- Active（激活）：always；
- Location Dependency：Delta location from coordinate；
- Coordinate Reference（参考坐标）：._FSAE_sus_front_torsion_400.ground.cfl_torsion_front_mount；
- Location：0，0，0；
- Location in：local；
- Closest Interface Node：left/5951，right/5951（5951 指有限元软件中创建的接口点）；
- Orientation Dependency：User entered values；
- Orient using：Euler Angles；

- Euler Angles：90，90，0。

（2）单击 OK 按钮，完成约束副 .FSAE_sus_front_torsion_400.jolrev_torsion_front 的创建。

至此悬架模型调试完成，如图 14-73 所示；后悬架按相同方式调试，此处不再重复，后悬架模型存储在章节电子文件中，创建好的后悬架模型如图 14-74 所示。

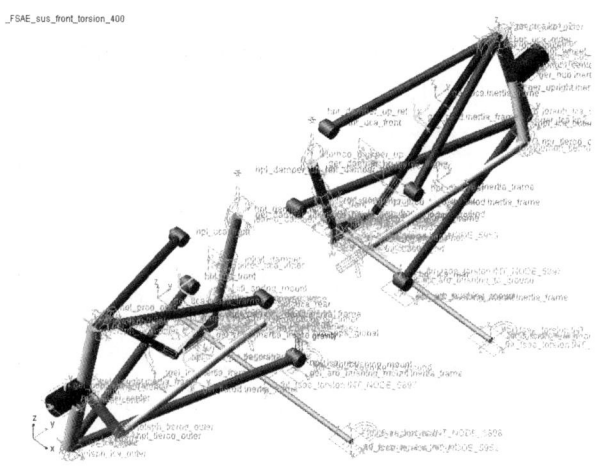

图 14-73　前悬架 FSAE_sus_front_torsion_400

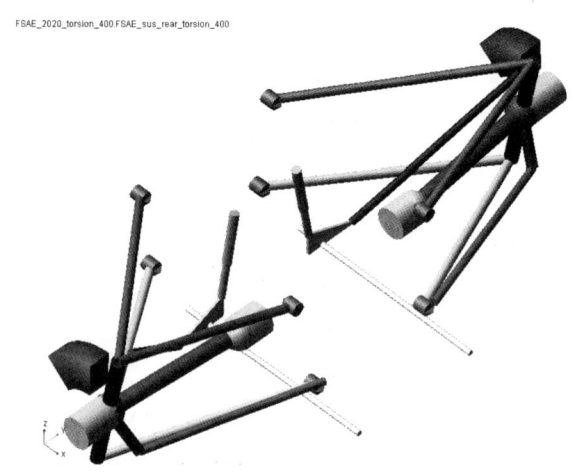

图 14-74　后悬架 FSAE_sus_rear_torsion_400

14.11　谐波脉冲转向仿真

1. 整车装配

在进行谐波脉冲转向仿真之前，需要完成整车的装配，替换上述整车 FSAE_2020_torsion 模型中的前后悬架，完成替换后整车模型另存为 FSAE_2020_torsion_400，此时整车模型如图 14-75 所示。

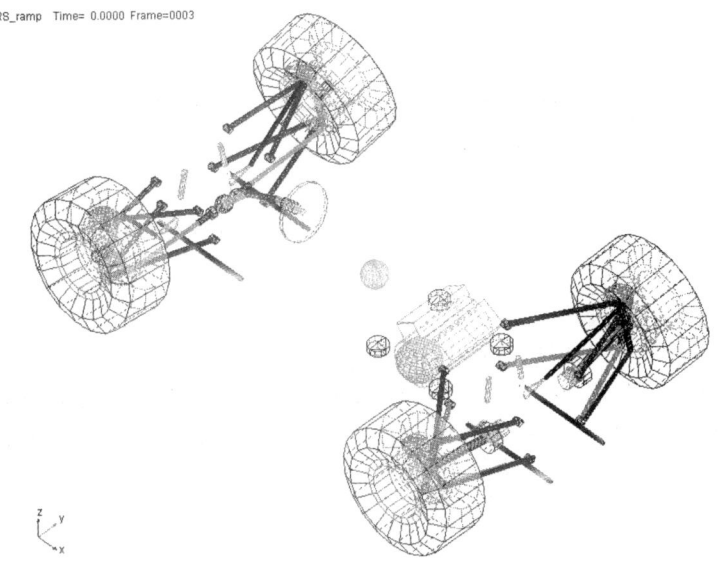

图 14-75　整车 FSAE_2020_torsion_400

2. 谐波脉冲转向仿真

（1）单击 Simulate > Full-Vehicle Analysis > Open-loop steering Events > Ramp Steer 命令，弹出阶跃仿真对话框，如图 14-76 所示，在下列对话框中输入相应的数据：

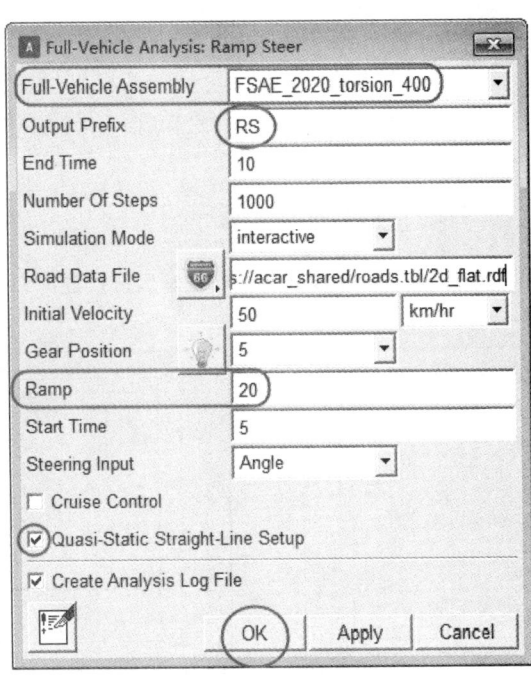

图 14-76　谐波脉冲转向参数设置

- Full-Vehicle Assembly：FSAE_2020_torsion_400；
- Output Prefix：RS；
- End Time：10；

- Number of steps：1000；
- Simulation Mode：interactive；
- Road Date File：mdids：//acar_shared/roads.tbl/2d_flat.rdf；
- Initial Velocity（单位：km/h）：50；
- Gear Position：5；
- Ramp（单位：度）：20，指反向盘转动的频率，每秒钟 20 度；
- Start Time（单位：s）：5；
- Steering Input：Angle；
- 勾选 Quasi-Static Straight-Line Setup。

（2）单击 OK 按钮，完成 FSAE_2020_torsion_400 赛车谐波脉冲转向仿真设置并提交运算，运算完成后，整车运行轨迹如图 14-77 所示。整车运行参数如图 14-78～图 14-83 所示。

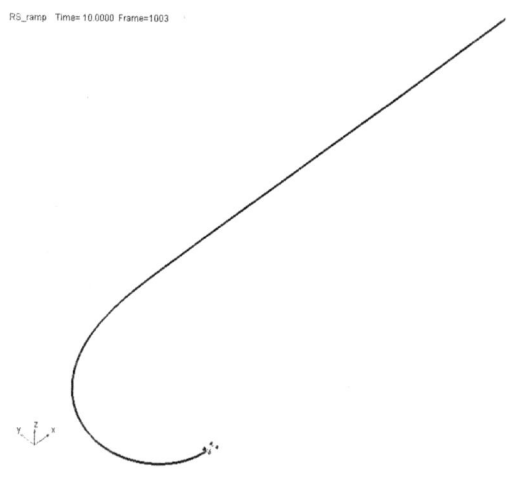

图 14-77 谐波脉冲转向整车运行轨迹

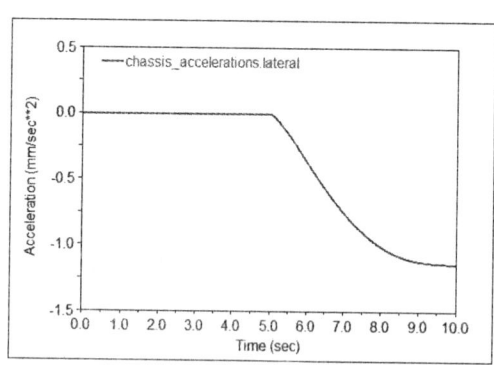

图 14-78 车身侧向加速度

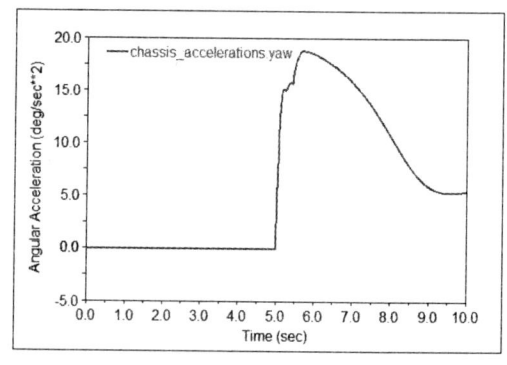

图 14-79 横摆角加速度

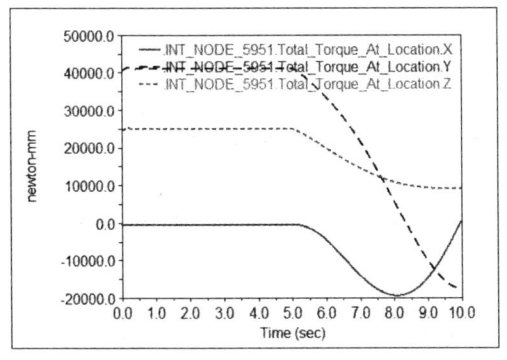

图 14-80　5951 节点 X/Y/Z 方向力矩

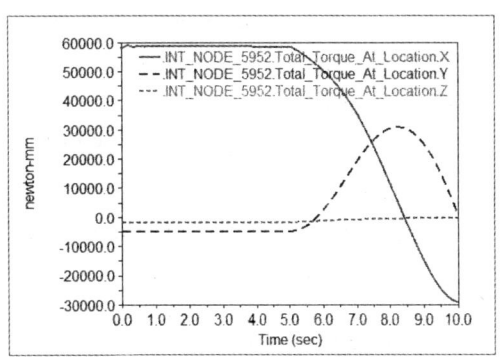

图 14-81　5952 节点 X/Y/Z 方向力矩

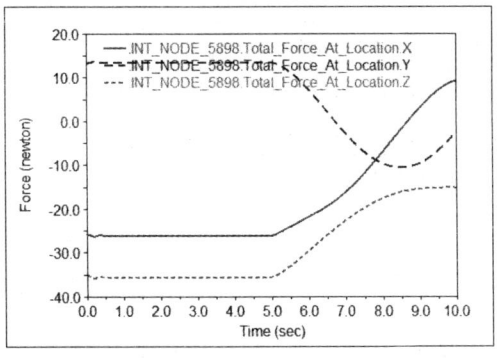

图 14-82　5898 节点 X/Y/Z 方向受力

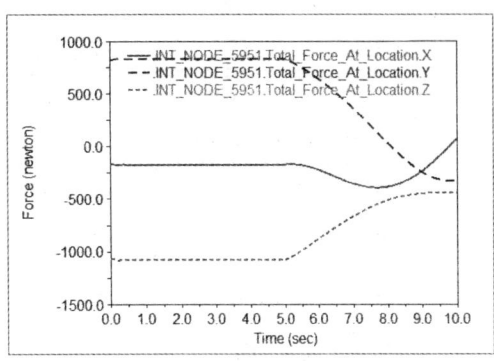

图 14-83　5951 节点 X/Y/Z 方向受力

14.12　考虑横向避震器特性

大多数情况下，垂向避震器完全可以满足要求，不用添加横向避震器；但考虑方程式赛车的特殊性，提升其起步抬头、刹车点头及过弯时的稳定性，就显得尤为必要。

在左右支架上添加横向避震器，此时，横向避震器可以限制支架摆臂在 X 方向的摆动，进而限制车身高速过弯时的振幅；车身的振动幅值并不是越小越好，允许其存在振幅，但必须在可控范围内，提升悬架的"韧度"，需要注意的是韧度与刚度是两个完全不同的概念。

添加避震器前，将模型._FSAE_sus_front_torsion_400 另存为._FSAE_sus_front_torsion_400_damper。

（1）单击 Build > Construction Frame > New 命令，在下列对话框中输入相应的数据：
- Construction Frame（结构框名称）：damper_line；
- Location Dependency：Delta location from coordinate；
- Coordinate Reference（参考坐标）：._FSAE_sus_front_torsion_400_damper.ground.cfl_zhijia_to_chassis_ref；
- Location：30，30，-50；
- Location in：local；

- Orientation Dependency：User-entered values；
- Orient using：Euler Angles；
- Euler Angles：0，0，0。

（2）单击 OK 按钮，完成._FSAE_sus_front_torsion_400_damper.ground.cfl_damper_line 结构框的创建。

（3）单击 Build > Force > Damper > New 命令，弹出避震器创建对话框，如图 14-84 所示，在下列对话框中输入相应的数据。

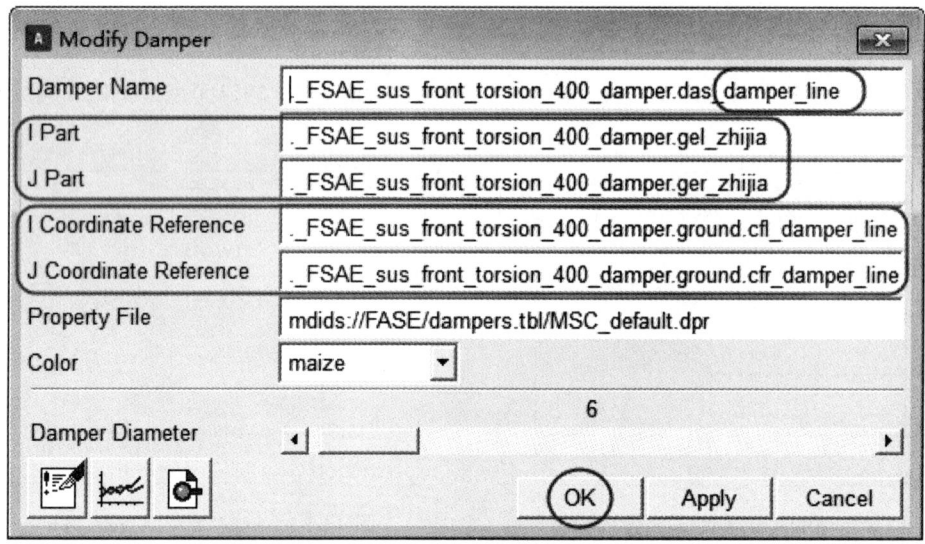

图 14-84　避震器_damper_line

- Damper Name（减震器名称）：damper_line；
- I Part：._FSAE_sus_front_third_spring.ges_damper_up；
- J Part：._FSAE_sus_front_torsion_400_damper.ger_zhijia；
- I Coordinate Reference（参考坐标）：._FSAE_sus_front_torsion_400_damper.ground.cfl_damper_line；
- J Coordinate Reference（参考坐标）：._FSAE_sus_front_torsion_400_damper.ground.cfr_damper_line；
- Property File（属性文件）：mdids：//FASE/dampers.tbl/MDI_default.dpr；
- Damper Diameter（避震器直径）：拖动滑块选择 6 mm；
- Color：maize；

（4）单击 OK 按钮，完成避震器._FSAE_sus_front_torsion_400_damper.das_damper_line 的创建。

仅需在左右支架部件上添加避震器即可，不需要添加约束，因为左右支架摆臂部件与副车架存在旋转约束，如果在部件上添加圆柱副约束或移动副约束，会产生过约束而导致悬架系统不能成功仿真，添加完成横向避震器后前悬架如图 14-85 所示，后悬架按相同方式添加横向避震器即可，如图 14-86 所示。

图 14-85　前悬架._FSAE_sus_front_torsion_400_damper

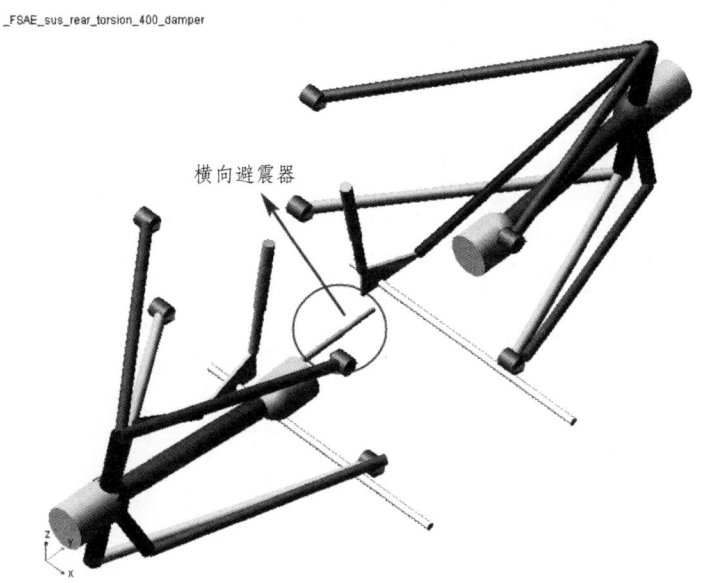

图 14-86　前悬架._FSAE_sus_front_torsion_400_damper

14.13　正弦扫频转向仿真

替换上述整车 FSAE_2020_torsion_400 模型中的前后悬架，完成替换后整车模型另存为 FSAE_2020_torsion_400_damper，此时整车模型如图 14-87 所示。

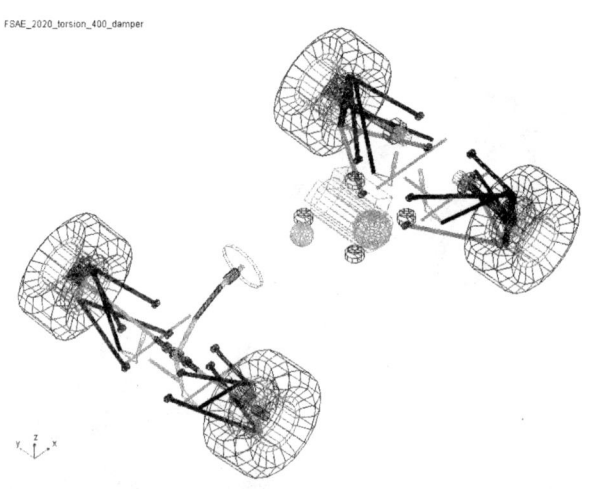

图 14-87　整车 FSAE_2020_torsion_400_damper

（1）单击 Simulate > Full-Vehicle Analysis > Open-loop steering Events > Swept-Sne Steer 命令，弹出正弦扫频转向仿真对话框，如图 14-88 所示，在下列对话框中输入相应的数据：

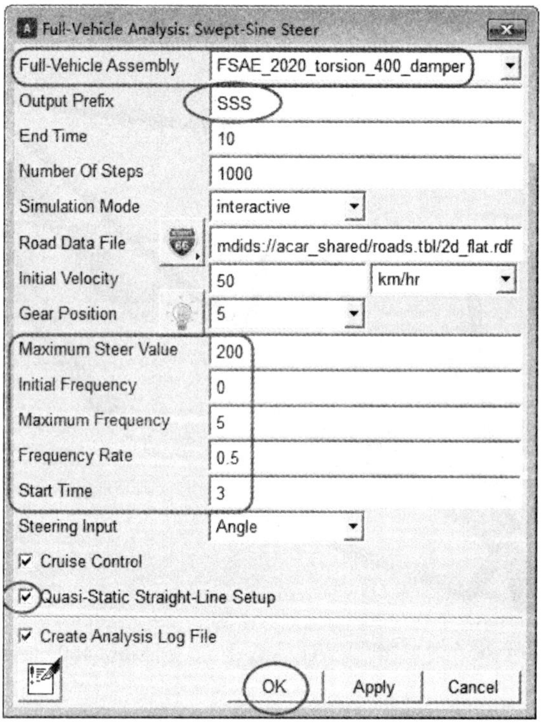

图 14-88　正弦扫频转向仿真设置参数

- Full-Vehicle Assembly：FSAE_2020_torsion_400_damper；
- Output Prefix：SSS；
- End Time：10；
- Number of steps：1000；
- Simulation Mode：interactive；

- 306 -

- Road Date File：mdids：//acar_shared/roads.tbl/2d_flat.rdf；
- Initial Velocity（单位：km/h）：50；
- Gear Position：5；
- Maximum Steer Value：200；
- Initial Frequency：0；
- Maximum Frequency：5；
- Frequency Rate：0.5；
- Start Time（单位：s）：3；
- Steering Input：Angle；
- 勾选 Quasi-Static Straight-Line Setup。

（2）单击 OK 按钮，完成 FSAE_2020_torsion_400_damper 赛车正弦扫频转向仿真设置并提交运算，运算完成后，整车运行轨迹如图 14-89 所示。整车运行参数如图 14-90~14-95 所示。

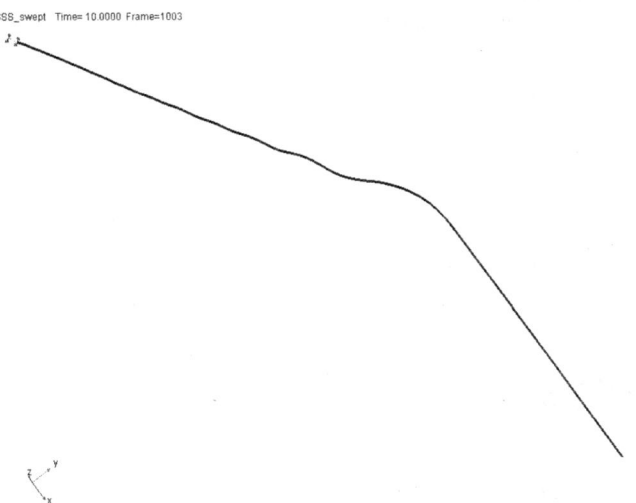

图 14-89 正弦扫频转向整车运行轨迹

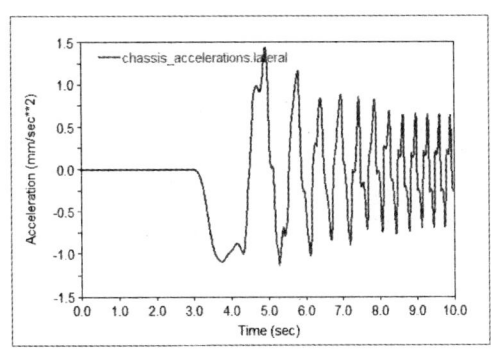

图 14-90 侧向加速度

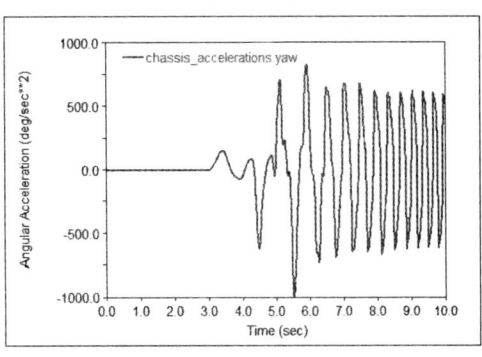

图 14-91 横摆角加速度

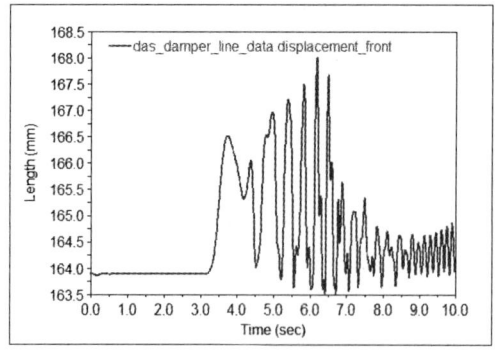

图 14-92　前横向避震器位移

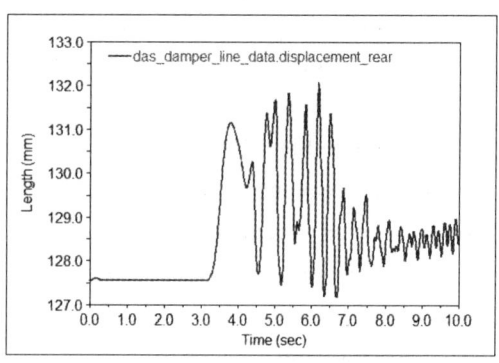

图 14-93　后横向避震器位移

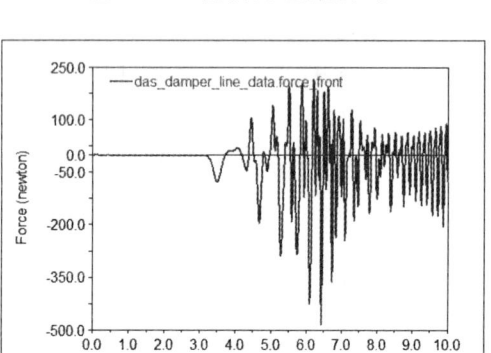

图 14-94　前横向避震器受力

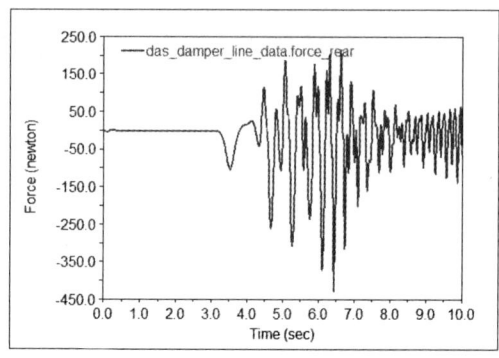

图 14-95　后横向避震器受力

14.14　稳定性参数对比

整车 FSAE_2020_torsion_400 与 FSAE_2020_torsion_400_damper 在 50 km/h 相同条件下进行蛇形绕桩仿真，从图 14-96 ~ 图 14-99 可以看出稳定性参数均有提升，因此证明 FSAE 赛车安装横向避震器的必要性。

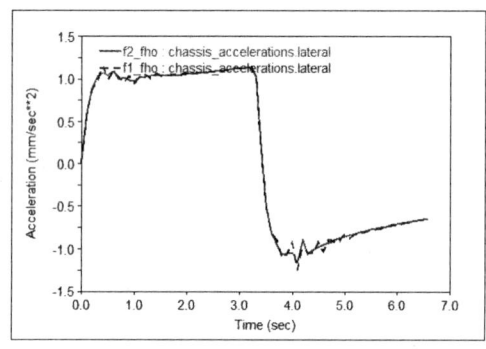

图 14-96　侧向加速度

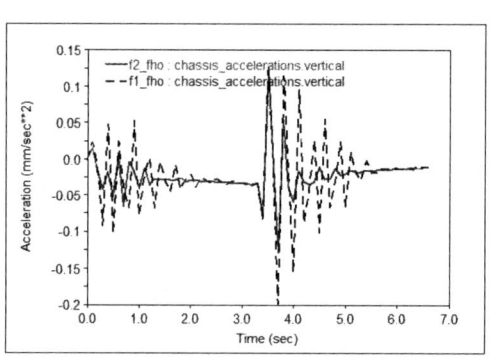

图 14-97　垂向加速度

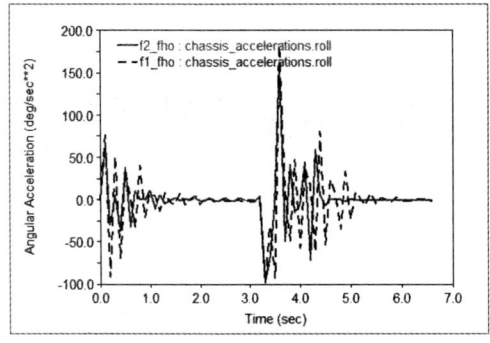
图 14-98 俯仰角加速度

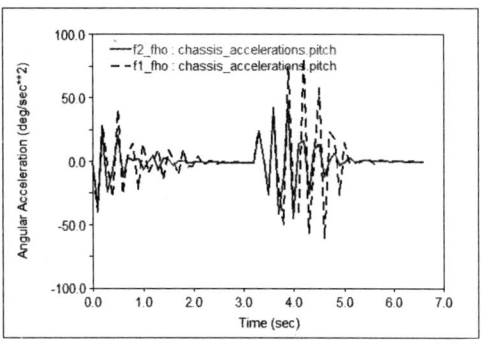
图 14-99 侧倾角加速度

第 15 章 拉杆式悬架模型 II

拉杆式悬架模型弹簧及横向稳定杆有多种布置方式,不同的布置方式主要考虑安装空间、车辆性能设计要求等。除了可以降低重心外,拉杆悬架的优势是可以在较大的裕度空间范围内布置拉杆,例如可以把拉杆布置到上下控制臂之外。对于推杆悬架来说,过长的推杆容易导致挠度变形过大,而拉杆不存此问题。拉杆悬架模型如图15-1所示,从图中可以看出,拉杆、弹簧、避震器及柔性体稳定杆均布置在上下控制臂之外,此种布置方式大大扩展了悬架模型的建模思路。为了不重复叙述建模过程,本章依然采用在后推杆悬架模型 FSAE_sus_rear_axle 上修改建模。

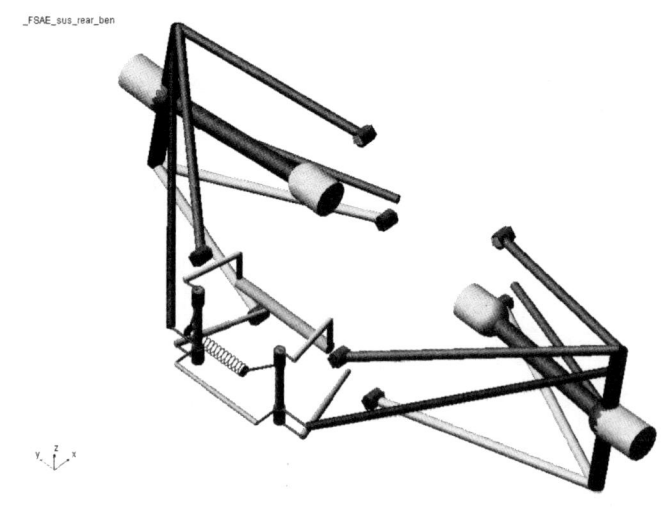

图 15-1 拉杆悬架模型

学习目标

(1)了解拉杆悬架思路。
(2)了解拉杆悬架模型。
(3)会收油门转弯仿真。
(4)了解纵置式扭杆弹簧拉杆悬架。
(5)会弯道收油门仿真。

15.1 拉杆悬架概述

车辆的悬架系统没有好坏优劣之分，并不是采用双 A 臂悬架的车辆一定比采用麦弗逊悬架的好；悬架好坏优劣的衡量最关键是悬架和车辆的匹配，如果匹配得好，低成本的悬架依然可以起到高技术含量悬架的性能。

2009 年，红牛 F1 车队 RB5 车型后悬架采用纵置扭杆弹簧式拉杆悬架后，拉杆悬架又一次进入方程式赛车的设计视野。拉杆悬架特有的降低车身的特性配合扭杆弹簧的韧度实属完美的结合，这在红牛 RB 系列车型（冠军车）及 W10 车型（冠军车）上得到了很好的印证，其具备弯道稳定性、高速下压力等优良特性。推杆在传力时的不稳定性，可能导致推杆发生折叠损害，因此推杆一般较短，大多在车轮质心附近小角度范围内布置。而拉杆悬架不存在此缺点，因此拉杆的长度可以设计得较长，甚至将拉杆放置到上下控制臂之外，合理调试拉杆的布置角度，可以减小车身重量传递到弹簧与避震器上的力，并可降低避震器能耗（如果采用主动悬置设备的话）。图 15-1 所示的拉杆悬架模型的建模思路来自 W10 赛车后悬架，对于刚性旋转轴，可以替换成柔性体旋转轴，即把垂置扭杆弹簧式推杆悬架颠倒过来，具体建模过程在垂置扭杆弹簧式拉杆悬架模型中叙述。

15.2 拉杆悬架模型

1. 模型导入

（1）启动 ADAMS/CAR，选择专家模块进入建模界面。

（2）单击 File > Open 命令，弹出打开模板对话框，如图 15-2 所示，在 Template Name 中输入：mdids://FASE/templates.tbl/_FSAE_sus_rear_axle.tpl。

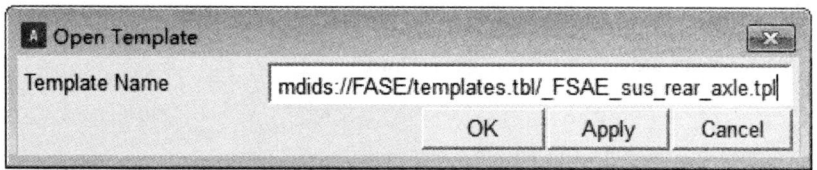

图 15-2　打开模板对话框

（3）单击 OK 按钮，导入模型。

2. 删除部件、弹簧、避震器、约束、结构框

通过删除部件，与部件相关联的约束、弹簧、避震器等都会全部删除，最后删除多余的硬点和结构框，此种方法删除速度较快，同时不容易出现错误。需要删除的信息如下：

① ._FSAE_sus_rear_axle.gel_prod；
② ._FSAE_sus_rear_axle.ger_prod；
③ ._FSAE_sus_rear_axle.gel_bellcrank；
④ ._FSAE_sus_rear_axle.ger_bellcrank；

⑤ ._FSAE_sus_rear_axle.gel_damper_chassis；
⑥ ._FSAE_sus_rear_axle.ger_damper_chassis；
⑦ ._FSAE_sus_rear_axle.gel_damper_bellcrank；
⑧ ._FSAE_sus_rear_axle.ger_damper_bellcrank；
⑨ ._FSAE_sus_rear_axle.nsl_spring；
⑩ ._FSAE_sus_rear_axle.nsr_spring；
⑪ ._FSAE_sus_rear_axle.dal_damper；
⑫ ._FSAE_sus_rear_axle.dar_damper；
⑬ ._FSAE_sus_rear_axle.ground.cfl_damper_chassis_orient；
⑭ ._FSAE_sus_rear_axle.ground.cfr_damper_chassis_orient；
⑮ ._FSAE_sus_rear_axle.ground.cfl_damper_bellcrank_orient；
⑯ ._FSAE_sus_rear_axle.ground.cfr_damper_bellcrank_orient；
⑰ ._FSAE_sus_rear_axle.ground.cfl_bellcrank_pivot；
⑱ ._FSAE_sus_rear_axle.ground.cfr_bellcrank_pivot；
⑲ ._FSAE_sus_rear_axle.ground.hpl_shock_to_chassis；
⑳ ._FSAE_sus_rear_axle.ground.hpr_shock_to_chassis；
㉑ ._FSAE_sus_rear_axle.ground.hpl_arblink_to_bellcrank；
㉒ ._FSAE_sus_rear_axle.ground.hpr_arblink_to_bellcrank；
㉓ ._FSAE_sus_rear_axle.ground.hpl_bellcrank_pivot；
㉔ ._FSAE_sus_rear_axle.ground.hpr_bellcrank_pivot；
㉕ ._FSAE_sus_rear_axle.ground.hpl_shock_to_bellcrank；
㉖ ._FSAE_sus_rear_axle.ground.hpr_shock_to_bellcrank；
㉗ ._FSAE_sus_rear_axle.ground.hpl_prod_to_bellcrank；
㉘ ._FSAE_sus_rear_axle.ground.hpr_prod_to_bellcrank；
㉙ ._FSAE_sus_rear_axle.ground.hpl_prod_outer；
㉚ ._FSAE_sus_rear_axle.ground.hpr_prod_outer。

3．保存模型

（1）单击 File > Save As 命令，弹出保存模板对话框，如图 15-3 所示，在下列对话框中输入相应的数据。

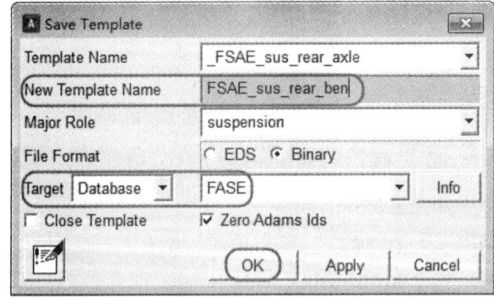

图 15-3　模型保存 FSAE_sus_rear_ben

- Template Name：FSAE_sus_rear_axle；
- New Template Name：FSAE_sus_rear_ben；
- Major Role（主特征）：suspension；
- File Format：Binary；
- Target：Datebase/FASE。

（2）单击 OK 按钮，完成拉杆式悬架模型模板 FSAE_sus_rear_ben 的保存。

4．添加硬点

（1）单击 Build > Hardpoind > New 命令，在下列对话框中输入相应的数据：
- Hardpoint：prod_to_zhijia；
- Typer：left；
- Location：1200.0，-150.0，204.8。

（2）单击 Apply 按钮，完成._FSAE_sus_rear_ben.ground.hpl_prod_to_zhijia 硬点的创建。

（3）重复上述创建硬点步骤，完成图 15-4 中标注线内的硬点创建及位置修改。

hpl_bellcrank_pivot	1547.8	-170.0	305.0
hpl_drive_shaft_inr	1550.0	-200.0	225.0
hpl_lca_front	1370.0	-127.0	127.0
hpl_lca_outer	1498.6	-482.6	101.6
hpl_lca_rear	1651.0	-127.0	127.0
hpl_prod_outer	1530.0	-477.2	320.0
hpl_prod_to_zhijia	1200.0	-150.0	204.8
hpl_tierod_inner	1676.4	-127.0	152.4
hpl_tierod_outer	1574.8	-457.2	152.4
hpl_uca_front	1270.0	-152.4	304.8
hpl_uca_outer	1549.4	-482.6	355.6
hpl_uca_rear	1625.6	-152.4	304.8
hpl_wheel_center	1524.0	-558.8	228.6
hpl_zhijia_to_torsion	1200.0	-90.0	204.8
hps_global	1524.0	0.0	0.0

图 15-4 硬点信息

5．部件 prod

（1）单击 Build > Part > General Part > New 命令，弹出创建部件对话框，如图 15-5 所示，在下列对话框中输入相应的数据：
- General Part：prod；
- Type：left；
- Location Dependency：Centered between coordinates；
- Centered between：Two Coordinates；
- Coordinate Reference #1（参考坐标）：._FSAE_sus_rear_ben.ground.hpl_prod_outer；
- Coordinate Reference #2（参考坐标）：._FSAE_sus_rear_ben.ground.hpl_prod_to_zhijia；
- Orientation Dependency：Orient axis along line；

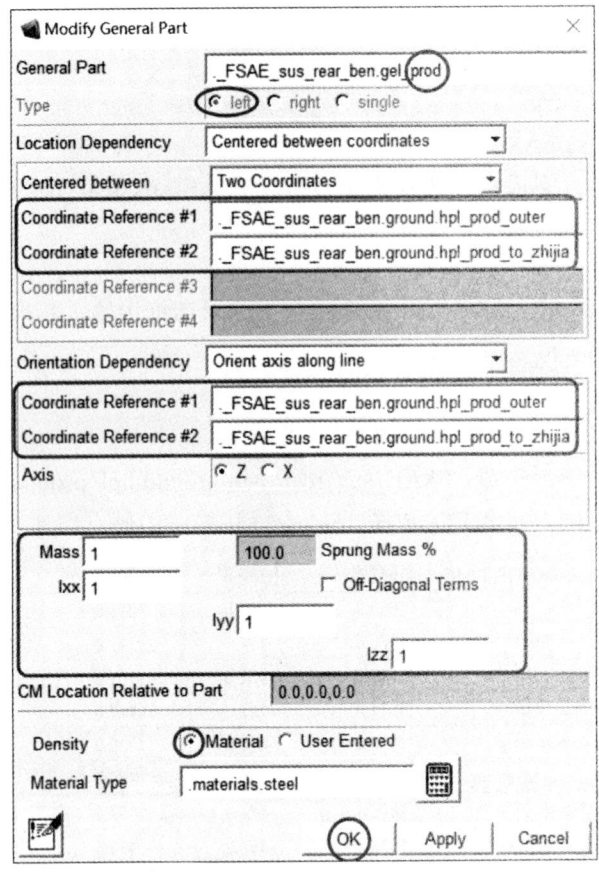

图 15-5 部件_prod

- Coordinate Reference #1（参考坐标）：._FSAE_sus_rear_ben.ground.hpl_prod_outer；
- Coordinate Reference #2（参考坐标）：._FSAE_sus_rear_ben.ground.hpl_prod_to_zhijia；
- Axis：Z；
- Mass：1；
- Ixx：1；
- Iyy：1；
- Izz：1；
- Density：Material；
- Material Type：.materials.steel。

（2）单击 OK 按钮，完成部件._FSAE_sus_rear_ben.gel_prod 的创建。

6. 几何体 prod

（1）单击 Build > Geometry > Link > New 命令，弹出创建几何体对话框，如图 15-6 所示，在下列对话框中输入相应的数据：

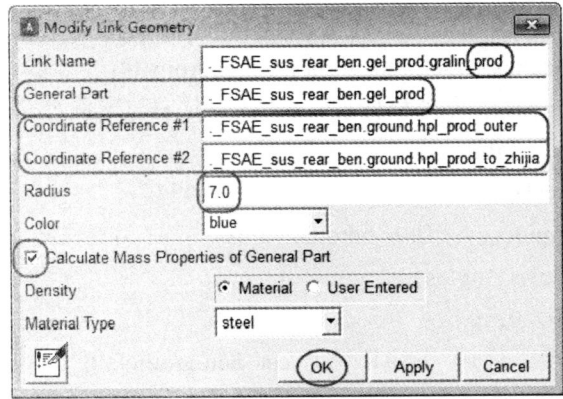

图 15-6 连杆几何体_prod

- Link Name（连杆名称）：prod；
- General Part：._FSAE_sus_rear_ben.gel_prod；
- Coordinate Reference #1（参考坐标）：._FSAE_sus_rear_ben.ground.hpl_prod_outer；
- Coordinate Reference #2（参考坐标）：._FSAE_sus_rear_ben.ground.hpl_prod_to_zhijia；
- Radius（半径）：7；
- Color：blue；
- 勾选 Calculate Mass Properties of General Part 复选框，当几何体建立好之后，会更新对应部件的质量和惯量参数；
- Density：Material；
- Material Type：steel。

（2）单击 OK 按钮，完成._FSAE_sus_rear_ben.gel_prod.gralin_prod 几何体的创建。

7. 结构框

（1）单击 Build > Construction Frame > New 命令，弹出创建结构框对话框，如图 15-7 所示，在下列对话框中输入相应的数据：

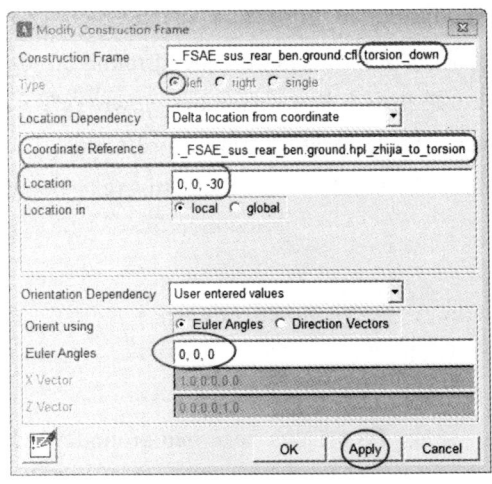

图 15-7 结构框对话框

- Construction Frame（结构框名称）：torsion_down；
- Location Dependency：Delta location from coordinate；
- Coordinate Reference（参考坐标）：._FSAE_sus_rear_ben.ground.hpl_zhijia_to_torsion；
- Location：0，0，-30；
- Location in：local；
- Orientation Dependency：User-entered values；
- Orient using：Euler Angles；
- Euler Angles：0，0，0。

（2）单击 Apply 按钮，完成._FSAE_sus_rear_ben.ground.cfl_torsion_down 结构框的创建。同理，创建其他结构框。

- Construction Frame（结构框名称）：torsion_up；
- Location Dependency：Delta location from coordinate；
- Coordinate Reference（参考坐标）：._FSAE_sus_rear_ben.ground.hpl_zhijia_to_torsion；
- Location：0，0，120；
- Location in：local；
- Orientation Dependency：User-entered values；
- Orient using：Euler Angles；
- Euler Angles：0，0，0。

（3）单击 Apply 按钮，完成._FSAE_sus_rear_ben.ground.cfl_torsion_up 结构框的创建。

- Construction Frame（结构框名称）：damper_down；
- Location Dependency：Delta location from coordinate；
- Coordinate Reference（参考坐标）：._FSAE_sus_rear_ben.ground.cfl_damper_up；
- Location：150，50，0；
- Location in：local；
- Orientation Dependency：User-entered values；
- Orient using：Euler Angles；
- Euler Angles：0，0，0。

（4）单击 Apply 按钮，完成._FSAE_sus_rear_ben.ground.cfl_damper_down 结构框的创建。

- Construction Frame（结构框名称）：damper1_down_ref；
- Location Dependency：Delta location from coordinate；
- Coordinate Reference（参考坐标）：._FSAE_sus_rear_ben.ground.cfl_damper_down；
- Location：50，0，0；
- Location in：local；
- Orientation Dependency：User-entered values；
- Orient using：Euler Angles；
- Euler Angles：0，0，0。

（5）单击 Apply 按钮，完成._FSAE_sus_rear_ben.ground.cfl_damper1_down_ref 结构框的创建。

- Construction Frame（结构框名称）：damper_up；

- Location Dependency：Delta location from coordinate；
- Coordinate Reference（参考坐标）：._FSAE_sus_rear_ben.ground.cfl_damper_down；
- Location：0，-50，0；
- Location in：local；
- Orientation Dependency：User-entered values；
- Orient using：Euler Angles；
- Euler Angles：0，0，0。

（6）单击 Apply 按钮，完成._FSAE_sus_rear_ben.ground.cfl_damper_up 结构框的创建。
- Construction Frame（结构框名称）：damper_front；
- Location Dependency：Delta location from coordinate；
- Coordinate Reference（参考坐标）：._FSAE_sus_rear_ben.ground.hpl_zhijia_to_torsion；
- Location：-50，0，0；
- Location in：local；
- Orientation Dependency：User-entered values；
- Orient using：Euler Angles；
- Euler Angles：0，0，0。

（7）单击 Apply 按钮，完成._FSAE_sus_rear_ben.ground.cfl_damper_front 结构框的创建；
- Construction Frame（结构框名称）：zhijia_up；
- Location Dependency：Delta location from coordinate；
- Coordinate Reference（参考坐标）：._FSAE_sus_rear_ben.ground.cfl_torsion_up；
- Location：0，0，-30；
- Location in：local；
- Orientation Dependency：User-entered values；
- Orient using：Euler Angles；
- Euler Angles：0，0，0。

（8）单击 Apply 按钮，完成._FSAE_sus_rear_ben.ground.cfl_zhijia_up 结构框的创建。
- Construction Frame（结构框名称）：spring_mount；
- Location Dependency：Delta location from coordinate；
- Coordinate Reference（参考坐标）：._FSAE_sus_rear_ben.ground.cfl_torsion_up；
- Location：-50，150，-30；
- Location in：local；
- Orientation Dependency：User-entered values；
- Orient using：Euler Angles；
- Euler Angles：0，0，0。

（9）单击 Apply 按钮，完成._FSAE_sus_rear_ben.ground.cfl_spring_mount 结构框的创建。
- Construction Frame（结构框名称）：zhijia_to_link；
- Location Dependency：Delta location from coordinate；
- Coordinate Reference（参考坐标）：._FSAE_sus_rear_ben.ground.cfl_torsion_up；
- Location：0，-30，0；

- Location in：local；
- Orientation Dependency：User-entered values；
- Orient using：Euler Angles；
- Euler Angles：0，0，0。

（10）单击 Apply 按钮，完成._FSAE_sus_rear_ben.ground.cfl_zhijia_to_link 结构框的创建。
- Construction Frame（结构框名称）：link_to_torsion；
- Location Dependency：Delta location from coordinate；
- Coordinate Reference（参考坐标）：._FSAE_sus_rear_ben.ground.cfl_torsion_up；
- Location：100，-10，-60；
- Location in：local；
- Orientation Dependency：User-entered values；
- Orient using：Euler Angles；
- Euler Angles：0，0，0。

（11）单击 Apply 按钮，完成._FSAE_sus_rear_ben.ground.cfl_link_to_torsion 结构框的创建。
- Construction Frame（结构框名称）：link_ref；
- Location Dependency：Delta location from coordinate；
- Coordinate Reference（参考坐标）：._FSAE_sus_rear_ben.ground.cfl_link_to_torsion；
- Location：0，0，60；
- Location in：local；
- Orientation Dependency：User-entered values；
- Orient using：Euler Angles；
- Euler Angles：0，0，0。

（12）单击 Apply 按钮，完成._FSAE_sus_rear_ben.ground.cfl_link_ref 结构框的创建。
- Construction Frame（结构框名称）：torsion_mid；
- Location Dependency：Centered between coordinates；
- Centered between：Two Coordinates；
- Coordinate Reference #1（参考坐标）：._FSAE_sus_rear_ben.ground.cfl_link_to_torsion；
- Coordinate Reference #2（参考坐标）：._FSAE_sus_rear_ben.ground.cfr_link_to_torsion；
- Orientation Dependency：User-entered values；
- Orient using：Euler Angles；
- Euler Angles：0，0，0。

（13）单击 Apply 按钮，完成._FSAE_sus_rear_ben.ground.cfs_torsion_mid 结构框的创建。
- Construction Frame（结构框名称）：bumpstop；
- Location Dependency：Delta location from coordinate；
- Coordinate Reference（参考坐标）：._FSAE_sus_rear_rigid_ben.ground.hpl_spring_mount；
- Location：0，80，0；
- Location in：local；
- Orientation Dependency：User-entered values；

- Orient using：Euler Angles；
- Euler Angles：0，0，0。

（14）单击 Apply 按钮，完成._FSAE_sus_rear_rigid_ben.ground.cfl_bumpstop 结构框的创建。

- Construction Frame（结构框名称）：rebumpstop；
- Location Dependency：Delta location from coordinate；
- Coordinate Reference（参考坐标）：._FSAE_sus_rear_rigid_ben.ground.cfl_spring_mount；
- Location：0，10，0；
- Location in：local；
- Orientation Dependency：User-entered values；
- Orient using：Euler Angles；
- Euler Angles：0，0，0。

（15）单击 OK 按钮，完成._FSAE_sus_rear_rigid_ben.ground.cfl_rebumpstop 结构框的创建。

8. 部件 axis_vertical

（1）单击 Build > Part > General Part > New 命令，在下列对话框中输入相应的数据：

- General Part：axis_vertical；
- Type：left；
- Location Dependency：Centered between coordinates；
- Centered between：Two Coordinates；
- Coordinate Reference #1（参考坐标）：._FSAE_sus_rear_rigid_ben.ground.cfl_torsion_up；
- Coordinate Reference #2（参考坐标）：._FSAE_sus_rear_rigid_ben.ground.cfl_torsion_down；
- Orientation Dependency：User-entered values；
- Orient using：Euler Angles；
- Euler Angles：0，0，0；
- Axis：Z；
- Mass：1；
- Ixx：1；
- Iyy：1；
- Izz：1；
- Density：Material；
- Material Type：.materials.steel。

（2）单击 OK 按钮，完成部件._FSAE_sus_rear_rigid_ben.gel_axis_vertical 的创建。

9. 几何体连杆 axis_vertical

（1）单击 Build > Geometry > Link > New 命令，在下列对话框中输入相应的数据：

- Link Name（连杆名称）：axis_vertical；

- General Part：._FSAE_sus_rear_rigid_ben.gel_axis_vertical；
- Coordinate Reference #1（参考坐标）：._FSAE_sus_rear_rigid_ben.ground.cfl_torsion_up；
- Coordinate Reference #2（参考坐标）：._FSAE_sus_rear_rigid_ben.ground.cfl_torsion_down；
- Radius（半径）：7；
- Color：green；
- 勾选 Calculate Mass Properties of General Part 复选框，当几何体建立好之后，会更新对应部件的质量和惯量参数；
- Density：Material；
- Material Type：steel。

（2）单击 OK 按钮，完成._FSAE_sus_rear_rigid_ben.gel_axis_vertical.gralin_axis_vertical 几何体的创建。

10. 部件 zhijia_down

（1）单击 Build > Part > General Part > New 命令，在下列对话框中输入相应的数据：
- General Part：zhijia_down；
- Type：left；
- Location Dependency：Centered between coordinates；
- Centered between：Two Coordinates；
- Coordinate Reference #1（参考坐标）：._FSAE_sus_rear_rigid_ben.ground.cfl_torsion_down；
- Coordinate Reference #2（参考坐标）：._FSAE_sus_rear_rigid_ben.ground.hpl_zhijia_to_torsion；
- Orientation Dependency：User-entered values；
- Orient using：Euler Angles；
- Euler Angles：0，0，0；
- Axis：Z；
- Mass：1；
- Ixx：1；
- Iyy：1；
- Izz：1；
- Density：Material；
- Material Type：.materials.steel。

（2）单击 OK 按钮，完成部件._FSAE_sus_rear_rigid_ben.gel_zhijia_down 的创建。

11. 几何体 zhijia_down 集

（1）单击 Build > Geometry > Link > New 命令，在下列结构框中输入相应的数据：
- Link Name（连杆名称）：zhijia_down_main；
- General Part：._FSAE_sus_rear_rigid_ben.gel_zhijia_down；

- Coordinate Reference #1（参考坐标）：._FSAE_sus_rear_rigid_ben.ground.hpl_zhijia_to_torsion；
- Coordinate Reference #2（参考坐标）：._FSAE_sus_rear_rigid_ben.ground.cfl_torsion_down；
- Radius（半径）：7；
- Color：red；
- 勾选 Calculate Mass Properties of General Part 复选框，当几何体建立好之后，会更新对应部件的质量和惯量参数；
- Density：Material；
- Material Type：steel。

（2）单击 Apply 按钮，完成._FSAE_sus_rear_rigid_ben.gel_zhijia_down.gralin_zhijia_down_main 几何体的创建。

- Link Name（连杆名称）：zhijia_down_to_damper_front；
- General Part：._FSAE_sus_rear_rigid_ben.gel_zhijia_down；
- Coordinate Reference #1（参考坐标）：._FSAE_sus_rear_rigid_ben.ground.hpl_zhijia_to_torsion；
- Coordinate Reference #2（参考坐标）：._FSAE_sus_rear_rigid_ben.ground.cfl_damper_front；
- Radius（半径）：2；
- Color：red；
- 勾选 Calculate Mass Properties of General Part 复选框，当几何体建立好之后，会更新对应部件的质量和惯量参数；
- Density：Material；
- Material Type：steel。

（3）单击 Apply 按钮，完成._FSAE_sus_rear_rigid_ben.gel_zhijia_down.gralin_zhijia_down_to_damper_front 几何体的创建。

- Link Name（连杆名称）：zhijia_down_to_prod；
- General Part：._FSAE_sus_rear_rigid_ben.gel_zhijia_down；
- Coordinate Reference #1（参考坐标）：._FSAE_sus_rear_rigid_ben.ground.hpl_zhijia_to_torsion；
- Coordinate Reference #2（参考坐标）：._FSAE_sus_rear_rigid_ben.ground.hpl_prod_to_zhijia；
- Radius（半径）：2；
- Color：red；
- 勾选 Calculate Mass Properties of General Part 复选框，当几何体建立好之后，会更新对应部件的质量和惯量参数；
- Density：Material；
- Material Type：steel。

（4）单击 Apply 按钮，完成._FSAE_sus_rear_rigid_ben.gel_zhijia_down.gralin_zhijia_

down_to_prod 几何体的创建。
- Link Name（连杆名称）：zhijia_down_to_damper；
- General Part：._FSAE_sus_rear_rigid_ben.gel_zhijia_down；
- Coordinate Reference #1（参考坐标）：._FSAE_sus_rear_rigid_ben.ground.cfl_torsion_down；
- Coordinate Reference #2（参考坐标）：._FSAE_sus_rear_rigid_ben.ground.cfl_damper_up；
- Radius（半径）：2；
- Color：red；
- 勾选 Calculate Mass Properties of General Part 复选框，当几何体建立好之后，会更新对应部件的质量和惯量参数；
- Density：Material；
- Material Type：steel。

（5）单击 OK 按钮，完成 ._FSAE_sus_rear_rigid_ben.gel_zhijia_down.gralin_zhijia_down_to_damper 几何体的创建。

12. 部件 damper_down

（1）单击 Build > Part > General Part > New 命令，在下列结构框中输入相应的数据：
- General Part：damper_down；
- Type：left；
- Location Dependency：Delta location from coordinate；
- Coordinate Reference（参考坐标）：._FSAE_sus_rear_rigid_ben.ground.cfl_damper_down；
- Location：0，0，0；
- Location in：local；
- Orientation Dependency：User-entered values；
- Orient using：Euler Angles；
- Euler Angles：0，0，0；
- Mass：1；
- Ixx：1；
- Iyy：1；
- Izz：1；
- Density：Material；
- Material Type：.materials.steel。

（2）单击 Apply 按钮，完成部件 ._FSAE_sus_rear_rigid_ben.gel_damper_down 的创建。

13. 部件 damper_up

（1）单击 Build > Part > General Part > New 命令，在下列结构框中输入相应的数据：
- General Part：damper_up；
- Type：left；

- Location Dependency：Delta location from coordinate；
- Coordinate Reference（参考坐标）：._FSAE_sus_rear_rigid_ben.ground.cfl_damper_up；
- Location：0，0，0；
- Location in：local；
- Orientation Dependency：User-entered values；
- Orient using：Euler Angles；
- Euler Angles：0，0，0；
- Mass：1；
- Ixx：1；
- Iyy：1；
- Izz：1；
- Density：Material；
- Material Type：.materials.steel。

（2）单击 OK 按钮，完成部件._FSAE_sus_rear_rigid_ben.gel_damper_up 的创建。

14. 部件 zhijia_up

（1）单击 Build > Part > General Part > New 命令，在下列结构框中输入相应的数据：
- General Part：zhijia_up；
- Type：left；
- Location Dependency：Centered between coordinates；
- Centered between：Two Coordinates；
- Coordinate Reference #1（参考坐标）：._FSAE_sus_rear_rigid_ben.ground.cfl_torsion_up；
- Coordinate Reference #2（参考坐标）：._FSAE_sus_rear_rigid_ben.ground.cfl_zhijia_up；
- Orientation Dependency：User-entered values；
- Orient using：Euler Angles；
- Euler Angles：0，0，0；
- Axis：Z；
- Mass：1；
- Ixx：1；
- Iyy：1；
- Izz：1；
- Density：Material；
- Material Type：.materials.steel。

（2）单击 OK 按钮，完成部件._FSAE_sus_rear_rigid_ben.gel_zhijia_up 的创建。

15. 几何体 zhijia_up 集

（1）单击 Build > Geometry > Link > New 命令，在下列对话框中输入相应的数据：
- Link Name（连杆名称）：zhijia_up_main；

- General Part：._FSAE_sus_rear_rigid_ben.gel_zhijia_up；
- Coordinate Reference #1（参考坐标）：._FSAE_sus_rear_rigid_ben.ground.cfl_torsion_up；
- Coordinate Reference #2（参考坐标）：._FSAE_sus_rear_rigid_ben.ground.cfl_zhijia_up；
- Radius（半径）：10；
- Color：red；
- 勾选 Calculate Mass Properties of General Part 复选框，当几何体建立好之后，会更新对应部件的质量和惯量参数；
- Density：Material；
- Material Type：steel。

（2）单击 Apply 按钮，完成._FSAE_sus_rear_rigid_ben.gel_zhijia_up.gralin_zhijia_up_main 几何体的创建。

- Link Name（连杆名称）：zhijia_to_link；
- General Part：._FSAE_sus_rear_rigid_ben.gel_zhijia_up；
- Coordinate Reference #1（参考坐标）：._FSAE_sus_rear_rigid_ben.ground.cfl_torsion_up；
- Coordinate Reference #2（参考坐标）：._FSAE_sus_rear_rigid_ben.ground.cfl_zhijia_to_link；
- Radius（半径）：2；
- Color：red；
- 勾选 Calculate Mass Properties of General Part 复选框，当几何体建立好之后，会更新对应部件的质量和惯量参数；
- Density：Material；
- Material Type：steel。

（3）单击 Apply 按钮，完成._FSAE_sus_rear_rigid_ben.gel_zhijia_up.gralin_zhijia_to_link 几何体的创建。

- Link Name（连杆名称）：zhijia_up_to_spring；
- General Part：._FSAE_sus_rear_rigid_ben.gel_zhijia_up；
- Coordinate Reference #1（参考坐标）：._FSAE_sus_rear_rigid_ben.ground.hpl_spring_mount；
- Coordinate Reference #2（参考坐标）：._FSAE_sus_rear_rigid_ben.ground.cfl_zhijia_up；
- Radius（半径）：2；
- Color：red；
- 勾选 Calculate Mass Properties of General Part 复选框，当几何体建立好之后，会更新对应部件的质量和惯量参数；
- Density：Material；
- Material Type：steel。

（4）单击 Apply 按钮，完成._FSAE_sus_rear_rigid_ben.gel_zhijia_up.gralin_zhijia_up_to_spring 几何体的创建。

16. 部件 link1

（1）单击 Build > Part > General Part > New 命令，在下列对话框中输入相应的数据：
- General Part：link1；
- Type：left；
- Location Dependency：Centered between coordinates；
- Centered between：Two Coordinates；
- Coordinate Reference #1（参考坐标）：._FSAE_sus_rear_rigid_ben.ground.cfl_zhijia_to_link；
- Coordinate Reference #2（参考坐标）：._FSAE_sus_rear_rigid_ben.ground.cfl_link_ref；
- Orientation Dependency：User-entered values；
- Orient using：Euler Angles；
- Euler Angles：0，0，0；
- Axis：Z；
- Mass：1；
- Ixx：1；
- Iyy：1；
- Izz：1；
- Density：Material；
- Material Type：.materials.steel。

（2）单击 OK 按钮，完成部件._FSAE_sus_rear_rigid_ben.gel_link1 的创建。

17. 几何体 link1

（1）单击 Build > Geometry > Link > New 命令，在下列对话框中输入相应的数据：
- Link Name（连杆名称）：link1；
- General Part：._FSAE_sus_rear_rigid_ben.gel_link1；
- Coordinate Reference #1（参考坐标）：._FSAE_sus_rear_rigid_ben.ground.cfl_zhijia_to_link；
- Coordinate Reference #2（参考坐标）：._FSAE_sus_rear_rigid_ben.ground.cfl_link_ref；
- Radius（半径）：5；
- Color：yellow；
- 勾选 Calculate Mass Properties of General Part 复选框；
- Density：Material；
- Material Type：steel。

（2）单击 OK 按钮，完成._FSAE_sus_rear_rigid_ben.gel_link1.gralin_link1 几何体的创建。

18. 部件 link2

（1）单击 Build > Part > General Part > New 命令，在下列对话框中输入相应的数据：
- General Part：link2；

- Type：left；
- Location Dependency：Centered between coordinates；
- Centered between：Two Coordinates；
- Coordinate Reference #1（参考坐标）：._FSAE_sus_rear_rigid_ben.ground.cfl_link_ref；
- Coordinate Reference #2（参考坐标）：._FSAE_sus_rear_rigid_ben.ground.cfl_link_to_torsion；
- Orientation Dependency：User-entered values；
- Orient using：Euler Angles；
- Euler Angles：0，0，0；
- Axis：Z；
- Mass：1；
- Ixx：1；
- Iyy：1；
- Izz：1；
- Density：Material；
- Material Type：.materials.steel。

(2) 单击 OK 按钮，完成部件._FSAE_sus_rear_rigid_ben.gel_link2 的创建。

19. 几何体 link2

(1) 单击 Build > Geometry > Link > New 命令，在下列对话框中输入相应的数据：

- Link Name（连杆名称）：link2；
- General Part 输入：._FSAE_sus_rear_rigid_ben.gel_link2；
- Coordinate Reference #1（参考坐标）：._FSAE_sus_rear_rigid_ben.ground.cfl_link_ref；
- Coordinate Reference #2（参考坐标）：._FSAE_sus_rear_rigid_ben.ground.cfl_link_to_torsion；
- Radius（半径）：5；
- Color：red；
- 勾选 Calculate Mass Properties of General Part 复选框；
- Density：Material；
- Material Type：steel。

(2) 单击 OK 按钮，完成._FSAE_sus_rear_rigid_ben.gel_link1.gralin_link2 几何体的创建。

20. 部件 torsion

(1) 单击 Build > Part > General Part > New 命令，在下列对话框中输入相应的数据：

- General Part：torsion；
- Type：single；
- Location Dependency：Delta location from coordinate；
- Coordinate Reference（参考坐标）：._FSAE_sus_rear_rigid_ben.ground.cfs_torsion_mid；
- Location：0，0，0；

- Location in：local；
- Orientation Dependency：User-entered values；
- Orient using：Euler Angles；
- Euler Angles：0，0，0；
- Axis：Z；
- Mass：1；
- Ixx：1；
- Iyy：1；
- Izz：1；
- Density：Material；
- Material Type：.materials.steel。

（2）单击 OK 按钮，完成部件._FSAE_sus_rear_rigid_ben.ges_torsion 的创建。

21. 几何体 torsion

（1）单击 Build > Geometry > Link > New 命令，在下列对话框中输入相应的数据：

- Link Name（连杆名称）：torsion；
- General Part：._FSAE_sus_rear_rigid_ben.ges_torsion；
- Coordinate Reference #1（参考坐标）：._FSAE_sus_rear_rigid_ben.ground.cfl_link_to_torsion；
- Coordinate Reference #2（参考坐标）：._FSAE_sus_rear_rigid_ben.ground.cfr_link_to_torsion；
- Radius（半径）：5；
- Color：skyblue；
- 勾选 Calculate Mass Properties of General Part 复选框；
- Density：Material；
- Material Type：steel。

（2）单击 OK 按钮，完成._FSAE_sus_rear_rigid_ben.ges_torsion.gralin_torsion 几何体的创建。

22. 部件 spring_mount

（1）单击 Build > Part > General Part > New 命令，在下列对话框中输入相应的数据：

- General Part：spring_mount；
- Type：left；
- Location Dependency：Delta location from coordinate；
- Coordinate Reference（参考坐标）：._FSAE_sus_rear_rigid_ben.ground.hpl_spring_mount；
- Location：0，0，0；
- Location in：local；
- Orientation Dependency：User-entered values；
- Orient using：Euler Angles；

- Euler Angles：0，0，0；
- Axis：Z；
- Mass：1；
- Ixx：1；
- Iyy：1；
- Izz：1；
- Density：Material；
- Material Type：.materials.steel。

（2）单击 Apply 按钮，完成部件._FSAE_sus_rear_rigid_ben.gel_spring_mount 的创建。

23. 部件 body_ref

（1）单击 Build > Part > General Part > New 命令，在下列对话框中输入相应的数据：

- General Part：body_ref；
- Type：left；
- Location Dependency：Delta location from coordinate；
- Coordinate Reference（参考坐标）：._FSAE_sus_rear_rigid_ben.ground.cfl_bumpstop；
- Location：0，0，0；
- Location in：local；
- Orientation Dependency：User-entered values；
- Orient using：Euler Angles；
- Euler Angles：0，0，0；
- Axis：Z；
- Mass：1；
- Ixx：1；
- Iyy：1；
- Izz：1；
- Density：Material；
- Material Type：.materials.steel。

（2）单击 OK 按钮，完成部件._FSAE_sus_rear_rigid_ben.gel_body_ref 的创建。

24. 部件 upright 与 prod 之间的 spherical 约束

（1）单击 Build > Attachments > Joint > New 命令，弹出创建约束件对话框，如图 15-8 所示。在下列话框中输入相应的数据：

- Joint Name（约束副名称）：prod_outer；
- Type：left；
- I Part：._FSAE_sus_rear_rigid_ben.ger_upright；
- J Part：._FSAE_sus_rear_rigid_ben.ger_prod；
- Joint Type（约束副类型）：spherical；

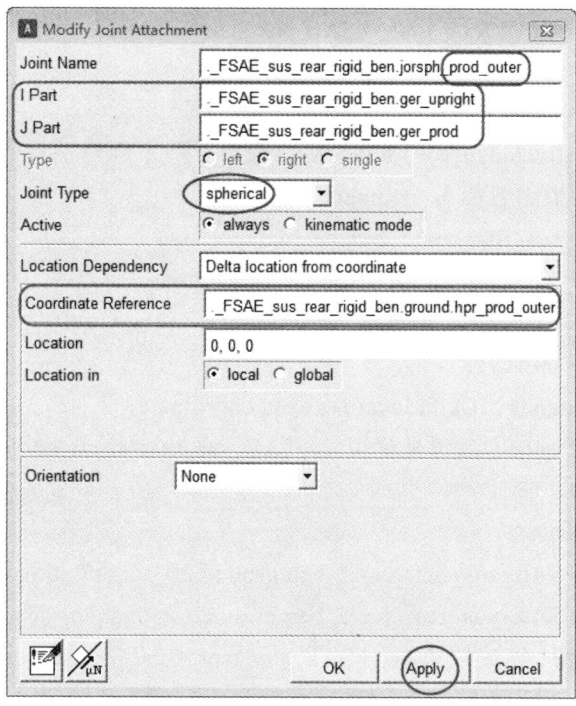

图 15-8 刚性约束对话框_ spherical

- Active（激活）：always；
- Location Dependency：Delta location from coordinate；
- Coordinate Reference（参考坐标）：._FSAE_sus_rear_rigid_ben.ground.hpr_prod_outer；
- Location：0，0，0；
- Location in：local。

（2）单击 Apply 按钮，完成约束副 ._FSAE_sus_rear_rigid_ben.jorsph_prod_outer 的创建。

25. 部件 prod 与 zhijia 之间的 hook 约束

（1）单击 Build > Attachments > Joint > New 命令，在下列对话框中输入相应的数据：
- Joint Name（约束副名称）：prod_to_zhijia；
- I Part：._FSAE_sus_rear_rigid_ben.gel_prod；
- J Part：._FSAE_sus_rear_rigid_ben.gel_zhijia_down；
- Joint Type（约束副类型）：hook；
- Active（激活）：always；
- Location Dependency：Delta location from coordinate；
- Coordinate Reference（参考坐标）：._FSAE_sus_rear_rigid_ben.ground.hpl_prod_to_zhijia；
- Location：0，0，0；
- Location in：local。
- I-Part Axis：._FSAE_sus_rear_rigid_ben.ground.hpl_prod_outer；
- J-Part Axis：._FSAE_sus_rear_rigid_ben.ground.hpl_zhijia_to_torsion；

（2）单击 Apply 按钮，完成约束副._FSAE_sus_rear_rigid_ben.jolhoo_prod_to_zhijia 的创建。

26. 部件 damper_up 与 zhijia_down 之间的 hook 约束

（1）单击 Build > Attachments > Joint > New 命令，在下列对话框中输入相应的数据：
- Joint Name（约束副名称）：damper1_up；
- I Part：._FSAE_sus_rear_rigid_ben.gel_damper_up；
- J Part：._FSAE_sus_rear_rigid_ben.gel_zhijia_down；
- Joint Type（约束副类型）：hook；
- Active（激活）：always；
- Location Dependency：Delta location from coordinate；
- Coordinate Reference（参考坐标）：._FSAE_sus_rear_rigid_ben.ground.cfl_damper_up；
- Location：0，0，0；
- Location in：local；
- I-Part Axis：._FSAE_sus_rear_rigid_ben.ground.cfl_torsion_down；
- J-Part Axis：._FSAE_sus_rear_rigid_ben.ground.cfl_damper_down。

（2）单击 Apply 按钮，完成约束副._FSAE_sus_rear_rigid_ben.jolhoo_damper1_up 的创建。

27. 部件 damper_down 与 suspension_to_chassis 之间的 hook 约束

（1）单击 Build > Attachments > Joint > New 命令，在下列对话框中输入相应的数据：
- Joint Name（约束副名称）：damper_down；
- I Part：._FSAE_sus_rear_rigid_ben.gel_damper_down；
- J Part：._FSAE_sus_rear_rigid_ben.mts_suspension_to_chassis；
- Joint Type（约束副类型）：hook；
- Active（激活）：always；
- Location Dependency：Delta location from coordinate；
- Coordinate Reference（参考坐标）：._FSAE_sus_rear_rigid_ben.ground.cfl_damper_down；
- Location：0，0，0；
- Location in：local；
- I-Part Axis：._FSAE_sus_rear_rigid_ben.ground.cfl_damper1_down_ref；
- J-Part Axis：._FSAE_sus_rear_rigid_ben.ground.cfl_damper_up。

（2）单击 Apply 按钮，完成约束副._FSAE_sus_rear_rigid_ben.jolhoo_damper_down 的创建。

28. 部件 damper_up 与 damper_down 之间的 cylindrical 约束

（1）单击 Build > Attachments > Joint > New 命令，在下列对话框中输入相应的数据：
- Joint Name（约束副名称）：damper1_mid；
- Type：left；
- I Part：._FSAE_sus_rear_rigid_ben.gel_damper_up；
- J Part：._FSAE_sus_rear_rigid_ben.gel_damper_down；

- Joint Type（约束副类型）：cylindrical；
- Active（激活）：always；
- Location Dependency：Centered between coordinates；
- Centered between：Two Coordinates；
- Coordinate Reference #1（参考坐标）：._FSAE_sus_rear_rigid_ben.ground.cfl_damper_up；
- Coordinate Reference #2（参考坐标）：._FSAE_sus_rear_rigid_ben.ground.cfl_damper_down；
- Location Dependency：Orient axis to point；
- Coordinate Reference（参考坐标）：._FSAE_sus_rear_rigid_ben.ground.cfl_damper_down。

（2）单击 Apply 按钮，完成约束副._FSAE_sus_rear_rigid_ben.jolcyl_damper1_mid 的创建。

29. 部件 axis_vertical 与安装件 suspension_to_chassis 之间的 revolute 约束

（1）单击 Build > Attachments > Joint > New 命令，在下列对话框中输入相应的数据：
- Joint Name（约束副名称）：axis_vertical；
- Type：left；
- I Part：._FSAE_sus_rear_rigid_ben.gel_axis_vertical；
- J Part：._FSAE_sus_rear_rigid_ben.mts_suspension_to_chassis；
- Joint Type（约束副类型）：revolute；
- Active（激活）：always；
- Location Dependency：Centered between coordinates；
- Centered between：Two Coordinates；
- Coordinate Reference #1（参考坐标）：._FSAE_sus_rear_rigid_ben.ground.cfl_torsion_up；
- Coordinate Reference #2（参考坐标）：._FSAE_sus_rear_rigid_ben.ground.cfl_torsion_down；
- Orientation Dependency：Orient axis to point；
- Coordinate Reference：._FSAE_sus_rear_rigid_ben.ground.cfl_torsion_up。

（2）单击 Apply 按钮，完成约束副._FSAE_sus_rear_rigid_ben.jolrev_axis_vertical 的创建。

30. 部件 zhijia_down 与 _axis_vertical 之间的 fixed 约束

（1）单击 Build > Attachments > Joint > New 命令，在下列对话框中输入相应的数据：
- Joint Name（约束副名称）：zhijia_down；
- Type：left；
- I Part：._FSAE_sus_rear_rigid_ben.gel_zhijia_down；
- J Part：._FSAE_sus_rear_rigid_ben.gel_axis_vertical；
- Joint Type（约束副类型）：fixed；
- Active（激活）：always；
- Location Dependency：Centered between coordinates；
- Centered between：Two Coordinates；

- Coordinate Reference #1（参考坐标）：._FSAE_sus_rear_rigid_ben.ground.cfl_torsion_down；
- Coordinate Reference #2（参考坐标）：._FSAE_sus_rear_rigid_ben.ground.hpl_zhijia_to_torsion。

（2）单击 Apply 按钮，完成约束副._FSAE_sus_rear_rigid_ben.jolfix_zhijia_down 的创建。

31. 部件 zhijia_up 与 _axis_vertical 之间的 fixed 约束

（1）单击 Build > Attachments > Joint > New 命令，在下列对话框中输入相应的数据：

- Joint Name（约束副名称）：zhijia_up；
- Type：left；
- I Part：._FSAE_sus_rear_rigid_ben.gel_zhijia_up；
- J Part：._FSAE_sus_rear_rigid_ben.gel_axis_vertical；
- Joint Type（约束副类型）：fixed；
- Active（激活）：always；
- Location Dependency：Centered between coordinates；
- Centered between：Two Coordinates；
- Coordinate Reference #1（参考坐标）：._FSAE_sus_rear_rigid_ben.ground.cfl_torsion_up；
- Coordinate Reference #2（参考坐标）：._FSAE_sus_rear_rigid_ben.ground.cfl_zhijia_up。

（2）单击 Apply 按钮，完成约束副._FSAE_sus_rear_rigid_ben.jolfix_zhijia_up 的创建。

32. 部件 zhijia_up 与 link1 之间的 hook 约束

（1）单击 Build > Attachments > Joint > New 命令，在下列对话框中输入相应的数据：

- Joint Name（约束副名称）：zhijia_up_to_link1；
- I Part：._FSAE_sus_rear_rigid_ben.gel_zhijia_up；
- J Part：._FSAE_sus_rear_rigid_ben.gel_link1；
- Joint Type（约束副类型）：hook；
- Active（激活）：always；
- Location Dependency：Delta location from coordinate；
- Coordinate Reference（参考坐标）：._FSAE_sus_rear_rigid_ben.ground.cfl_zhijia_to_link；
- Location：0, 0, 0；
- Location in：local；
- I-Part Axis：._FSAE_sus_rear_rigid_ben.ground.cfl_torsion_up；
- J-Part Axis：._FSAE_sus_rear_rigid_ben.ground.cfl_link_ref。

（2）单击 Apply 按钮，完成约束副._FSAE_sus_rear_rigid_ben.jolhoo_zhijia_up_to_link1 的创建。

33. 部件 link1 与 link2 之间的 hook 约束

（1）单击 Build > Attachments > Joint > New 命令，在下列对话框中输入相应的数据：

- Joint Name（约束副名称）：link_ref；
- I Part：._FSAE_sus_rear_rigid_ben.gel_link1；
- J Part：._FSAE_sus_rear_rigid_ben.gel_link2；
- Joint Type（约束副类型）：hook；
- Active（激活）：always；
- Location Dependency：Delta location from coordinate；
- Coordinate Reference（参考坐标）：._FSAE_sus_rear_rigid_ben.ground.cfl_link_ref；
- Location：0，0，0；
- Location in：local；
- I-Part Axis：._FSAE_sus_rear_rigid_ben.ground.cfl_zhijia_to_link；
- J-Part Axis：._FSAE_sus_rear_rigid_ben.ground.cfl_link_to_torsion。

（2）单击 Apply 按钮，完成约束副._FSAE_sus_rear_rigid_ben.jolhoo_link_ref 的创建。

34. 部件 link2 与 torsion 之间的 fixed 约束

（1）单击 Build > Attachments > Joint > New 命令，在下列对话框中输入相应的数据：

- Joint Name（约束副名称）：link2_to_torsion；
- Type：left；
- I Part：._FSAE_sus_rear_rigid_ben.gel_link2；
- J Part：._FSAE_sus_rear_rigid_ben.ges_torsion；
- Joint Type（约束副类型）：fixed；
- Active（激活）：always；
- Location Dependency：Delta location from coordinate；
- Coordinate Reference（参考坐标）：._FSAE_sus_rear_rigid_ben.ground.cfl_link_to_torsion；
- Location：0，0，0；
- Location in：local。

（2）单击 Apply 按钮，完成约束副._FSAE_sus_rear_rigid_ben.jolfix_link2_to_torsion 的创建。

35. 部件 spring_mount 与 zhijia_up 之间的 convel 约束

（1）单击 Build > Attachments > Joint > New 命令，在下列对话框中输入相应的数据：

- Joint Name（约束副名称）：spirng_mount；
- I Part：._FSAE_sus_rear_rigid_ben.gel_spring_mount；
- J Part：._FSAE_sus_rear_rigid_ben.gel_zhijia_up；
- Joint Type（约束副类型）：convel；
- Active（激活）：always；
- Location Dependency：Delta location from coordinate；
- Coordinate Reference（参考坐标）：._FSAE_sus_rear_rigid_ben.ground.hpl_spring_mount；
- Location：0，0，0；
- Location in：local；

- I-Part Axis：._FSAE_sus_rear_rigid_ben.ground.hpr_spring_mount；
- J-Part Axis：._FSAE_sus_rear_rigid_ben.ground.cfl_zhijia_up。

（2）单击 Apply 按钮，完成约束副._FSAE_sus_rear_rigid_ben.jolcon_spirng_mount 的创建。

36. 部件 spring_mount（左侧）与 spring_mount（右侧）之间的 cylindrical 约束

（1）单击 Build > Attachments > Joint > New 命令，在下列对话框中输入相应的数据：

- Joint Name（约束副名称）：sping_mid；
- Type：single；
- I Part：._FSAE_sus_rear_rigid_ben.gel_spring_mount；
- J Part：._FSAE_sus_rear_rigid_ben.ger_spring_mount；
- Joint Type（约束副类型）：cylindrical；
- Active（激活）：always；
- Location Dependency：Centered between coordinates；
- Centered between：Two Coordinates；
- Coordinate Reference #1（参考坐标）：._FSAE_sus_rear_rigid_ben.ground.hpl_spring_mount；
- Coordinate Reference #2（参考坐标）：._FSAE_sus_rear_rigid_ben.ground.hpr_spring_mount；
- Location Dependency：Orient axis to point；
- Coordinate Reference（参考坐标）：._FSAE_sus_rear_rigid_ben.ground.hpl_spring_mount。

（2）单击 Apply 按钮，完成约束副._FSAE_sus_rear_rigid_ben.joscyl_sping_mid 的创建。

37. 部件 body_ref 与 suspension_to_chassis 之间的 fixed 约束

（1）单击 Build > Attachments > Joint > New 命令，在下列对话框中输入相应的数据：

- Joint Name（约束副名称）：body_ref；
- Type：left；
- I Part：._FSAE_sus_rear_rigid_ben.gel_body_ref；
- J Part：._FSAE_sus_rear_rigid_ben.mts_suspension_to_chassis；
- Joint Type（约束副类型）：fixed；
- Active（激活）：always；
- Location Dependency：Delta location from coordinate；
- Coordinate Reference（参考坐标）：._FSAE_sus_rear_rigid_ben.ground.cfl_bumpstop；
- Location：0，0，0；
- Location in：local。

（2）单击 Apply 按钮，完成约束副._FSAE_sus_rear_rigid_ben.jolfix_body_ref 的创建。

38. 避震器

（1）单击 Build > Force > Damper > New 命令，弹出避震器创建对话框，如图 15-9 所示，在下列对话框中输入相应的数据：

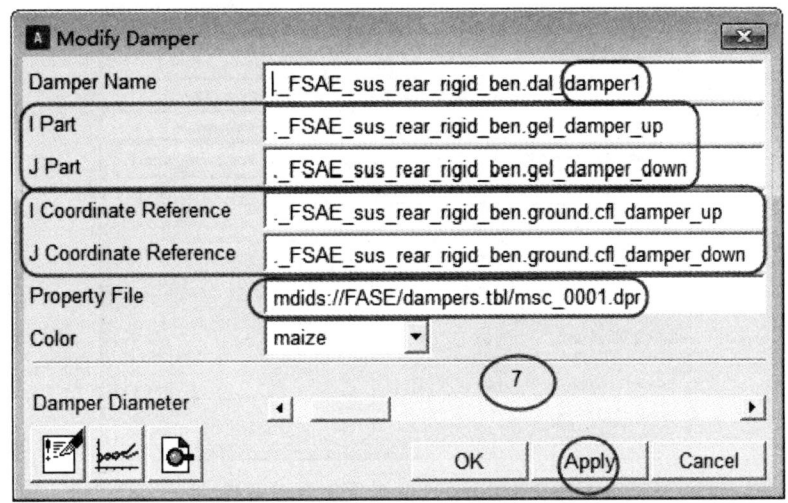

图 15-9　damper 避震器创建对话框

- Damper Name（减震器名称）：damper1；
- I Part：._FSAE_sus_rear_rigid_ben.gel_damper_up；
- J Part：._FSAE_sus_rear_rigid_ben.gel_damper_down；
- I Coordinate Reference（参考坐标）：._FSAE_sus_rear_rigid_ben.ground.cfl_damper_up；
- J Coordinate Reference（参考坐标）：._FSAE_sus_rear_rigid_ben.ground.cfl_damper_down；
- Property File（属性文件）：mdids：//FASE/dampers.tbl/msc_0001.dpr；
- Damper Diameter（避震器直径）：拖动滑块选择 7 mm；
- Color：maize。

（2）单击 Apply 按钮，完成避震器._FSAE_sus_rear_rigid_ben.dal_damper1 的创建。同理，创建其他避震器。

- Damper Name（减震器名称）：damper_front；
- I Part：._FSAE_sus_rear_rigid_ben.gel_zhijia_down；
- J Part：._FSAE_sus_rear_rigid_ben.ger_zhijia_down；
- I Coordinate Reference（参考坐标）：._FSAE_sus_rear_rigid_ben.ground.cfl_damper_front；
- J Coordinate Reference（参考坐标）：._FSAE_sus_rear_rigid_ben.ground.cfr_damper_front；
- Property File（属性文件）：mdids：//FASE/dampers.tbl/MSC_default.dpr；
- Damper Diameter（避震器直径）：拖动滑块选择 5 mm；
- Color：maize。

（3）单击 OK 按钮，完成避震器._FSAE_sus_rear_rigid_ben.das_damper_front 的创建。

39. 弹　簧

（1）单击 Build > Force > Spring > New 命令，弹出弹簧创建对话框，如图 15-10 所示，在下列对话框中输入相应的数据：

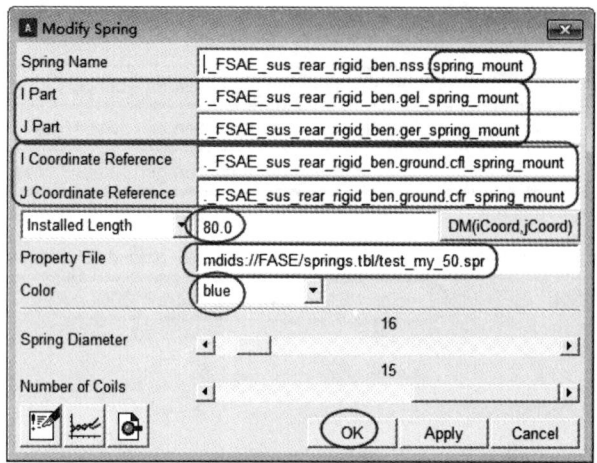

图 15-10　spring 弹簧创建对话框

- Spring Name（减震器名称）：spring_mount；
- I Part：._FSAE_sus_rear_rigid_ben.gel_spring_mount；
- J Part：._FSAE_sus_rear_rigid_ben.ger_spring_mount；
- I Coordinate Reference（参考坐标）：._FSAE_sus_rear_rigid_ben.ground.cfl_spring_mount；
- J Coordinate Reference（参考坐标）：._FSAE_sus_rear_rigid_ben.ground.cfr_spring_mount；
- Installed Length（安装长度）：单击 DM（iCoord，jCoord），自动计算弹簧的安装长度并填入到方框中，此模型的安装长度为 80；
- Property File（属性文件）：mdids：//FASE/springs.tbl/test_my_50.spr；
- Color：blue；
- Spring Diameter（弹簧直径）：拖动滑块选择 16 mm；
- Spring of Coils（弹簧圈数）：拖动滑块选择 15。

（2）单击 OK 按钮，完成弹簧._FSAE_sus_rear_rigid_ben.nss_spring_mount 的创建。

40. 上限位缓冲块

（1）单击 Build > Force > Bumpstop > New 命令，弹出创建上限位缓冲块对话框，如图 15-11 所示。在下列对话框中输入相应的数据：

- Bumpstop Name（上限位缓冲块名称）：bumpstop；
- I Part：._FSAE_sus_rear_rigid_ben.gel_spring_mount；
- J Part：._FSAE_sus_rear_rigid_ben.gel_body_ref；
- I Coordinate Reference（参考坐标）：._FSAE_sus_rear_rigid_ben.ground.cfl_spring_mount；

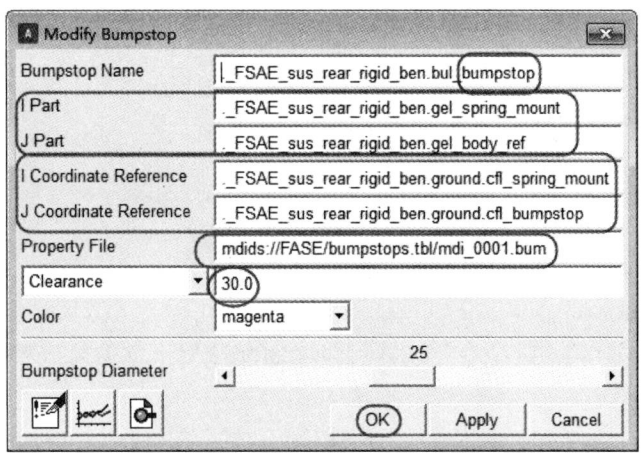

图 15-11　上限位缓冲块对话框

- J Coordinate Reference（参考坐标）：._FSAE_sus_rear_rigid_ben.ground.cfl_bumpstop；
- Property File（属性文件）：mdids：//FASE/bumpstops.tbl/mdi_0001.bum，属性文件为缓冲块的刚度，刚度曲线如图 15-12 所示，数据信息如下；
- Clearance（空行程）：30。

（2）单击 OK 按钮，完成上限位缓冲块 ._FSAE_sus_rear_rigid_ben.bul_bumpstop 的创建。

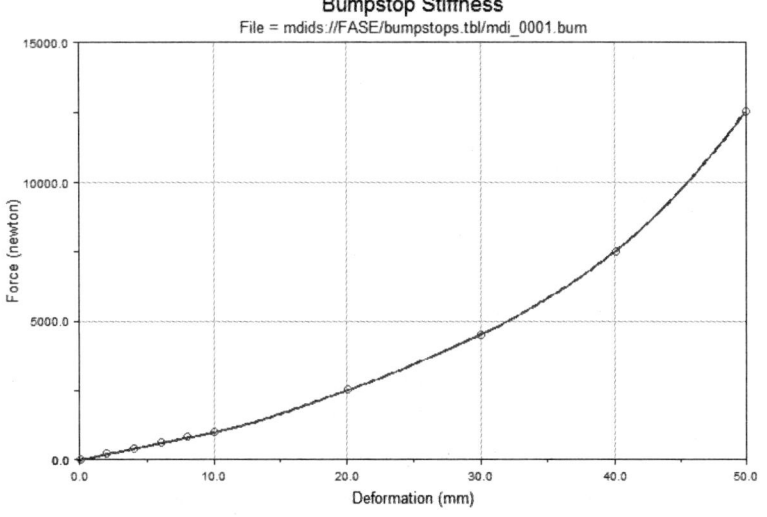

图 15-12　上限位缓冲块刚度

缓冲块属性文件信息

$---MDI_HEADER
[MDI_HEADER]
 FILE_TYPE   = 'bum'
 FILE_VERSION   = 4.0
 FILE_FORMAT   = 'ASCII'

```
$----------------------------------------UNITS
[UNITS]
 LENGTH  =  'mm'
 ANGLE   =  'degrees'
 FORCE   =  'newton'
 MASS    =  'kg'
 TIME    =  'second'
$--------------------------------CURVE %以下数据可以根据实验修改
[CURVE]
{ disp        force}
 0.0          0.0
 2.0          200.0
 4.0          400.0
 6.0          600.0
 8.0          800.0
 10.0         1000.0
 20.0         2500.0
 30.0         4500.0
 40.0         7500.0
 50.0         12500.0
```

41. 下限位缓冲块

（1）单击 Build > Force > Bumpstop > New 命令，弹出创建下限位缓冲块对话框，如图 15-13 所示，在下列对话框中输入相应的数据：

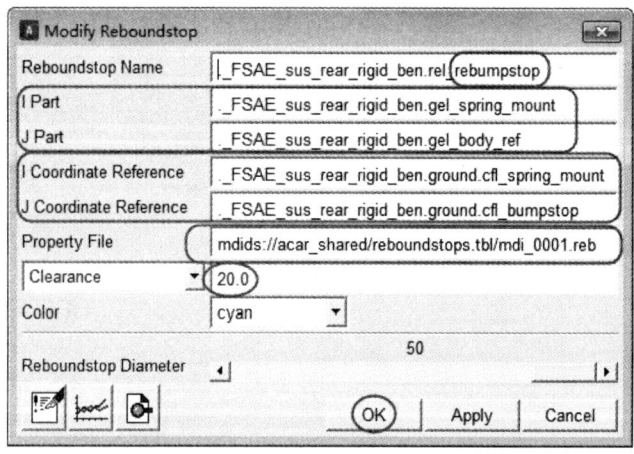

图 15-13 下限位缓冲块对话框

- Rebumpstop Name（下限位缓冲块名称）：rebumpstop；
- I Part：._FSAE_sus_rear_rigid_ben.gel_spring_mount；
- J Part：._FSAE_sus_rear_rigid_ben.gel_body_ref；
- I Coordinate Reference（参考坐标）：._FSAE_sus_rear_rigid_ben.ground.cfl_spring_mount；
- J Coordinate Reference（参考坐标）：._FSAE_sus_rear_rigid_ben.ground.cfl_bumpstop；
- Property File（属性文件）：mdids：//acar_shared/reboundstops.tbl/mdi_0001.reb；
- Clearance（空行程）：20。

（2）单击 OK 按钮，完成下限位缓冲块 ._FSAE_sus_rear_rigid_ben.rel_rebumpstop 的创建。

42. 部件 torsion 与 suspension_to_chassis 之间的 bushing 约束

（1）单击 Build > Attachments > Bushing > New 命令，弹出创衬套件对话框，如图 15-14 所示，在下列对话框中输入相应的数据：

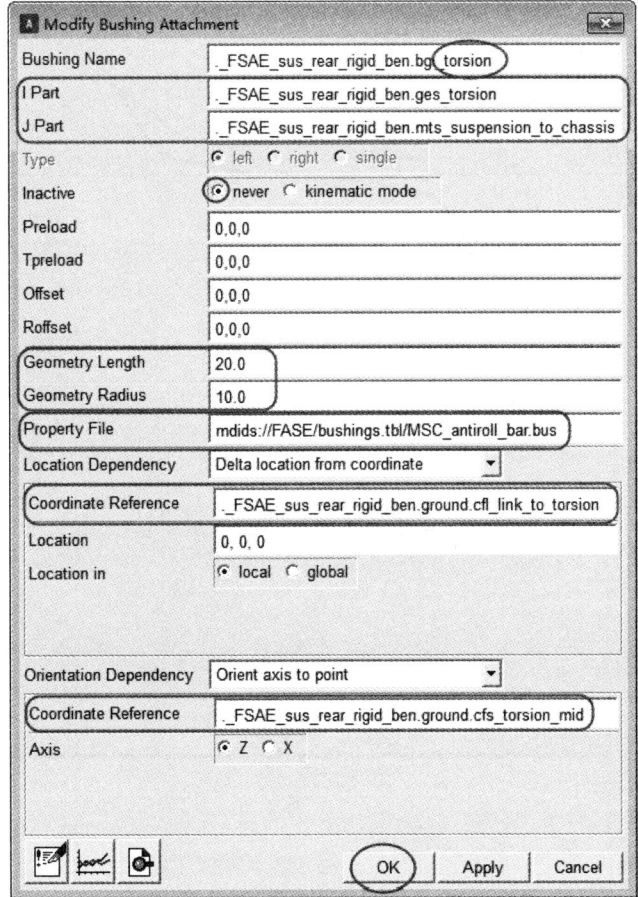

图 15-14 衬套约束对话框_torsion

- Bushing Name（约束副名称）：torsion；
- I Part：._FSAE_sus_rear_rigid_ben.ges_torsion；
- J Part：._FSAE_sus_rear_rigid_ben.mts_suspension_to_chassis；

- Inactive（抑制）：never；
- Preload：0，0，0；
- Tpreload：0，0，0；
- Offset：0，0，0；
- Roffset：0，0，0；
- Geometry Length：20；
- Geometry Radius：10；
- Property File：mdids：//FASE/bushings.tbl/MSC_antiroll_bar.bus，用记事本文件打开衬套属性文件，用 MATLAB 软件绘制 X、Y、Z 方向的垂向刚度及扭转刚度，如图 15-15 和图 15-16 所示，属性文件数据信息如下；

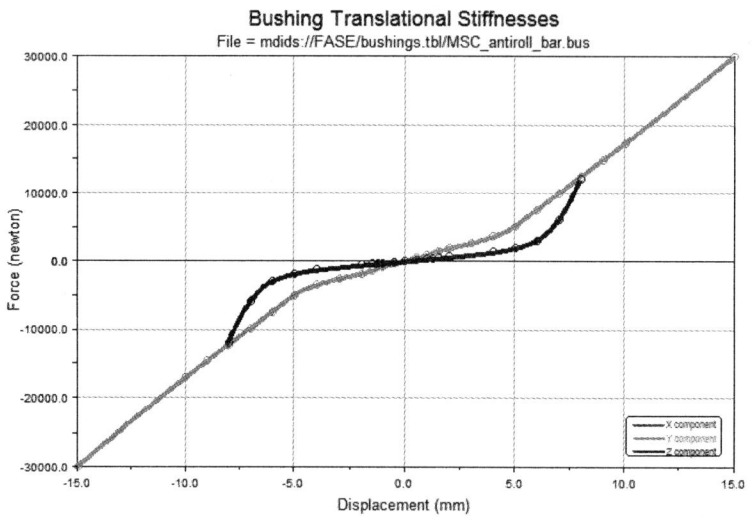

图 15-15　衬套垂向刚度

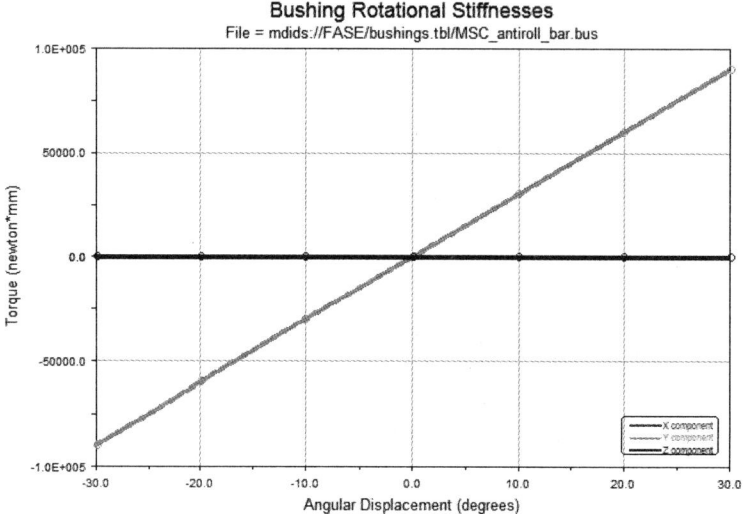

图 15-16　衬套扭转刚度

- Location Dependency：Delta location from coordinate；
- Coordinate Reference（参考坐标）：._FSAE_sus_rear_rigid_ben.ground.cfl_link_to_torsion；
- Location：0，0，0；
- Location in：local；
- Orientation Dependency：Orient axis to point；
- Coordinate Reference（参考坐标）：._FSAE_sus_rear_rigid_ben.ground.cfs_torsion_mid；
- Axis：Z。

（2）单击 Apply 按钮，完成衬套._FSAE_sus_rear_rigid_ben.bgl_torsion 的创建。至此，拉杆式悬架模型建立完成，如图 15-1 所示。

```
衬套属性文件，数据可以通过实验的方式或者有限元法分析获取：
$--------------------------------------------MDI_HEADER
[MDI_HEADER]
  FILE_TYPE      =  'bus'
  FILE_VERSION   =  4.0
  FILE_FORMAT    =  'ASCII'
$--------------------------------------------UNITS
[UNITS]
  LENGTH   =  'mm'
  ANGLE    =  'degrees'
  FORCE    =  'newton'
  MASS     =  'kg'
  TIME     =  'second'
$--------------------------------------------DAMPING
[DAMPING]
  FX_DAMPING  =  10.0
  FY_DAMPING  =  10.0
  FZ_DAMPING  =  10.0
  TX_DAMPING  =  0.0
  TY_DAMPING  =  50.0
  TZ_DAMPING  =  0.0
$--------------------------------------------FX_CURVE
[FX_CURVE]
{x                fx}
 -15.0            -30000.0
 -10.0            -17182.0
```

-9.0	-14762.0
-8.0	-12342.0
-7.0	-9922.0
-6.0	-7502.0
-5.0	-5082.0
-4.0	-3569.5
-3.0	-2613.6
-2.0	-1875.5
-1.5	-1427.8
-1.0	-847.0
-0.5	-423.5
0.0	0.0
0.5	423.5
1.0	847.0
1.5	1427.8
2.0	1875.5
3.0	2613.6
4.0	3569.5
5.0	5082.0
6.0	7502.0
7.0	9922.0
8.0	12342.0
9.0	14762.0
10.0	17182.0
15.0	30000.0

$--FY_CURVE

[FY_CURVE]

{y	fy}
-15.0	-30000.0
-10.0	-17182.0
-9.0	-14762.0
-8.0	-12342.0
-7.0	-9922.0
-6.0	-7502.0
-5.0	-5082.0

-4.0	-3569.5
-3.0	-2613.6
-2.0	-1875.5
-1.5	-1427.8
-1.0	-847.0
-0.5	-423.5
0.0	0.0
0.5	423.5
1.0	847.0
1.5	1427.8
2.0	1875.5
3.0	2613.6
4.0	3569.5
5.0	5082.0
6.0	7502.0
7.0	9922.0
8.0	12342.0
9.0	14762.0
10.0	17182.0
15.0	30000.0

$---FZ_CURVE

[FZ_CURVE]

{z	fz}
-8.0	-12000.0
-7.0	-6000.0
-6.0	-3000.0
-5.0	-1900.0
-4.0	-1300.0
-2.0	-600.0
-1.5	-450.0
-1.25	-375.0
-1.0	-300.0
-0.5	-150.0
0.0	0.0
0.5	150.0

1.0	300.0
1.25	375.0
1.5	450.0
2.0	600.0
4.0	1300.0
5.0	1900.0
6.0	3000.0
7.0	6000.0
8.0	12000.0

$---TX_CURVE

[TX_CURVE]

{ax	tx}
-30.0	-90000.0
-20.0	-60000.0
-10.0	-30000.0
0.0	0.0
10.0	30000.0
20.0	60000.0
30.0	90000.0

$---TY_CURVE

[TY_CURVE]

{ay	ty}
-30.0	-90000.0
-20.0	-60000.0
-10.0	-30000.0
0.0	0.0
10.0	30000.0
20.0	60000.0
30.0	90000.0

$---TZ_CURVE

[TZ_CURVE]

{az	tz}
-30.0	-30.0
-20.0	-20.0
-10.0	-10.0

0.0	0.0
10.0	10.0
20.0	20.0
30.0	30.0

15.3 收油门转弯仿真

替换整车模型 FSAE_2020_torsion_vertical_damper.asy 的后悬架后,此时由于替换后的后悬架模型 FSAE_sus_rear_rigid_ben 将拉杆设置到上下控制臂之外,这样会导致发动机和拉杆产生重合(实际上并不影响整车仿真,多体动力学和三维软件有本质的区别),把发动机向前移动 200 mm,将发动机模型另存为 FSAE_powertrain_2020(最好另存为,否则会导致其他采用发动机的车型在打开装配模型时也会向前移动 200 mm,此时发动机模型为替换后整车模型的单独用发动机,不与其他整车模型共享)。隐藏发动机,此时后悬架模型在整车视图下显得更加清楚,完成整车模型并另存为 FSAE_2020_rigid_Ben,如图 15-17 所示。

图 15-17 整车 FSAE_2020_rigid_Ben

收油门仿真分为两个阶段:① 在圆形车道上将整车模型加速到预设的侧向加速度,侧向加速度满足条件为收油门转弯仿真的前提;② 整车在达到预设的侧向加速度后,关闭油门(此时分离合器分离与离合器结合两种模式),然后按设定的角速度转动方向盘,直到整车停止运动完成仿真。

(1)单击 Simulate > Full-Vehicle Analysis > Cornering Events > Lift-Off-Turn-In 命令,仿真参数设置如图 15-18 和图 15-19 所示,在下列对话框中输入相应的数据:

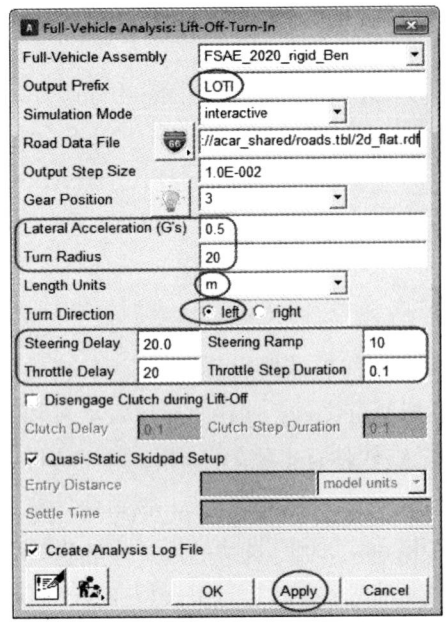 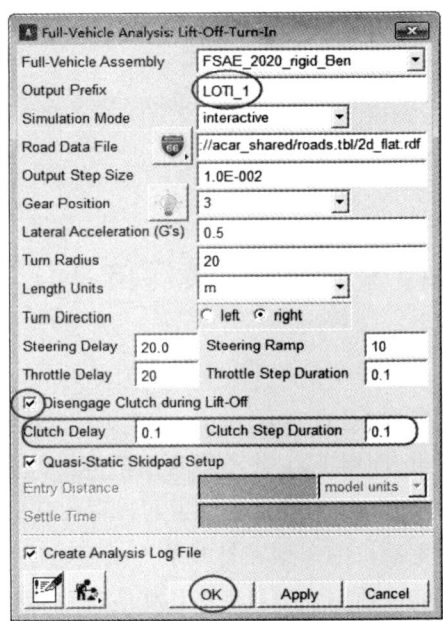

图 15-18　收油门转弯仿真（离合器结合）　　图 15-19　收油门转弯仿真（离合器分离）

- Full-Vehicle Assembly：FSAE_2020_rigid_Ben；
- Output Prefix：LOTI；
- Simulation Mode：interactive；
- Road Date File：mdids：//acar_shared/roads.tbl/2d_flat.rdf；
- Output Step step：0.01；
- Gear Position：3；
- Lateral Acceleration（G's）：0.5；
- Turn Radius：20；
- Length Units：m；
- Turn Direction：left；
- Steering Delay：20；
- Steering Ramp：10；
- Throttle Delay：20；
- Throttle Step Duration：0.1；
- 勾选 Disengage Clutch during Lift-Off，使离合器分离，如图 15-19 所示；
- Clutch Delay：0.1；
- Clutch Step Duration：0.1；
- 勾选 Quasi-Static Straight-Line Setup。

（2）单击 OK 按钮，完成 FSAE_2020_rigid_Ben 收油门转弯仿真设置并提交运算，运算完成后，整车运行轨迹如图 15-20 和图 15-21 所示；参数曲线如图 15-22～图 15-24 所示，从图中可以看出，20 s 附近方向盘开始放手。

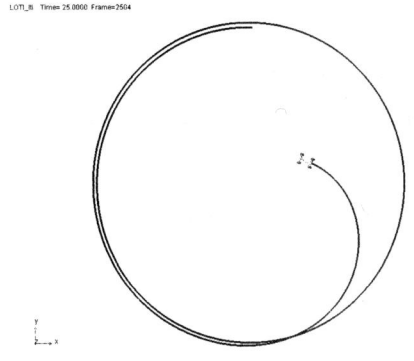

图 15-20　整车运动轨迹（离合器结合）

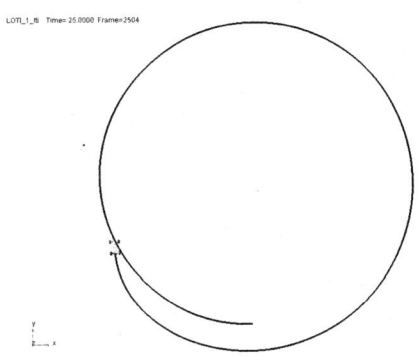

图 15-21　整车运动轨迹（离合器分离）

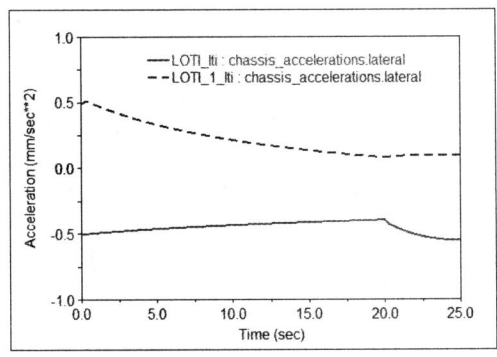

图 15-22　侧向加速度

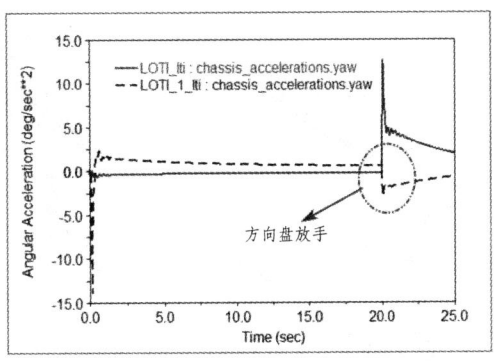

图 15-23　横摆角加速度

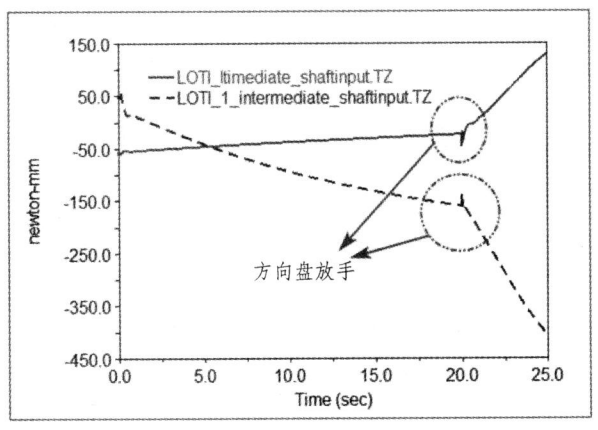

图 15-24　方向盘中间传动轴 Z 方向扭矩

15.4 纵置扭杆弹簧式拉杆悬架

纵置扭杆弹簧式拉杆悬架模型如图 15-25 所示,扭杆弹簧横截面半径为 5 mm,长度为 200 mm;横向稳定杆也采用柔性体,横截面半径为 10 mm,长度为 200 mm。纵置扭杆弹簧式拉杆悬架与上述悬架模型 FSAE_sus_rear_rigid_ben 较为相似,刚性扭转轴用垂向扭转弹簧替换,删除上部旋转支架、螺旋弹簧及上下限位块后添加对应的约束,该小节不再叙述详细建模步骤,读者可打开电子文件中存储的悬架模型_FSAE_sus_rear_torsion_flex_ben.tpl,逐步建立对应的部件、柔性体、硬点、结构框、约束等关系。同理,前悬架模型_FSAE_sus_front_torsion_flex_ben.tpl 如图 15-26 所示。

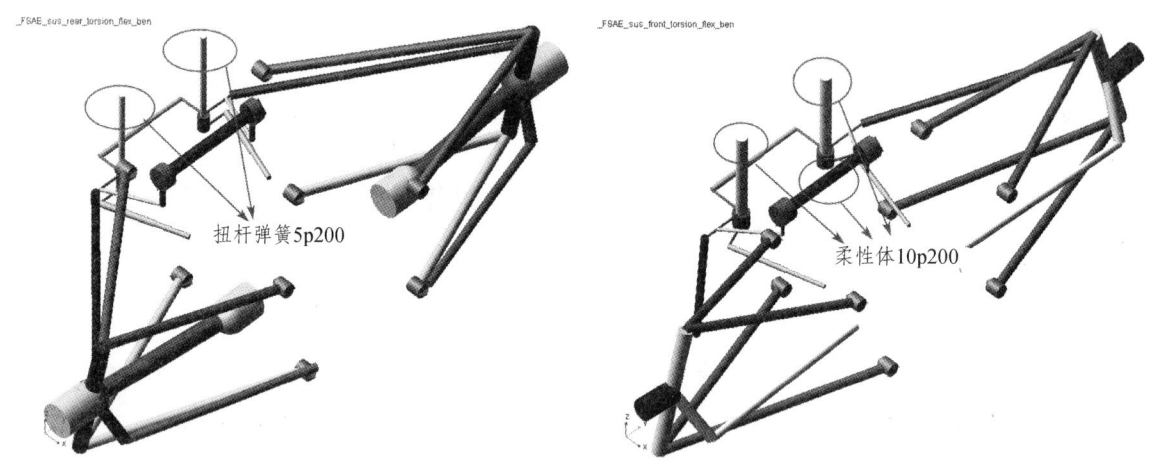

图 15-25　后纵置扭杆弹簧式拉杆悬架　　　　图 15-26　前纵置扭杆弹簧式拉杆悬架

15.5 弯道收油门仿真

稳态回转时突然关闭油门,此时分两种模式:① 方向盘锁定;② 由驱动器调节转向盘角度维持原有的转弯半径。用悬架模型_FSAE_sus_front_torsion_flex_ben.tpl 与 _FSAE_sus_rear_torsion_flex_ben.tpl 替换整车 FSAE_2020_rigid_Ben 中的前后悬架,此时整车模型另存为 FSAE_2020 _Ben,如图 15-27 所示。

图 15-27 整车 FSAE_2020_Ben

（1）单击 Simulate > Full-Vehicle Analysis > Cornering Events > Power-Off Cornering 命令，仿真参数设置如图 15-28 和 15-29 所示，在下列对话框中输入相应的数据：

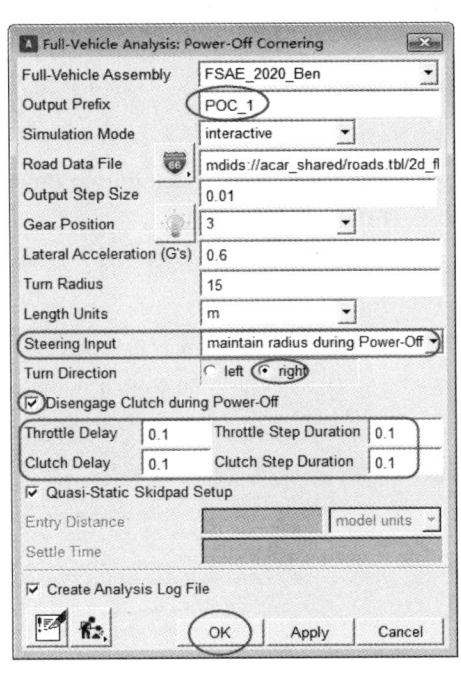

图 15-28 弯道收油门仿真（锁定方向盘）　　图 15-29 弯道收油门仿真（保持转弯半径）

- Full-Vehicle Assembly：FSAE_2020_Ben；
- Output Prefix：POC；
- Simulation Mode：interactive；
- Road Date File：mdids：//acar_shared/roads.tbl/2d_flat.rdf；

- Output Step step: 0.01;
- Gear Position: 3;
- Lateral Acceleration (G's): 0.6;
- Turn Radius: 15;
- Length Units: m;
- Steering Input: ① lock steering during Power-Off（锁定方向盘）；② maintain radius during Power-Off（保持转向半径不变），从图 15-30 和图 15-31 中可以看出在两种不同模式下车辆运行的轨迹；

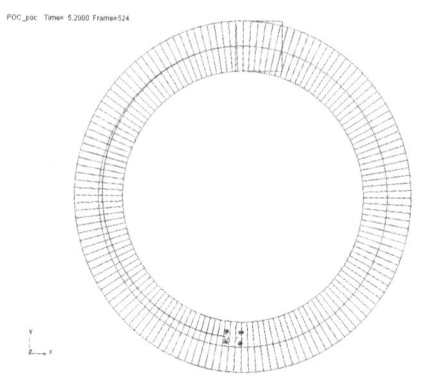

图 15-30　整车运动轨迹（锁定方向盘）　　图 15-31　整车运动轨迹（保持转弯半径）

- Turn Direction: left;
- 勾选 Disengage Clutch during Power-Off;
- Throttle Delay: 0.1;
- Throttle Step Duration: 0.1;
- Clutch Delay: 0.1;
- Clutch Step Duration: 0.1;
- 勾选 Quasi-Static Straight-Line Setup。

（2）单击 OK 按钮，完成 FSAE_2020_Ben 弯道收油门仿真设置并提交运算，运算完成后，参数曲线如图 15-32～15-36 所示。

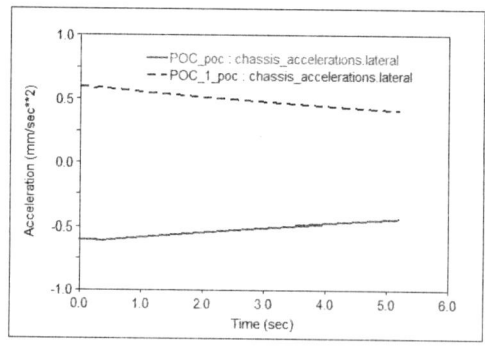

图 15-32　车身侧向加速度

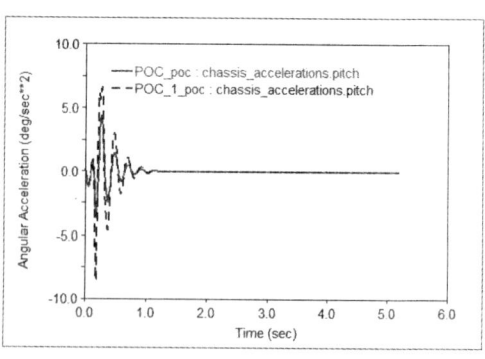

图 15-33　侧倾角加速度

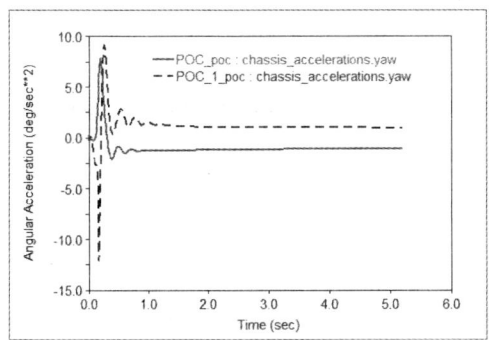
图 15-34 横摆角加速度

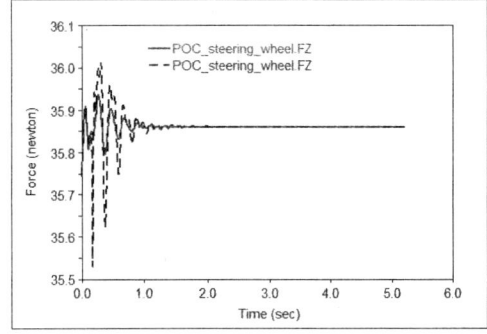
图 15-35 方向盘力

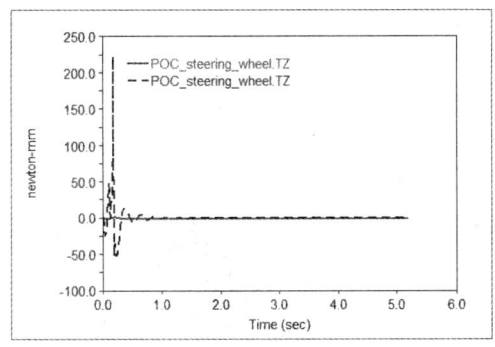
图 15-36 方向盘力矩

第 16 章 扭力梁悬架

方程式赛车悬架具有多种悬架形式,目前方程式赛车大多采用推杆式独立悬架。整车操控性能取决于车辆的悬架,悬架精髓在于调试,并非独立悬架一定比非独立悬架的操控性好。扭力梁悬架为非独立悬架,结构简单,占用空间小,尤其是拖拽臂的延伸可以缩短方程式赛车车身的长度,减轻整车质量。同时,对扭力梁悬架适当调试,可以实现随动转向,进一步提升整车的操控性并兼顾平顺性。本章通过介绍扭力梁悬架模型的建立,替换方程式赛车后推杆式独立悬架,完成整车模型的建立,如图 16-1 所示。

图 16-1 FSAE 整车模型

学习目标

(1)了解扭力梁悬架。
(2)了解 FSAE 整车模型。
(3)会定常半径转弯仿真。

16.1 扭力梁悬架

(1) 启动 ADAMS/CAR, 选择专家模块进入建模界面。
(2) 单击 File > New 命令, 弹出建模对话框; 在模板名称中输入: _torsion_beam_sus_FSAE, 主特征选择 suspension, 单击 OK 按钮。
(3) 单击 Build > Hardpoind > New 命令, 弹出创建硬点对话框;
在硬点名称里输入: wheel_center, 类型选择 left; 在位置文本框输入: 1265, -550.0, 3755.0。
(4) 单击 Apply 按钮, 完成 wheel_center 硬点的创建。重复上述步骤完成图 16-2 中硬点的创建, 创建完成后单击 OK 按钮。

	loc x	loc y	loc z
hpl_rca_outer	1265.0	-370.0	375.0
hpl_rca_outer_ref	1365.0	-370.0	375.0
hpl_rca_pivot	889.37	-370.0	375.0
hpl_rca_pivot_ref	799.37	-370.0	375.0
hpl_shaft_inner	1250.0	-100.0	350.0
hpl_shock_lower	1300.0	-320.0	375.0
hpl_shock_lower_ref	1300.0	-320.0	275.0
hpl_shock_up	1365.0	-320.0	641.0
hpl_shock_up_ref	1365.0	-320.0	841.0
hpl_spring_lower	1170.42	-320.0	375.0
hpl_spring_up	1170.42	-320.0	551.0
hpl_torsion_beam_location	989.37	-370.0	375.0
hpl_wheel_center	1265.0	-550.0	375.0

图 16-2 扭力梁悬架硬点

16.1.1 下拖拽臂部件

(1) 单击 Build > Part > General Part > New 命令, 弹出创建部件对话框, 如图 16-3 所示, 在下列对话框中输入相应的数据:
- General Part: rca;
- Location Dependency: Centered between coordinates;
- Centered between: Two Coordinates, 拖拽臂部件 rca 位于两点坐标的中心位置;
- Coordinate Reference #1 (参考坐标): ._torsion_beam_sus_FSAE.ground.hpl_rca_pivot;
- Coordinate Reference #2 (参考坐标): ._torsion_beam_sus_FSAE.ground.hpl_rca_outer;
- Orient using: Euler Angles, 部件定向采用欧拉角模式;
- Euler Angles: 0, 0, 0;
- Mass: 1;
- Ixx: 1;

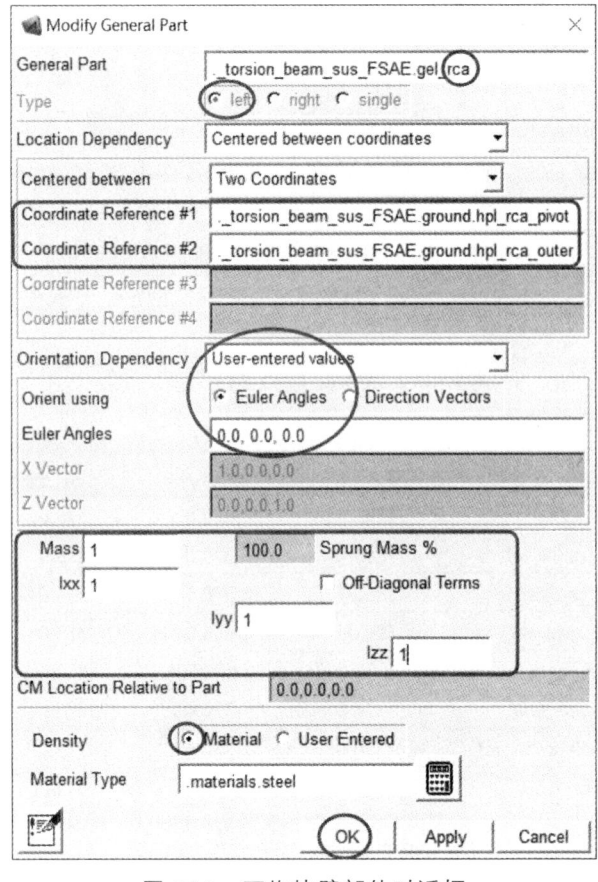

图 16-3　下拖拽臂部件对话框

- Iyy：1；
- Izz：1；
- Density：Material；
- Material Type：.materials.steel。

（2）单击 OK 按钮，完成部件 ._torsion_beam_sus_FSAE.gel_rca 的创建。

（3）单击 Build > Geometry > Link > New 命令，弹出创建部件对话框，如图 16-4 所示，在下列对话框中输入相应的数据：

- Link Name（连杆名称）：link；
- General Part：._torsion_beam_sus_FSAE.gel_rca；
- Coordinate Reference #1（参考坐标）：._torsion_beam_sus_FSAE.ground.hpl_rca_pivot；
- Coordinate Reference #2（参考坐标）：._torsion_beam_sus_FSAE.ground.hpl_rca_outer；
- Radius（半径）：15；
- Color：green；

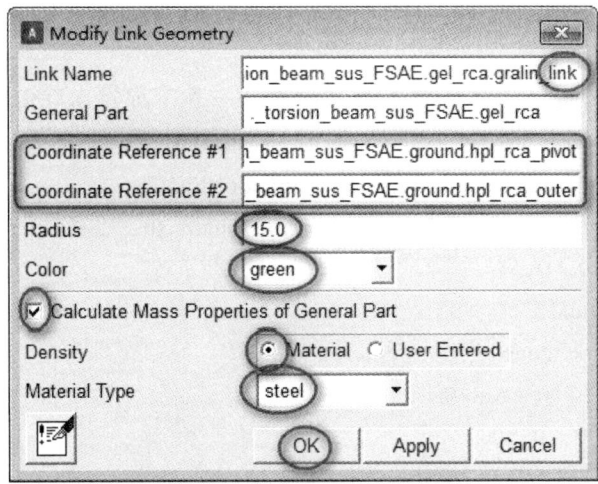

图 16-4　下拖拽臂 link 几何体

- 选择 Calculate Mass Properties of General Part 复选框，当几何体建立好之后，会更新对应部件的质量和惯量参数；
  - Density：Material；
  - Material Type：steel。

（4）单击 OK 按钮，完成 ._torsion_beam_sus_FSAE.gel_rca.gralin_link 几何体的创建。

16.1.2　部件 Strut_up

（1）单击 Build > Part > General Part > New 命令，在下列对话框中输入相应的数据：
- General Part：strut_up；
- Location Dependency：Delta location from coordinate；
- Coordinate Reference（参考坐标）：._torsion_beam_sus_FSAE.ground.hpl_shock_up；
- Location：0，0，0；
- Location in：local；
- Orientation Dependency：User-entered values；
- Orient using：Euler Angles；
- Euler Angles：0，0，0；
- Mass：1；
- Ixx：1；
- Iyy：1；
- Izz：1；
- Density：Material；
- Material Type：.materials.steel。

（2）单击 OK 按钮，完成部件 ._torsion_beam_sus_FSAE.gel_strut_up 的创建。

16.1.3 部件 Strut_down

(1) 单击 Build > Part > General Part > New 命令,在下列对话框中输入相应的数据:
- General Part:strut_up;
- Location Dependency:Delta location from coordinate;
- Coordinate Reference(参考坐标):._torsion_beam_sus_FSAE.ground.hpl_shock_lower;
- Location:0,0,0;
- Location in:local;
- Orientation Dependency:User-entered values;
- Orient using:Euler Angles;
- Euler Angles:0,0,0;
- Mass:1;
- Ixx:1;
- Iyy:1;
- Izz:1;
- Density:Material;
- Material Type:.materials.steel。

(2) 单击 OK 按钮,完成部件._torsion_beam_sus_FSAE.gel_strut_down 的创建。

16.1.4 轮毂 spindle 部件

(1) 单击 Build > Suspension Parameters > Toe/Camber Values > Set 命令,弹出悬架参数对话框,如图 16-5 所示。前束角输入:0;外倾角输入:0;单击 OK 按钮,完成参数创建。与此同时,系统自动建立两个输出通信器:col[r]_toe_angle、col[r]_camber_angle。

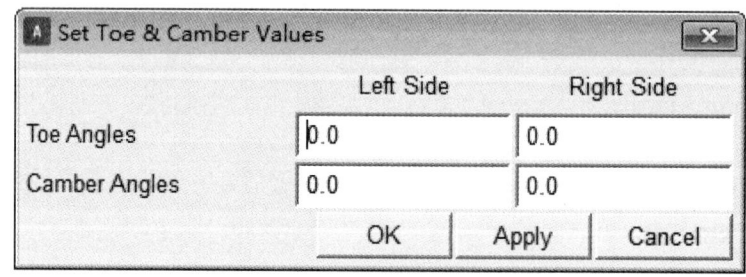

图 16-5 悬架参数

(2) 单击 Build > Construction Frame > New 命令,弹出创建结构框对话框,如图 16-6 所示,在下列对话框中输入相应的数据:
- Construction Frame(结构框名称):wheel_center;
- Coordinate Reference(参考坐标):._torsion_beam_sus_FSAE.ground.hpl_wheel_center;
- Location:0,0,0;
- Location in:local;

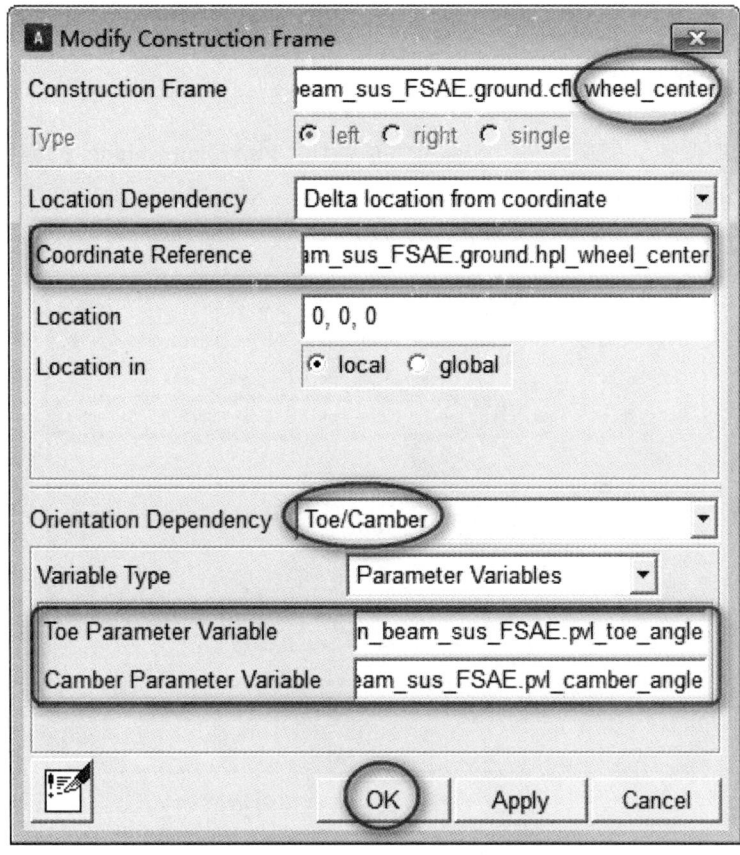

图 16-6 wheel_center 结构框

- Orientation Dependency：User-entered values；
- Variable Type（变量类型）：Parameter Variable（参数变量）；
- Toe Parameter Values（前束变量值）：._torsion_beam_sus_FSAE.pvl_toe_angle；
- Camber Parameter Values（外倾变量值）：._torsion_beam_sus_FSAE.pvl_camber_angle。

（3）单击 OK 按钮，完成 ._torsion_beam_sus_FSAE.ground.cfl_wheel_center 结构框的创建。
（4）单击 Build > Part > General Part > New 命令，在下列对话框中输入相应的数据：

- General Part：spindle；
- Location Dependency：Delta location from coordinate；
- Coordinate Reference（参考坐标）：._torsion_beam_sus_FSAE.ground.hpl_wheel_center；
- Location：0，0，0；
- Location in：local；
- Orientation Dependency：Toe/Camber；
- Variable Type（变量类型）：Parameter Variable；
- Toe Parameter Values（前束变量值）：._torsion_beam_sus_FSAE.pvl_toe_angle；
- Camber Parameter Values（外倾变量值）：._torsion_beam_sus_FSAE.pvl_camber_angle；
- Ixx：1；
- Iyy：1；

- Izz：1；
- Density：Material；
- Material Type：.materials.steel。

（5）单击 OK 按钮，完成部件._torsion_beam_sus_FSAE.gel_spindle 的创建。

（6）单击 Build > Geometry > Cylinder（圆柱体）> New 命令，弹出创建部件对话框，如图 16-7 所示，在下列对话框中输入相应的数据：

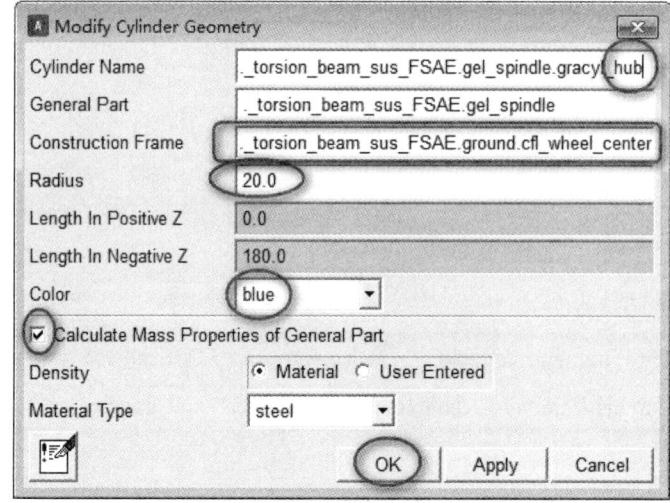

图 16-7 轮毂几何体创建对话框

- Cylinder Name（连杆名称）：hub；
- General Part：._torsion_beam_sus_FSAE.gel_spindle；
- Radius（半径）：20；
- Length In Positive Z（Z 轴正方向长度）：0；
- Length In Negative Z（Z 轴负方向长度）：180；
- Color（圆柱体几何体颜色）：blue；
- 选择 Calculate Mass Properties of General Part 复选框。

（7）单击 OK 按钮，完成轮毂圆柱体._torsion_beam_sus_FSAE.gel_spindle.gracyl_hub 几何体的创建。

16.1.5 部件 torsion_beam

（1）单击 Build > Part > General Part > New 命令，在下列对话框中输入相应的数据：
- General Part：torsion_beam；
- Type：single；
- Location Dependency：Centered between coordinates；
- Centered between：Two Coordinates；
- Coordinate Reference #1（参考坐标）：._torsion_beam_sus_FSAE.ground.hpl_torsion_beam_location；

- Coordinate Reference #2（参考坐标）：._torsion_beam_sus_FSAE.ground.hpr_torsion_beam_location；
- Orient using：Euler Angles；
- Euler Angles：0，0，0；
- Mass：1；
- Ixx：1；
- Iyy：1；
- Izz：1；
- Density：Material；
- Material Type：.materials.steel。

（2）单击 OK 按钮，完成部件._torsion_beam_sus_FSAE.ges_torsion_beam 的创建。

（3）单击 Build > Geometry > Link > New 命令，在下列对话框中输入相应的数据：
- Link Name（连杆名称）：link；
- General Part：._torsion_beam_sus_FSAE.ges_torsion_beam；
- Coordinate Reference #1（参考坐标）：._torsion_beam_sus_FSAE.ground.hpl_torsion_beam_location；
- Coordinate Reference #2（参考坐标）：._torsion_beam_sus_FSAE.ground.hpr_torsion_beam_location；
- Radius（半径）：10；
- Color（杆件几何体颜色）：yellow；
- 选择 Calculate Mass Properties of General Part 复选框，当几何体建立好之后，会更新对应部件的质量和惯量参数；
- Density：Material；
- Material Type：steel。

（4）单击 OK 按钮，完成._torsion_beam_sus_FSAE.ges_torsion_beam.gralin_link 几何体的创建。

16.1.6 部件 tripot

（1）单击 Build > Parameter Variable > New 命令，弹出参数变量对话框，如图 16-8 所示，在下列对话框中输入相应的数据：
- Parameter Variable Name：drive_shaft_offset；
- Real Value（实数值）：180；
- Units：length；
- Hide from standard user（是否从标准界面隐藏）：no。

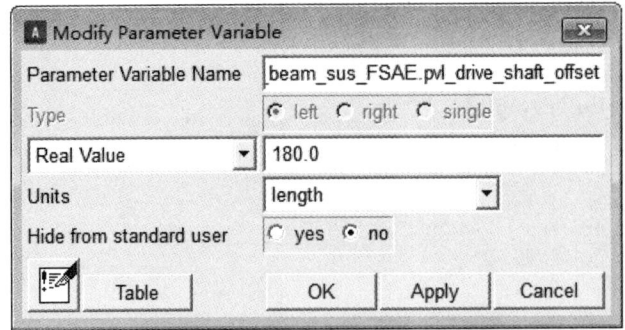

图 16-8 drive_shaft_offset 变量

（2）单击 OK 按钮，完成变量._torsion_beam_sus_FSAE.pvl_drive_shaft_offset 的创建。

（3）单击 Build > Construction Frame > New 命令，在下列对话框中输入相应的数据：
- Construction Frame（结构框名称）：drive_shaft_otr；
- Location Dependency：Delta location from coordinate；
- Coordinate Reference（参考坐标）：._torsion_beam_sus_FSAE.ground.cfl_wheel_center；
- Location：0.0，0.0，(-1.0 * ._torsion_beam_sus_FSAE.pvl_drive_shaft_offset)；
- Location in：local；
- Orientation Dependency：Orient axis to point；
- Coordinate Reference（参考坐标）：._torsion_beam_sus_FSAE.ground.hpl_wheel_center；
- Axis：Z。

（4）单击 OK 按钮，完成._fsae_suspension_rear_axle.ground.cfl_drive_shaft_otr 结构框的创建。

（5）单击 Build > Construction Frame > New 命令，在下列对话框中输入相应的数据：
- Construction Frame（结构框名称）：drive_shaft_inr；
- Location Dependency：Delta location from coordinate；
- Coordinate Reference（参考坐标）：._torsion_beam_sus_FSAE.ground.hpl_shaft_inner；
- Location：0，0，0；
- Location in：local；
- Orientation Dependency：Orient in plane；
- Coordinate Reference #1（参考坐标）：._torsion_beam_sus_FSAE.ground.hpl_shaft_inner；
- Coordinate Reference #2（参考坐标）：._torsion_beam_sus_FSAE.ground.hpr_shaft_inner；
- Coordinate Reference #3（参考坐标）：._torsion_beam_sus_FSAE.ground.cfl_drive_shaft_otr；
- Axis：ZX。

（6）单击 OK 按钮，完成._torsion_beam_sus_FSAE.ground.cfl_drive_shaft_inr 结构框的创建。

（7）单击 Build > Part > General Part > New 命令，在下列对话框中输入相应的数据：
- General Part：tripot；

- Location Dependency：Delta location from coordinate；
- Coordinate Reference（参考坐标）：._torsion_beam_sus_FSAE.ground.hpl_shaft_inner；
- Location：0，0，0；
- Location in：local；
- Orientation Dependency：Orient to zpoint-xpoint；
- Coordinate Reference #1（参考坐标）：._torsion_beam_sus_FSAE.ground.hpr_shaft_inner；
- Coordinate Reference #2（参考坐标）：._torsion_beam_sus_FSAE.ground.cfl_drive_shaft_otr；
- Axes：ZX；
- Mass：1；
- Ixx：1；
- Iyy：1；
- Izz：1；
- Density：Material；
- Material Type：.materials.steel。

（8）单击 OK 按钮，完成部件._torsion_beam_sus_FSAE.gel_tripot 的创建。

（9）单击 Build > Geometry > Cylinder（圆柱体）> New 命令，在下列对话框中输入相应的数据：
- Cylinder Name（连杆名称）：tripot_housing_extention；
- General Part：._torsion_beam_sus_FSAE.gel_tripot；
- Radius（半径）：20；
- Length In Positive Z（Z轴正方向长度）：50；
- Length In Negative Z（Z轴负方向长度）：0；
- Color（圆柱体几何体颜色）：red；
- 选择 Calculate Mass Properties of General Part 复选框。

（10）单击 OK 按钮，完成轮毂圆柱体._torsion_beam_sus_FSAE.gel_tripot.gracyl_tripot_housing_extention 几何体的创建。

16.1.7　部件 drive_shaft

（1）单击 Build > Part > General Part > New 命令，在下列对话框中输入相应的数据：
- General Part：drive_shaft；
- Location Dependency：Delta location from coordinate；
- Coordinate Reference（参考坐标）：._torsion_beam_sus_FSAE.ground.hpl_shaft_inner；
- Location：0，0，0；
- Location in：local；
- Orientation Dependency：Orient in plane；
- Coordinate Reference #1（参考坐标）：._torsion_beam_sus_FSAE.ground.cfl_drive_shaft_otr；

- Coordinate Reference #2（参考坐标）：._torsion_beam_sus_FSAE.ground.hpl_shaft_inner；
- Coordinate Reference#3（参考坐标）：._torsion_beam_sus_FSAE.ground.hpl_wheel_center；
- Axis：ZX；
- Mass：1；
- Ixx：1；
- Iyy：1；
- Izz：1；
- Density：Material；
- Material Type：.materials.steel。

（2）单击 OK 按钮，完成部件._torsion_beam_sus_FSAE.gel_drive_shaft 的创建。

（3）单击 Build > Geometry > Link > New 命令，在下列对话框中输入相应的数据：
- Link Name（连杆名称）：drive_shaft；
- General Part 输入：._torsion_beam_sus_FSAE.gel_drive_shaft；
- Coordinate Reference #1（参考坐标）：._torsion_beam_sus_FSAE.ground.hpl_shaft_inner；
- Coordinate Reference #2（参考坐标）：._torsion_beam_sus_FSAE.ground.cfl_drive_shaft_otr；
- Radius（半径）：10；
- Color（杆件几何体颜色）：skyblue；
- 选择 Calculate Mass Properties of General Part 复选框，当几何体建立好之后，会更新对应部件的质量和惯量参数；
- Density：Material；
- Material Type：steel。

（4）单击 OK 按钮，完成._torsion_beam_sus_FSAE.gel_drive_shaft.gralin_drive_shaft 几何体的创建。

（5）单击 Build > Geometry > Ellipsoid > New 命令，在下列对话框中输入相应的数据：
- Ellipsoid Name（连杆名称）：tripot_housing；
- Coordinate Reference（参考坐标）：._torsion_beam_sus_FSAE.ground.hpl_shaft_inner；
- Link：._torsion_beam_sus_FSAE.gel_drive_shaft.gralin_drive_shaft；
- X Scale：2；
- Y Scale：2；
- Z Scale：2；
- Color（杆件几何体颜色）：red；
- 选择 Calculate Mass Properties of General Part 复选框，当几何体建立好之后，会更新对应部件的质量和惯量参数；
- Density：Material；

- Material Type：steel。

（6）单击 Apply 按钮，完成._torsion_beam_sus_FSAE.gel_drive_shaft.graell_tripot_housing 几何体的创建。

（7）单击 Build > Geometry > Ellipsoid > New 命令，在下列对话框中输入相应的数据：
- Ellipsoid Name（连杆名称）：otr_cv_housing；
- Coordinate Reference（参考坐标）：._torsion_beam_sus_FSAE.ground.cfl_drive_ shaft_otr；
- Link：._torsion_beam_sus_FSAE.gel_drive_shaft.gralin_drive_shaft；
- X Scale：2；
- Y Scale：2；
- Z Scale：2；
- Color（杆件几何体颜色）：skyblue；
- 选择 Calculate Mass Properties of General Part 复选框，当几何体建立好之后，会更新对应部件的质量和惯量参数；
- Density：Material；
- Material Type：steel。

（8）单击 OK 按钮，完成._torsion_beam_sus_FSAE.gel_drive_shaft.graell_otr_cv_housing 几何体的创建。

16.1.8 安装部件

（1）单击 Build > Part > Mount > New 命令，弹出创建部件对话框，如图 16-9 所示，在下列对话框中输入相应的数据：

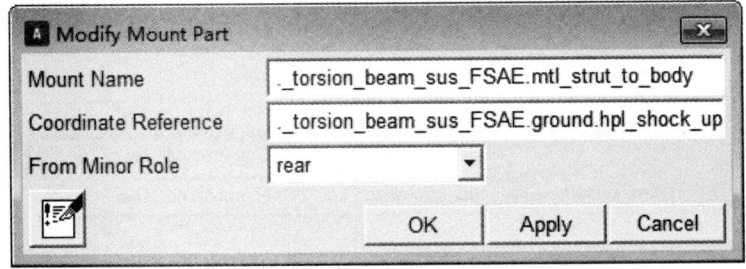

图 16-9 安装部件对话框

- Mount name（安装件名称）：strut_to_body；
- Coordinate Reference（参考坐标）：._torsion_beam_sus_FSAE.ground.hpl_shock_up；
- From Minor Role：rear。

（2）单击 Apply 按钮，完成._torsion_beam_sus_FSAE.mtl_strut_to_body 安装部件的创建。同理，创建其他部件。
- Mount name（安装件名称）：tripot_to_differential；
- Coordinate Reference（参考坐标）：._torsion_beam_sus_FSAE.ground.hpl_shaft_inner；

- From Minor Role：inherit（继承特性）；

（3）单击 Apply 按钮，完成._torsion_beam_sus_FSAE.mtl_strut_to_body 安装部件的创建。
- Mount name（安装件名称）：spring_to_body；
- Coordinate Reference（参考坐标）：._torsion_beam_sus_FSAE.ground.hpl_spring_up；
- From Minor Role：inherit（继承特性）。

（4）单击 Apply 按钮，完成._torsion_beam_sus_FSAE.mtl_spring_to_body 安装部件的创建。

16.1.9 刚性约束

1. 部件 strut_up 与安装件 strut_to_body 之间的 hooke 约束

（1）单击 Build > Attachments > Joint > New 命令，弹出创建约束件对话框，如图 16-10 所示，在下列对话框中输入相应的数据：

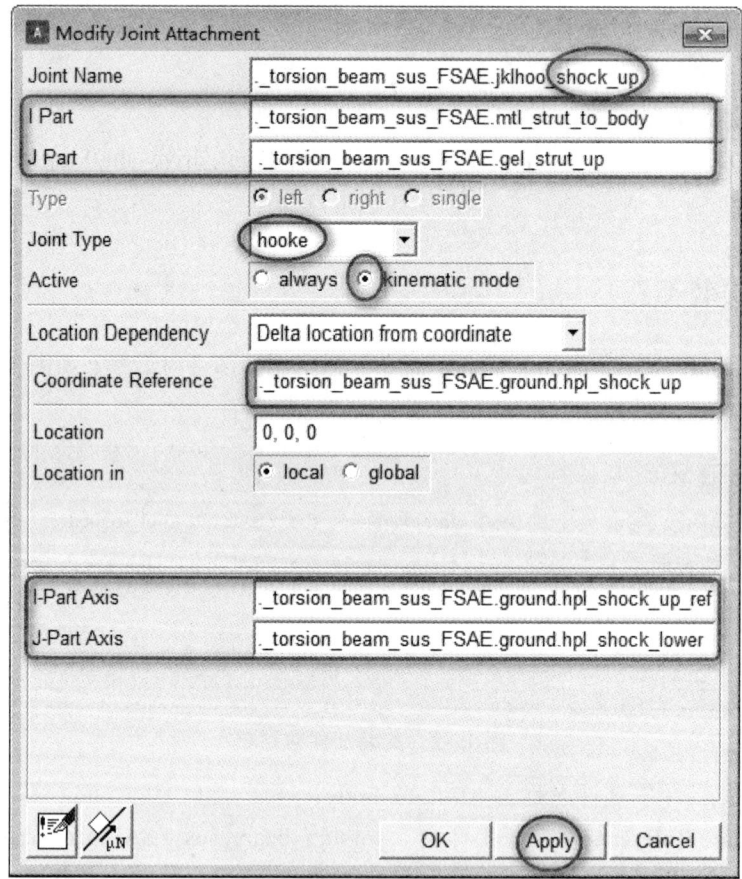

图 16-10 刚性约束对话框- hooke

- Joint Name（约束副名称）：shock_up；
- I Part：._torsion_beam_sus_FSAE.mtl_strut_to_body；

- J Part：._torsion_beam_sus_FSAE.gel_strut_up；
- Joint Type（约束副类型）：hooke；
- Active（激活）：kinematic mode（运动学模式）；
- Location Dependency：Delta location from coordinate；
- Coordinate Reference（参考坐标）：._torsion_beam_sus_FSAE.ground.hpl_shock_up；
- Location：0，0，0；
- Location in：local；
- I-Part Axis：._torsion_beam_sus_FSAE.ground.hpl_shock_up_ref；
- J-Part Axis：._torsion_beam_sus_FSAE.ground.hpl_shock_lower。

（2）单击 Apply 按钮，完成约束副._torsion_beam_sus_FSAE.jklhoo_shock_up 的创建。

2. 部件 strut_down 与 rca 之间的 hooke 约束

（1）单击 Build > Attachments > Joint > New 命令，在下列对话框中输入相应的数据：
- Joint Name（约束副名称）：shock_up；
- I Part：._torsion_beam_sus_FSAE.gel_strut_down；
- J Part：._torsion_beam_sus_FSAE.gel_rca；
- Joint Type（约束副类型）：hooke；
- Active（激活）：kinematic mode（运动学模式）；
- Location Dependency：Delta location from coordinate；
- Coordinate Reference（参考坐标）：._torsion_beam_sus_FSAE.ground.hpl_shock_lower；
- Location：0，0，0；
- Location in：local；
- I-Part Axis：._torsion_beam_sus_FSAE.ground.hpl_shock_up；
- J-Part Axis：._torsion_beam_sus_FSAE.ground.hpl_shock_lower_ref。

（2）单击 Apply 按钮，完成约束副._torsion_beam_sus_FSAE.jklhoo_shock_down 的创建。

3. 部件 strut_down 与 strut_up 之间的 cylindrical 约束

（1）单击 Build > Attachments > Joint > New 命令，在下列对话框中输入相应的数据：
- Joint Name（约束副名称）：strut；
- I Part：._torsion_beam_sus_FSAE.gel_strut_down；
- J Part：._torsion_beam_sus_FSAE.gel_strut_up；
- Joint Type（约束副类型）：cylindrical；
- Active（激活）：always；
- Location Dependency：Centered between coordinates；
- Centered between：Two Coordinates；
- Coordinate Reference #1（参考坐标）：._torsion_beam_sus_FSAE.ground.hpl_shock_lower；
- Coordinate Reference #2（参考坐标）：._torsion_beam_sus_FSAE.ground.hpl_shock_up；
- Orientation Dependency：Orient axis along line；

- Coordinate Reference #1（参考坐标）：._torsion_beam_sus_FSAE.ground.hpl_shock_lower；
- Coordinate Reference #2（参考坐标）：._torsion_beam_sus_FSAE.ground.hpl_shock_up；
- Axis：Z。

（2）单击 Apply 按钮，完成 ._torsion_beam_sus_FSAE.jolcyl_strut 圆柱副的创建。

4. 部件 spindle 与 rca 之间的 revolute 约束

（1）单击 Build > Attachments > Joint > New 命令，在下列对话框中输入相应的数据：
- Joint Name（约束副名称）：hub；
- I Part：._torsion_beam_sus_FSAE.gel_spindle；
- J Part：._torsion_beam_sus_FSAE.gel_rca；
- Joint Type（约束副类型）：revolute；
- Active（激活）：always；
- Location Dependency：Delta location from coordinate；
- Coordinate Reference（参考坐标）：._torsion_beam_sus_FSAE.ground.hpl_wheel_center；
- Location：0, 0, 0；
- Location in：local；
- Orientation Dependency：Toe/Camber；
- Variable Type（变量类型）：Parameter Variable；
- Toe Parameter Values（前束变量值）：._torsion_beam_sus_FSAE.pvl_toe_angle；
- Camber Parameter Values（外倾变量值）：._torsion_beam_sus_FSAE.pvl_camber_angle。

（2）单击 Apply 按钮，完成 ._torsion_beam_sus_FSAE.jolrev_hub 转动副的创建。

5. 部件 tripot 与 tripot_to_differential 之间的 translational 约束

（1）单击 Build > Attachments > Joint > New 命令，在下列对话框中输入相应的数据：
- Joint Name（约束副名称）：_tripot_to_differential；
- I Part：._torsion_beam_sus_FSAE.gel_tripot；
- J Part：._torsion_beam_sus_FSAE.mtl_tripot_to_differential；
- Joint Type（约束副类型）：translational；
- Active（激活）：always；
- Location Dependency：Delta location from coordinate；
- Coordinate Reference（参考坐标）：._torsion_beam_sus_FSAE.ground.hpl_shaft_inner；
- Location：0, 0, 0；
- Location in：local；
- Orientation Dependency：Orient axis to point；
- Coordinate Reference：._torsion_beam_sus_FSAE.ground.cfr_drive_shaft_inr；
- Axis：Z。

（2）单击 Apply 按钮，完成 ._torsion_beam_sus_FSAE.joltra_tripot_to_differential 移动副的创建。

6. 部件 tripot 与 drive_shaft 之间的 convel 约束

（1）单击 Build > Attachments > Joint > New 命令，在下列对话框中输入相应的数据：
- Joint Name（约束副名称）：drive_sft_int_jt；
- I Part：._torsion_beam_sus_FSAE.gel_tripot；
- J Part：._torsion_beam_sus_FSAE.gel_drive_shaft；
- Joint Type（约束副类型）：convel；
- Active（激活）：always；
- Location Dependency：Delta location from coordinate；
- Coordinate Reference（参考坐标）：._torsion_beam_sus_FSAE.ground.hpl_shaft_inner；
- Location：0，0，0；
- Location in：local；
- I-Part Axis：._torsion_beam_sus_FSAE.ground.cfr_drive_shaft_inr；
- J-Part Axis：._torsion_beam_sus_FSAE.ground.cfl_drive_shaft_otr。

（2）单击 Apply 按钮，完成约束副 ._torsion_beam_sus_FSAE.jolcon_drive_sft_int_jt 的创建。

7. 部件 spindle 与 drive_shaft 之间的 convel 约束

（1）单击 Build > Attachments > Joint > New 命令，在下列对话框中输入相应的数据：
- Joint Name（约束副名称）：drive_sft_otr；
- I Part：._torsion_beam_sus_FSAE.gel_drive_shaft；
- J Part：._torsion_beam_sus_FSAE.gel_spindle；
- Joint Type（约束副类型）：convel；
- Active（激活）：always；
- Location Dependency：Delta location from coordinate；
- Coordinate Reference（参考坐标）：._torsion_beam_sus_FSAE.ground.cfl_drive_shaft_otr；
- Location：0，0，0；
- Location in：local；
- I-Part Axis：._torsion_beam_sus_FSAE.ground.hpl_shaft_inner；
- J-Part Axis：._torsion_beam_sus_FSAE.ground.hpl_wheel_center。

（2）单击 Apply 按钮，完成约束副 ._torsion_beam_sus_FSAE.jolcon_drive_sft_otr 的创建。

8. 部件 torsion_beam 与 rca 之间的 fixed 约束

（1）单击 Build > Attachments > Joint > New 命令，在下列对话框中输入相应的数据：
- Joint Name（约束副名称）：torsion_beam；
- I Part：._torsion_beam_sus_FSAE.ges_torsion_beam；
- J Part：._torsion_beam_sus_FSAE.gel_rca；
- Joint Type（约束副类型）：fixed；
- Active（激活）：always；
- Location Dependency：Delta location from coordinate；

- Coordinate Reference(参考坐标):._torsion_beam_sus_FSAE.ground.hpl_torsion_beam_location;
- Location:0,0,0;
- Location in:local。

(2)单击 Apply 按钮,完成约束副._torsion_beam_sus_FSAE.jolfix_torsion_beam 的创建;

9. 部件 tripot 与 tripot_to_differential 之间的 translational 约束

(1)单击 Build > Attachments > Joint > New 命令,在下列对话框中输入相应的数据:

- Joint Name(约束副名称):rca_pivot;
- I Part:._torsion_beam_sus_FSAE.gel_rca;
- J Part:._torsion_beam_sus_FSAE.mts_subframe_to_body;
- Joint Type(约束副类型):revolute;
- Active(激活):kinematic mode(运动学模式);
- Location Dependency:Delta location from coordinate;
- Coordinate Reference(参考坐标):._torsion_beam_sus_FSAE.ground.hpl_rca_pivot;
- Location:0,0,0;
- Location in:local;
- Orientation Dependency:Orient axis to point;
- Coordinate Reference:._torsion_beam_sus_FSAE.ground.hpr_rca_pivot;
- Axis:Z。

(2)单击 OK 按钮,完成._torsion_beam_sus_FSAE.jklrev_rca_pivotl 转动副的创建。

16.1.10 柔性约束

1. 部件 rca 与 subframe_to_body 之间的 bushing 约束

(1)单击 Build > Attachments > Bushing > New 命令,弹出创建衬套件对话框,如图 16-11 所示,在下列对话框中输入相应的数据:

- Bushing Name(约束副名称):rca_pivot;
- I Part:._torsion_beam_sus_FSAE.gel_rca;
- J Part:._torsion_beam_sus_FSAE.mts_subframe_to_body;
- Inactive(抑制):kinematic mode(运动学模式);
- Preload:0,0,0;
- Tpreload:0,0,0;
- Offset:0,0,0;
- Roffset:0,0,0;
- Geometry Length:20;
- Geometry Radius:30;

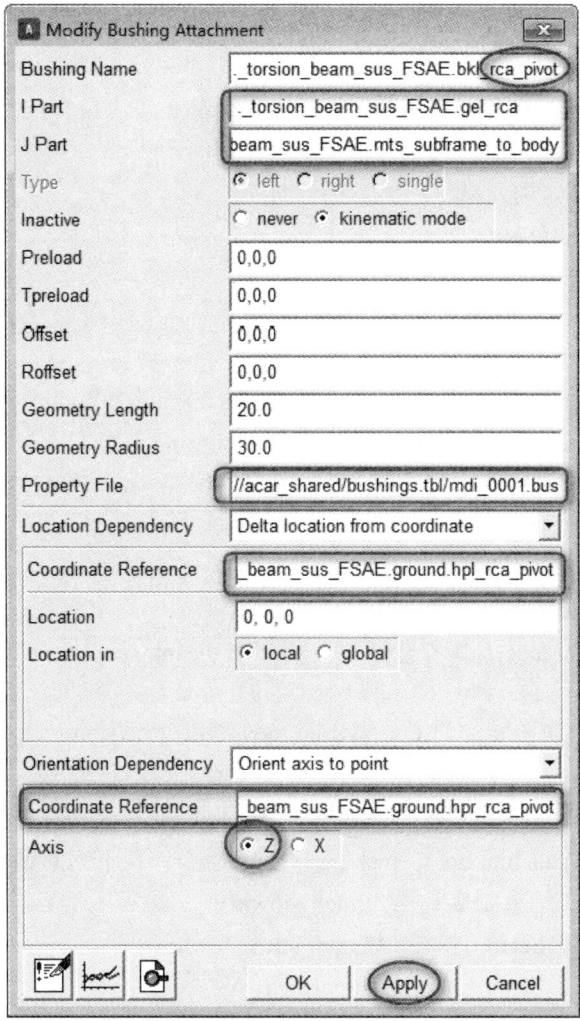

图 16-11 衬套约束对话框-bushing

- Property File：mdids：//acar_shared/bushings.tbl/mdi_0001.bus；
- Location Dependency：Delta location from coordinate；
- Coordinate Reference（参考坐标）：._torsion_beam_sus_FSAE.ground.hpl_rca_pivot；
- Location：0，0，0；
- Location in：local；
- Orientation Dependency：Orient axis to point；
- Coordinate Reference（参考坐标）：._torsion_beam_sus_FSAE.ground.hpr_rca_pivot；
- Axis：Z。

（2）单击 Apply 按钮，完成衬套._torsion_beam_sus_FSAE.bkl_rca_pivot 的创建。

2. 部件 strut_up 与 strut_to_body 之间的 bushing 约束

（1）单击 Build > Attachments > Bushing > New 命令，在下列对话框中输入相应的数据：

- Bushing Name（约束副名称）：shock_up；

- I Part：._torsion_beam_sus_FSAE.gel_strut_up；
- J Part：._torsion_beam_sus_FSAE.mtl_strut_to_body；
- Inactive（抑制）：kinematic mode（运动学模式）；
- Preload：0，0，0；
- Tpreload：0，0，0；
- Offset：0，0，0；
- Roffset：0，0，0；
- Geometry Length：20；
- Geometry Radius：30；
- Property File：mdids：//acar_shared/bushings.tbl/mdi_0001.bus；
- Location Dependency：Delta location from coordinate；
- Coordinate Reference（参考坐标）：._torsion_beam_sus_FSAE.ground.hpl_shock_up；
- Location：0，0，0；
- Location in：local；
- Orientation Dependency：Orient axis to point；
- Coordinate Reference（参考坐标）：._torsion_beam_sus_FSAE.ground.hpr_shock_up；
- Axis：Z。

（2）单击 Apply 按钮，完成衬套._torsion_beam_sus_FSAE.bkl_shock_up 的创建。

3．部件 strut_down 与 rca 之间的 bushing 约束

（1）单击 Build > Attachments > Bushing > New 命令，在下列对话框中输入相应的数据：
- Bushing Name（约束副名称）：shock_dowm；
- I Part：._torsion_beam_sus_FSAE.gel_strut_down；
- J Part：._torsion_beam_sus_FSAE.gel_rca；
- Inactive（抑制）：kinematic mode（运动学模式）；
- Preload：0，0，0；
- Tpreload：0，0，0；
- Offset：0，0，0；
- Roffset：0，0，0；
- Geometry Length：20；
- Geometry Radius：30；
- Property File：mdids：//acar_shared/bushings.tbl/mdi_0001.bus；
- Location Dependency：Delta location from coordinate；
- Coordinate Reference（参考坐标）：._torsion_beam_sus_FSAE.ground.hpl_shock_lower；
- Location：0，0，0；
- Location in：local；
- Orientation Dependency：Orient axis to point；
- Coordinate Reference（参考坐标）：._torsion_beam_sus_FSAE.ground.hpr_shock_lower；
- Axis：Z。

（2）单击 OK 按钮，完成衬套._torsion_beam_sus_FSAE.bkl_shock_dowm 的创建。

16.1.11 扭力梁悬架变量参数

（1）单击 Build > Parameter Variable > New 命令，弹出参数变量对话框，如图 16-12 所示，在下列对话框中输入相应的数据：

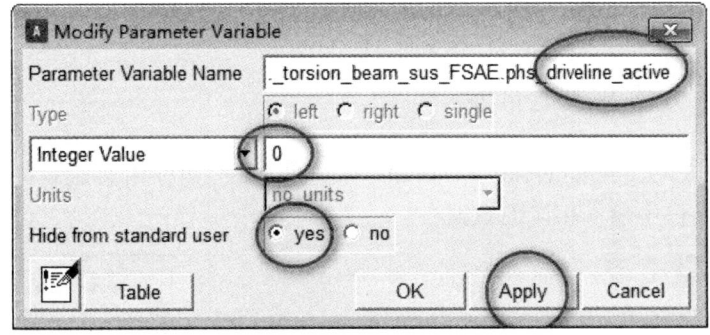

图 16-12 参数变量对话框

- Parameter Variable Name：driveline_active；
- Integer Value（实数值）：0；
- Units：length；
- Hide from standard user（是否从标准界面隐藏）：yes。

（2）单击 Apply 按钮，完成变量._torsion_beam_sus_FSAE.phs_driveline_active 的创建。同理，创建其他变量参数。

- Parameter Variable Name：kinematic_flag；
- Integer Value（实数值）：0；
- Units：length；
- Hide from standard user（是否从标准界面隐藏）：yes。

（3）单击 OK 按钮，完成变量._torsion_beam_sus_FSAE.phs_kinematic_flag 的创建。

（4）单击 Build > Suspension Parameters > Characteristics Array > Set 命令，弹出悬架参数变量对话框，如图 16-13 所示，在下列对话框中输入相应的数据：

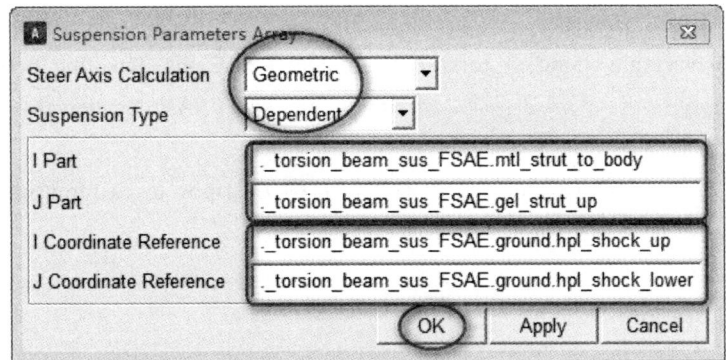

图 16-13 悬架参数变量对话框

- Steer Axis Calculation：Geometric；
- Suspension Type：Dependent（非独立悬架）；
- I Part：._torsion_beam_sus_FSAE.mtl_strut_to_body；
- J Part：._torsion_beam_sus_FSAE.gel_strut_up；
- I Coordinate Reference：._torsion_beam_sus_FSAE.ground.hpl_shock_up；
- J Coordinate Reference：._torsion_beam_sus_FSAE.ground.hpl_shock_lower。

（5）单击 OK 按钮，完成悬架参数变量设置。

16.1.12 扭力梁悬架通信器

（1）单击 Build > Communicator > Output > New 命令，弹出输出通信器对话框，如图 16-14 所示，在下列对话框中输入相应的数据：

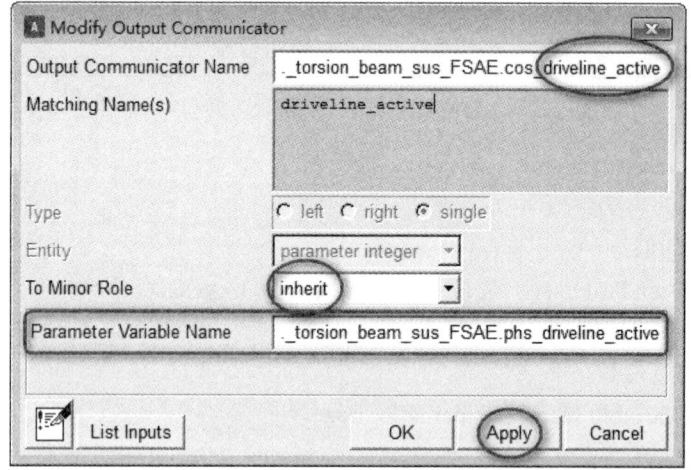

图 16-14 通信器设置对话框

- Output Communicator Name（输出通信器名称）：driveline_active；
- Matching Name（s）：driveline_active；
- Type：single；
- Entity：parameter integer；
- To Minor Role：inherit；
- Parameter Variable Name：._torsion_beam_sus_FSAE.phs_driveline_active。

（2）单击 Apply 按钮，完成通信器._torsion_beam_sus_FSAE.cos_driveline_active 的创建。同理，创建其他通信器。

- Output Communicator Name（输出通信器名称）：tripot_to_differential；
- Matching Name（s）：tripot_to_differential；
- Type：left；
- Entity：Location；
- To Minor Role：inherit；
- Coordinate Reference Name：._torsion_beam_sus_FSAE.ground.hpl_shaft_inner。

（3）单击 Apply 按钮，完成通信器._torsion_beam_sus_FSAE.col_tripot_to_differential 的创建。
- Output Communicator Name（输出通信器名称）：suspension_mount；
- Matching Name（s）：suspension_mount；
- Type：left；
- Entity：mount；
- To Minor Role：rear；
- Part Name：._torsion_beam_sus_FSAE.gel_spindle。

（4）单击 Apply 按钮，完成通信器._torsion_beam_sus_FSAE.col_suspension_mount 的创建。
- Output Communicator Name（输出通信器名称）：wheel_center；
- Matching Name（s）：wheel_center；
- Type：left；
- Entity：Location；
- To Minor Role：rear；
- Coordinate Reference Name：._torsion_beam_sus_FSAE.ground.hpl_wheel_center。

（5）单击 Apply 按钮，完成通信器._torsion_beam_sus_FSAE.col_wheel_center 的创建。
- Output Communicator Name（输出通信器名称）：suspension_upright；
- Matching Name（s）：suspension_upright；
- Type：left；
- Entity：mount；
- To Minor Role：rear；
- Part Name：._torsion_beam_sus_FSAE.gel_rca。

（6）单击 OK 按钮，完成通信器._torsion_beam_sus_FSAE.col_suspension_upright 的创建。

16.1.13　扭力梁驱动轴显示组件

（1）在模型树栏点击 Group 菜单，在 New Group 上右击鼠标，弹出创建组件对话框，如图 16-15 所示，在下列对话框中输入相应的数据：

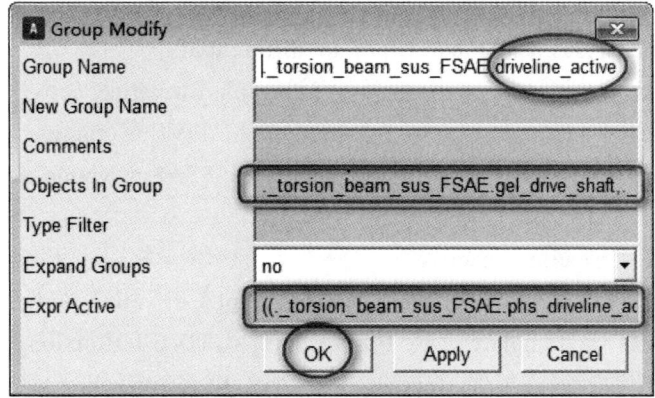

图 16-15　驱动轴显示组件对话框

- Group Name：driveline_active。
- Object In Group（显示组件包括的部件、几何体、约束等对象），顺序输入 1～26 对象的如下信息：

[1] ._torsion_beam_sus_FSAE.gel_drive_shaft；

[2] ._torsion_beam_sus_FSAE.gel_tripot；

[3] ._torsion_beam_sus_FSAE.ger_drive_shaft；

[4] ._torsion_beam_sus_FSAE.ger_tripot；

[5] ._torsion_beam_sus_FSAE.mtl_tripot_to_differential；

[6] ._torsion_beam_sus_FSAE.mtr_tripot_to_differential；

[7] ._torsion_beam_sus_FSAE.gel_drive_shaft.gralin_drive_shaft；

[8] ._torsion_beam_sus_FSAE.gel_drive_shaft.graell_otr_cv_housing；

[9] ._torsion_beam_sus_FSAE.gel_drive_shaft.graell_tripot_housing；

[10] ._torsion_beam_sus_FSAE.gel_tripot.gracyl_tripot_housing_extention；

[11] ._torsion_beam_sus_FSAE.ger_drive_shaft.gralin_drive_shaft；

[12] ._torsion_beam_sus_FSAE.ger_drive_shaft.graell_otr_cv_housing；

[13] ._torsion_beam_sus_FSAE.ger_drive_shaft.graell_tripot_housing；

[14] ._torsion_beam_sus_FSAE.ger_tripot.gracyl_tripot_housing_extention；

[15] ._torsion_beam_sus_FSAE.jolcon_drive_sft_int_jt；

[16] ._torsion_beam_sus_FSAE.jolcon_drive_sft_otr；

[17] ._torsion_beam_sus_FSAE.joltra_tripot_to_differential；

[18] ._torsion_beam_sus_FSAE.jorcon_drive_sft_int_jt；

[19] ._torsion_beam_sus_FSAE.jorcon_drive_sft_otr；

[20] ._torsion_beam_sus_FSAE.jortra_tripot_to_differential；

[21] ._torsion_beam_sus_FSAE.mtl_fixed_2；

[22] ._torsion_beam_sus_FSAE.mtr_fixed_2；

[23] ._torsion_beam_sus_FSAE.cil_tripot_to_differential；

[24] ._torsion_beam_sus_FSAE.cir_tripot_to_differential；

[25] ._torsion_beam_sus_FSAE.col_tripot_to_differential；

[26] ._torsion_beam_sus_FSAE.cor_tripot_to_differential。

- Expr Active：((._torsion_beam_sus_FSAE.phs_driveline_active || ._torsion_beam_sus_FSAE.model_class == "template" ? 1：0) && DB_ACTIVE(._torsion_beam_sus_FSAE))。

（2）单击 Apply 按钮，完成组件._torsion_beam_sus_FSAE.driveline_active 的创建。同理，创建其他组件。

- Group Name：driveline_inactive；
- Expr Active：((!._torsion_beam_sus_FSAE.phs_driveline_active || ._torsion_beam_sus_FSAE.model_class == "template" ? 1：0) && DB_ACTIVE(._torsion_beam_sus_FSAE))。

（3）单击 OK 按钮，完成组件._torsion_beam_sus_FSAE.driveline_inactive 的创建。此时扭力梁悬架模型创建完成，如图 16-16 所示。

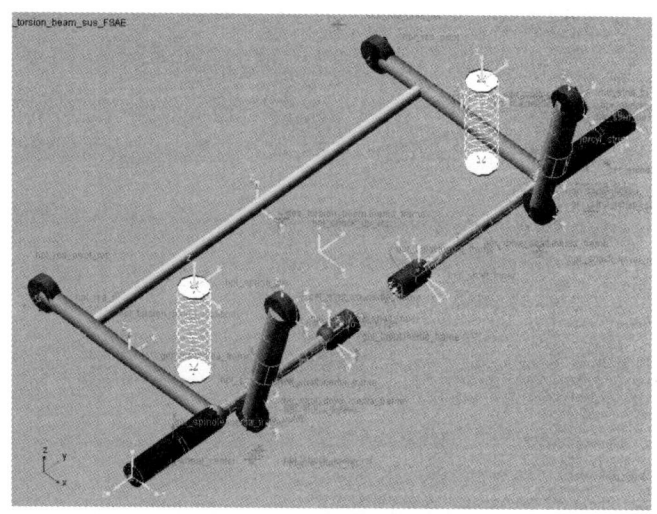

图 16-16　扭力梁悬架模型保存

（4）单击 File > Save As 命令，弹出保存模板对话框，如图 16-17 所示，在下列对话框中输入相应的数据：
- Major Role（主特征）：suspension；
- File Format：Binary；
- Target：Directory。

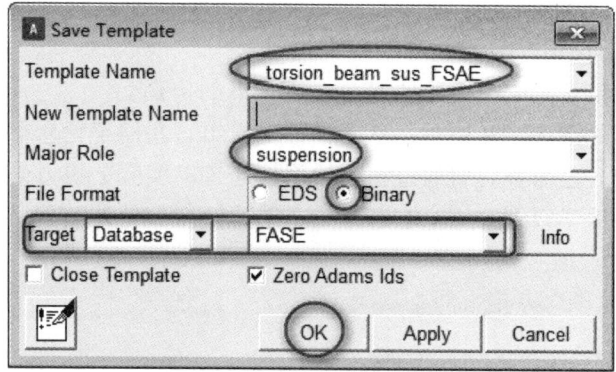

图 16-17　扭力梁悬架模型保存

（5）单击 Select 按钮，选择存储路径为 D：/fsae_MD_2010.cdb/templates.tbl。
（6）单击 OK 按钮，完成推杆式悬架模型模板._torsion_beam_sus_FSAE 的保存。

16.1.14　反向激振测试验证模型

（1）单击 Simulate > Suspension Analysis > Opposite Travel 命令，弹出反向激振对话框，如图 16-18 所示，在下列对话框中输入相应的数据：
- Output Prefix：OT_torsion；
- Number of Steps（仿真步数）：1000；
- Mode of Simulation：interactive；

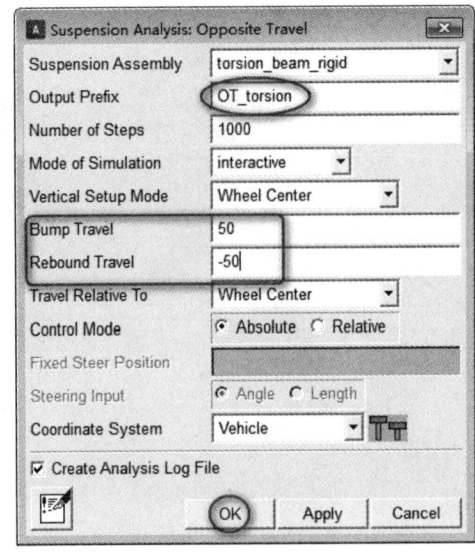

图 16-18　反向激振对话框

- Vertical Setup Mode：Wheel Center；
- Bump Travel：50；
- Rebound Travel：-50；
- Travel Relative To：Wheel Center；
- Control Mode：Absolute；
- Coordinate System：Vehicle。

（2）单击 OK 按钮，完成扭力梁悬架在 C 模式下的仿真。

扭力梁激振过程如图 16-19 所示，从图中可以看出，左侧车轮悬空，主要原因在于扭力梁悬架为非对立悬架，左右拖拽臂通过扭力梁刚性连接。实际上扭力梁为柔性梁，柔性梁内容在后续柔性系统动力学篇章介绍。左右车轮主销后倾与车轮外倾角如图 16-20 和图 16-21 所示。

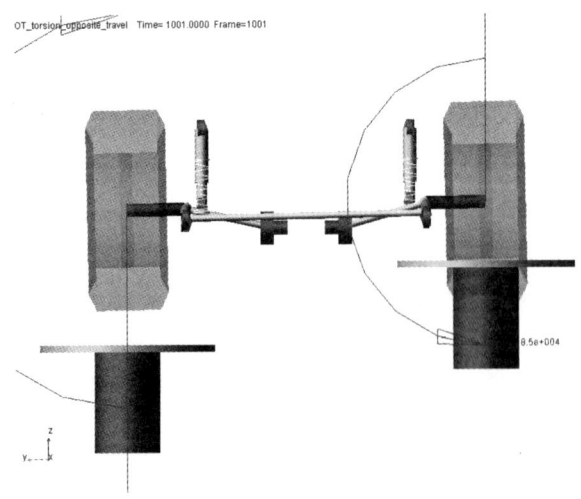

图 16-19　扭力梁激振图

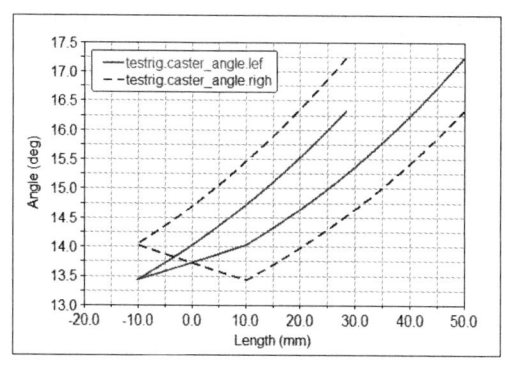 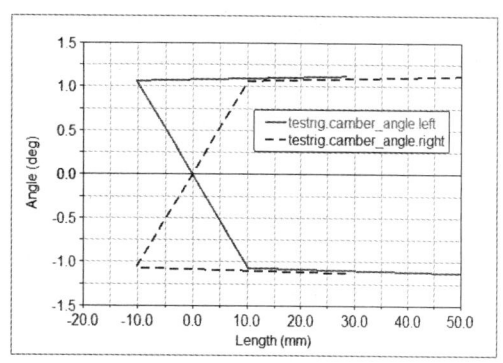

图 16-20 主销后倾角　　　　　　　　图 16-21 车轮外倾角

16.2　FSAE 整车模型

替换 FSAE 赛车后推力杆式双 A 臂悬架系统为扭力梁悬架系统后,移动后悬架位置,保持整车轴距为 1 524 mm。调试完成后,整车包含 58 个自由度,整车模型如图 16-22 所示。整车装配模型保存为 fsae_full_2018.asy。

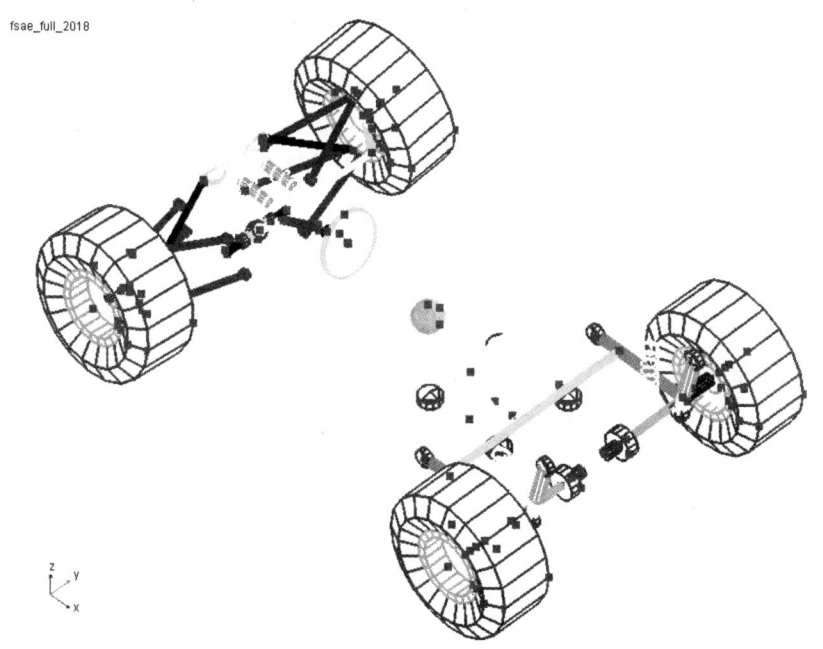

图 16-22　FSAE 整车模型

16.3　定常半径转弯仿真

定常半径转弯主要用于验证整车的转向特性,包含过渡转向、不足转向、中性转向。

（1）单击 Simulate > Full-Vehicle Analysis > Cornering Events > Constant Radius Cornering 命令，弹出定常半径转弯仿真对话框，如图 16-23 所示，在下列对话框中输入相应的数据：

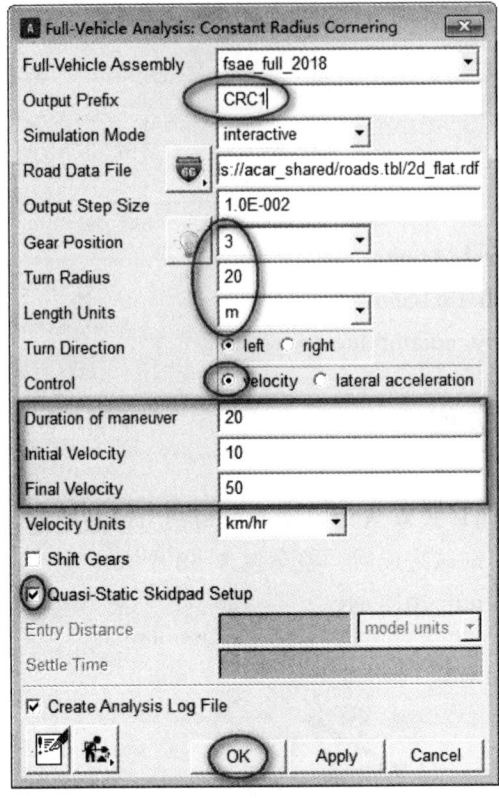

图 16-23　定常半径转弯设置

- Output Prefix：CRC1；
- Mode of Simulation：interactive；
- Road Date File：mdids：//FASE/roads.tbl/2d_flat.rdf；
- Output Step Size（仿真步数）：0.01；
- Gear Position：3；
- Turn Radius：20；
- Length Units：m；
- Control：velocity；
- Duration of maneuver：20；
- Initial Velocity：10；
- Final Velocity：50；
- Velocity Units：km/hr；
- 勾选 Quasi-Static Straight-Line Setup：整车模型包含发动机运行准静态平衡。

（2）单击 OK 按钮，完成定常半径仿真设置并提交运算。

仿真结束后，FSAE 整车运行轨迹如图 16-24 所示；整车的各参数输出如图 16-25～图 16-30 所示。

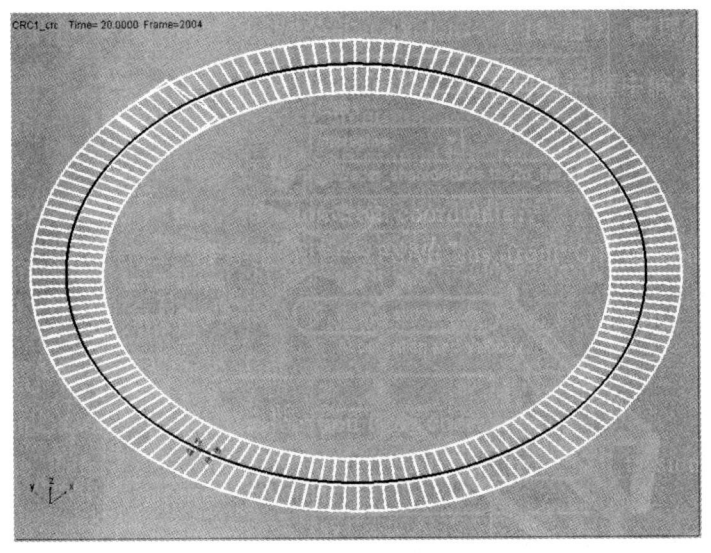

图 16-24 定常半径转弯运行轨迹

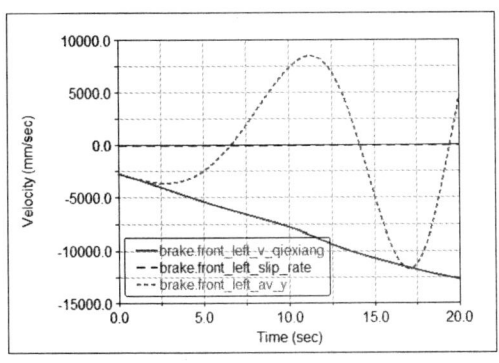

图 16-25 左前车辆切向、滑移、角速度参数

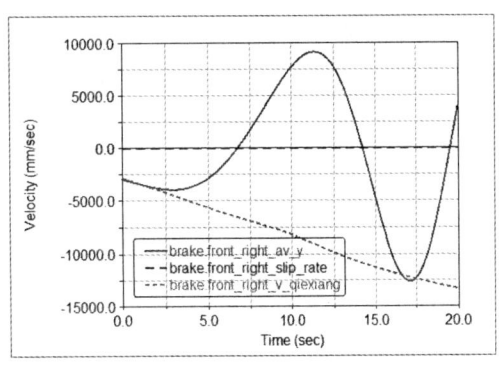

图 16-26 右前车辆切向、滑移、角速度参数

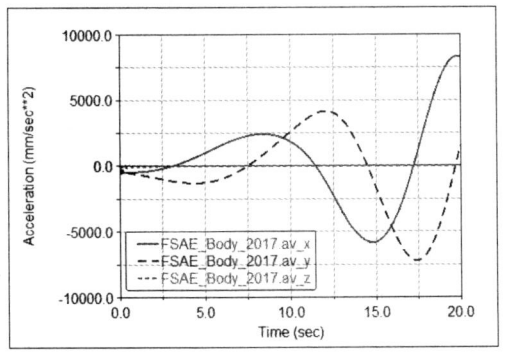

图 16-27 车身加速度 X/Y/Z

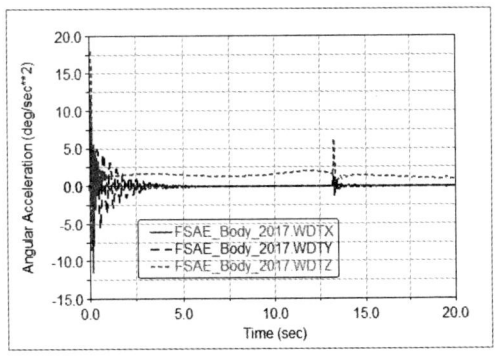

图 16-28 车身角加速度 X/Y/Z

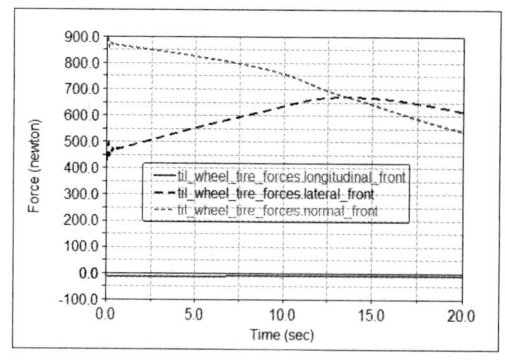
图 16-29 前轮胎 X/Y/Z 受力

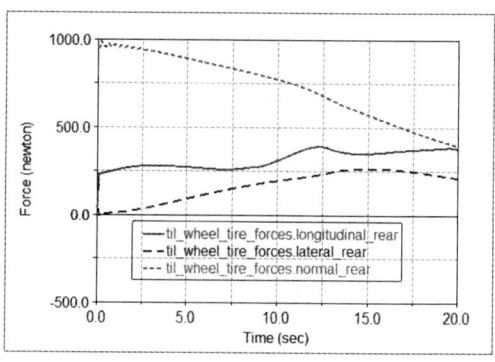
图 16-30 后轮胎 X/Y/Z 受力

第 17 章 扭杆弹簧式推杆悬架

扭杆弹簧的布置有多种方式，可以配合不同的悬架类型。本章介绍一种扭杆弹簧垂置及纵置方向布置的推杆悬架，扭杆垂置式推杆悬架如图 17-1 所示。此种方式布置可以将扭杆弹簧与推杆悬架的优势结合起来。对于支架摆臂的转动特性，可以考虑增加横向避震器限制及减小摆臂的旋转特性，进而减少车轮或车身上下振动的幅值特性，提升赛车的性能。为减少建模过程的重复特性，依然在推杆悬架模型_FSAE_sus_front.tpl 上修改建模。

图 17-1 扭杆垂置式推杆悬架

学习目标

（1）了解扭杆弹簧 MNF。
（2）了解垂置式扭杆弹簧推杆悬架。
（3）了解纵置式扭杆弹簧推杆悬架。
（4）了解横向避震器。
（5）会弯道制动仿真。
（6）会回正性仿真。

17.1 扭杆弹簧 MNF

扭杆弹簧截面半径为 10 mm，长度为 200 mm；在 ABAQUS 中建立有限元模型 torsion_10p200，如图 17-2 所示。RP1 与 RP2 之间的距离为 30 mm，RP2 与 RP3 之间的距离为 170 mm，模型包含 11 600 个六面体单元。RP1、RP2、RP3 与其同心圆两层厚度单元内所有的节点采用 MPC 多点约束，即 RP 点与节点采用梁单元连接；提取扭杆弹簧的前 20 阶模态，扭杆弹簧 MNF 制作过程不再重复，请参考"拉杆悬架模型"章节中的扭杆弹簧调试部分。扭杆弹簧的 20 阶模态如图 17-3 所示，已经建立好的扭杆弹簧 torsion_10p200、torsion_5p200 存储在章节电子文件夹中，请读者自行查阅。

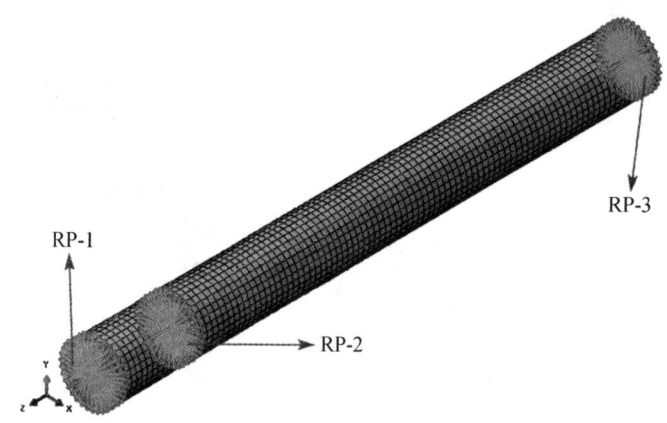

图 17-2　扭杆弹簧 torsion_10p200

```
1     Mode    1: Value =  2.25351E+08 Freq =  2389.2  (cycles/time)
2     Mode    2: Value =  2.25798E+08 Freq =  2391.6  (cycles/time)
3     Mode    3: Value =  1.50878E+09 Freq =  6182.1  (cycles/time)
4     Mode    4: Value =  1.51157E+09 Freq =  6187.8  (cycles/time)
5     Mode    5: Value =  2.77946E+09 Freq =  8390.7  (cycles/time)
6     Mode    6: Value =  5.12527E+09 Freq =  11394.  (cycles/time)
7     Mode    7: Value =  5.13406E+09 Freq =  11404.  (cycles/time)
8     Mode    8: Value =  7.51549E+09 Freq =  13797.  (cycles/time)
9     Mode    9: Value =  1.09343E+10 Freq =  16642.  (cycles/time)
10    Mode   10: Value =  1.22502E+10 Freq =  17615.  (cycles/time)
11    Mode   11: Value =  1.22696E+10 Freq =  17629.  (cycles/time)
12    Mode   12: Value =  2.36243E+10 Freq =  24462.  (cycles/time)
13    Mode   13: Value =  2.36593E+10 Freq =  24481.  (cycles/time)
14    Mode   14: Value =  2.42159E+10 Freq =  24767.  (cycles/time)
```

图 17-3　扭杆弹簧 20 阶频率

17.2 扭杆弹簧垂置式推杆悬架

1. 模型导入

（1）启动 ADAMS/CAR，选择专家模块进入建模界面。

（2）单击 File > Open 命令，弹出打开模板对话框，如图 17-4 所示，在 Template Name 中输入：mdids：//FASE/templates.tbl/_FSAE_sus_front.tpl。

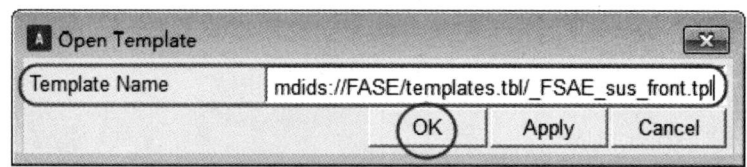

图 17-4　模板对话框

（3）单击 OK 按钮，导入模型。

2. 删除部件、弹簧、避震器、约束、结构框

通过删除部件，与部件相关联的约束、弹簧、避震器等都会全部删除，最后删除多余的硬点和结构框，此种方法删除速度较快，同时不容易出现错误。需要删除的信息如下：

① ._FSAE_sus_front.gel_bellcrank；
② ._FSAE_sus_front.ger_bellcrank；
③ ._FSAE_sus_front.gel_damper_chassis；
④ ._FSAE_sus_front.ger_damper_chassis；
⑤ ._FSAE_sus_front.gel_damper_bellcrank；
⑥ ._FSAE_sus_front.ger_damper_bellcrank；
⑦ ._FSAE_sus_front.ground.hpl_shock_to_bellcrank；
⑧ ._FSAE_sus_front.ground.hpr_shock_to_bellcrank；
⑨ ._FSAE_sus_front.ground.hpl_bellcrank_pivot；
⑩ ._FSAE_sus_front.ground.hpr_bellcrank_pivot；
⑪ ._FSAE_sus_front.ground.hpl_arblink_to_bellcrank；
⑫ ._FSAE_sus_front.ground.hpr_arblink_to_bellcrank；
⑬ ._FSAE_sus_front.ground.hpl_shock_to_chassis；
⑭ ._FSAE_sus_front.ground.hpr_shock_to_chassis；
⑮ ._FSAE_sus_front.ground.hpl_bellcrank_pivot_orient；
⑯ ._FSAE_sus_front.ground.hpr_bellcrank_pivot_orient；
⑰ ._FSAE_sus_front.ground.cfl_damper_bellcrank_orient；
⑱ ._FSAE_sus_front.ground.cfr_damper_bellcrank_orient；
⑲ ._FSAE_sus_front.ground.cfl_bellcrank_pivot；
⑳ ._FSAE_sus_front.ground.cfr_bellcrank_pivot；
㉑ ._FSAE_sus_front.ground.cfl_damper_chassis_orient；

㉒ ._FSAE_sus_front.ground.cfr_damper_chassis_orient；
㉓ ._FSAE_sus_front.ground.cfl_shock_to_bellcrank；
㉔ ._FSAE_sus_front.ground.cfr_shock_to_bellcrank；
㉕ ._FSAE_sus_front.ground.cfl_spring_to_bellcrank；
㉖ ._FSAE_sus_front.ground.cfr_spring_to_bellcrank。

3. 保存模型

（1）单击 File > Save As 命令，弹出保存模板对话框，如图 17-5 所示，在下列对话框中输入相应的数据：

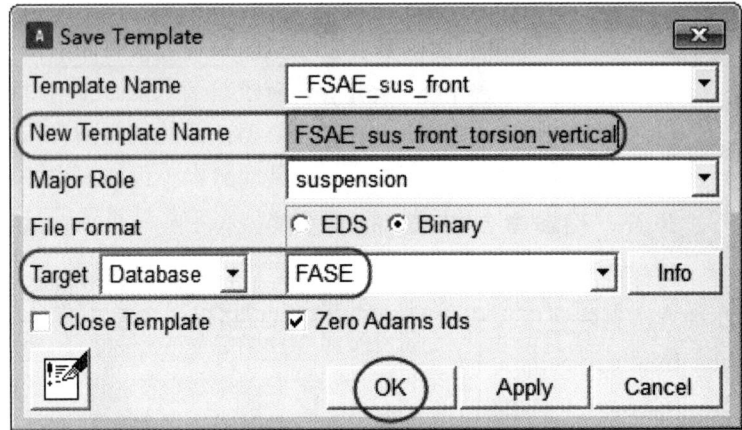

图 17-5 模型保存_FSAE_sus_front_torsion_vertical

- Template Name：_FSAE_sus_front；
- New Template Name：FSAE_sus_front_torsion_vertical；
- Major Role（主特征）：suspension；
- File Format：Binary；
- Target：Datebase/FASE。

（2）单击 OK 按钮，完成推杆式悬架模型模板 FSAE_sus_front_torsion_vertical 的保存。

4. 添加硬点

（1）单击 Build > Hardpoind > New 命令，在下列对话框中输入相应的数据：
- Hardpoint：zhijia_to_damper2；
- Typer：left；
- Location：-80.0，-90.8，381.0。

（2）单击 Apply 按钮，完成._FSAE_sus_front_torsion_vertical.ground.hpl_zhijia_to_damper2 硬点的创建。

（3）重复上述创建硬点步骤，完成图 17-6 中标注线内的硬点创建及位置修改。

	loc x	loc y	loc z
hpl_arb_bushing_mour	127.0	-127.0	101.6
hpl_lca_front	-127.0	-127.0	114.3
hpl_lca_outer	0.0	-546.1	120.65
hpl_lca_rear	127.0	-127.0	114.3
hpl_prod_outer	0.0	-457.2	139.7
hpl_prod_to_zhijia	-50.8	-127.0	381.0
hpl_tierod_inner	50.8	-127.0	152.4
hpl_tierod_outer	63.5	-482.6	152.4
hpl_uca_front	-101.6	-177.8	279.4
hpl_uca_outer	0.0	-482.6	355.6
hpl_uca_rear	101.6	-177.8	279.4
hpl_wheel_center	0.0	-558.8	241.3
hpl_zhijia_to_damper1	0.0	-50.8	381.0
hpl_zhijia_to_damper2	-80.0	-90.8	381.0
hpl_zhijia_to_torsion	15.4	-90.0	381.0
hps_hps_global	0.0	0.0	0.0

图 17-6　硬点信息

5. 部件 zhijia

（1）单击 Build > Part > General Part > New 命令，弹出创建部件对话框，如图 17-7 所示，在下列对话框中输入相应的数据：

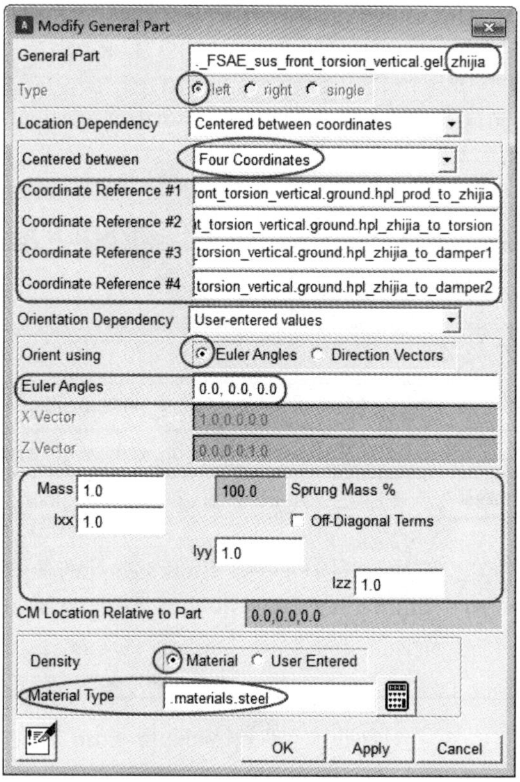

图 17-7　部件_zhijia

- General Part 输入：zhijia；
- Type：left；
- Location Dependency：Centered between coordinates；
- Centered between：Four Coordinates；
- Coordinate Reference #1（参考坐标）：._FSAE_sus_front_torsion_vertical.ground.hpl_prod_to_zhijia；
- Coordinate Reference #2（参考坐标）：._FSAE_sus_front_torsion_vertical.ground.hpl_zhijia_to_torsion；
- Coordinate Reference #3（参考坐标）：._FSAE_sus_front_torsion_vertical.ground.hpl_zhijia_to_damper1；
- Coordinate Reference #4（参考坐标）：._FSAE_sus_front_torsion_vertical.ground.hpl_zhijia_to_damper2；
- Orientation Dependency：User-entered values；
- Orient using：Euler Angles；
- Euler Angles：0，0，0；
- Mass：1；
- Ixx：1；
- Iyy：1；
- Izz：1；
- Density：Material；
- Material Type：.materials.steel。

（2）单击 OK 按钮，完成部件._FSAE_sus_front_torsion_vertical.gel_zhijia 的创建。

6. 轮廓几何体 zhijia

（1）单击 Build > Geometry > Outline > New 命令，弹出创建轮廓几何体对话框，如图 17-8 所示，在下列对话框中输入相应的数据：

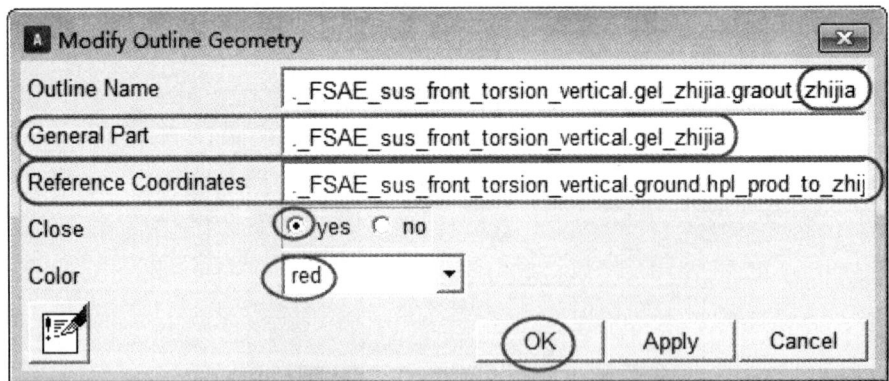

图 17-8　轮廓线_zhijia

- Outline Name：zhijia。
- General Part：._FSAE_sus_front_torsion_vertical.gel_zhijia。

- Reference Coordinates，顺序选取以下硬点，次序不能乱：
① ._FSAE_sus_front_torsion_vertical.ground.hpl_prod_to_zhijia；
② ._FSAE_sus_front_torsion_vertical.ground.hpl_zhijia_to_torsion；
③ ._FSAE_sus_front_torsion_vertical.ground.hpl_zhijia_to_damper1；
④ ._FSAE_sus_front_torsion_vertical.ground.hpl_zhijia_to_damper2；
⑤ ._FSAE_sus_front_torsion_vertical.ground.hpl_prod_to_zhijia。
- Close（封闭，所选的硬点或参考点形成一个闭合曲线的点集合，上述第一个与第五个点为同一个点，即所选的点集合封闭）：yes。
- Color：red。

（2）单击 OK 按钮，完成._FSAE_sus_front_torsion_vertical.gel_zhijia.graout_zhijia 轮廓线的创建。

7. 结构框

（1）单击 Build > Construction Frame > New 命令，弹出创建结构框对话框，如图 17-9 所示，在下列对话框中输入相应的数据：

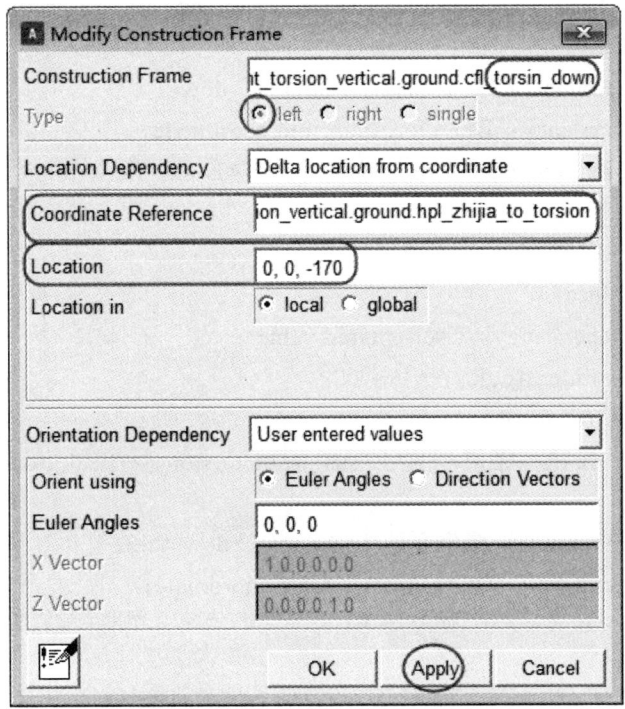

图 17-9　结构框_torsin_down

- Construction Frame（结构框名称）：torsin_down；
- Location Dependency：Delta location from coordinate；
- Coordinate Reference（参考坐标）：._FSAE_sus_front_torsion_vertical.ground.hpl_zhijia_to_torsion；
- Location：0，0，-170；

- Location in：local；
- Orientation Dependency：User-entered values；
- Orient using：Euler Angles；
- Euler Angles：0，0，0。

（2）单击 Apply 按钮，完成 ._FSAE_sus_front_torsion_vertical.ground.cfl_torsin_down 结构框的创建。同理，创建其他结构框。

- Construction Frame（结构框名称）：torsion_up；
- Location Dependency：Delta location from coordinate；
- Coordinate Reference（参考坐标）：._FSAE_sus_front_torsion_vertical.ground.hpl_zhijia_to_torsion；
- Location：0，0，30；
- Location in：local；
- Orientation Dependency：User-entered values；
- Orient using：Euler Angles；
- Euler Angles：0，0，0。

（3）单击 Apply 按钮，完成 ._FSAE_sus_front_torsion_vertical.ground.cfl_torsion_up 结构框的创建。

- Construction Frame（结构框名称）：damper1_down；
- Location Dependency：Delta location from coordinate；
- Coordinate Reference（参考坐标）：._FSAE_sus_front_torsion_vertical.ground.hpl_zhijia_to_damper1；
- Location：150，0，0；
- Location in：local；
- Orientation Dependency：User-entered values；
- Orient using：Euler Angles；
- Euler Angles：0，0，0。

（4）单击 Apply 按钮，完成 ._FSAE_sus_front_torsion_vertical.ground.cfl_damper1_down 结构框的创建。

- Construction Frame（结构框名称）：damper1_down_ref；
- Location Dependency：Delta location from coordinate；
- Coordinate Reference（参考坐标）：._FSAE_sus_front_torsion_vertical.ground.cfl_damper1_down；
- Location：60，0，0；
- Location in：local；
- Orientation Dependency：User-entered values；
- Orient using：Euler Angles；
- Euler Angles：0，0，0。

（5）单击 OK 按钮，完成 ._FSAE_sus_front_torsion_vertical.ground.cfl_damper1_down_ref 结构框的创建。

8. 部件 damper1_up

（1）单击 Build > Part > General Part > New 命令，在下列对话框中输入相应的数据：
- General Part：zhijia；
- Type：left；
- Location Dependency：Delta location from coordinate；
- Coordinate Reference（参考坐标）：._FSAE_sus_front_torsion_vertical.ground.hpl_zhijia_to_damper1；
- Location：0，0，0；
- Location in：local；
- Orientation Dependency：User-entered values；
- Orient using：Euler Angles；
- Euler Angles：0，0，0；
- Mass：1；
- Ixx：1；
- Iyy：1；
- Izz：1；
- Density：Material；
- Material Type：.materials.steel。

（2）单击 Apply 按钮，完成部件._FSAE_sus_front_torsion_vertical.gel_damper1_up 的创建。

9. 部件 dmper1_down

（1）单击 Build > Part > General Part > New 命令，在下列对话框中输入相应的数据：
- General Part：dmper1_down；
- Type：left；
- Location Dependency：Delta location from coordinate；
- Coordinate Reference（参考坐标）：._FSAE_sus_front_torsion_vertical.ground.cfl_damper1_down；
- Location：0，0，0；
- Location in：local；
- Orientation Dependency：User-entered values；
- Orient using：Euler Angles；
- Euler Angles：0，0，0；
- Mass：1；
- Ixx：1；
- Iyy：1；
- Izz：1；
- Density：Material；
- Material Type：.materials.steel。

（2）单击 OK 按钮，完成部件._FSAE_sus_front_torsion_vertical.gel_dmper1_down 的创建。

10. 扭杆弹簧 FSAE_torsion

（1）单击 Build > Part > Flexible Body > New 命令，弹出创建部件对话框，如图 17-10 所示，在下列对话框中输入相应的数据：

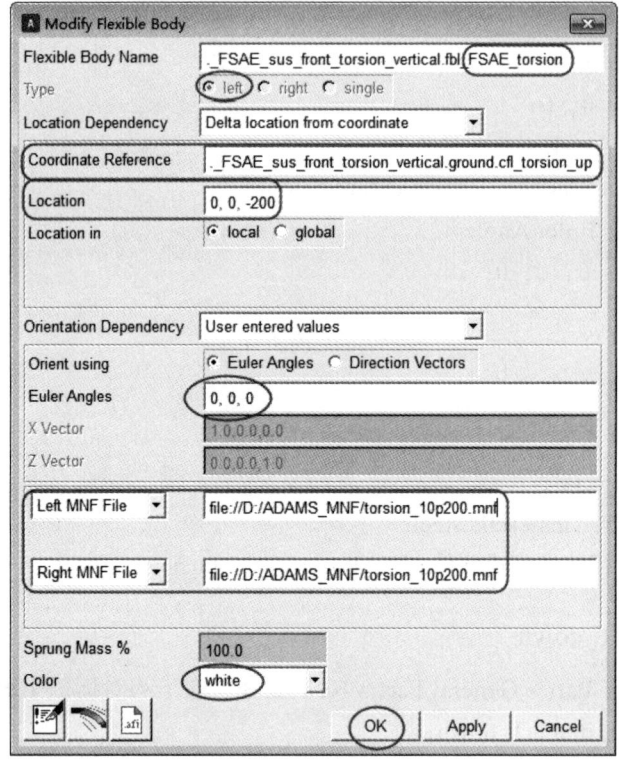

图 17-10　扭杆弹簧部件

- Flexible Body Name：FSAE_torsion；
- Type：left；
- Location Dependency：Delta location from coordinate；
- Coordinate Reference（参考坐标）：._FSAE_sus_front_torsion_vertical.ground.cfl_torsion_up；
- Location：0，0，-200；
- Location in：local；
- Orientation Dependency：User-entered values；
- Orient using：Euler Angles；
- Euler Angles：0，0，0；
- Left MNF File：file：//D：/ADAMS_MNF/torsion_10p200.mnf；
- Right MNF File：file：//D：/ADAMS_MNF/torsion_10p200.mnf；
- Color：white。

（2）单击 Apply 按钮，完成部件._FSAE_sus_front_torsion_vertical.fbl_FSAE_torsion 的创建。

11. 部件 FSAE_torsion 与安装件 suspension_to_chassis 之间的 revolute 约束

（1）单击 Build > Attachments > Joint > New 命令，弹出创建约束件对话框，如图 17-11 所示，在下列对话框中输入相应的数据：

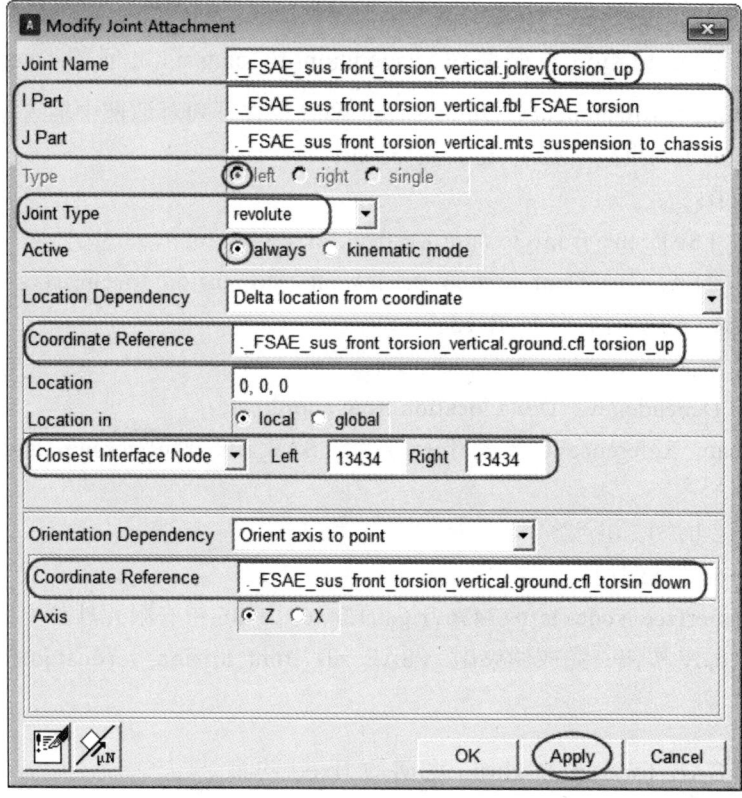

图 17-11 刚性约束对话框_ torsion_up

- Joint Name（约束副名称）：torsion_up；
- Type：left；
- I Part：._FSAE_sus_front_torsion_vertical.fbl_FSAE_torsion；
- J Part：._FSAE_sus_front_torsion_vertical.mts_suspension_to_chassis；
- Joint Type（约束副类型）：revolute；
- Active（激活）：always；
- Location Dependency：Delta location from coordinate；
- Coordinate Reference（参考坐标）：._FSAE_sus_front_torsion_vertical.ground.cfl_torsion_up；
- Location：0，0，0；
- Location in：local；
- Closest Interface Node（指参考点：._FSAE_sus_front_torsion.ground.cfl_zhijia_to_chassis 附近的节点，注意节点指的是有限元节点或者是在有限元软件中创建的接口点）：left/13434，right/13434（13434 指有限元软件中创建的接口点）；

- 391 -

- Orientation Dependency：Orient axis to point；
- Coordinate Reference：._FSAE_sus_front_torsion_vertical.ground.cfl_torsin_down。

（2）单击 Apply 按钮，完成约束副._FSAE_sus_front_torsion_vertical.jolrev_torsion_up 的创建。

12. 部件 FSAE_torsion 与安装件 suspension_to_chassis 之间的 fixed 约束

（1）单击 Build > Attachments > Joint > New 命令，在下列对话框中输入相应的数据：
- Joint Name（约束副名称）：torsion_down；
- Type：left；
- I Part：._FSAE_sus_front_torsion_vertical.fbl_FSAE_torsion；
- J Part：._FSAE_sus_front_torsion_vertical.mts_suspension_to_chassis；
- Joint Type（约束副类型）：fixed；
- Active（激活）：always；
- Location Dependency：Delta location from coordinate；
- Coordinate Reference（参考坐标）：._FSAE_sus_front_torsion_vertical.ground.cfl_torsin_down；
- Location：0，0，0；
- Location in：local；
- Closest Interface Node：left/13436，right/13436（13436 指有限元软件中创建的接口点）；

（2）单击 Apply 按钮，完成约束副._FSAE_sus_front_torsion_vertical.jolfix_torsion_down 的创建。

13. 部件 FSAE_torsion 与 zhijia 之间的 fixed 约束

（1）单击 Build > Attachments > Joint > New 命令，在下列对话框中输入相应的数据：
- Joint Name（约束副名称）：torsion_to_zhijia；
- Type：left；
- I Part：._FSAE_sus_front_torsion_vertical.fbl_FSAE_torsion；
- J Part：._FSAE_sus_front_torsion_vertical.gel_zhijia；
- Joint Type（约束副类型）：fixed；
- Active（激活）：always；
- Location Dependency：Delta location from coordinate；
- Coordinate Reference（参考坐标）：._FSAE_sus_front_torsion_vertical.ground.hpl_zhijia_to_torsion；
- Location：0，0，0；
- Location in：local；
- Closest Interface Node：left/13435，right/13435。

（2）单击 Apply 按钮,完成约束副._FSAE_sus_front_torsion_vertical.jolfix_torsion_to_zhijia 的创建。

14. 部件 zhijia 与 prod 之间的 convel 约束

（1）单击 Build > Attachments > Joint > New 命令，在下列对话框中输入相应的数据：

- Joint Name（约束副名称）：prod_to_zhijia；
- I Part：._FSAE_sus_front_torsion_vertical.gel_prod；
- J Part：._FSAE_sus_front_torsion_vertical.gel_zhijia；
- Joint Type（约束副类型）：convel；
- Active（激活）：always；
- Location Dependency：Delta location from coordinate；
- Coordinate Reference（参考坐标）：._FSAE_sus_front_torsion_vertical.ground.hpl_prod_to_zhijia；
- Location：0，0，0；
- Location in：local；
- I-Part Axis：._FSAE_sus_front_torsion_vertical.ground.hpl_prod_outer；
- J-Part Axis：._FSAE_sus_front_torsion_vertical.ground.hpl_zhijia_to_torsion；

（2）单击 Apply 按钮，完成约束副._FSAE_sus_front_torsion_vertical.jolcon_prod_to_zhijia 的创建。

15. 部件 damper1_up 与 zhijia 之间的 hook 约束

（1）单击 Build>Attachments>Joint>New 命令，在下列对话框中输入相应的数据：

- Joint Name（约束副名称）：damper1_up；
- I Part：._FSAE_sus_front_torsion_vertical.gel_damper1_up；
- J Part：._FSAE_sus_front_torsion_vertical.gel_zhijia；
- Joint Type（约束副类型）：hook；
- Active（激活）：always；
- Location Dependency：Delta location from coordinate；
- Coordinate Reference（参考坐标）：._FSAE_sus_front_torsion_vertical.ground.hpl_zhijia_to_damper1；
- Location：0，0，0；
- Location in：local；
- I-Part Axis：._FSAE_sus_front_torsion_vertical.ground.hpl_zhijia_to_torsion；
- J-Part Axis：._FSAE_sus_front_torsion_vertical.ground.cfl_damper1_down。

（2）单击 Apply 按钮，完成约束副._FSAE_sus_front_torsion_vertical.jolhoo_damper1_up 的创建。

16. 部件 damper1_up 与 zhijia 之间的 hook 约束

（1）单击 Build > Attachments > Joint > New 命令，在下列对话框中输入相应的数据：

- Joint Name（约束副名称）：damper_down；
- I Part：._FSAE_sus_front_torsion_vertical.gel_dmper1_down；

- J Part：._FSAE_sus_front_torsion_vertical.mts_suspension_to_chassis；
- Joint Type（约束副类型）：hook；
- Active（激活）：always；
- Location Dependency：Delta location from coordinate；
- Coordinate Reference（参考坐标）：._FSAE_sus_front_torsion_vertical.ground.cfl_damper1_down；
- Location：0，0，0；
- Location in：local；
- I-Part Axis：._FSAE_sus_front_torsion_vertical.ground.hpl_zhijia_to_damper1；
- J-Part Axis：._FSAE_sus_front_torsion_vertical.ground.cfl_damper1_down_ref。

（2）单击 Apply 按钮，完成约束副._FSAE_sus_front_torsion_vertical.jolhoo_damper_down 的创建。

17. 部件 damper_up 与 damper_down 之间的 cylindrical 约束

（1）单击 Build > Attachments > Joint > New 命令，在下列对话框中输入相应的数据：
- Joint Name（约束副名称）：damper1_mid；
- Type：left；
- I Part：._FSAE_sus_front_torsion_vertical.gel_damper1_up；
- J Part：._FSAE_sus_front_torsion_vertical.gel_dmper1_down；
- Joint Type（约束副类型）：cylindrical；
- Active（激活）：always；
- Location Dependency：Centered between coordinates；
- Centered between：Two Coordinates；
- Coordinate Reference #1（参考坐标）：._FSAE_sus_front_torsion_vertical.ground.hpl_zhijia_to_damper1；
- Coordinate Reference #2（参考坐标）：._FSAE_sus_front_torsion_vertical.ground.cfl_damper1_down；
- Location Dependency：Orient axis to point；
- Coordinate Reference（参考坐标）：._FSAE_sus_front_torsion_vertical.ground.cfl_damper1_down。

（2）单击 OK 按钮，完成约束副._FSAE_sus_front_torsion_vertical.jolcyl_damper1_mid 的创建；至此，前扭杆弹簧垂置式悬架模型建立完成，如图 17-1 所示，保存模型，参数保持默认。

18. 后悬架硬点信息

同理，参考前扭杆弹簧垂置式推杆悬架的创建完成后悬架模型的建立，模型建立完成后，切换到标准模板，建立前后悬架子系统，建立好的后悬架子系统如图 17-12 所示。

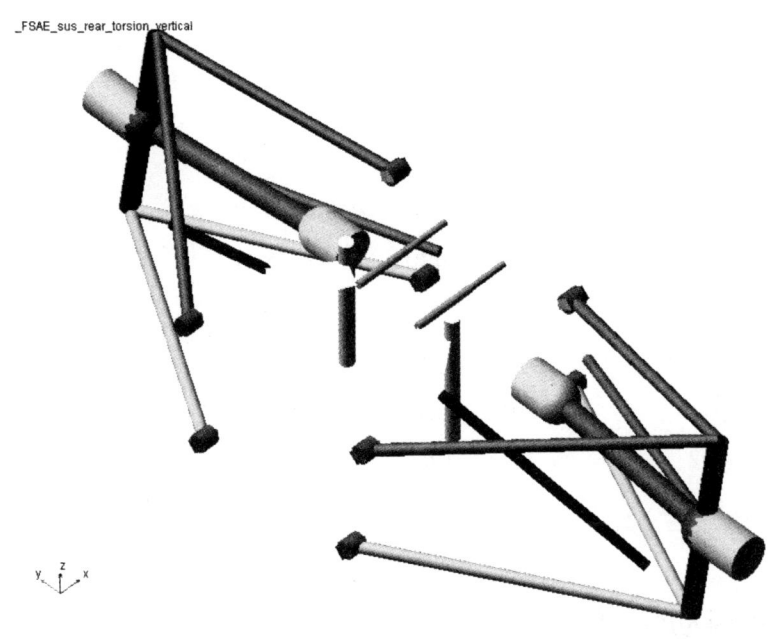

图 17-12　后扭杆弹簧垂置式推杆悬架

```
Info for subsystem:    FSAE_sus_rear_torsion_vertical

   File Name      :   <FASE>/subsystems.tbl/FSAE_sus_rear_torsion_vertical.sub
   Template                                                                    :
mdids://FASE/templates.tbl/_FSAE_sus_rear_torsion_vertical.tpl
   Comments       :   *no comments found*
   Major Role     :   suspension
   Minor Role     :   rear


   HARDPOINTS:
```

hardpoint name	symmetry	x_value	y_value	z_value
global	single	1524.0	0.0	0.0
bellcrank_pivot	left/right	1547.8	-170.0	305.0
drive_shaft_inr	left/right	1550.0	-200.0	225.0
lca_front	left/right	1270.0	-127.0	127.0
lca_outer	left/right	1498.6	-482.6	101.6
lca_rear	left/right	1651.0	-127.0	127.0
prod_outer	left/right	1498.6	-407.2	127.0

prod_to_zhijia	left/right	1400.0	-150.0	304.8
tierod_inner	left/right	1676.4	-127.0	152.4
tierod_outer	left/right	1574.8	-457.2	152.4
uca_front	left/right	1270.0	-152.4	304.8
uca_outer	left/right	1549.4	-482.6	355.6
uca_rear	left/right	1625.6	-152.4	304.8
wheel_center	left/right	1524.0	-558.8	228.6
zhijia_to_damper1	left/right	1459.7	-50.0	304.8
zhijia_to_damper2	left/right	1359.7	-100.0	304.8
zhijia_to_torsion	left/right	1479.7	-90.0	304.8

17.3 弯道制动仿真

整车模型信息如下，其中包含各子系统及对应的模板，读者可以通过前面建立的任意整车模型替换前后悬架或者重新装配整车模型 FSAE_2020_torsion_vertical 完成整车装配，如图 17-13 所示。

```
************************  ASSEMBLY INFO  ****************************
  Assembly Name   :  FSAE_2020_torsion_vertical
  Assembly Class  :  full_vehicle
  File Name       :  <FASE>/assemblies.tbl/FSAE_2020_torsion_vertical.asy

  SUBSYSTEM NAME                         MAJOR ROLE              MINOR ROLE
  front_tire                             wheel                   front
  rear_tire                              wheel                   rear
  FSAE_brake                             brake_system            any
  FSAE_body                              body                    any
  FSAE_steering_2017                     steering                front
  FSAE_powertrain                        powertrain              rear
  FSAE_sus_front_torsion_vertical        suspension              front
  FSAE_sus_rear_torsion_vertical         suspension              rear
```

```
*************************   SUBSYSTEM INFO   *************************

HARDPOINTS:
hardpoint name              symmetry        x_value    y_value    z_value
---------------             --------        -------    -------    -------

path_error_reference        single          0.0        0.0        0.0
upright_reference           left/right      0.0        0.0        0.0


Info for subsystem:  front_tire


File Name        :   <FASE>/subsystems.tbl/front_tire.sub
Template         :   mdids://FASE/templates.tbl/_handling_tire.tpl
Comments         :   *no comments found*
Major Role       :   wheel
Minor Role       :   front


Info for subsystem:  rear_tire


File Name        :   <FASE>/subsystems.tbl/rear_tire.sub
Template         :   mdids://FASE/templates.tbl/_handling_tire.tpl
Comments         :   *no comments found*
Major Role       :   wheel
Minor Role       :   rear


Info for subsystem:  FSAE_brake


File Name        :   <FASE>/subsystems.tbl/FSAE_brake.sub
Template         :   mdids://FASE/templates.tbl/_brake_system_4Wdisk.tpl
Comments         :
Template      :  4 Wheel Disk Brake System
Subsystem     :  *no subsystem comments found*
Major Role       :   brake_system
Minor Role       :   any
```

```
Info for subsystem:   FSAE_body

File Name       :   <FASE>/subsystems.tbl/FSAE_body.sub
Template        :   mdids://FASE/templates.tbl/_fsae_chassis.tpl
Comments        :
Template        :   Simple One Part Rigid Chassis
Subsystem       :   *no subsystem comments found*
Major Role      :   body
Minor Role      :   any
```

图 17-13　整车模型_FSAE_2020_torsion_vertical

弯道制动是较危险的驾驶情况之一，真实实验时安全隐患极大，采用仿真技术可以大大降低成本周期和保证人员安全等；弯道制动仿真时，软件中的程序驱动使车辆从直线引道进入固定半径的圆周车道，当车辆的侧向加速度与设定的值相同时，此时驱动器保持转弯半径并制动车辆，制动减速度为参数设定的值，在规定的制动时间内，使车辆速度降低到 2.5 km/h 以下。

1. 单线移动超车仿真（锁定方向盘）

（1）单击 Simulate > Full-Vehicle Analysis > Cornering Events > Braking-In-Turn 命令，弹出弯道制动仿真对话框，如图 17-14 所示，在下列对话框中输入相应的数据：
- Full-Vehicle Assembly：FSAE_2020_torsion_vertical；

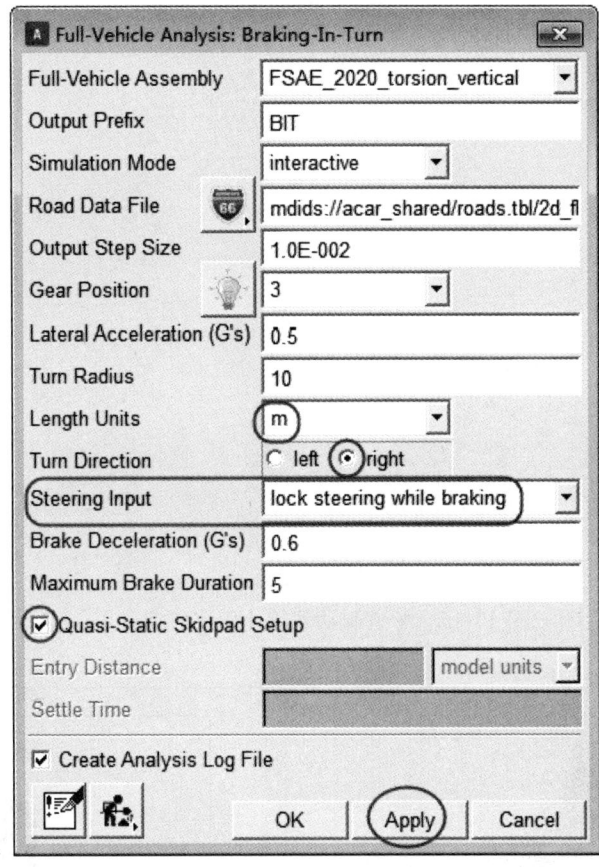

图 17-14 弯道制动仿真（锁定方向盘）对话框

- Output Prefix：BIT；
- Output Step step：0.01；
- Simulation Mode：interactive；
- Road Date File：mdids：//acar_shared/roads.tbl/2d_flat.rdf；
- Gear Position：3；
- Lateral Acceleration（G's）：0.5；
- Turn Radius：10；
- Length Units：m；
- Turn Direction：right；
- Steering Input：lock steering while braking；
- Brake Deceleration（G's）：0.6；
- Maximum Brake Duration：5；
- 不勾选 Quasi-Static Straight-Line Setup。

（2）单击 OK 按钮，完成 FSAE_2020_torsion_vertical 弯道制动仿真设置并提交运算，运算完成后，整车运行轨迹如图 17-15 所示，从图中可以看出，在方向盘锁定情况下，车身质心基本上沿路面中心线移动，在结束制动时，整车有滑出中心线的倾向。仿真完成，计算参数如图 17-16~图 17-18 所示，从滑移率看，前轮已经失稳。

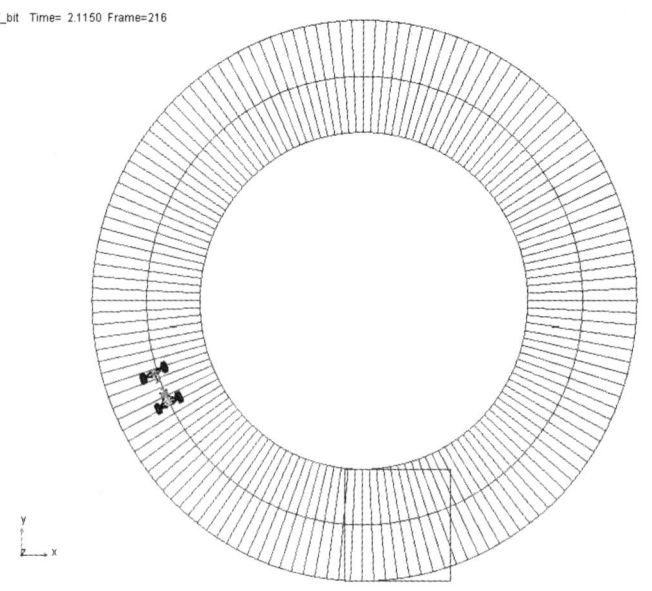

图 17-15 弯道制动轨迹（锁定方向盘）

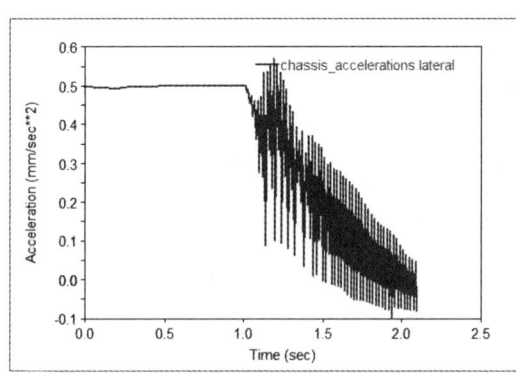

图 17-16 车身侧向加速度

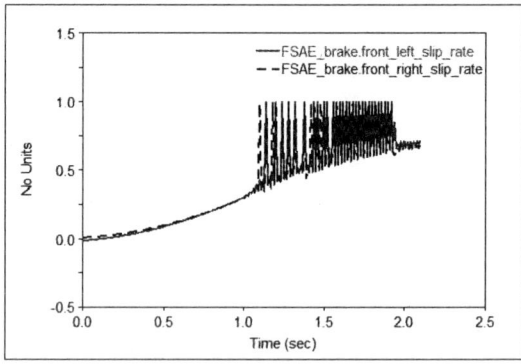

图 17-17 前左右轮滑移率

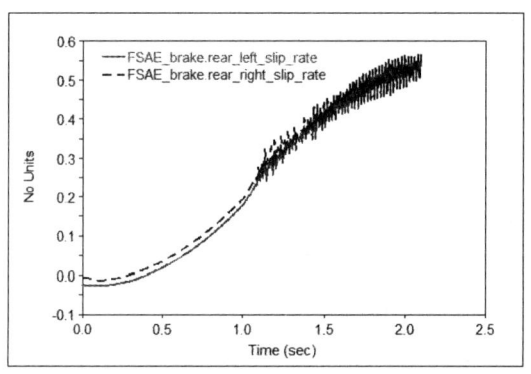

图 17-18 后左右轮滑移率

2. 单线移动超车仿真（修正方向盘转向）

（1）单击 Simulate > Full-Vehicle Analysis > Cornering Events > Braking-In-Turn 命令，在下列对话框中输入相应的数据：

- Full-Vehicle Assembly：FSAE_2020_torsion_vertical；
- Output Prefix：BIT1；
- Output Step step：0.01；
- Simulation Mode：interactive；
- Road Date File：mdids：//acar_shared/roads.tbl/2d_flat.rdf；
- Gear Position：3；
- Lateral Acceleration（G's）：0.4；
- Turn Radius：20；
- Length Units：m；
- Turn Direction：left；
- Steering Input：maintain radius while braking；
- Brake Deceleration（G's）：0.5；
- Maximum Brake Duration：3；
- 勾选 Quasi-Static Straight-Line Setup；
- Entry Distance（引道长度，单位：米）：10；
- Settle Time（引道长度内驱动程序计算时间，单位：秒）：3。

（2）单击 OK 按钮，完成 FSAE_2020_torsion_vertical 弯道制动仿真设置并提交运算，运算完成后，整车运行轨迹如图 17-19 所示，从图中可以看出，整车始终按圆形路面中心线运动。

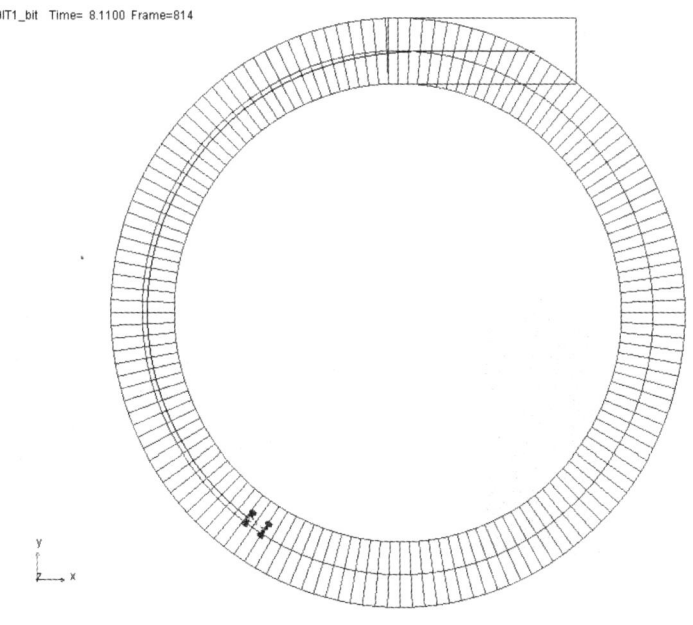

图 17-19　弯道制动轨迹（方向盘修正）

（3）切换到后处理模块，单击 Plot > Cerate Plots 命令，导入驱动绘图配置文件，如图 17-20 所示，在下列对话框中输入相应的数据：

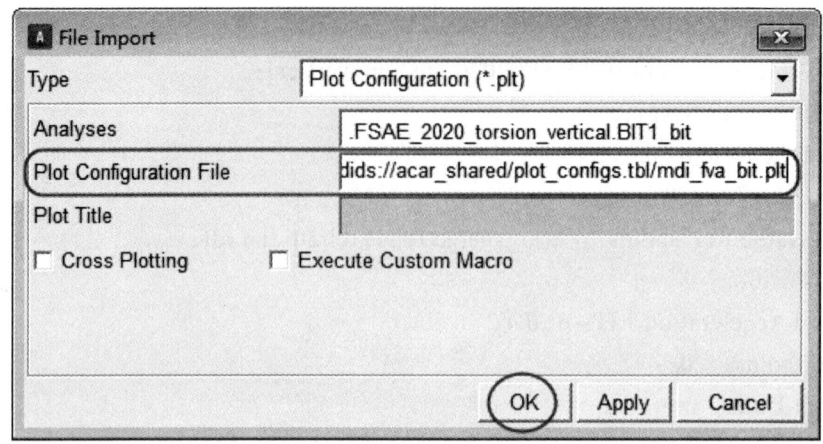

图 17-20　绘图文件

- Analyses：.FSAE_2020_torsion_vertical.BIT1_bit；
- Plot Configuration File：mdids：//acar_shared/plot_configs.tbl/mdi_fva_bit.plt。

（4）单击 OK 按钮，完成绘图配置文件的导入。

绘制文件导入成功后，绘制整车参数曲线，如图 17-21～图 17-25 所示。

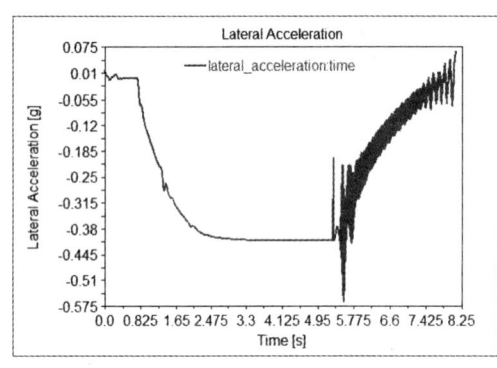

图 17-21　侧向加速度

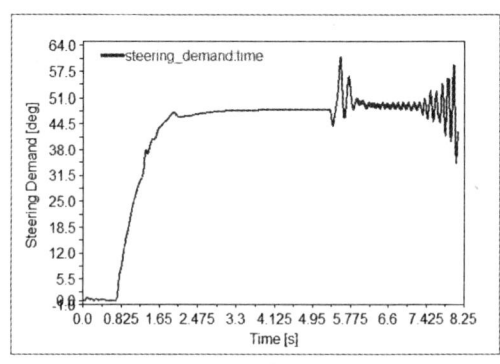

图 17-22　转向盘角度

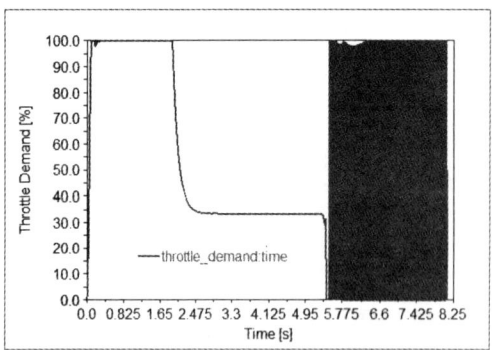

图 17-23　油门开度

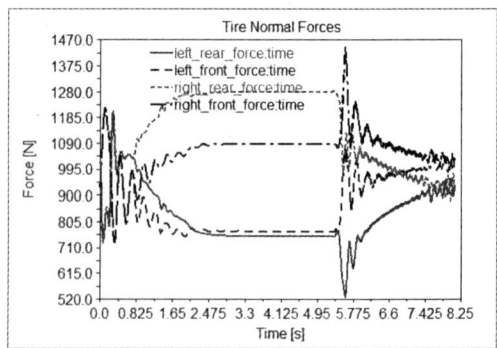

图 17-24　轮胎受力

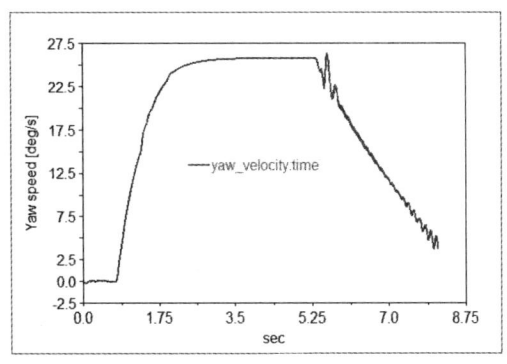

图 17-25　横摆角速度

17.4　扭杆弹簧纵置式推杆悬架

纵置式扭杆弹簧悬架与垂置式扭杆弹簧悬架相似，不需要重新建模，约束关系也不用调试，只需要对硬点的位置和结构框进行修改即可得到纵置式扭杆弹簧式推杆悬架。

1. 保存模型

（1）单击 File > Save As 命令，弹出保存模板对话框，如图 17-26 所示，在下列对话框中输入相应的数据：
- Template Name：FSAE_sus_front_torsion_vertical；
- New Template Name：FSAE_sus_front_torson_longtitudinal；
- Major Role（主特征）：suspension；
- File Format：Binary；
- Target：Datebase/FASE。

（2）单击 OK 按钮，完成推杆式悬架模型模板 FSAE_sus_front_torson_longtitudinal 的保存。

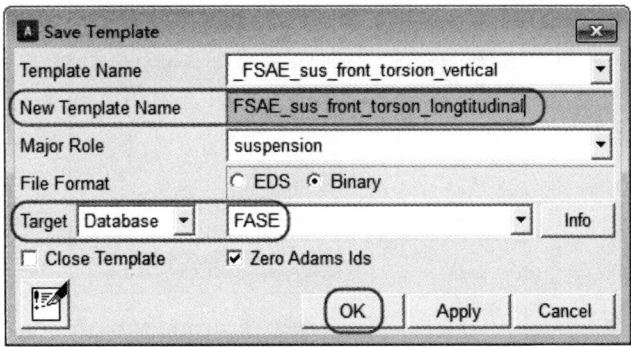

图 17-26　模型保存 FSAE_sus_front_torson_longtitudinal

2. 调试硬点

（1）单击 Build > Hardpoind > Modify 命令，在下列对话框中输入相应的数据：

- Hardpoint：zhijia_to_damper1；
- Typer：left；
- Location：-50.8，-50.8，311.0。

（2）单击 Apply 按钮，完成 ._FSAE_sus_front_torson_longtitudinal.ground.hpl_zhijia_to_damper1 硬点的创建。

（3）重复上述创建硬点步骤，完成图 17-27 中标注线内的硬点创建及位置修改。

硬点	X	Y	Z
hpl_arb_bushing_mour	127.0	-127.0	101.6
hpl_lca_front	-127.0	-127.0	114.3
hpl_lca_outer	0.0	-546.1	120.65
hpl_lca_rear	127.0	-127.0	114.3
hpl_prod_outer	0.0	-457.2	139.7
hpl_prod_to_zhijia	-50.8	-127.0	381.0
hpl_tierod_inner	50.8	-127.0	152.4
hpl_tierod_outer	63.5	-482.6	152.4
hpl_uca_front	-101.6	-177.8	279.4
hpl_uca_outer	0.0	-482.6	355.6
hpl_uca_rear	101.6	-177.8	279.4
hpl_wheel_center	0.0	-558.8	241.3
hpl_zhijia_to_damper1	-50.8	-50.8	311.0
hpl_zhijia_to_damper2	-50.8	-90.8	401.0
hpl_zhijia_to_torsion	-50.8	-120.0	321.0
hps_hps_global	0.0	0.0	0.0

图 17-27 硬点信息

3. 调试结构框

（1）单击 Build > Construction Frame > Modify 命令，弹出结构框对话框，如图 17-28 所示，该结构框调试主要修改结构框的方向，对扭杆弹簧布置形式与约束点的接触进行调试。

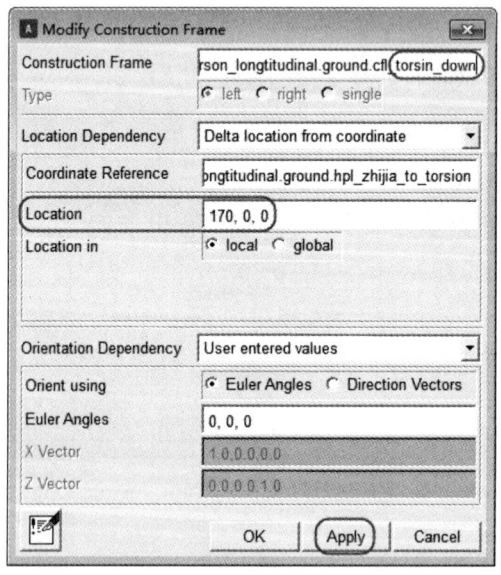

图 17-28 结构框 cfl_torsin_down 对话框

- Construction Frame（结构框名称，右键快捷寻找输入）：torsin_down；
- Location Dependency：Delta location from coordinate；
- Coordinate Reference（参考坐标）：._FSAE_sus_front_torsion_vertical.ground.hpl_zhijia_to_torsion；
- Location：170，0，0；
- Location in：local；
- Orientation Dependency：User-entered values；
- Orient using：Euler Angles；
- Euler Angles：0，0，0。

（2）单击 Apply 按钮，完成._FSAE_sus_front_torsion_vertical.ground.cfl_torsin_down 结构框的调试。同理，调试其他结构框。

- Construction Frame（结构框名称）：torsion_up；
- Location Dependency：Delta location from coordinate；
- Coordinate Reference（参考坐标）：._FSAE_sus_front_torsion_vertical.ground.hpl_zhijia_to_torsion；
- Location：-30，0，0；
- Location in：local；
- Orientation Dependency：User-entered values；
- Orient using：Euler Angles；
- Euler Angles：0，0，0。

（3）单击 Apply 按钮，完成._FSAE_sus_front_torsion_vertical.ground.cfl_torsion_up 结构框的调试。

- Construction Frame（结构框名称）：damper1_down；
- Location Dependency：Delta location from coordinate；
- Coordinate Reference（参考坐标）：._FSAE_sus_front_torsion_vertical.ground.hpl_zhijia_to_damper1；
- Location：0，0，-150；
- Location in：local；
- Orientation Dependency：User-entered values；
- Orient using：Euler Angles；
- Euler Angles：0，0，0。

（4）单击 Apply 按钮，完成._FSAE_sus_front_torsion_vertical.ground.cfl_damper1_down 结构框的调试。

- Construction Frame（结构框名称）：damper1_down_ref；
- Location Dependency：Delta location from coordinate；
- Coordinate Reference（参考坐标）：._FSAE_sus_front_torsion_vertical.ground.cfl_damper1_down；
- Location：0，0，-50；
- Location in：local；

- Orientation Dependency：User-entered values；
- Orient using：Euler Angles；
- Euler Angles：0，0，0。

（5）单击 OK 按钮，完成._FSAE_sus_front_torsion_vertical.ground.cfl_damper1_down_ref 结构框的调试。

4．调试扭杆弹簧 FSAE_torsion

（1）单击 Build > Part > Flexible Body > Modify 命令，弹出修改部件对话框，如图 17-29 所示，在下列对话框中输入相应的数据：

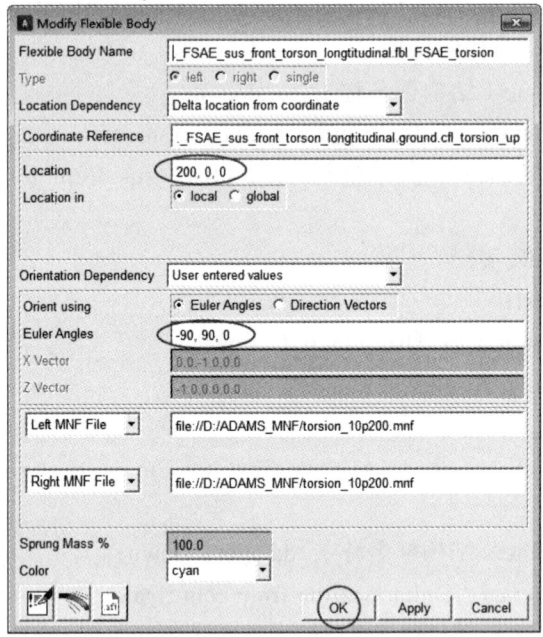

图 17-29 扭杆弹簧部件

- Flexible Body Name：FSAE_torsion；
- Type：left；
- Location Dependency：Delta location from coordinate；
- Coordinate Reference（参考坐标）：._FSAE_sus_front_torsion_vertical.ground.cfl_torsion_up；
- Location：200，0，0；
- Location in：local；
- Orientation Dependency：User-entered values；
- Orient using：Euler Angles；
- Euler Angles：-90，90，0；
- Left MNF File：file：//D：/ADAMS_MNF/torsion_10p200.mnf；
- Right MNF File：file：//D：/ADAMS_MNF/torsion_10p200.mnf；
- Color：cyan。

（2）单击 OK 按钮，完成部件._FSAE_sus_front_torsion_vertical.fbl_FSAE_torsion 的调试；至此前纵置扭杆弹簧式推杆悬架建立完成，如图 17-30 所示。

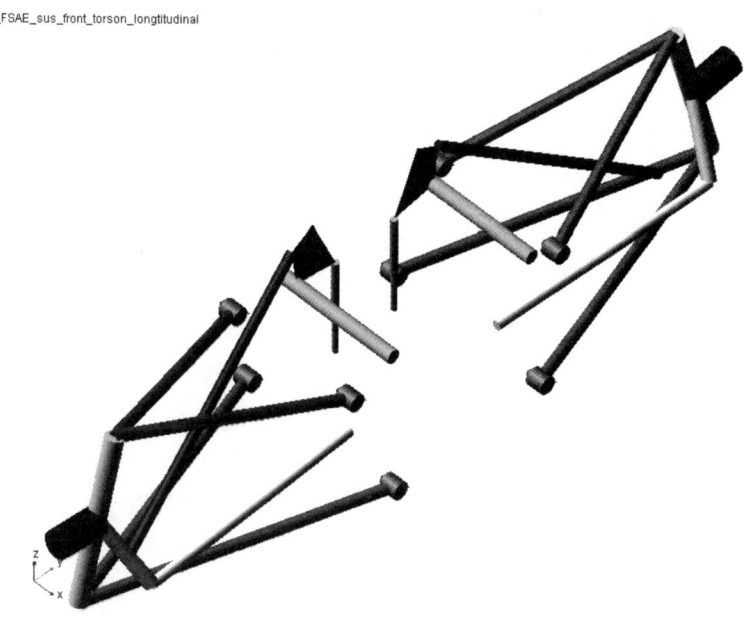

图 17-30　前纵置扭杆弹簧式推杆悬架

5. 后纵置扭杆弹簧式推杆悬架

后纵置扭杆弹簧式推杆悬架硬点信息如下，调试方式与前纵置扭杆弹簧式推杆悬架相同，调试完毕后纵置扭杆弹簧式推杆悬架模型如图 17-31 所示。

```
Info for subsystem:  FSAE_sus_front_torson_longtitudinal

File Name    : <FASE>/subsystems.tbl/FSAE_sus_front_torson_longtitudinal.sub
Template     : mdids://FASE/templates.tbl/_FSAE_sus_front_torson_longtitudinal.tpl
Comments     : *no comments found*
Major Role   : suspension
Minor Role   : front


HARDPOINTS:

hardpoint name          symmetry        x_value     y_value     z_value
--------------          --------        -------     -------     -------

hps_global              single             0.0         0.0         0.0
arb_bushing_mount       left/right       127.0      -127.0       101.6
lca_front               left/right      -127.0      -127.0       114.3
```

lca_outer	left/right	0.0	-546.1	120.65
lca_rear	left/right	127.0	-127.0	114.3
prod_outer	left/right	0.0	-457.2	139.7
prod_to_zhijia	left/right	-50.8	-127.0	381.0
tierod_inner	left/right	50.8	-127.0	152.4
tierod_outer	left/right	63.5	-482.6	152.4
uca_front	left/right	-101.6	-177.8	279.4
uca_outer	left/right	0.0	-482.6	355.6
uca_rear	left/right	101.6	-177.8	279.4
wheel_center	left/right	0.0	-558.8	241.3
zhijia_to_damper1	left/right	-50.8	-50.8	301.0
zhijia_to_damper2	left/right	-50.8	-90.8	401.0
zhijia_to_torsion	left/right	-50.8	-120.0	301.0

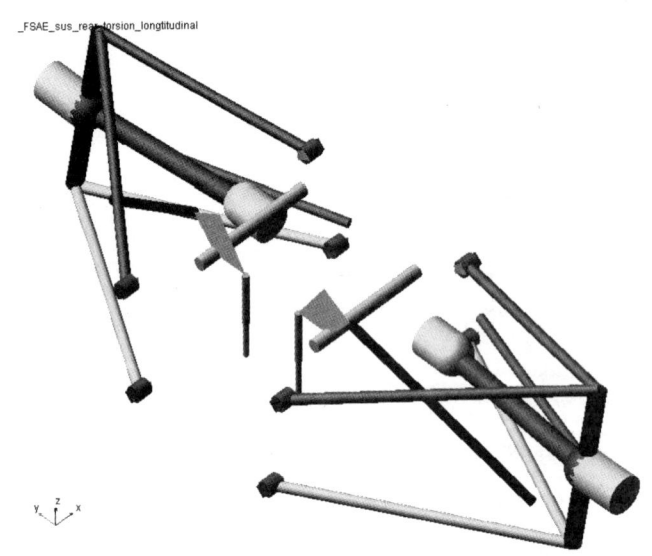

图 17-31 后纵置扭杆弹簧式推杆悬架

17.5 回正性仿真

回正性实验是模拟驾驶员在开车时瞬间松开方向盘，根据车辆的底盘参数（四轮定位参数等因素）设定使汽车在回正力矩的作用下检测响应，仿真时分为两种条件模式：① 转弯半径与速度；② 加速度与速度。仿真之前，装配整车 FSAE_2020_torson_longtitudinal 模型如图 17-32 所示。

图 17-32　整车 FSAE_2020_torson_longtitudinal 模型

1. 回正性仿真（加速度与速度）

（1）单击 Simulate > Full-Vehicle Analysis > Cornering Events > Cornering Steer Release 命令，弹出仿真参数设置对话框，如图 17-33 所示，在下列对话框中输入相应的数据：

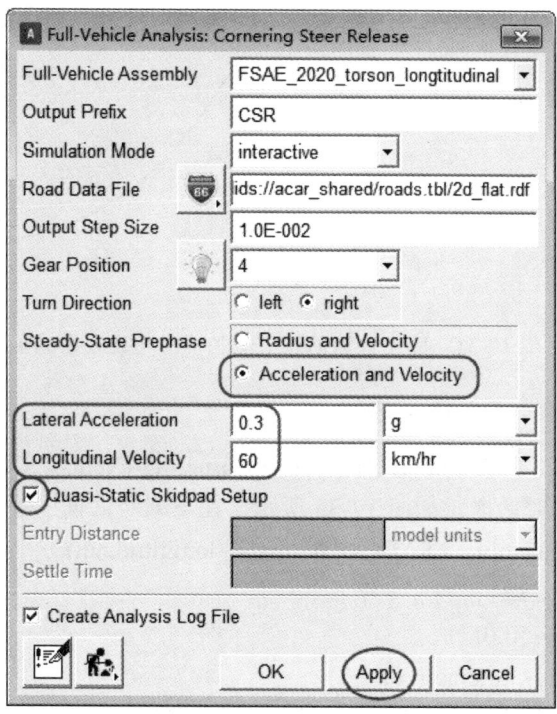

图 17-33　回正性仿真（加速度与速度模式）

- Full-Vehicle Assembly：FSAE_2020_torson_longtitudinal；
- Output Prefix：CSR；
- Output Step step：0.01；

- Simulation Mode：interactive；
- Road Date File：mdids：//acar_shared/roads.tbl/2d_flat.rdf；
- Gear Position：4；
- Turn Direction：right；
- Steady-Statc Prephase：Acceleration and Velocity；
- Lateral Acceleration（G's）：0.3，g；
- Longtitudinal Velocity：60，km/hr。
- 勾选 Quasi-Static Straight-Line Setup。

（2）单击 OK 按钮，完成 FSAE_2020_torson_longtitudinal 回正性仿真设置并提交运算，运算完成后，整车运行轨迹如图 17-34 所示。

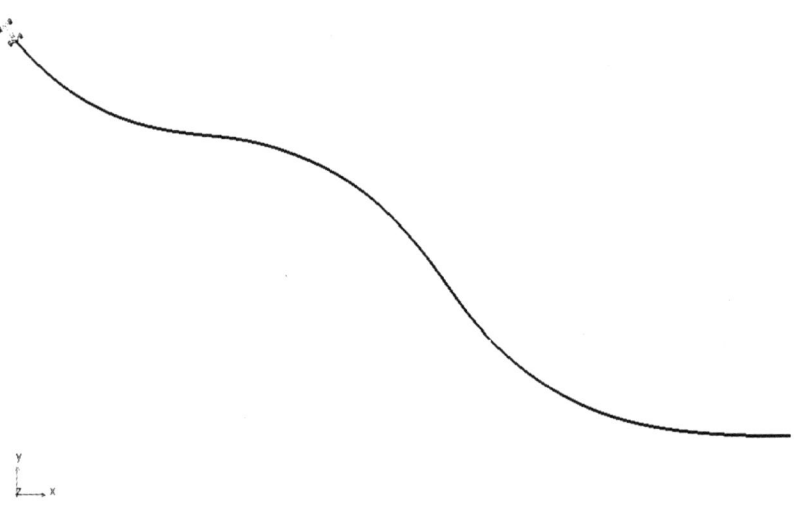

图 17-34　回正性运动轨迹（加速度与速度模式）

2. 回正性仿真（转弯半径与速度）

（1）单击 Simulate > Full-Vehicle Analysis > Cornering Events > Cornering Steer Release 命令，弹出仿真参数设置对话框，如图 17-35 所示，在下列对话框中输入相应的数据：

- Full-Vehicle Assembly：FSAE_2020_torson_longtitudinal；
- Output Prefix：CSR1；
- Output Step step：0.01；
- Simulation Mode：interactive；
- Road Date File：mdids：//acar_shared/roads.tbl/2d_flat.rdf；
- Gear Position：4；
- Turn Direction：right；
- Steady-State Prephase：Radius and Velocity；

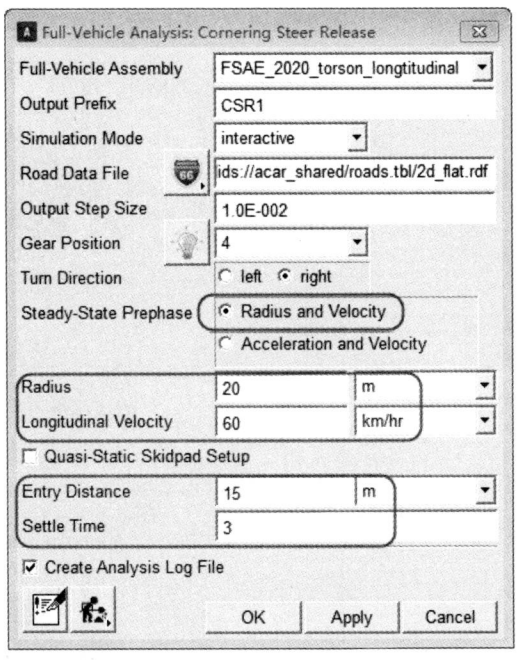

图 17-35 回正性仿真(半径与速度模式)

- Radius:20,m;
- Longtitudinal Velocity:60,km/hr;
- 勾选 Quasi-Static Straight-Line Setup;
- Entry Distance(引道长度):15,m;
- Settle Time(引道长度内驱动程序计算时间,单位:秒):3。

(2)单击 OK 按钮,完成 FSAE_2020_torson_longtitudinal 回正性仿真设置并提交运算,运算完成后,整车运行轨迹如图 17-36 所示。两种模式下整车的运行参数如图 17-37~图 17-42 所示。

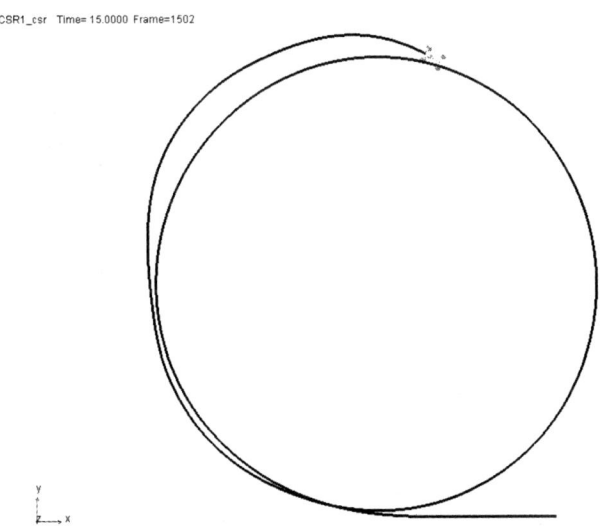

图 17-36 回正性运动轨迹(半径与速度模式)

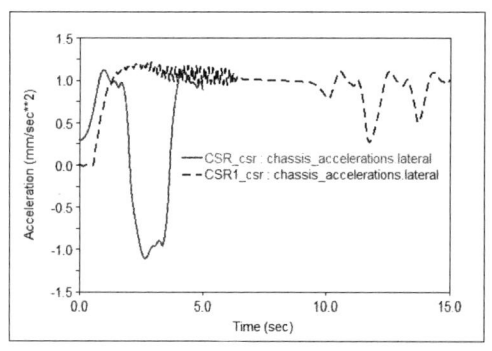

图 17-37 车身侧向加速度

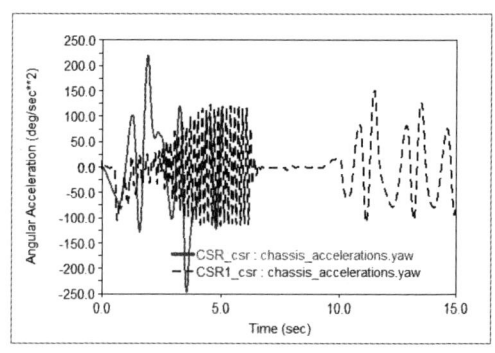

图 17-38 车身横摆角加速度

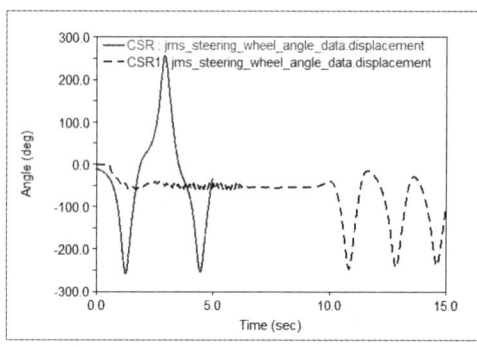

图 17-39 方向盘转动角度

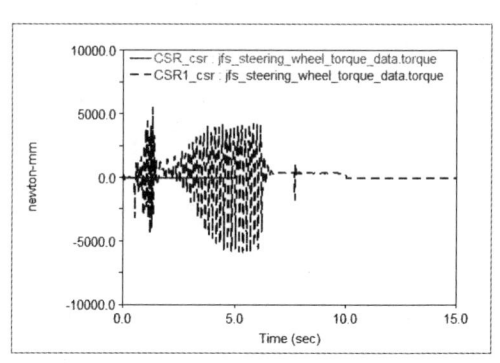

图 17-40 方向盘力矩

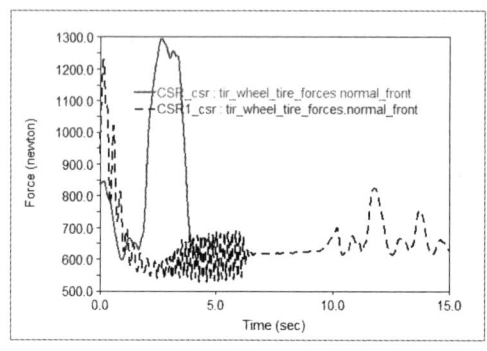

图 17-41 前轮胎法向受力

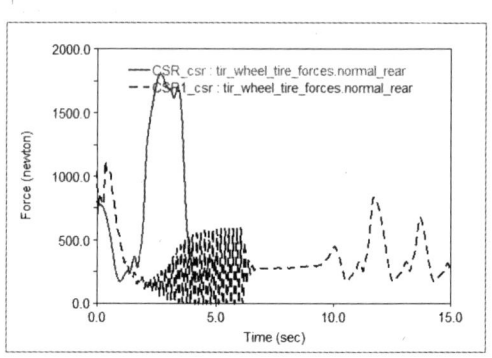

图 17-42 后轮胎法向受力

17.6 考虑横向避震器

对于整车 FSAE_sus_front_torsion_vertical 与整车 FSAE_2020_torson_longtitudinal，均可考虑横向避震器，横向避震器既可限制支架摆臂运动，又可缓冲支架摆臂的震动，而支架摆臂运动的大小与车身和轮胎的垂向振幅相关，因此加装横向避震器，可以提升整车的性能。

1. 整车 FSAE_2020_vertical_damper

加装避震器前，将前扭杆弹簧垂置式推杆悬架 FSAE_sus_front_torsion_vertical 另存为 FSAE_sus_front_torsion_vertical_damper；将后扭杆弹簧垂置式推杆悬架 FSAE_sus_front_torson_longtitudinal 另存为 FSAE_sus_front_torson_longtitudinal_damper。

（1）单击 Build > Force > Damper > New 命令，弹出避震器创建对话框，如图 17-43 所示，在下列对话框中输入相应的数据：

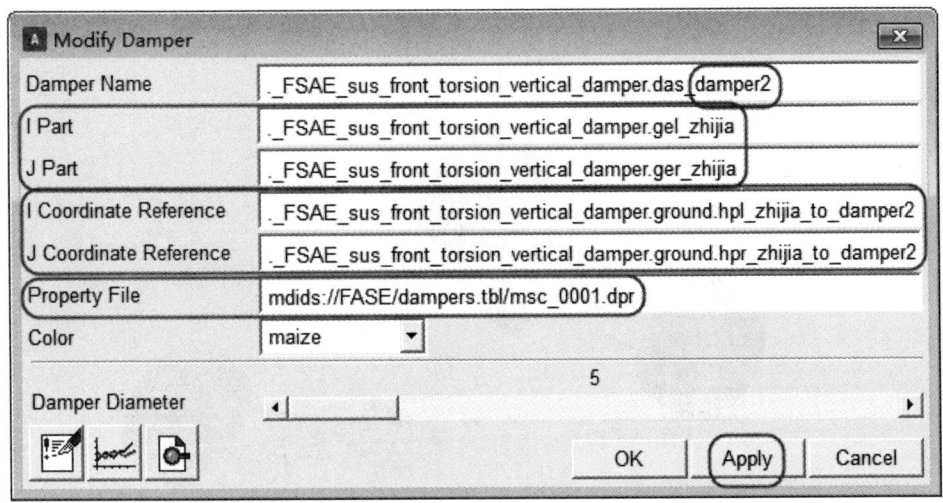

图 17-43　避震器_ damper2

- Damper Name（减震器名称）：damper2；
- I Part：._FSAE_sus_front_torsion_vertical_damper.gel_zhijia；
- J Part：._FSAE_sus_front_torsion_vertical_damper.ger_zhijia；
- I Coordinate Reference（参考坐标）：._FSAE_sus_front_torsion_vertical_damper.ground.hpl_zhijia_to_damper2；
- J Coordinate Reference（参考坐标）：._FSAE_sus_front_torsion_vertical_damper. ground.hpr_zhijia_to_damper2；
- Property File（属性文件）：mdids：//FASE/dampers.tbl/msc_0001.dpr；
- Damper Diameter（避震器直径）：拖动滑块选择 5 mm；
- Color：maize。

（2）单击 OK 按钮，完成避震器._FSAE_sus_front_torsion_vertical_damper.das_damper2 的创建。同理，完成对应后悬架避震器的添加，此处不再重复，加装完成后前推杆悬架如图 17-44 所示，装配好的整车如图 17-45 所示。

图 17-44 前扭杆弹簧垂置式推杆悬架
（考虑横向避震器）

图 17-45 整车 FSAE_2020_vertical_damper 模型

2. 整车 FSAE_2020_ longtitudinal _damper

加装避震器前，将前扭杆弹簧垂置式推杆悬架 FSAE_sus_front_torson_longtitudinal 另存为 FSAE_sus_front_torson_longtitudinal_damper；将后扭杆弹簧垂置式推杆悬架 FSAE_sus_rear_torsion_longtitudinal 另存为 FSAE_sus_rear_torsion_longtitudinal_damper。

（1）单击 Build > Force > Damper > New 命令，在下列对话框中输入相应的数据：
- Damper Name（减震器名称）：damper_line；
- I Part：._FSAE_sus_rear_torsion_longtitudinal_damper.gel_zhijia；
- J Part：._FSAE_sus_rear_torsion_longtitudinal_damper.ger_zhijia；
- I Coordinate Reference（参考坐标）：._FSAE_sus_rear_torsion_longtitudinal_damper.ground.hpl_zhijia_to_damper2；
- J Coordinate Reference（参考坐标）：._FSAE_sus_rear_torsion_longtitudinal_damper.ground.hpr_zhijia_to_damper2；

- Property File（属性文件）：mdids：//FASE/dampers.tbl/MSC_default.dpr；
- Damper Diameter（避震器直径）：拖动滑块选择 5 mm；
- Color：maize；

（2）单击 OK 按钮，完成避震器._FSAE_sus_rear_torsion_longtitudinal_damper.das_damper_line 的创建。同理，完成对应前悬架避震器的添加，此处不再重复，加装完成后的后推杆悬架如图 17-46 所示，装配好的整车如图 17-47 所示。

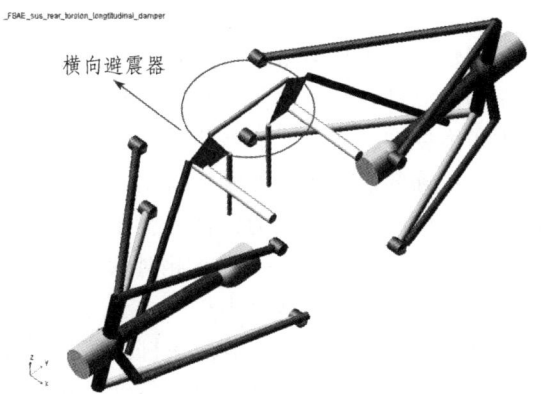

图 17-46　后扭杆弹簧垂置式推杆悬架
（考虑横向避震器）

图 17-47　整车 FSAE_2020_longtitudinal_damper 模型

第18章 空间斜置扭杆弹簧推杆式悬架

不论是家用轿车还是商用轿车，当悬架系统的弹簧刚度与避震器参数确定后，整车底盘的基本性能就确定了，整车的操纵性与平顺性只能折中。整车加速起步、紧急制动时，希望弹簧的刚度大，避免车身大角度俯仰；转弯时，希望弹簧刚度小（多数人认为转弯时弹簧刚度应过大，这是错误的理解，正确的设定是弹簧刚度应稍小，弹簧刚度小可以充分保证车轮与地面的接触面积，进而提供足够的侧向抓地力，保证整车不产生侧向滑移及摆动现象；如果弹簧刚度过大或者刚性把轮胎与车身连接在一起，此时轮胎与地面的接触力是不均衡的，大概率会导致某一小区域范围内侧向与垂向接触力极大，超过极限而产生侧滑或摆尾）。空间斜置扭杆弹簧推杆式解耦悬架如图 18-1 所示，通过扭杆弹簧与螺旋弹簧两种不同刚度的设定，在起步、制动、高速行驶时采用扭转弹簧的大刚度特性；在转弯及低速行驶时采用螺旋弹簧提升整车的平顺性及弯道稳定性；在静止、低速、高速时可以通过滑阀避震器改变车身的高度。此种悬架设定与半主动悬架有本质的区别，半主动悬架是通过改变避震器的特性来改变悬架系统的特性，半主动悬架的弹簧刚度是不可调节的。另外，此种悬架通过液压作用器改变扭杆弹簧与螺旋弹簧的工作特性，避震器可用滑阀特性改变车身高度，也可用变阻尼特性（如磁流变避震器）改变车身高度。建模过程不再重复，本章直接在模型_FSAE_sus_front_white.tpl 上通过添加扭杆弹簧、螺旋弹簧等部件完成建模，完成后的悬架模型存放在章节文件夹中，读者请自行查阅。

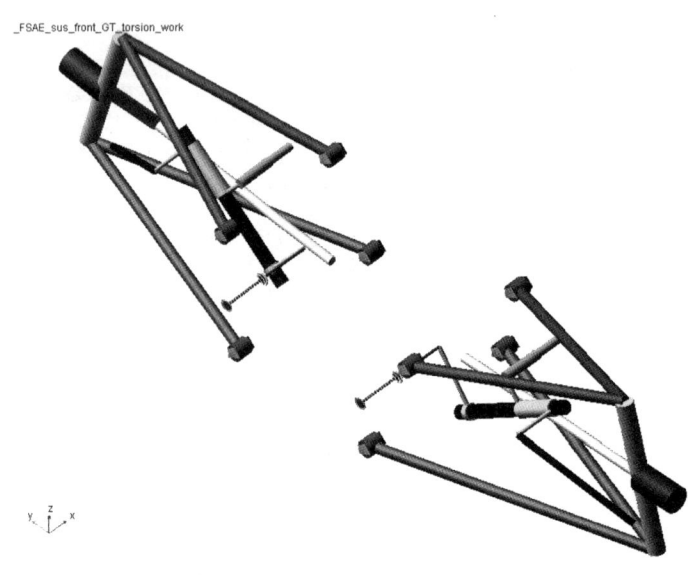

图 18-1 扭杆弹簧斜置式推杆悬架

学习目标

（1）了解前扭杆弹簧斜置式推杆悬架。
（2）会刚度匹配。
（3）会阻尼匹配。
（4）会四轮定位参数对标。
（5）会加速仿真。
（6）了解后扭杆弹簧斜置式推杆悬架。
（7）会制动仿真。

18.1 前扭杆弹簧斜置式推杆悬架

1. 模型导入

（1）启动 ADAMS/CAR，选择专家模块进入建模界面。
（2）单击 File > Open 命令，弹出打开模板对话框，如图 18-2 所示，在 Template Name 中输入：mdids：//FASE/templates.tbl/_FSAE_sus_front_white.tpl。
（3）单击 OK 按钮，完成模型导入。

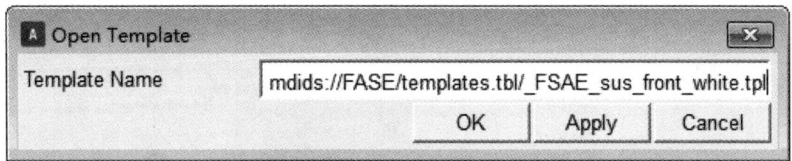

图 18-2　模板对话框

2. 推杆部件硬点

（1）单击 Build > Hardpoind > New 命令，在下列对话框中输入相应的数据：
- Hardpoint：prod_outer；
- Type：left；
- Location：0.0，-500.0，140.65。

（2）单击 Apply 按钮，完成 prod_outer 硬点的创建。同理，创建其他硬件点。
- Hardpoint：prod_to_bellcrank；
- Type：left；
- Location：0.0，-350.0，250.0。

（3）单击 OK 按钮，完成 prod_to_bellcrank 硬点的创建。
（4）单击 File > Save As 命令，弹出保存模板对话框，如图 18-3 所示，在下列对话框中输入相应的数据：

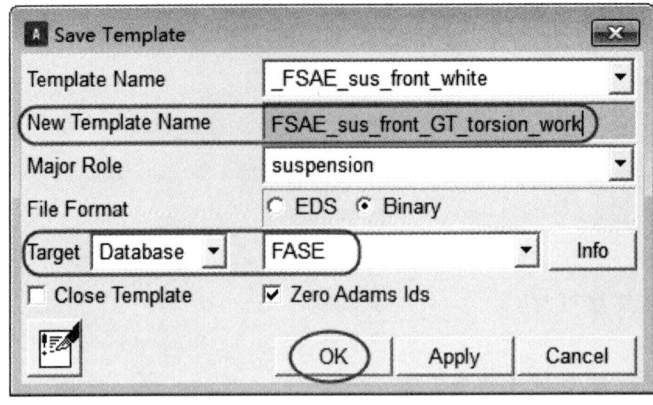

图 18-3 模型保存

- Template Name：_FSAE_sus_front_wite；
- New Template Name：FSAE_sus_front_GT_torsion_work；
- Major Role（主特征）：suspension；
- File Format：Binary；
- Target：Datebase/FASE。

(5) 单击 OK 按钮，完成推杆式悬架模型模板 FSAE_sus_front_GT_torsion_work 的保存。

3. 结构框

(1) 单击 Build > Construction Frame > New 命令，弹出创建结构框对话框，如图 18-4 所示，在下列对话框中输入相应的数据：

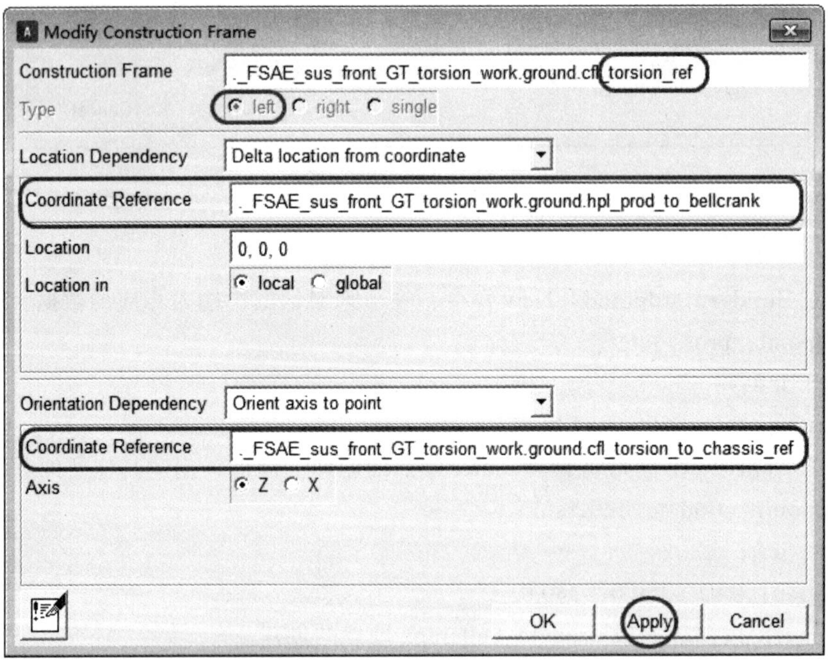

图 18-4 结构框 torsion_ref

- Construction Frame（结构框名称）：torsion_ref；
- Location Dependency：Delta location from coordinate；
- Coordinate Reference（参考坐标）：._FSAE_sus_front_GT_torsion_work.ground.hpl_prod_to_bellcrank；
- Location：0，0，0；
- Location in：local；
- Orientation Dependency：Orient axis to point；
- Coordinate Reference（参考坐标）：._FSAE_sus_front_GT_torsion_work.ground.cfl_torsion_to_chassis_ref；
- Axis：Z。

（2）单击 Apply 按钮，完成._FSAE_sus_front_GT_torsion_work.ground.cfl_torsion_ref 结构框的创建，同理，创建其他结构框。

- Construction Frame（结构框名称）：torsion_front；
- Location Dependency：Delta location from coordinate；
- Coordinate Reference（参考坐标）：._FSAE_sus_front_GT_torsion_work.ground.cfl_torsion_ref；
- Location：70，0，0；
- Location in：local；
- Orientation Dependency：User entered values；
- Orient using：Euler Angles；
- Euler Angles：0，0，0。

（3）单击 Apply 按钮，完成._FSAE_sus_front_GT_torsion_work.ground.cfl_torsion_front 结构框的创建。

- Construction Frame（结构框名称）：torsion_to_damper；
- Location Dependency：Delta location from coordinate；
- Coordinate Reference（参考坐标）：._FSAE_sus_front_GT_torsion_work.ground.cfl_torsion_ref；
- Location：0，0，80；
- Location in：local；
- Orientation Dependency：Delta location from coordinate；
- Coordinate Reference（参考坐标）：._FSAE_sus_front_GT_torsion_work.ground.cfl_torsion_ref；
- Orientation：0，0，0。

（4）单击 Apply 按钮，完成._FSAE_sus_front_GT_torsion_work.ground.cfl_torsion_to_damper 结构框的创建。

- Construction Frame（结构框名称）：torsion_ref_up；
- Location Dependency：Delta location from coordinate；
- Coordinate Reference（参考坐标）：._FSAE_sus_front_GT_torsion_work.ground.cfl_torsion_ref；

- Location：0，50，0；
- Location in：local；
- Orientation Dependency：Delta location from coordinate；
- Coordinate Reference（参考坐标）：._FSAE_sus_front_GT_torsion_work.ground.cfl_torsion_ref；
- Orientation：0，0，0。

（5）单击 Apply 按钮，完成._FSAE_sus_front_GT_torsion_work.ground.cfl_torsion_ref_up 结构框的创建。

- Construction Frame（结构框名称）：torsion_to_damper_up；
- Location Dependency：Delta location from coordinate；
- Coordinate Reference（参考坐标）：._FSAE_sus_front_GT_torsion_work.ground.cfl_torsion_to_damper；
- Location：0，50，0；
- Location in：local；
- Orientation Dependency：Delta location from coordinate；
- Coordinate Reference（参考坐标）：._FSAE_sus_front_GT_torsion_work.ground.cfl_torsion_to_damper；
- Orientation：0，0，0。

（6）单击 Apply 按钮，完成._FSAE_sus_front_GT_torsion_work.ground.cfl_torsion_to_damper_up 结构框的创建。

- Construction Frame（结构框名称）：torsion_to_chassis_up；
- Location Dependency：Delta location from coordinate；
- Coordinate Reference（参考坐标）：._FSAE_sus_front_GT_torsion_work.ground.cfl_torsion_ref；
- Location：0，100，200；
- Location in：local；
- Orientation Dependency：Delta location from coordinate；
- Coordinate Reference（参考坐标）：._FSAE_sus_front_GT_torsion_work.ground.cfl_torsion_ref；
- Orientation：0，0，0。

（7）单击 Apply 按钮，完成._FSAE_sus_front_GT_torsion_work.ground.cfl_torsion_to_chassis_up 结构框的创建。

- Construction Frame（结构框名称）：torsion_to_chassis；
- Location Dependency：Delta location from coordinate；
- Coordinate Reference（参考坐标）：._FSAE_sus_front_GT_torsion_work.ground.cfl_torsion_ref；
- Location：0，0，200；
- Location in：local；
- Orientation Dependency：Delta location from coordinate；

- Coordinate Reference（参考坐标）：._FSAE_sus_front_GT_torsion_work.ground.cfl_torsion_ref；
- Orientation：0，0，0。

（8）单击 Apply 按钮，完成._FSAE_sus_front_GT_torsion_work.ground.cfl_torsion_to_chassis 结构框的创建。

- Construction Frame（结构框名称）：spring_down；
- Location Dependency：Delta location from coordinate；
- Coordinate Reference（参考坐标）：._FSAE_sus_front_GT_torsion_work.ground.cfl_torsion_to_chassis_up；
- Location：150，0，0；
- Location in：local；
- Orientation Dependency：Delta location from coordinate；
- Coordinate Reference（参考坐标）：._FSAE_sus_front_GT_torsion_work.ground.cfl_torsion_to_chassis_up；
- Orientation：90，90，0。

（9）单击 Apply 按钮，完成._FSAE_sus_front_GT_torsion_work.ground.cfl_spring_down 结构框的创建。

- Construction Frame（结构框名称）：damper_down；
- Location Dependency：Delta location from coordinate；
- Coordinate Reference（参考坐标）：._FSAE_sus_front_GT_torsion_work.ground.cfl_torsion_to_damper_up；
- Location：-100，0，0；
- Location in：local；
- Orientation Dependency：Delta location from coordinate；
- Coordinate Reference（参考坐标）：._FSAE_sus_front_GT_torsion_work.ground.cfl_torsion_to_damper_up；
- Orientation：0，0，0。

（10）单击 Apply 按钮，完成._FSAE_sus_front_GT_torsion_work.ground.cfl_damper_down 结构框的创建。

- Construction Frame（结构框名称）：damper_down_ref；
- Location Dependency：Delta location from coordinate；
- Coordinate Reference（参考坐标）：._FSAE_sus_front_GT_torsion_work.ground.cfl_damper_down；
- Location：-30，0，0；
- Location in：local；
- Orientation Dependency：Delta location from coordinate；
- Coordinate Reference（参考坐标）：._FSAE_sus_front_GT_torsion_work.ground.cfl_damper_down；
- Orientation：0，0，0。

（11）单击 Apply 按钮，完成._FSAE_sus_front_GT_torsion_work.ground.cfl_damper_down_ref 结构框的创建。

- Construction Frame（结构框名称）：zhijia_up_to_torsion；
- Location Dependency：Centered between coordinates；
- Centered between：Two Coordinates；
- Coordinate Reference #1（参考坐标）：._FSAE_sus_front_GT_torsion_work.ground.cfl_torsion_ref；
- Coordinate Reference #2（参考坐标）：._FSAE_sus_front_GT_torsion_work.ground.cfl_torsion_to_damper；
- Orientation Dependency：Delta location from coordinate；
- Coordinate Reference（参考坐标）：._FSAE_sus_front_GT_torsion_work.ground.cfl_torsion_ref；
- Orientation：0，0，0。

（12）单击 Apply 按钮，完成._FSAE_sus_front_GT_torsion_work.ground.cfl_zhijia_up_to_torsion 结构框的创建。

- Construction Frame（结构框名称）：chassis_base；
- Location Dependency：Delta location from coordinate；
- Coordinate Reference（参考坐标）：._FSAE_sus_front_GT_torsion_work.ground.cfl_torsion_to_chassis_up；
- Location：80，0，0；
- Location in：local；
- Orientation Dependency：Delta location from coordinate；
- Coordinate Reference（参考坐标）：._FSAE_sus_front_GT_torsion_work.ground.cfl_spring_down；
- Orientation：0，0，0。

（13）单击 Apply 按钮，完成._FSAE_sus_front_GT_torsion_work.ground.cfl_chassis_base 结构框的创建。

- Construction Frame（结构框名称）：chassis_ref；
- Location Dependency：Delta location from coordinate；
- Coordinate Reference（参考坐标）：._FSAE_sus_front_GT_torsion_work.ground.cfl_chassis_base；
- Location：0，0，30；
- Location in：local；
- Orientation Dependency：Delta location from coordinate；
- Coordinate Reference（参考坐标）：._FSAE_sus_front_GT_torsion_work.ground.cfl_chassis_base；
- Orientation：0，0，0。

（14）单击 Apply 按钮，完成._FSAE_sus_front_GT_torsion_work.ground.cfl_chassis_ref 结构框的创建。

- Construction Frame（结构框名称）：torsion_down_to_chassis；
- Location Dependency：Delta location from coordinate；
- Coordinate Reference（参考坐标）：._FSAE_sus_front_GT_torsion_work.ground.cfl_torsion_to_chassis；
- Location：0，0，25；
- Location in：local；
- Orientation Dependency：Delta location from coordinate；
- Coordinate Reference（参考坐标）：._FSAE_sus_front_GT_torsion_work.ground.cfl_torsion_to_chassis；
- Orientation：0，0，0。

（15）单击 OK 按钮，完成._FSAE_sus_front_GT_torsion_work.ground.cfl_torsion_down_to_chassis 结构框的创建。

4. 部件 prod

（1）单击 Build > Part > General Part > New 命令，弹出创建部件对话框，如图 18-5 所示，在下列对话框中输入相应的数据：

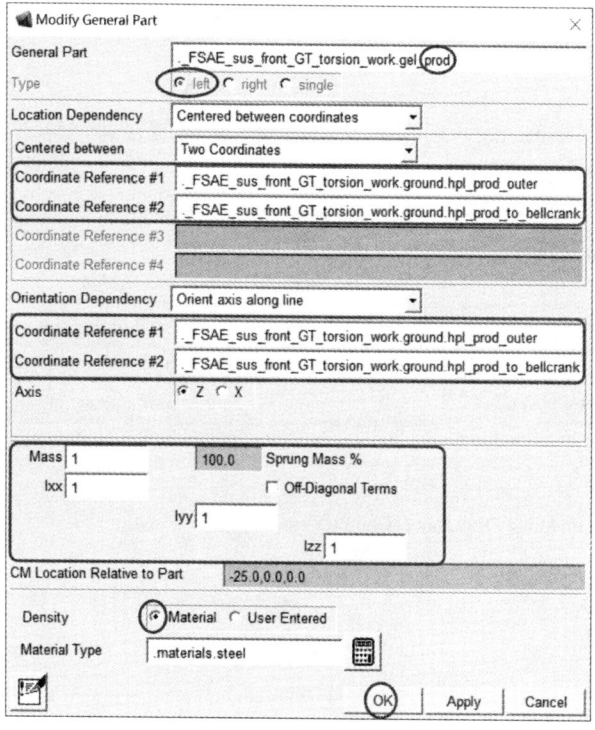

图 18-5 部件 prod

- General Part：prod；
- Type：left；
- Location Dependency：Centered between coordinates；
- Centered between：Two Coordinates；

- Coordinate Reference #1（参考坐标）：._FSAE_sus_front_GT_torsion_work.ground.hpl_prod_outer；
- Coordinate Reference #2（参考坐标）：._FSAE_sus_front_GT_torsion_work.ground.hpl_prod_to_bellcrank；
- Orientation Dependency：Orient axis along line；
- Coordinate Reference #1（参考坐标）：._FSAE_sus_front_GT_torsion_work.ground.hpl_prod_outer；
- Coordinate Reference #2（参考坐标）：._FSAE_sus_front_GT_torsion_work.ground.hpl_prod_to_bellcrank；
- Axis：Z；
- Mass：1；
- Ixx：1；
- Iyy：1；
- Izz：1；
- Density：Material；
- Material Type：.materials.steel。

（2）单击 OK 按钮，完成部件._FSAE_sus_front_GT_torsion_work.gel_prod 的创建。

5. 几何体 prod

（1）单击 Build > Geometry > Link > New 命令，弹出创建几何体对话框，如图 18-6 所示，在下列对话框中输入相应的数据：

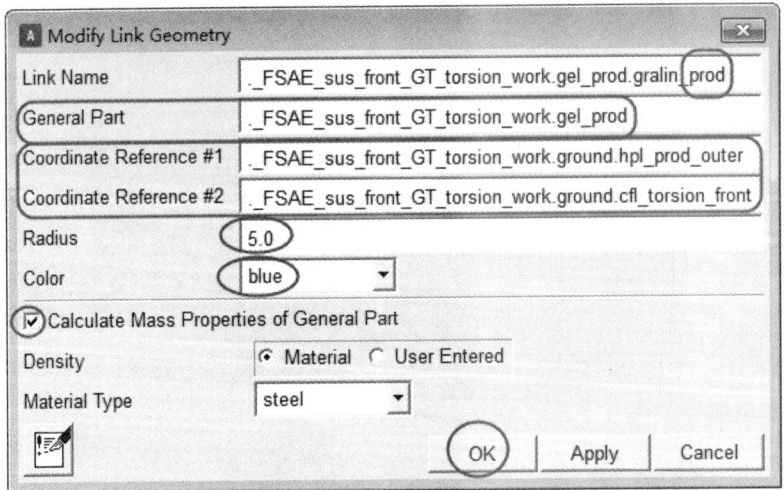

图 18-6　连杆几何体_prod

- Link Name（连杆名称）：prod；
- General Part：._FSAE_sus_front_GT_torsion_work.gel_prod；
- Coordinate Reference #1（参考坐标）：._FSAE_sus_front_GT_torsion_work.ground.hpl_prod_outer；

- Coordinate Reference #2（参考坐标）：._FSAE_sus_front_GT_torsion_work.ground.cfl_torsion_front；
- Radius（半径）：5；
- Color：blue；
- 勾选 Calculate Mass Properties of General Part 复选框，当几何体建立好之后，会更新对应部件的质量和惯量参数；
- Density：Material；
- Material Type：steel。

（2）单击 OK 按钮，完成._FSAE_sus_front_GT_torsion_work.gel_prod.gralin_prod 几何体的创建。

6. 部件 zhijia_up

（1）单击 Build > Part > General Part > New 命令，在下列对话框中输入相应的数据：
- General Part：zhijia_up；
- Type：left；
- Location Dependency：Centered between coordinates；
- Centered between：Two Coordinates；
- Coordinate Reference #1（参考坐标）：._FSAE_sus_front_GT_torsion_work.ground.cfl_torsion_ref；
- Coordinate Reference #2（参考坐标）：._FSAE_sus_front_GT_torsion_work.ground.cfl_torsion_to_damper；
- Orientation Dependency：Delta location from coordinate；
- Coordinate Reference（参考坐标）：._FSAE_sus_front_GT_torsion_work.ground.cfl_torsion_to_damper；
- Orientation：0，0，0；
- Mass：1；
- Ixx：1；
- Iyy：1；
- Izz：1；
- Density：Material；
- Material Type：.materials.steel。

（2）单击 OK 按钮，完成部件._FSAE_sus_front_GT_torsion_work.gel_zhijia_up 的创建。

7. 几何体 zhijia_up

（1）单击 Build > Geometry > Link > New 命令，在下列对话框中输入相应的数据：
- Link Name（连杆名称）：zhijia_up_main；
- General Part：._FSAE_sus_front_GT_torsion_work.gel_zhijia_up；
- Coordinate Reference #1（参考坐标）：._FSAE_sus_front_GT_torsion_work.ground.cfl_torsion_ref；

- Coordinate Reference #2（参考坐标）：._FSAE_sus_front_GT_torsion_work.ground.cfl_torsion_to_damper；
- Radius（半径）：12；
- Color：yellow；
- 勾选 Calculate Mass Properties of General Part 复选框，当几何体建立好之后，会更新对应部件的质量和惯量参数；
- Density：Material；
- Material Type：steel。

（2）单击 Apply 按钮，完成 ._FSAE_sus_front_GT_torsion_work.gel_zhijia_up.gralin_zhijia_up_main 几何体的创建。同理，创建其他几何体。

- Link Name（连杆名称）：zhijia_up_to_poshrod；
- General Part：._FSAE_sus_front_GT_torsion_work.gel_zhijia_up；
- Coordinate Reference #1（参考坐标）：._FSAE_sus_front_GT_torsion_work.ground.cfl_torsion_ref；
- Coordinate Reference #2（参考坐标）：._FSAE_sus_front_GT_torsion_work.ground.cfl_torsion_front；
- Radius（半径）：3；
- Color：yellow；
- 勾选 Calculate Mass Properties of General Part 复选框，当几何体建立好之后，会更新对应部件的质量和惯量参数；
- Density：Material；
- Material Type：steel。

（3）单击 Apply 按钮，完成 ._FSAE_sus_front_GT_torsion_work.gel_zhijia_up.gralin_zhijia_up_to_poshrod 几何体的创建。

- Link Name（连杆名称）：zhijia_up_to_damper；
- General Part：._FSAE_sus_front_GT_torsion_work.gel_zhijia_up；
- Coordinate Reference #1（参考坐标）：._FSAE_sus_front_GT_torsion_work.ground.cfl_torsion_to_damper；
- Coordinate Reference #2（参考坐标）：._FSAE_sus_front_GT_torsion_work.ground.cfl_torsion_to_damper_up；
- Radius（半径）：3；
- Color：yellow；
- 勾选 Calculate Mass Properties of General Part 复选框，当几何体建立好之后，会更新对应部件的质量和惯量参数；
- Density：Material；
- Material Type：steel。

（4）单击 OK 按钮，完成 ._FSAE_sus_front_GT_torsion_work.gel_zhijia_up.gralin_zhijia_up_to_damper 几何体的创建。

8. 部件 zhijia_down

（1）单击 Build > Part > General Part > New 命令，在下列对话框中输入相应的数据：

- General Part：zhijia_down；
- Type：left；
- Location Dependency：Delta location from coordinate；
- Coordinate Reference（参考坐标）：._FSAE_sus_front_GT_torsion_work.ground.cfl_torsion_to_chassis；
- Location：0，0，0；
- Location in：local；
- Orientation Dependency：Delta location from coordinate；
- Coordinate Reference（参考坐标）：._FSAE_sus_front_GT_torsion_work.ground.cfl_torsion_to_chassis；
- Orientation：0，0，0；
- Mass：1；
- Ixx：1；
- Iyy：1；
- Izz：1；
- Density：Material；
- Material Type：.materials.steel。

（2）单击 OK 按钮，完成部件._FSAE_sus_front_GT_torsion_work.gel_zhijia_dwon 的创建。

9. 几何体 zhijia_down

（1）单击 Build > Geometry > Cylinder> New 命令，弹出创建圆柱体对话框，如图 18-7 所示，在下列对话框中输入相应的数据：

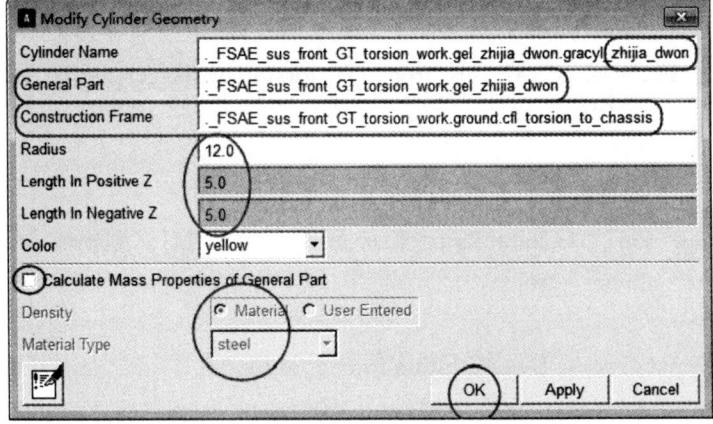

图 18-7　圆柱几何体 zhijia_down

- Cylinder Name（连杆名称）：zhijia_down；
- General Part：._FSAE_sus_front_GT_torsion_work.gel_zhijia_down；

- Coordinate Frame（参考坐标）：._FSAE_sus_front_GT_torsion_work.ground.cfl_torsion_to_chassis；
- Radius（半径）：12；
- Length in Positive Z（Z轴正方向）：5，圆柱几何体创建要求必须在结构框的Z轴正负方向拉伸；
- Length in Negative Z（Z轴负方向）：5；
- Color：yellow；
- 勾选Calculate Mass Properties of General Part复选框，当几何体建立好之后，会更新对应部件的质量和惯量参数；
- Density：Material；
- Material Type：steel。

（2）单击OK按钮，完成._FSAE_sus_front_GT_torsion_work.gel_zhijia_down.gracyl_zhijia_down几何体的创建。

（3）单击Build > Geometry > Link > New命令，在下列对话框中输入相应的数据：
- Link Name（连杆名称）：zhijia_down_to_spring；
- General Part：._FSAE_sus_front_GT_torsion_work.gel_zhijia_down；
- Coordinate Reference #1（参考坐标）：._FSAE_sus_front_GT_torsion_work.ground.cfl_torsion_to_chassis；
- Coordinate Reference #2（参考坐标）：._FSAE_sus_front_GT_torsion_work.ground.cfl_torsion_to_chassis_up；
- Radius（半径）：3；
- Color：yellow；
- 勾选Calculate Mass Properties of General Part复选框，当几何体建立好之后，会更新对应部件的质量和惯量参数；
- Density：Material；
- Material Type：steel。

（4）单击OK按钮，完成._FSAE_sus_front_GT_torsion_work.gel_zhijia_down.gralin_zhijia_down_to_spring几何体的创建。

10. 部件damper_front

（1）单击Build > Part > General Part > New命令，在下列对话框中输入相应的数据：
- General Part：damper_front；
- Type：left；
- Location Dependency：Delta location from coordinate；
- Coordinate Reference（参考坐标）：._FSAE_sus_front_GT_torsion_work.ground.cfl_torsion_to_damper_up；
- Location：0，0，0；
- Location in：local；
- Orientation Dependency：Delta location from coordinate；

- Coordinate Reference（参考坐标）：._FSAE_sus_front_GT_torsion_work.ground.cfl_torsion_to_damper_up；
- Orientation：0，0，0；
- Mass：1；
- Ixx：1；
- Iyy：1；
- Izz：1；
- Density：Material；
- Material Type：.materials.steel。

（2）单击 OK 按钮，完成部件._FSAE_sus_front_GT_torsion_work.gel_damper_front 的创建。

11. 部件 damper_rear

- 单击 Build > Part > General Part > New 命令，在下列对话框中输入相应的数据：
- General Part：damper_rear；
- Type：left；
- Location Dependency：Delta location from coordinate；
- Coordinate Reference（参考坐标）：._FSAE_sus_front_GT_torsion_work.ground.cfl_damper_down；
- Location：0，0，0；
- Location in：local；
- Orientation Dependency：Delta location from coordinate；
- Coordinate Reference（参考坐标）：._FSAE_sus_front_GT_torsion_work.ground.cfl_damper_down；
- Orientation：0，0，0；
- Mass：1；
- Ixx：1；
- Iyy：1；
- Izz：1；
- Density：Material；
- Material Type：.materials.steel。

（2）单击 OK 按钮，完成部件._FSAE_sus_front_GT_torsion_work.gel_damper_rear 的创建。

12. 部件 spring_base

ADAMS 中的弹簧拉压均可，实际上悬架系统中的弹簧仅受压力，如果在 ADAMS 中参考物理模型建立出来的弹簧为受拉特性，此处通过部件位置的反响布置实现螺旋弹簧的受压特性。

（1）单击 Build > Part > General Part > New 命令，在下列对话框中输入相应的数据：
- General Part：spring_base；
- Type：left；

- Location Dependency：Delta location from coordinate；
- Coordinate Reference（参考坐标）：._FSAE_sus_front_GT_torsion_work.ground.cfl_spring_down；
- Location：0，0，0；
- Location in：local；
- Orientation Dependency：Delta location from coordinate；
- Coordinate Reference（参考坐标）：._FSAE_sus_front_GT_torsion_work.ground.cfl_spring_down；
- Orientation：0，0，0；
- Mass：1；
- Ixx：1；
- Iyy：1；
- Izz：1；
- Density：Material；
- Material Type：.materials.steel。

（2）单击 OK 按钮，完成部件._FSAE_sus_front_GT_torsion_work.gel_spring_base 的创建。

13. 几何体 spring_base

（1）单击 Build > Geometry > Cylinder> New 命令，在下列对话框中输入相应的数据：
- Cylinder Name（连杆名称）：spring_base；
- General Part：._FSAE_sus_front_GT_torsion_work.gel_spring_base；
- Coordinate Frame（参考坐标）：._FSAE_sus_front_GT_torsion_work.ground.cfl_spring_down；
- Radius（半径）：7；
- Length in Positive Z（Z 轴正方向）：1；
- Length in Negative Z（Z 轴负方向）：1；
- Color：red；
- 勾选 Calculate Mass Properties of General Part 复选框，当几何体建立好之后，会更新对应部件的质量和惯量参数；
- Density：Material；
- Material Type：steel。

（2）单击 OK 按钮，完成._FSAE_sus_front_GT_torsion_work.gel_spring_base.gracyl_spring_base 几何体的创建。

（3）单击 Build > Geometry > Link > New 命令，在下列对话框中输入相应的数据：
- Link Name（连杆名称）：sping_pllrod；
- General Part：._FSAE_sus_front_GT_torsion_work.gel_spring_base；
- Coordinate Reference #1（参考坐标）：._FSAE_sus_front_GT_torsion_work.ground.cfl_spring_down；
- Coordinate Reference #2（参考坐标）：._FSAE_sus_front_GT_torsion_work.ground.cfl_

torsion_to_chassis_up；
- Radius（半径）：2；
- Color：red；
- 勾选 Calculate Mass Properties of General Part 复选框，当几何体建立好之后，会更新对应部件的质量和惯量参数；
- Density：Material；
- Material Type：steel。

（4）单击 OK 按钮，完成 ._FSAE_sus_front_GT_torsion_work.gel_spring_base.gralin_sping_pllrod 几何体的创建。

14. 部件 chassic_base

（1）单击 Build > Part > General Part > New 命令，在下列对话框中输入相应的数据：
- General Part：chassic_base；
- Type：left；
- Location Dependency：Delta location from coordinate；
- Coordinate Reference（参考坐标）：._FSAE_sus_front_GT_torsion_work.ground.cfl_torsion_to_chassis_up；
- Location：0，0，0；
- Location in：local；
- Orientation Dependency：Delta location from coordinate；
- Coordinate Reference（参考坐标）：._FSAE_sus_front_GT_torsion_work.ground.cfl_spring_down；
- Orientation：0，0，0；
- Mass：1；
- Ixx：1；
- Iyy：1；
- Izz：1；
- Density：Material；
- Material Type：.materials.steel。

（2）单击 OK 按钮，完成部件 ._FSAE_sus_front_GT_torsion_work.gel_chassic_base 的创建。

15. 几何体 chassis_base

（1）单击 Build > Geometry > Cylinder> New 命令，在下列对话框中输入相应的数据：
- Cylinder Name（连杆名称）：chassis_base；
- General Part：._FSAE_sus_front_GT_torsion_work.gel_chassic_base；
- Coordinate Frame（参考坐标）：._FSAE_sus_front_GT_torsion_work.ground.cfl_chassis_base；
- Radius（半径）：7；
- Length in Positive Z（Z 轴正方向）：1；

- Length in Negative Z（Z 轴负方向）：1；
- Color：red；
- 勾选 Calculate Mass Properties of General Part 复选框，当几何体建立好之后，会更新对应部件的质量和惯量参数；
- Density：Material；
- Material Type：steel。

（2）单击 OK 按钮，完成 ._FSAE_sus_front_GT_torsion_work.gel_chassic_base.gracyl_chassis_base 几何体的创建。

16. 柔性扭杆弹簧 torsion

扭杆弹簧制作不再重复，请读者参考相关章节自己制作练习。本模型已经制作好的扭杆弹簧 MNF 中性文件：torsion_10p250.mnf 存在章节文件中，请读者自行调阅查看。

（1）单击 Build > Part > Flexible Body > New 命令，弹出创建部件对话框，如图 18-8 所示，在下列对话框中输入相应的数据：

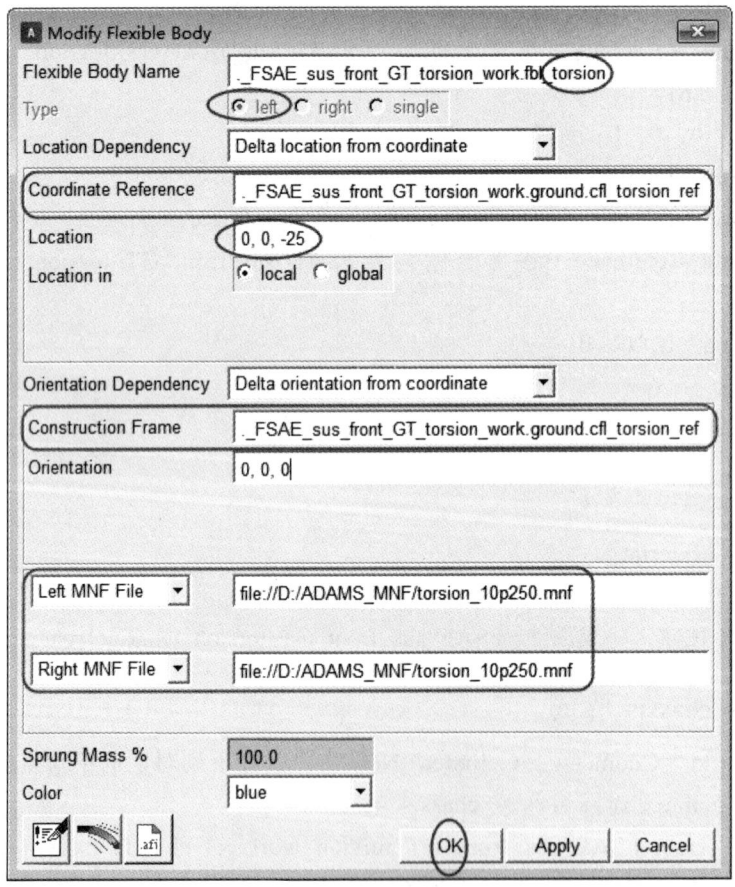

图 18-8　扭杆弹簧部件

- Flexible Body Name：torsion；
- Type：left；

- Location Dependency：Delta location from coordinate；
- Coordinate Reference（参考坐标）：._FSAE_sus_front_GT_torsion_work.ground.cfl_torsion_ref；
- Location：0，0，-25；
- Orientation Dependency：Delta location from coordinate；
- Coordinate Reference（参考坐标）：._FSAE_sus_front_GT_torsion_work.ground.cfl_torsion_ref；
- Orientation：0，0，0；
- Left MNF File：file：//D：/ADAMS_MNF/torsion_10p250.mnf；
- Right MNF File：file：//D：/ADAMS_MNF/torsion_10p250.mnf；
- Color：blue。

（2）单击 Apply 按钮，完成部件._FSAE_sus_front_GT_torsion_work.fbl_torsion 的创建。

17. 弹　簧

（1）单击 Build > Force > Spring > New 命令，弹出创建部件对话框，如图 18-9 所示，在下列对话框中输入相应的数据：

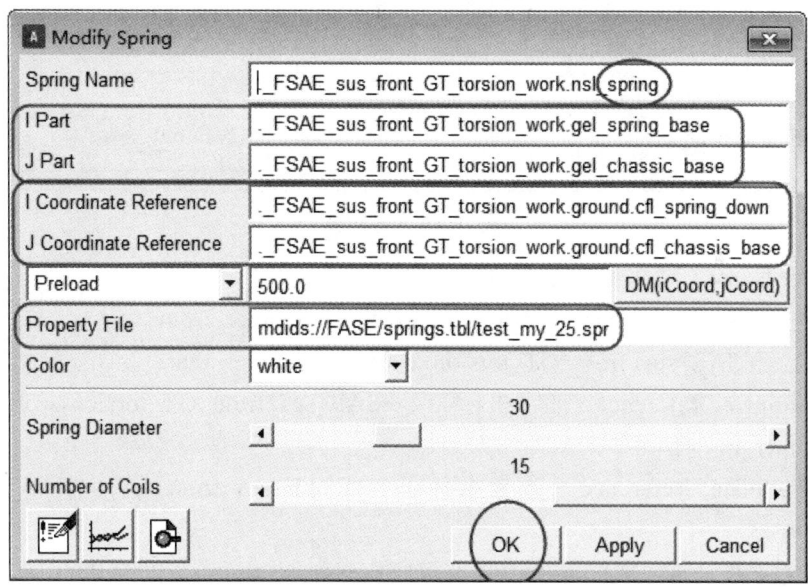

图 18-9　弹　簧

- Spring Name（减震器名称）：spring；
- I Part：._FSAE_sus_front_GT_torsion_work.gel_spring_base；
- J Part：._FSAE_sus_front_GT_torsion_work.gel_chassic_base；
- I Coordinate Reference（参考坐标）：._FSAE_sus_front_GT_torsion_work.ground.cfl_spring_down；
- J Coordinate Reference（参考坐标）：._FSAE_sus_front_GT_torsion_work.ground.cfl_chassis_base；

- Preload（弹簧预载荷）：500；
- Property File（属性文件）：mdids：//FASE/springs.tbl/test_my_25.spr；
- Spring Diameter（弹簧直径）：拖动滑块选择 30 mm；
- Spring of Coils（弹簧圈数）：拖动滑块选择 15。

（2）单击 OK 按钮，完成弹簧._FSAE_sus_front_GT_torsion_work.nsl_spring 的创建。

18．避震器

（1）单击 Build > Force > Damper > New 命令，弹出避震器创建对话框，如图 18-10 所示，在下列对话框中输入相应的数据：

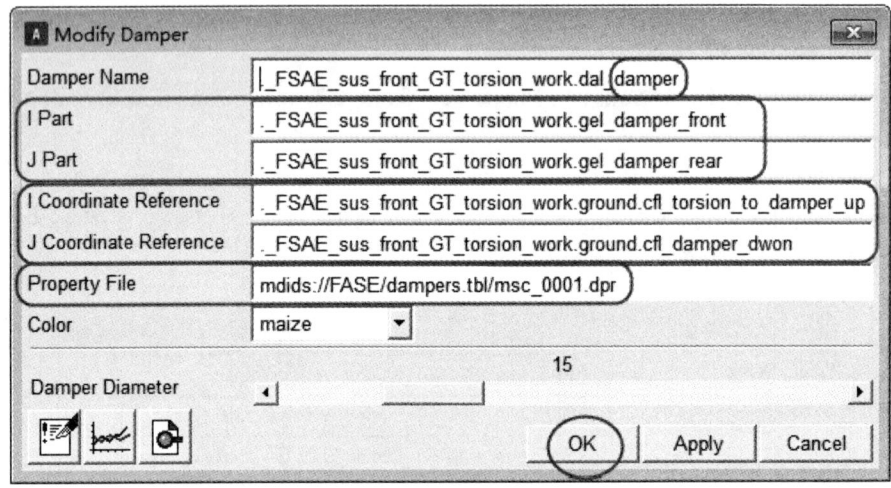

图 18-10　避震器

- Damper Name（减震器名称）：damper；
- I Part：._FSAE_sus_front_GT_torsion_work.gel_damper_front；
- J Part：._FSAE_sus_front_GT_torsion_work.gel_damper_rear；
- I Coordinate Reference（参考坐标）：._FSAE_sus_front_GT_torsion_work.ground.cfl_torsion_to_damper_up；
- J Coordinate Reference（参考坐标）：._FSAE_sus_front_GT_torsion_work.ground.cfl_damper_down；
- Property File（属性文件）：mdids：//FASE/dampers.tbl/msc_0001.dpr；
- Damper Diameter（避震器直径）：拖动滑块选择 15 mm；
- Color：maize。

（2）单击 OK 按钮，完成避震器._FSAE_sus_front_GT_torsion_work.dal_damper 的创建。

19．部件 lca 与 prod 之间的 spherical 约束

（1）单击 Build > Attachments > Joint > New 命令，弹出创建约束件对话框，如图 18-11 所示，在下列对话框中输入相应的数据：

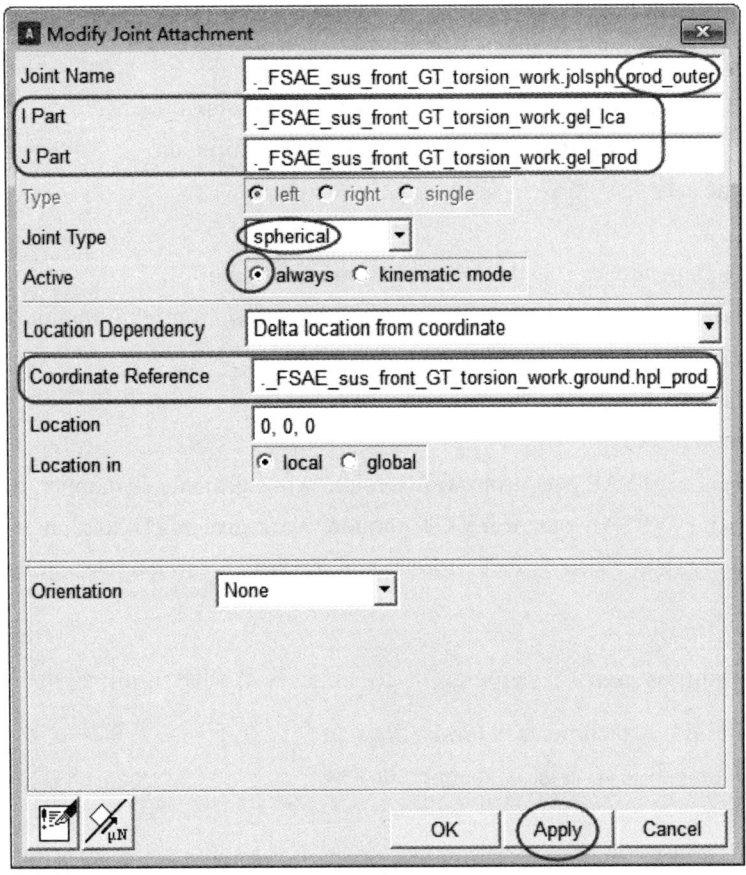

图 18-11 刚性约束-spherical

- Joint Name（约束副名称）：prod_outer；
- Type：left；
- I Part：._FSAE_sus_front_GT_torsion_work.gel_lca；
- J Part：._FSAE_sus_front_GT_torsion_work.gel_prod；
- Joint Type（约束副类型）：spherical；
- Active（激活）：always；
- Location Dependency：Delta location from coordinate；
- Coordinate Reference（参考坐标）：._FSAE_sus_front_GT_torsion_work.ground.hpl_prod_outer；
- Location：0，0，0；
- Location in：local。

（2）单击 Apply 按钮，完成约束副._FSAE_sus_front_GT_torsion_work.jolsph_prod_outer 的创建。

20. 部件 damper_front 与 zhijia_up 之间的 hook 约束

（1）单击 Build > Attachments > Joint > New 命令，在下列对话框中输入相应的数据：

- Joint Name（约束副名称）：damper_front；
- Type：left；
- I Part：._FSAE_sus_front_GT_torsion_work.gel_damper_front；
- J Part：._FSAE_sus_front_GT_torsion_work.gel_zhijia_up；
- Joint Type（约束副类型）：hook；
- Active（激活）：always；
- Location Dependency：Delta location from coordinate；
- Coordinate Reference（参考坐标）：._FSAE_sus_front_GT_torsion_work.ground.cfl_torsion_to_damper_up；
- Location：0，0，0；
- Location in：local；
- I-Part Axis：._FSAE_sus_front_GT_torsion_work.ground.cfl_damper_down；
- J-Part Axis：._FSAE_sus_front_GT_torsion_work.ground.cfl_torsion_to_damper。

（2）单击 Apply 按钮，完成约束副._FSAE_sus_front_GT_torsion_work.jolhoo_damper_front 的创建。

21. 部件 damper_rear 与 suspension_to_chassis 之间的 hook 约束

（1）单击 Build > Attachments > Joint > New 命令，在下列对话框中输入相应的数据：
- Joint Name（约束副名称）：damper_down；
- Type：left；
- I Part：._FSAE_sus_front_GT_torsion_work.gel_damper_rear；
- J Part：._FSAE_sus_front_GT_torsion_work.mts_suspension_to_chassis；
- Joint Type（约束副类型）：hook；
- Active（激活）：always；
- Location Dependency：Delta location from coordinate；
- Coordinate Reference（参考坐标）：._FSAE_sus_front_GT_torsion_work.ground.cfl_damper_down；
- Location：0，0，0；
- Location in：local；
- I-Part Axis：._FSAE_sus_front_GT_torsion_work.ground.cfl_torsion_to_damper_up；
- J-Part Axis：._FSAE_sus_front_GT_torsion_work.ground.cfl_damper_down_ref。

（2）单击 Apply 按钮，完成约束副._FSAE_sus_front_GT_torsion_work.jolhoo_damper_down 的创建。

22. 部件 damper_front 与 damper_rear 之间的 cylindrical 约束

（1）单击 Build > Attachments > Joint > New 命令，在下列对话框中输入相应的数据：
- Joint Name（约束副名称）：damper_mid；
- Type：left；
- I Part：._FSAE_sus_front_GT_torsion_work.gel_damper_front；

- J Part：._FSAE_sus_front_GT_torsion_work.gel_damper_rear；
- Joint Type（约束副类型）：cylindrical；
- Active（激活）：always；
- Location Dependency：Centered between coordinates；
- Centered between：Two Coordinates；
- Coordinate Reference #1（参考坐标）：._FSAE_sus_front_GT_torsion_work.ground.cfl_torsion_to_damper_up；
- Coordinate Reference #2（参考坐标）：._FSAE_sus_front_GT_torsion_work.ground.cfl_damper_dwon；
- Location Dependency：Orient axis to point；
- Coordinate Reference（参考坐标）：._FSAE_sus_front_GT_torsion_work.ground.cfl_damper_down。

（2）单击 Apply 按钮，完成约束副._FSAE_sus_front_GT_torsion_work.jolcyl_damper_mid 的创建。

23. 部件 zhijia_up 与 suspension_to_chassis 之间的 revolute 约束

（1）单击 Build > Attachments > Joint > New 命令，在下列对话框中输入相应的数据：
- Joint Name（约束副名称）：zhijia_up_to_prod；
- Type：left；
- I Part：._FSAE_sus_front_GT_torsion_work.gel_zhijia_up；
- J Part：._FSAE_sus_front_GT_torsion_work.mts_suspension_to_chassis；
- Joint Type（约束副类型）：revolute；
- Active（激活）：always；
- Location Dependency：Centered between coordinates；
- Centered between：Two Coordinates；
- Coordinate Reference #1（参考坐标）：._FSAE_sus_front_GT_torsion_work.ground.cfl_torsion_ref；
- Coordinate Reference #2（参考坐标）：._FSAE_sus_front_GT_torsion_work.ground.cfl_torsion_to_damper；
- Location Dependency：Orient axis to point；
- Coordinate Reference（参考坐标）：._FSAE_sus_front_GT_torsion_work.ground.cfl_torsion_to_damper；

（2）单击 Apply 按钮，完成约束副._FSAE_sus_front_GT_torsion_work.jolrev_zhijia_up_to_prod 的创建。

24. 部件 zhijia_up 与 torsion 之间的 fixed 约束

（1）单击 Build > Attachments > Joint > New 命令，在下列对话框中输入相应的数据：
- Joint Name（约束副名称）：zhijia_up_to_damper；
- Type：left；

- I Part：._FSAE_sus_front_GT_torsion_work.gel_zhijia_up；
- J Part：._FSAE_sus_front_GT_torsion_work.fbl_torsion；
- Joint Type（约束副类型）：fixed；
- Active（激活）：always；
- Location Dependency：Delta location from coordinate；
- Coordinate Reference（参考坐标）：._FSAE_sus_front_GT_torsion_work.ground.cfl_zhijia_up_to_torsion；
- Location：0，0，0；
- Location in：local；
- Closest Interface Node：left/4569，right/4569（4569 指有限元软件中创建的接口点）。

（2）单击 Apply 按钮，完成约束副 ._FSAE_sus_front_GT_torsion_work.jolfix_zhijia_up_to_damper 的创建。

25. 部件 zhijia_up 与 prod 之间的 hook 约束

（1）单击 Build > Attachments > Joint > New 命令，在下列对话框中输入相应的数据：
- Joint Name（约束副名称）：zhijia_to_prod；
- Type：left；
- I Part：._FSAE_sus_front_GT_torsion_work.gel_zhijia_up；
- J Part：._FSAE_sus_front_GT_torsion_work.gel_prod；
- Joint Type（约束副类型）：hook；
- Active（激活）：always；
- Location Dependency：Delta location from coordinate；
- Coordinate Reference（参考坐标）：._FSAE_sus_front_GT_torsion_work.ground.cfl_torsion_front；
- Location：0，0，0；
- Location in：local；
- I-Part Axis：._FSAE_sus_front_GT_torsion_work.ground.cfl_torsion_ref；
- J-Part Axis：._FSAE_sus_front_GT_torsion_work.ground.hpl_prod_outer。

（2）单击 Apply 按钮，完成约束副 ._FSAE_sus_front_GT_torsion_work.jolhoo_zhijia_to_prod 的创建。

26. 部件 zhijia_down 与 torsion 之间的 fixed 约束

（1）单击 Build > Attachments > Joint > New 命令，在下列对话框中输入相应的数据：
- Joint Name（约束副名称）：zhijia_down_to_torsion；
- Type：left；
- I Part：._FSAE_sus_front_GT_torsion_work.gel_zhijia_down；
- J Part：._FSAE_sus_front_GT_torsion_work.fbl_torsion；
- Joint Type（约束副类型）：fixed；
- Active（激活）：always；

- Location Dependency：Delta location from coordinate；
- Coordinate Reference（参考坐标）：._FSAE_sus_front_GT_torsion_work.ground.cfl_torsion_to_chassis；
- Location：0，0，0；
- Location in：local；
- Closest Interface Node：left/4567，right/4567（4567 指有限元软件中创建的接口点）。

（2）单击 Apply 按钮，完成约束副._FSAE_sus_front_GT_torsion_work.jolfix_zhijia_down_to_torsion 的创建。

27. 部件 zhijia_up 与 suspension_to_chassis 之间的 revolute 约束

（1）单击 Build > Attachments > Joint > New 命令，在下列对话框中输入相应的数据：
- Joint Name（约束副名称）：zhijia_down_to_chassic；
- Type：left；
- I Part：._FSAE_sus_front_GT_torsion_work.gel_zhijia_down；
- J Part：._FSAE_sus_front_GT_torsion_work.mts_suspension_to_chassis；
- Joint Type（约束副类型）：revolute；
- Active（激活）：always；
- Location Dependency：Delta location from coordinate；
- Coordinate Reference（参考坐标）：._FSAE_sus_front_GT_torsion_work.ground.cfl_torsion_to_chassis；
- Location：0，0，0；
- Location in：local；
- Location Dependency：Orient axis to point；
- Coordinate Reference（参考坐标）：._FSAE_sus_front_GT_torsion_work.ground.cfl_torsion_to_damper。

（2）单击 Apply 按钮，完成约束副._FSAE_sus_front_GT_torsion_work.jolrev_zhijia_down_to_chassic 的创建。

28. 部件 spring_base 与 zhijia_down 之间的 spherical 约束

（1）单击 Build > Attachments > Joint > New 命令，在下列对话框中输入相应的数据：
- Joint Name（约束副名称）：spring_to_zhijia_down；
- Type：left；
- I Part：._FSAE_sus_front_GT_torsion_work.gel_spring_base；
- J Part：._FSAE_sus_front_GT_torsion_work.gel_zhijia_down；
- Joint Type（约束副类型）：spherical；
- Active（激活）：always；
- Location Dependency：Delta location from coordinate；
- Coordinate Reference（参考坐标）：._FSAE_sus_front_GT_torsion_work.ground.cfl_torsion_to_chassis_up；

- Location：0，0，0；
- Location in：local。

（2）单击 Apply 按钮，完成约束副._FSAE_sus_front_GT_torsion_work.jolsph_spring_to_zhijia_down 的创建。

29. 部件 chassic_base 与 suspension_to_chassis 之间的 convel 约束

（1）单击 Build > Attachments > Joint > New 命令，在下列对话框中输入相应的数据：
- Joint Name（约束副名称）：chassis_base；
- I Part：._FSAE_sus_front_GT_torsion_work.gel_chassic_base；
- J Part：._FSAE_sus_front_GT_torsion_work.mts_suspension_to_chassis；
- Joint Type（约束副类型）：convel；
- Active（激活）：always；
- Location Dependency：Delta location from coordinate；
- Coordinate Reference（参考坐标）：._FSAE_sus_front_GT_torsion_work.ground.cfl_chassis_base；
- Location：0，0，0；
- Location in：local；
- I-Part Axis：._FSAE_sus_front_GT_torsion_work.ground.cfl_spring_down；
- J-Part Axis：._FSAE_sus_front_GT_torsion_work.ground.cfl_torsion_to_chassis_up。

（2）单击 Apply 按钮，完成约束副._FSAE_sus_front_GT_torsion_work.jolcon_chassis_base 的创建。

30. 部件 spring_base 与 chassic_base 之间的 translational 约束

（1）单击 Build > Attachments > Joint > New 命令，在下列对话框中输入相应的数据：
- Joint Name（约束副名称）：spring；
- Type：left；
- I Part：._FSAE_sus_front_GT_torsion_work.gel_spring_base；
- J Part：._FSAE_sus_front_GT_torsion_work.gel_chassic_base；
- Joint Type（约束副类型）：translational；
- Active（激活）：always；
- Location Dependency：Delta location from coordinate；
- Coordinate Reference（参考坐标）：._FSAE_sus_front_GT_torsion_work.ground.cfl_chassis_base；
- Location：0，0，0；
- Location in：local；
- Location Dependency：Orient axis to point；
- Coordinate Reference（参考坐标）：._FSAE_sus_front_GT_torsion_work.ground.cfl_spring_down。

（2）单击 OK 按钮，完成约束副._FSAE_sus_front_GT_torsion_work.joltra_spring 的创建。

31. 部件 zhijia_down 与 torsion 之间的 fixed 约束

（1）单击 Build > Attachments > Joint > New 命令，在下列对话框中输入相应的数据：
- Joint Name（约束副名称）：torsion_down；
- Type：left；
- I Part：._FSAE_sus_front_GT_torsion_work.fbl_torsion；
- J Part：._FSAE_sus_front_GT_torsion_work.mts_suspension_to_chassis；
- Joint Type（约束副类型）：fixed；
- Active（激活）：always；
- Location Dependency：Delta location from coordinate；
- Coordinate Reference（参考坐标）：._FSAE_sus_front_GT_torsion_work.ground.cfl_torsion_down_to_chassis；
- 四轮 Location：0，0，0；
- Location in：local；
- Closest Interface Node：left/4566，right/4566（4566 指有限元软件中创建的接口点）。

（2）单击 OK 按钮，完成约束副._FSAE_sus_front_GT_torsion_work.jolfix_torsion_down 的创建。

18.2　四轮定位参数对标

空间斜置扭杆弹簧推杆式悬架模型如图 18-1 所示，建成的模型工作模式为扭杆弹簧工作模式；将约束._FSAE_sus_front_GT_torsion_work.jolfix_torsion_down 更改为旋转约束，此时模型转化为螺旋弹簧工作模式，实际车辆是通过液压作动器根据车速的特性自动切换的。图 18-12～图 18-14 为悬架四轮定位参数变化曲线，图中 p1 为螺旋弹簧工作模式，p2 为扭杆弹簧工作模式。

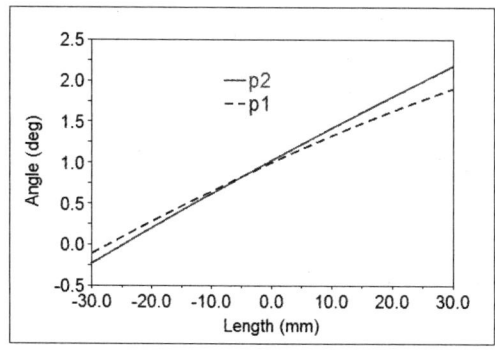

图 18-12　前束角

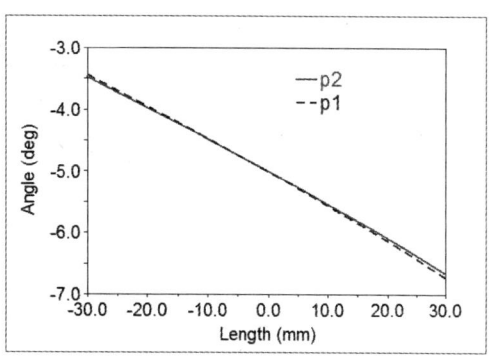

图 18-13　主销内倾角与车辆外倾角

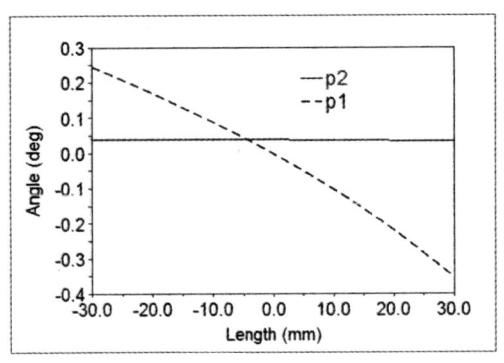

图 18-14　主销后倾角

18.3　刚度阻尼匹配

空间斜置扭杆弹簧推杆式悬架模型建立好之后，需要对弹簧的刚度与整车进行匹配，悬架传力模型如图 18-15 所示，推杆空间位置受力如图 18-16 所示。

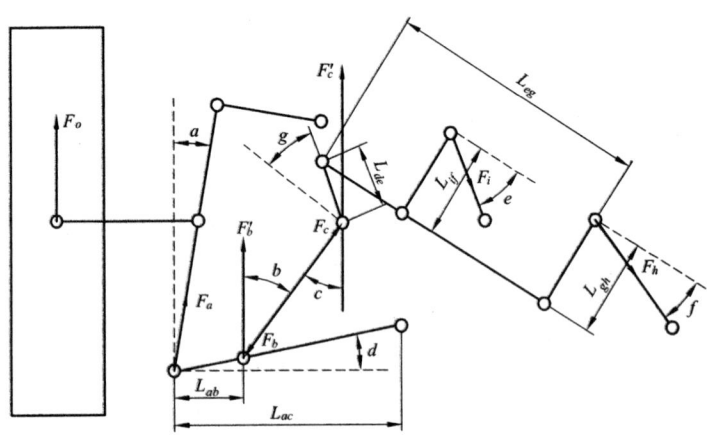

图 18-15　悬架传力模型

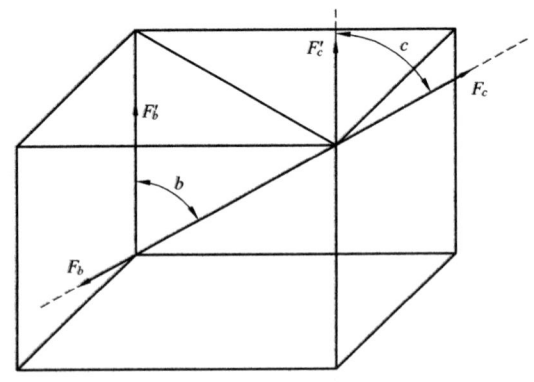

图 18-16　扭杆弹簧空间受力

根据受力分析，悬架传力数学描述如公式（18-1）~（18-13），整理公式后弹簧刚度公式为（18-14），阻尼系数公式为（18-15）：

$$F_o = mg \tag{18-1}$$

$$F_a \cos a = F_o \tag{18-2}$$

$$-F_{b'}(L_{ac} - L_{ab})\cos d = F_a \cos a L_{ac} \cos d \tag{18-3}$$

$$-F_b \cos b = F_{b'} \tag{18-4}$$

$$F_c = -F_b \tag{18-5}$$

$$F_c \cos c = F_{c'} \tag{18-6}$$

$$b = c \tag{18-7}$$

$$F_{c'} L_{de} \cos g = F_i L_{if} \cos e + F_h L_{gh} \cos f \tag{18-8}$$

$$F_i = kx \tag{18-9}$$

$$F_{o'} = m\ddot{z} \tag{18-10}$$

$$m_1 L_{de}^2 \phi = F_{c'} L_{de} \cos g \tag{18-11}$$

$$v_1 = \int L_{de} \phi \mathrm{d}t \tag{18-12}$$

$$\frac{v_1}{v} = \frac{L_{de}}{L_{if}} \tag{18-13}$$

$$K = -\frac{mgL_{ac}L_{de}\cos g}{xL_{if}(L_{ac} - L_{ab})\cos e} \tag{18-14}$$

$$\delta = m_1 L_{de} \frac{m\ddot{z}L_{ac}L_{de}\cos c \cos g + F_i L_{if} \cos b \cos e(L_{ac} - L_{ab})}{m\dot{z}L_{ac}L_{gh}L_{if}\cos c \cos f \cos g} \tag{18-15}$$

式中：F_o 为地面对轮胎的支撑力；m 为簧上质量；g 为重力加速度；F_a 为下控制臂外点所受的拉力；F_b 为下控制臂与推杆连接处所承受的压力；F_c 为推杆与上部支架摆臂连接处的推力；$F_{b'}$、$F_{c'}$ 为推杆在空间的垂向分力；F_i 为弹簧刚度；F_h 为弹簧受力；k 为避震器阻尼力；x 为弹簧的安装压缩长度；a 为转向节与垂向线之间的夹角；b、c 为推杆与垂向线之间的夹角；d 为下控制臂与水平线之间的夹角；e 为避震器与水平面之间的夹角；f 为螺旋弹簧与水平面之间的夹角；g 为上部支架摆臂和推杆连接处与水平面之间的夹角；L_{ac} 为下控制臂在水平面投影的长度；L_{ab} 为下控制臂到推杆与下控制臂连接处在水平面上的投影长度；L_{de} 为上部支架摆臂与推杆连接处之间的长度；L_{if} 为避震器摆臂长度；L_{gh} 为弹簧摆臂长度；$F_{o'}$ 为簧上质量在加速度作用下的惯性力；$\ddot{z}$ 为车身在垂向方向的加速度；$\dot{z}$ 为车身在垂向方向的速度；m_1 为扭杆弹簧的质量；ϕ 为扭杆弹簧转动的角加速度；v_1 为上部支架摆臂与推杆连接处的切向速度；v 为避震器于上部支架摆臂连接处的切向速度。

通过测量获取悬架参数，如表 18-1 所示，根据公式（18-14）计算出：前悬架弹簧力为 -984.72 N，后悬架弹簧力为 -1 086.34 N，负号表示弹簧受拉，实际物理模型为受压；此时可以根据弹簧力来确定刚度及安装长度，此处需要强调的是，方程式赛车要求弹簧的安装长度（或弹簧行程）较短，因此弹簧刚度相对于轿车要大很多；此处不具体确定，读者可根据设计的车辆确定刚度与长度。根据公式（18-15），用 MATLAB 计算出前后悬架的阻尼参数，如图 18-17 和图 18-18 所示。

表 18-1 悬架参数

参数	前悬架	后悬架	单位符号
L_{ac}	0.419 1	0.355 6	m
L_{ab}	0.046 1	0.002 6	m
L_{de}	0.070	0.080	m
L_{gh}	0.100	0.100	m
L_{if}	0.050	0.050	m
a	15.1	0	°
b	56.56	50.08	°
e	0	0	°
g	0	0	°
f	0	0	°
m	62.6	67.4	kg
m_1	0.612 7	0.612 7	kg

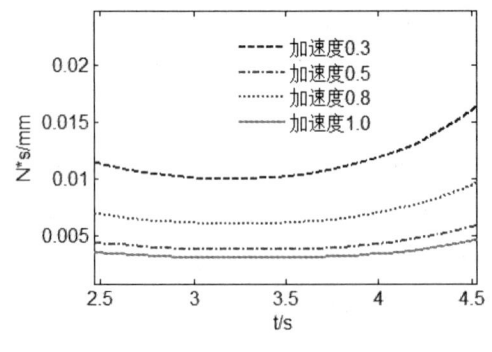

图 18-17 前悬架阻尼匹配

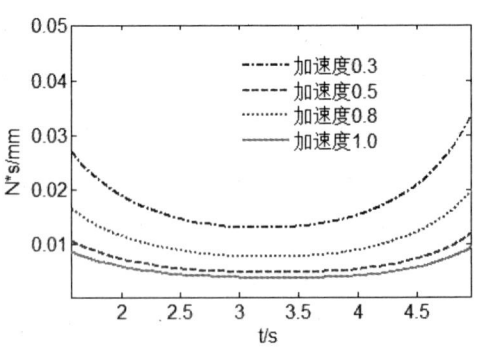

图 18-18 后悬架阻尼匹配

18.4 后扭杆弹簧斜置式推杆悬架

后空间扭杆弹簧斜置式推杆悬架模型与前空间扭杆弹簧斜置式推杆悬架模型框架完全一样，不同之处在于参数的定位点不同，其建模过程一致，因此建模不再详细展开，请读者参阅前空间扭杆弹簧斜置式推杆悬架模型建模过程，其建立好的后悬架模型如图 18-19 所示。后悬架的硬点、变量参数信息如下：

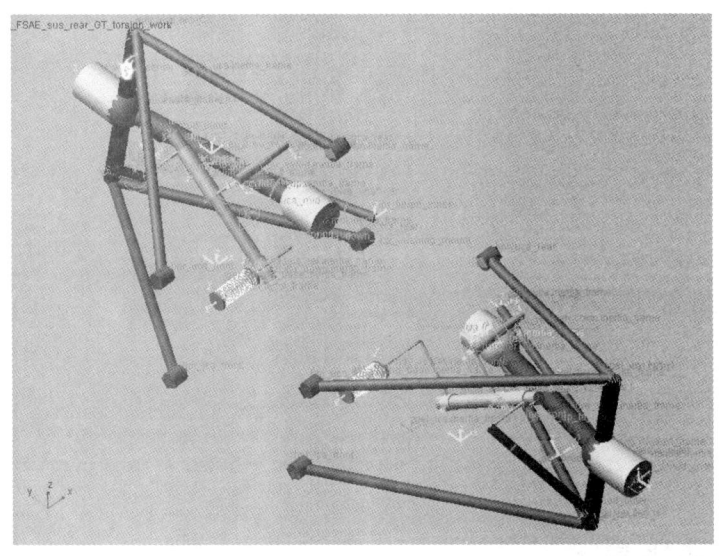

图 18-19 后空间扭杆弹簧斜置式推杆悬架

```
File Name      : <FASE>/subsystems.tbl/FSAE_sus_rear_GT_torsion_work.sub
Template       : mdids://FASE/templates.tbl/_FSAE_sus_rear_GT_torsion_work.tpl
Comments       : *no comments found*
Major Role     : suspension
Minor Role     : rear


HARDPOINTS:
hardpoint name          symmetry         x_value       y_value       z_value
--------------          --------         -------       -------       -------

global                  single           1524.0          0.0           0.0
arb_bushing_mount       left/right       1651.0        -127.0         101.6
drive_shaft_inr         left/right       1600.0        -200.0         185.0
lca_front               left/right       1270.0        -127.0         127.0
lca_outer               left/right       1498.6        -482.6         101.6
lca_rear                left/right       1651.0        -127.0         127.0
prod_outer              left/right       1498.6        -480.0         127.0
```

prod_to_bellcrank	left/right	1510.0	-350.0	250.0
tierod_inner	left/right	1676.4	-127.0	152.4
tierod_outer	left/right	1574.8	-457.2	152.4
uca_front	left/right	1270.0	-152.4	304.8
uca_outer	left/right	1549.4	-482.6	355.6
uca_rear	left/right	1625.6	-152.4	304.8
wheel_center	left/right	1524.0	-558.8	228.6

PARAMETERS:

parameter name	symmetry	type	value
driveline_active	single	integer	1
kinematic_flag	single	integer	0
camber_angle	left/right	real	-1.5
drive_shaft_offset	left/right	real	75.0
toe_angle	left/right	real	0.0

18.5 加速仿真

整车模型通过替换前后悬架模型获取，请读者参阅前述章节；建立好的整车模型如图 18-20 所示。

图 18-20　整车模型_FSAE_2020_GT_spring_work_preload（弹簧工作模式）

1. 加速仿真（闭环模式）

（1）单击 Simulate > Full-Vehicle Analysis > Straight-Line Events > Acceleration 命令，弹出阶跃仿真对话框，如图 18-21 所示，在下列对话框中输入相应的数据：

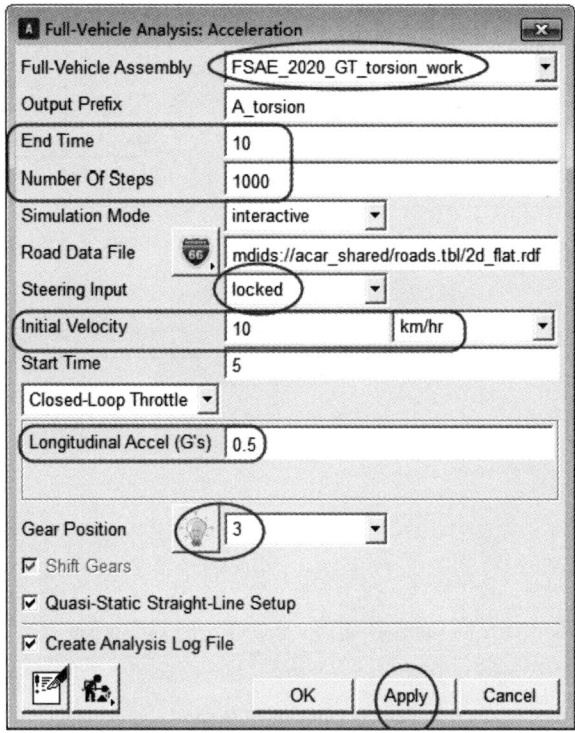

图 18-21　加速仿真设置（闭环）

- Full-Vehicle Assembly：FSAE_2020_GT_torsion_work；
- Output Prefix：A_torsion；
- End Time：10；
- Number of steps：1000；
- Mode of Simulation：interactive；
- Road Date File：mdids：//acar_shared/roads.tbl/2d_flat.rdf；
- Steering Input：locked，仿真过程中方向盘锁定；
- Initial Velocity（单位：km/h）：10；
- Start Time：5；
- Closed-Loop Throttle：工作模式选择闭环工作模式；
- Gear Position：3；
- 勾选 Quasi-Static Straight-Line Setup。

（2）单击 Apply 按钮，完成 FSAE_2020_GT_torsion_work 赛车加速仿真设置并提交运算；运算完成后，仿真参数的方法不变，按同样的方法完成整车 FSAE_2020_GT_spring_work_preload 仿真；整车参数如图 18-22 ~ 图 18-24 所示，从图中可以看出，采用扭杆弹簧工作模式时参数变化比较稳定。

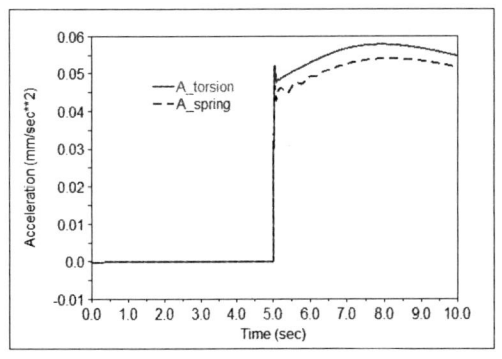
图 18-22 纵向加速度

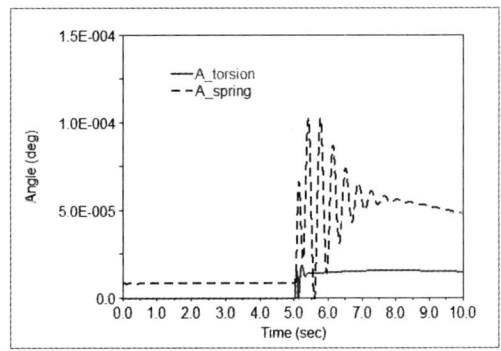
图 18-23 俯仰角位移

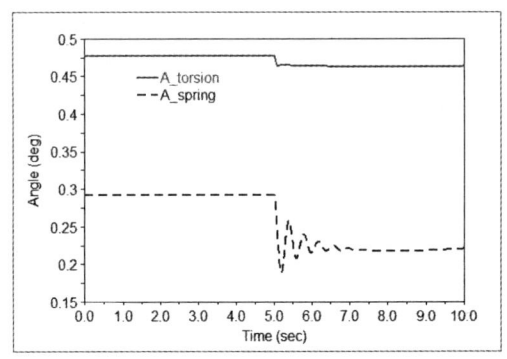
图 18-24 侧倾角位移

2. 加速仿真（开环模式）

（1）单击 Simulate > Full-Vehicle Analysis > Straight-Line Events > Acceleration 命令，弹出阶跃仿真对话框，如图 18-15 所示，在下列对话框中输入相应的数据。

- Full-Vehicle Assembly：FSAE_2020_GT_spring_work_preload；
- Output Prefix：A_spring；
- End Time：10；
- Number of steps：1000；
- Mode of Simulation：interactive；
- Road Date File：mdids：//acar_shared/roads.tbl/2d_flat.rdf；
- Steering Input：locked，仿真过程中方向盘锁定；
- Initial Velocity（单位：km/h）：10；
- Start Time：5；
- Open-Loop Throttle：工作模式选择闭环工作模式；
- Final Throttle：100，最终油门开度；
- Duration of Step：1，油门开度持续时间；
- Gear Position：3；

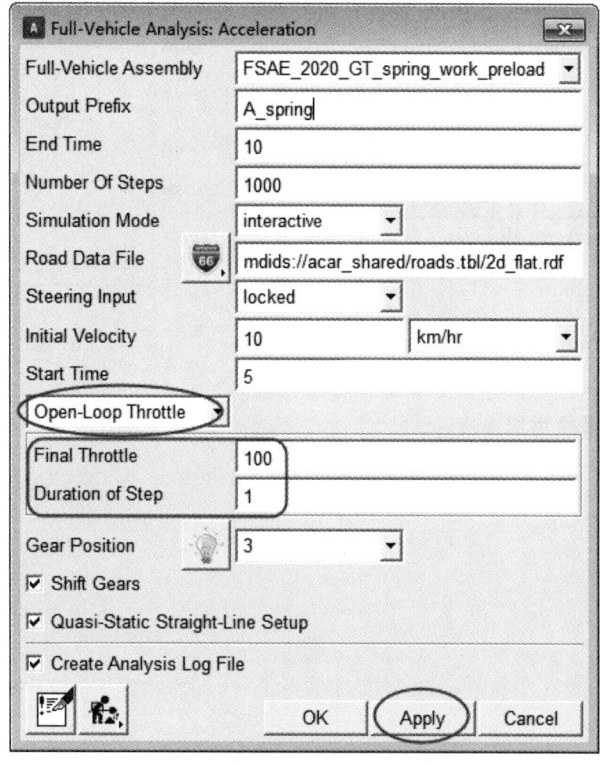

图 18-25 加速仿真设置（开环）

- 勾选 Shift Gears，仿真过程中可以根据整车运行情况自动换挡；
- 勾选 Quasi-Static Straight-Line Setup。

（2）单击 Apply 按钮，完成 FSAE_2020_GT_spring_work_preload 赛车加速仿真设置并提交运算；运算完成后，仿真参数设置不变，按同样的方法完成整车 FSAE_2020_GT_torsion_work 仿真；整车参数如图 18-26～图 18-29 所示，从图中可以看出，采用扭杆弹簧工作模式相对于螺旋弹簧模式稳定性得到极大提升；同时对于平顺性指标，螺旋弹簧工作模式优于扭杆弹簧工作模式，因此 FSAE 赛车底盘是解耦的，即平顺性指标与稳定性指标互不干涉。

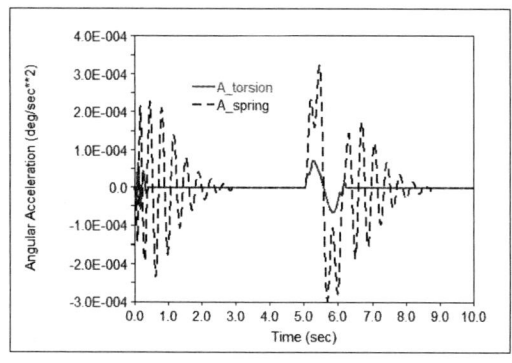

图 18-26 俯仰角加速度

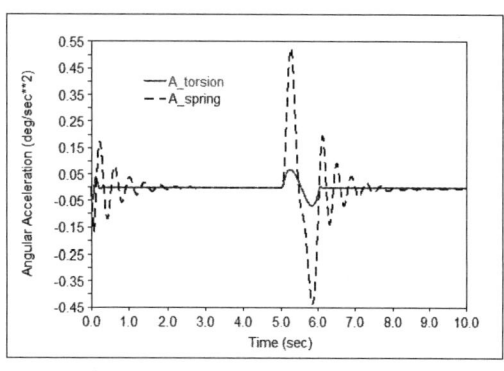

图 18-27 侧倾角加速度

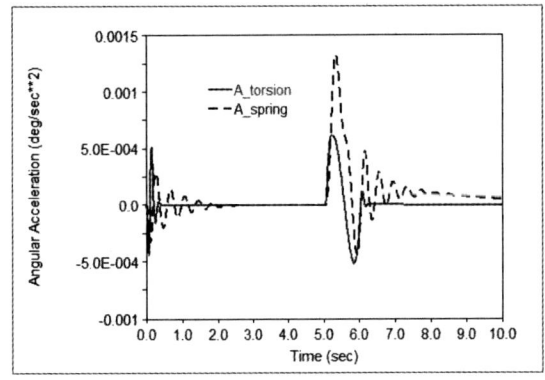 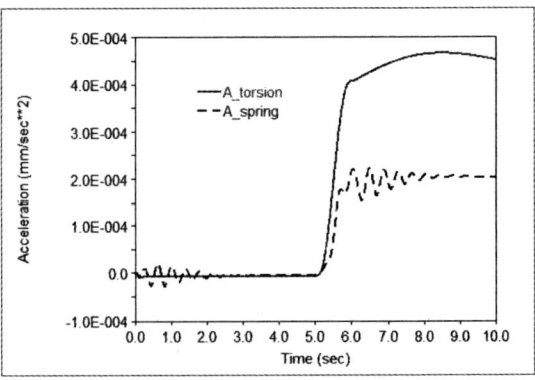

图 18-28　横摆角加速度　　　　　　　图 18-29　垂向加速度

第 19 章 上横置板簧悬架模型

上横置板簧悬架模型如图 19-1 所示，悬架左右上控制臂用单根横置板簧替代，板簧中心段部分可以通过不同的连接孔与车身连接，不同孔连接对应横置板簧的不同刚度，即横置板簧的刚度可调节。横置板簧既可以起到横向拉杆的作用，也可以起到不同弹簧刚度的作用，同时还能兼顾横向稳定杆的特性。悬架的性能在于与整车的匹配及调校，而非采用技术含量高的悬架性能。避震器有多种布置方式，可以通过推杆（自由度大）布置到较远位置，此处布置在转向主销的前锋，底部与下控制臂支架连接，上部与车身连接。本章直接在模型_FSAE_sus_front_white.tpl 上添加其他部件完成模型的建立，完成后的前上横置板簧悬架模型：_FSAE_sus_front_leaf_up.tpl 存放在章节文件夹中，读者请自行查阅。

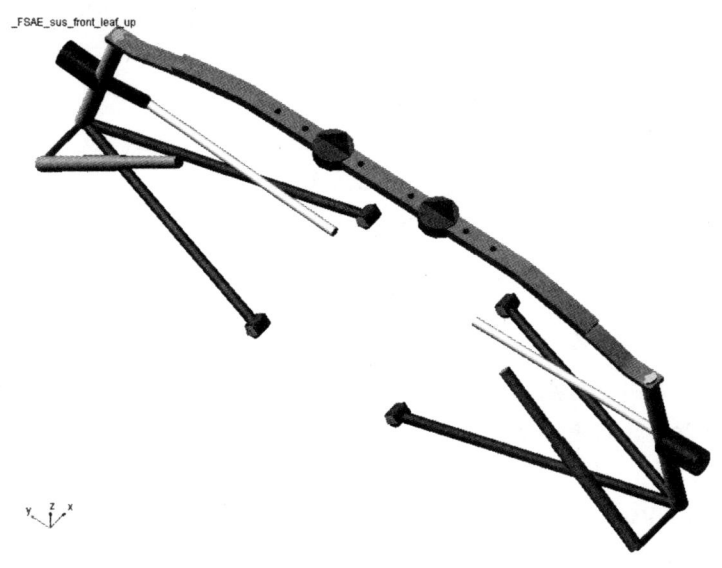

图 19-1　前上横置板簧悬架模型

学习目标

（1）了解前上横置板簧悬架。
（2）了解后上横置板簧悬架。
（3）会文件驱动仿真。

19.1 前上横置板簧悬架

1. 模型导入

（1）启动 ADAMS/CAR，选择专家模块进入建模界面。

（2）单击 File > Open 命令，弹出打开模板对话框，如图 19-2 所示，在 Template Name 中输入：mdids：//FASE/templates.tbl/_FSAE_sus_front_white.tpl。

（3）单击 OK 按钮，完成模型导入。

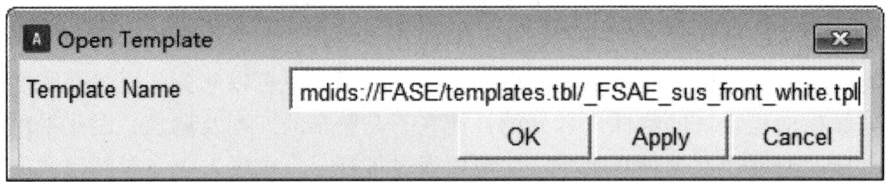

图 19-2　模板对话框

2. 删除上控制臂几何体

删除对应的上控制臂前后拉杆，但要注意不能删除上控制部件，删除后会导致模型错误，原因在于悬架的数组参数（即转向主销）由上下控制臂部件确定。

删除悬架以上控制臂前后拉杆几何体：

① ._FSAE_sus_front_white.gel_lca.gralin_uca_link_front；

② ._FSAE_sus_front_white.ger_lca.gralin_uca_link_front；

③ ._FSAE_sus_front_white.gel_lca.gralin_uca_link_rear；

④ ._FSAE_sus_front_white.ger_lca.gralin_uca_link_rear。

3. 修改部件位置

上控制臂部件位置由三个点确定，删除前后拉杆后，需要重新确定上控制臂的部件位置；将上控制臂部件放置在转向主销下位置点。

修改信息如下：

（1）单击 Build > Part > General Part > Modify 命令，在下列对话框中输入相应的数据：

- Location Dependency：Delta location from coordinate；
- Coordinate Reference（参考坐标）：._FSAE_sus_front_white.ground.hpl_uca_outer；
- Location：0，0，0。

（2）其余信息保持默认，单击 OK 按钮，完成下控制臂部件的修改。

（3）单击 File > Save As 命令，弹出保存模板对话框，如图 19-3 所示，在下列对话框中输入相应的数据：

- Template Name：_FSAE_sus_front_white；
- New Template Name：FSAE_sus_front_leaf_up；

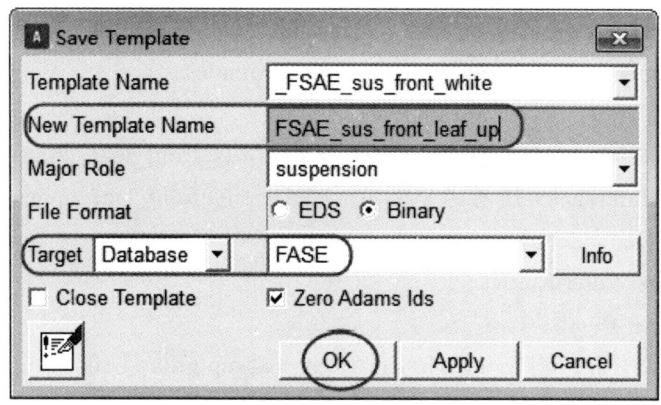

图 9-3　模型保存

- Major Role（主特征）：suspension；
- File Format：Binary；
- Target：Datebase/FASE。

（4）单击 OK 按钮，完成推杆式悬架模型模板 FSAE_sus_front_leaf_up 的保存。

4．结构框

（1）单击 Build > Construction Frame > New 命令，弹出创建结构框对话框，如图 19-4 所示，在下列对话框中输入相应的数据：

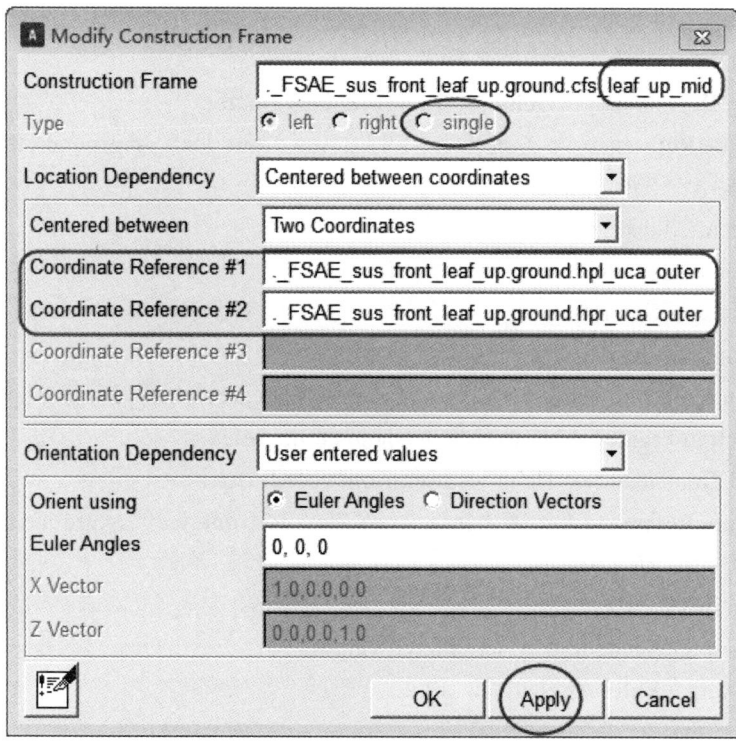

图 19-4　结构框 leaf_up_mid

- Construction Frame（结构框名称）：leaf_up_mid；
- Location Dependency：Centered between coordinates；
- Centered between：Two Coordinates；
- Coordinate Reference #1（参考坐标）：._FSAE_sus_front_leaf_up.ground.hpl_uca_outer；
- Coordinate Reference #2（参考坐标）：._FSAE_sus_front_leaf_up.ground.hpr_uca_outer；
- Orientation Dependency：User entered values；
- Orient using：Euler Angles；
- Euler Angles：0，0，0。

（2）单击 Apply 按钮，完成._FSAE_sus_front_leaf_up.ground.cfs_leaf_up_mid 结构框的创建。同理，创建其他结构框。

- Construction Frame（结构框名称）：damper_down；
- Location Dependency：Delta location from coordinate；
- Coordinate Reference（参考坐标）：._FSAE_sus_front_leaf_up.ground.hpl_lca_outer；
- Location：-100，0，0；
- Location in：local；
- Orientation Dependency：User entered values；
- Orient using：Euler Angles；
- Euler Angles：0，0，0。

（3）单击 Apply 按钮，完成._FSAE_sus_front_leaf_up.ground.cfl_damper_down 结构框的创建。

- Construction Frame（结构框名称）：damper_up；
- Location Dependency：Delta location from coordinate；
- Coordinate Reference（参考坐标）：._FSAE_sus_front_leaf_up.ground.cfl_damper_down；
- Location：0，250，200；
- Location in：local；
- Orientation Dependency：User entered values；
- Orient using：Euler Angles；
- Euler Angles：0，0，0。

（4）单击 Apply 按钮，完成._FSAE_sus_front_leaf_up.ground.cfl_damper_up 结构框的创建。

- Construction Frame（结构框名称）：damper_up_ref；
- Location Dependency：Delta location from coordinate；
- Coordinate Reference（参考坐标）：._FSAE_sus_front_leaf_up.ground.cfl_damper_up；
- Location：0，40，0；
- Location in：local；
- Orientation Dependency：User entered values；
- Orient using：Euler Angles；
- Euler Angles：0，0，0。

（5）单击 Apply 按钮，完成._FSAE_sus_front_leaf_up.ground.cfl_damper_up_ref 结构框的创建。

5. 部件 damper_up

（1）单击 Build > Part > General Part > New 命令，弹出创建部件对话框，如图 19-5 所示，在下列对话框中输入相应的数据：

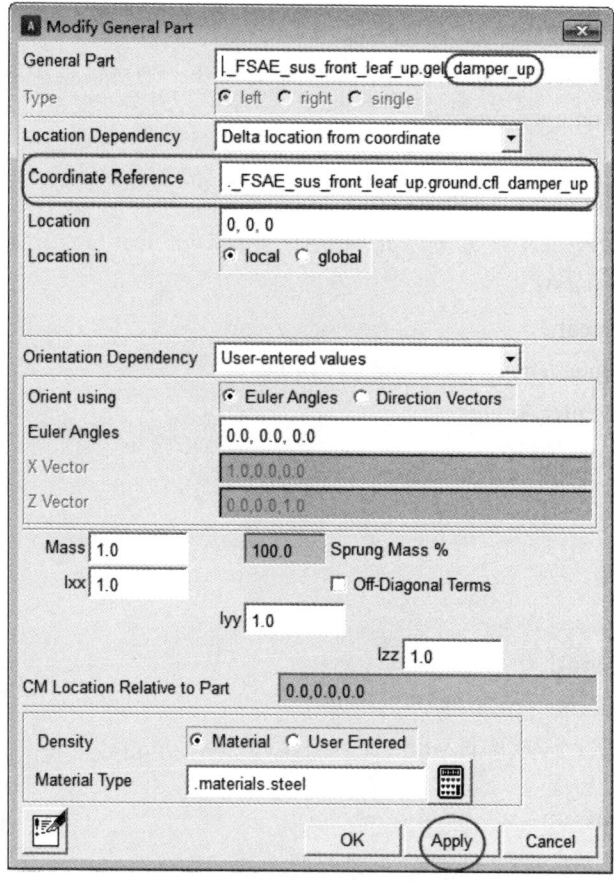

图 19-5　部件 damper_up

- General Part：damper_up；
- Type：left；
- Location Dependency：Delta location from coordinate；
- Coordinate Reference（参考坐标）：._FSAE_sus_front_leaf_up.ground.cfl_damper_up；
- Location：0，0，0；
- Location in：local；
- Orientation Dependency：User entered values；
- Orient using：Euler Angles；
- Euler Angles：0，0，0；
- Mass：1；
- Ixx：1；
- Iyy：1；
- Izz：1；

- Density：Material；
- Material Type：.materials.steel。

（2）单击 OK 按钮，完成部件._FSAE_sus_front_leaf_up.gel_damper_up 的创建。

6. 部件 damper_down

（1）单击 Build > Part > General Part > New 命令，在下列对话框中输入相应的数据：

- General Part：damper_down；
- Type：left；
- Location Dependency：Delta location from coordinate；
- Coordinate Reference（参考坐标）：._FSAE_sus_front_leaf_up.ground.cfl_damper_down；
- Location：0，0，0；
- Location in：local；
- Orientation Dependency：User entered values；
- Orient using：Euler Angles；
- Euler Angles：0，0，0；
- Mass：1；
- Ixx：1；
- Iyy：1；
- Izz：1；
- Density：Material；
- Material Type：.materials.steel。

（2）单击 OK 按钮，完成部件._FSAE_sus_front_leaf_up.gel_damper_down 的创建。

7. 板簧 fsae_leaf_up

板簧 MNF 中性文件：fsae_leaf_up.mnf 存放在章节文件中，请读者自行调阅查看。

（1）单击 Build > Part > Flexible Body > New 命令，弹出创建部件对话框，如图 19-6 所示。在下列对话框中输入相应的数据：

- Flexible Body Name：fsae_leaf_up；
- Type：single；
- Location Dependency：Delta location from coordinate；
- Coordinate Reference（参考坐标）：._FSAE_sus_front_leaf_up.ground.cfs_leaf_up_mid；
- Location：-15，0，0；
- Orientation Dependency：User entered values；
- Orient using：Euler Angles；
- Euler Angles：90，90，0；
- MNF File：file：//D：/ADAMS_MNF/fsae_leaf_up.mnf；
- Color：green。

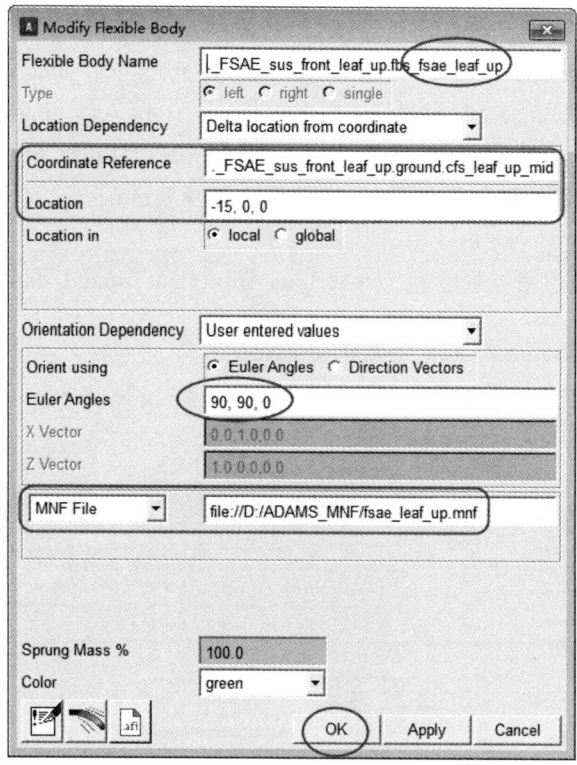

图 19-6　板簧 fsae_leaf_up

（2）单击 OK 按钮，完成部件._FSAE_sus_front_leaf_up.fbs_fsae_leaf_up 的创建。

8．避震器

（1）单击 Build > Force > Damper > New 命令，弹出避震器创建对话框，如图 19-7 所示，在下列对话框中输入相应的数据：

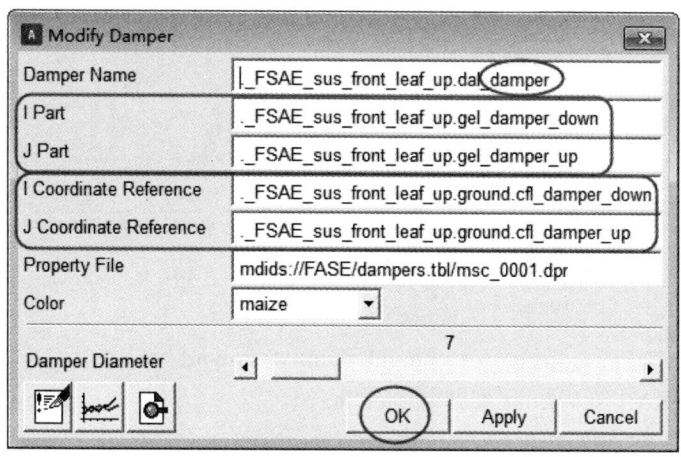

图 19-7　避震器

- Damper Name（减震器名称）：damper；
- I Part：._FSAE_sus_front_leaf_up.gel_damper_down；

- J Part：._FSAE_sus_front_leaf_up.gel_damper_up；
- I Coordinate Reference（参考坐标）：._FSAE_sus_front_leaf_up.ground.cfl_damper_down；
- J Coordinate Reference（参考坐标）：._FSAE_sus_front_leaf_up.ground.cfl_damper_up；
- Property File（属性文件）：mdids：//FASE/dampers.tbl/msc_0001.dpr；
- Damper Diameter（避震器直径）：拖动滑块选择 7 mm；
- Color：maize。

（2）单击 OK 按钮，完成避震器._FSAE_sus_front_leaf_up.dal_damper 的创建。

9. 部件 damper_down 与 lca 之间的 convel 约束

（1）单击 Build > Attachments > Joint > New 命令，弹出创建约束件对话框，如图 19-8 所示，在下列对话框中输入相应的数据：

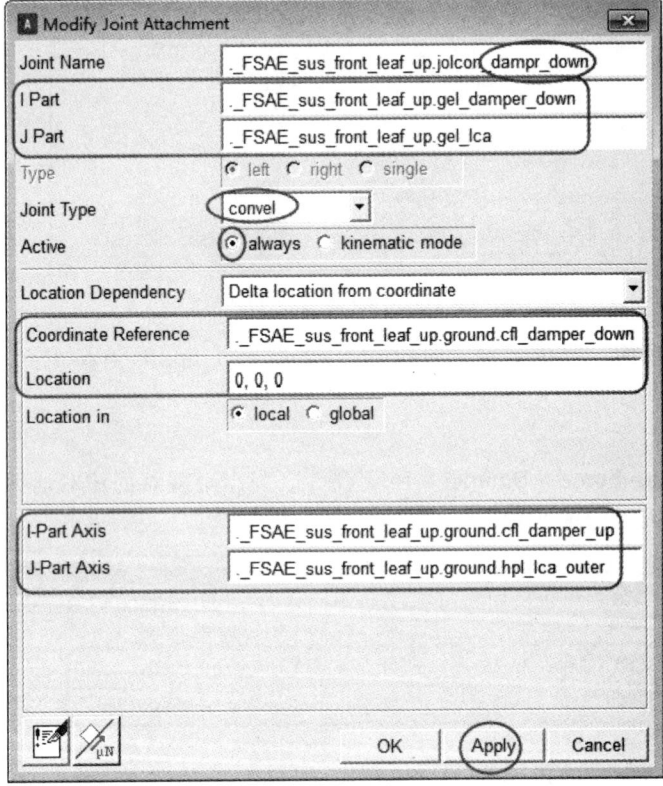

图 19-8　约束 convel

- Joint Name（约束副名称）：dampr_down；
- I Part：._FSAE_sus_front_leaf_up.gel_damper_down；
- J Part：._FSAE_sus_front_leaf_up.gel_lca；
- Joint Type（约束副类型）：convel；
- Active（激活）：always；
- Location Dependency：Delta location from coordinate；
- Coordinate Reference（参考坐标）：._FSAE_sus_front_leaf_up.ground.cfl_damper_down；

- Location：0，0，0；
- Location in：local；
- I-Part Axis：._FSAE_sus_front_leaf_up.ground.cfl_damper_up；
- J-Part Axis：._FSAE_sus_front_leaf_up.ground.hpl_lca_outer。

（2）单击 Apply 按钮，完成约束副._FSAE_sus_front_leaf_up.jolcon_dampr_down 的创建。

10. 部件 damper_up 与 suspension_to_chassis 之间的 convel 约束

（1）单击 Build > Attachments > Joint > New 命令，在下列对话框中输入相应的数据：
- Joint Name（约束副名称）：damper_up；
- I Part：._FSAE_sus_front_leaf_up.gel_damper_up；
- J Part：._FSAE_sus_front_leaf_up.mts_suspension_to_chassis；
- Joint Type（约束副类型）：convel；
- Active（激活）：always；
- Location Dependency：Delta location from coordinate；
- Coordinate Reference（参考坐标）：._FSAE_sus_front_leaf_up.ground.cfl_damper_up；
- Location：0，0，0；
- Location in：local；
- I-Part Axis：._FSAE_sus_front_leaf_up.ground.cfl_damper_down；
- J-Part Axis：._FSAE_sus_front_leaf_up.ground.cfl_damper_up_ref。

（2）单击 Apply 按钮，完成约束副._FSAE_sus_front_leaf_up.jolcon_damper_up 的创建。

11. 部件 damper_up 与 damper_down 之间的 cylindrical 约束

（1）单击 Build > Attachments > Joint > New 命令，在下列对话框中输入相应的数据：
- Joint Name（约束副名称）：damp_mid；
- Type：left；
- I Part：._FSAE_sus_front_leaf_up.gel_damper_down；
- J Part：._FSAE_sus_front_leaf_up.gel_damper_up；
- Joint Type（约束副类型）：cylindrical；
- Active（激活）：always；
- Location Dependency：Centered between coordinates；
- Centered between：Two Coordinates；
- Coordinate Reference #1（参考坐标）：._FSAE_sus_front_leaf_up.ground.cfl_damper_down；
- Coordinate Reference #2（参考坐标）：._FSAE_sus_front_leaf_up.ground.cfl_damper_up；
- Location Dependency：Orient axis to point；
- Coordinate Reference（参考坐标）：._FSAE_sus_front_leaf_up.ground.cfl_damper_up。

（2）单击 Apply 按钮，完成约束副._FSAE_sus_front_leaf_up.jolcyl_damp_mid 的创建。

12. 部件 leaf_front 与 lca 之间的 fixed 约束

（1）单击 Build > Attachments > Joint > New 命令，在下列对话框中输入相应的数据：
- Joint Name（约束副名称）：leaf_to_uca；
- Type：left；
- I Part：._FSAE_sus_front_leaf_up.fbs_fsae_leaf_up；
- J Part：._FSAE_sus_front_leaf_up.gel_uca；
- Joint Type（约束副类型）：fixed；
- Active（激活）：always；
- Location Dependency：Delta location from coordinate；
- Coordinate Reference（参考坐标）：._FSAE_sus_front_leaf_up.ground.hpl_uca_outer；
- Location：0，0，0；
- Location in：local；
- Closest Interface Node：left/8776，right/8776（8776 指有限元软件中创建的接口点）。

（2）单击 OK 按钮，完成约束副._FSAE_sus_front_leaf_up.jolfix_leaf_to_uca 的创建。

13. 衬套连接硬点

（1）单击 Build > Hardpoind > New 命令，在下列对话框中输入相应的数据：
- Hardpoint：leaf_mount；
- Type：left；
- Location：0.0，-95.4163132，407.8916489，此参数通过选取有限元节点获取。

（2）单击 OK 按钮，完成._FSAE_sus_front_leaf_up.ground.hpl_leaf_mount 硬点的创建。

14. 部件 leaf_up_mount 与 suspension_to_chassis 之间的 bushing 约束

（1）单击 **Build > Attachments > Bushing > New** 命令，弹出创建衬套对话框，如图 19-9 所示，在下列对话框中输入相应的数据：
- Bushing Name（约束副名称）：leaf_up_mount；
- I Part：._FSAE_sus_front_leaf_up.fbs_fsae_leaf_up；
- J Part：._FSAE_sus_front_leaf_up.mts_suspension_to_chassis；
- Inactive（抑制）：never；
- Preload：0，0，0；
- Tpreload：0，0，0；
- Offset：0，0，0；
- Roffset：0，0，0；
- Geometry Length：20；
- Geometry Radius：30；
- Property File：mdids://FASE/bushings.tbl/fsae_control_arm_bushing.bus；
- Location Dependency：Delta location from coordinate；

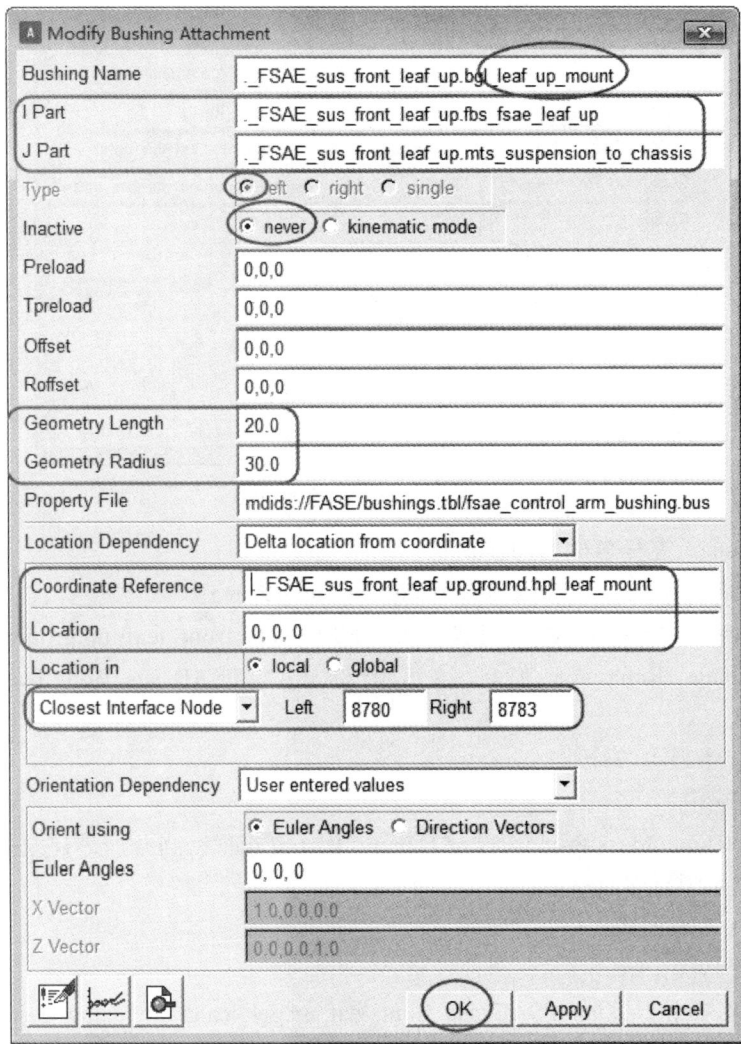

图 19-9 衬套 leaf_up_mount

- Coordinate Reference（参考坐标）：._FSAE_sus_front_leaf_up.ground.hpl_leaf_mount；
- Location：0，0，0；
- Location in：local；
- Closest Interface Node：Left：8780 / Right：8780；
- Orientation Dependency：User entered values；
- Orient using：Euler Angles；
- Euler Angles：0，0，0。

（2）单击 OK 按钮，完成轴套._FSAE_sus_front_leaf_up.bgl_leaf_up_mount 的创建。

15. 几何体 damp_base

（1）单击 Build > Geometry > Link > New 命令，弹出创建几何体对话框，如图 19-10 所示，在下列对话框中输入相应的数据：

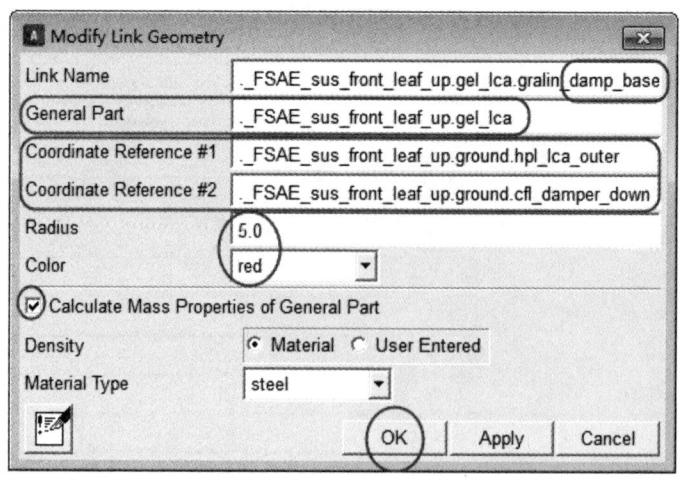

图 19-10 连杆几何体_link

- Link Name（连杆名称）：damp_base；
- General Part 输入：._FSAE_sus_front_leaf_up.gel_lca；
- Coordinate Reference #1（参考坐标）：._FSAE_sus_front_leaf_up.ground.hpl_lca_outer；
- Coordinate Reference #2（参考坐标）：._FSAE_sus_front_leaf_up.ground.cfl_damper_down；
- Radius（半径）：5；
- Color：red；
- 勾选 Calculate Mass Properties of General Part 复选框，当几何体建立好之后，会更新对应部件的质量和惯量参数；
- Density：Material；
- Material Type：steel。

（2）单击 OK 按钮，完成 ._FSAE_sus_front_leaf_up.gel_lca.gralin_damp_base 几何体的创建。

至此，前上横置板簧悬架模型建立完成，如图 19-1 所示。更改衬套与车身的连接位置，可以改变整个悬架的刚度，悬架与车身两点间的连接距离越大，连接段所起的横向稳定杆的作用力矩越大。

19.2 后上横置板簧悬架

后上横置板簧悬架模型建模不再展开，请读者参阅前上横置板簧悬架模型的建立；建立好的后上横置板簧悬架如图 19-11 所示。相关信息如下：

后上横置板簧悬架硬点信息：

Info for subsystem: FSAE_sus_rear_leaf_up

File Name ： <FASE>/subsystems.tbl/FSAE_sus_rear_leaf_up.sub

```
Template      : mdids://FASE/templates.tbl/_FSAE_sus_rear_leaf_up.tpl
Comments      : *no comments found*
Major Role    : suspension
Minor Role    : rear
```

HARDPOINTS:

hardpoint name	symmetry	x_value	y_value	z_value
global	single	1524.0	0.0	0.0
arb_bushing_mount	left/right	1651.0	-127.0	101.6
drive_shaft_inr	left/right	1550.0	-200.0	225.0
lca_front	left/right	1270.0	-127.0	127.0
lca_outer	left/right	1498.6	-482.6	101.6
lca_rear	left/right	1651.0	-127.0	127.0
leaf_up_mount	left/right	1549.4	-95.4	407.89
tierod_inner	left/right	1676.4	-127.0	152.4
tierod_outer	left/right	1574.8	-457.2	152.4
uca_outer	left/right	1549.4	-482.6	355.6
wheel_center	left/right	1524.0	-558.8	228.6

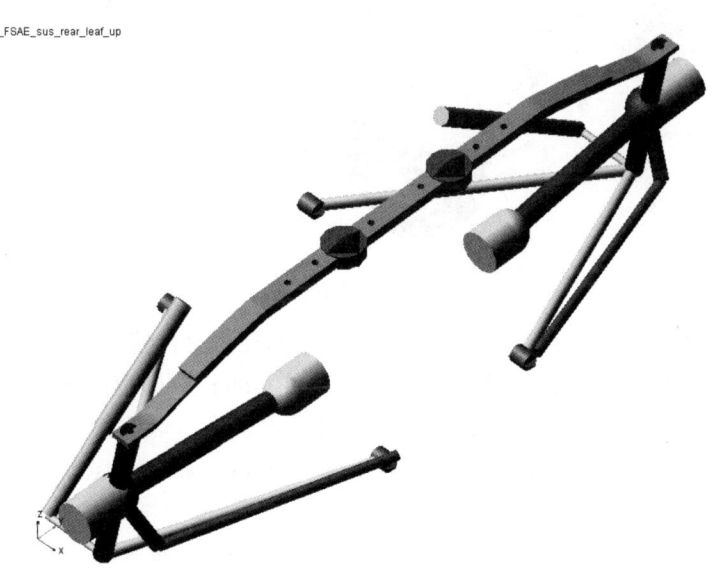

图 19-11　后上横置板簧悬架模型

后上横置板簧悬架变量信息：
PARAMETERS:

parameter name	symmetry	type	value
driveline_active	single	integer	1
kinematic_flag	single	integer	0
camber_angle	left/right	real	-1.5
drive_shaft_offset	left/right	real	75.0

19.3 文件驱动仿真

通过替换前后悬架完成整车 FSAE_2020_leaf_up 模型的建立，如图 19-12 所示。

图 19-12　整车模型 FSAE_2020_leaf_up

通过事件构造器（Event Builder）可以构造整车的任意事件特性仿真；同时仿真事件后在仿真目录中会有对应的存根，通过构造器（Event Builder）可以对存根文件进行修改；阶跃转向驱动文件：Step_steer_file_step.xml 存储在章节文件中，请读者自行查阅；

（1）单击 Simulate > Full-Vehicle Analysis > Event Builder 命令，构造驱动文件，如图 19-13 所示。

（2）单击 Event > Open，通过弹出的菜单在仿真目录中选取文件：file：//C：/Users/Administrator/Step_steer_file_step.xml。

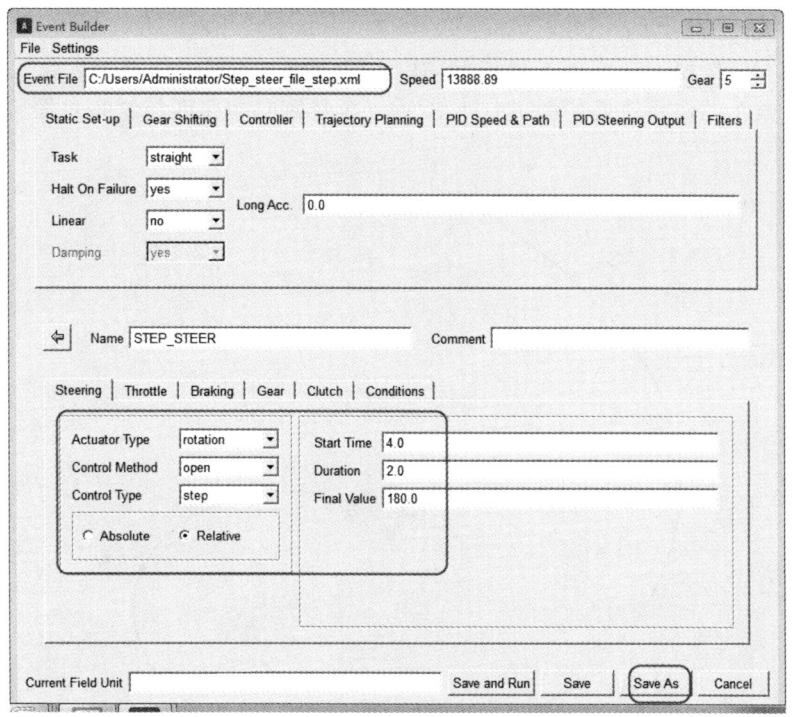

图 19-13　驱动构造文件

（3）在图 19-13 中可以构造转向角度、时间、转向持续时间等参数；建议先通过仿真文件获取存根，然后通过构造器对存根文件进行修改；修改完成后，保存修改。

（4）单击 Simulate > Full-Vehicle Analysis > File Driven Event 命令，弹出驱动控制文件仿真对话框，如图 19-14 所示，在下列对话框中输入相应的数据：

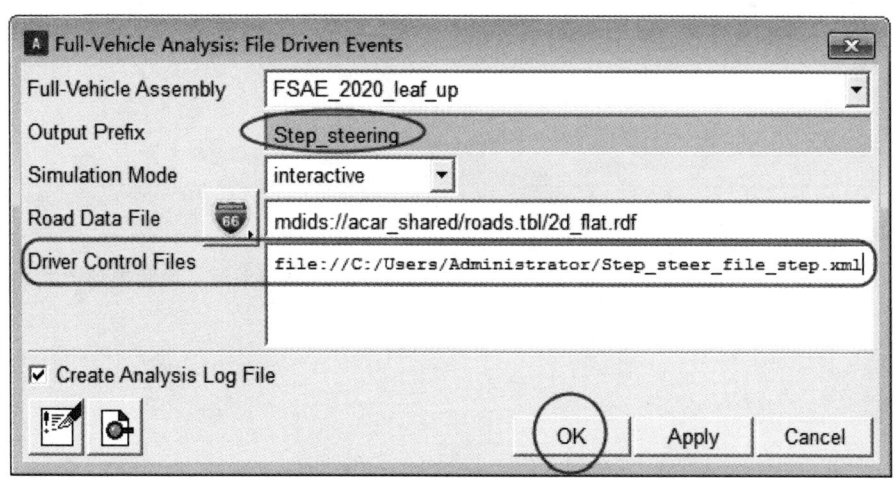

图 19-14　阶跃转向驱动文件仿真

- Output Prefix（输出别名）：Step_steering；
- Simulation Mode（仿真类型）：interactive；
- Road Date File：mdids：//acar_shared/roads.tbl/2d_flat.rdf；

- Driver Control Files：file：//C：/Users/Administrator/Step_steer_file_step.xml。

（5）单击 OK 按钮，完成直线制动 FSAE_2020_leaf_up 驱动控制仿真设置并提交计算。运算完成后，整车的运行轨迹如图 19-15 所示，车身侧向加速度与侧倾角加速度如图 19-16 和 19-17 所示。

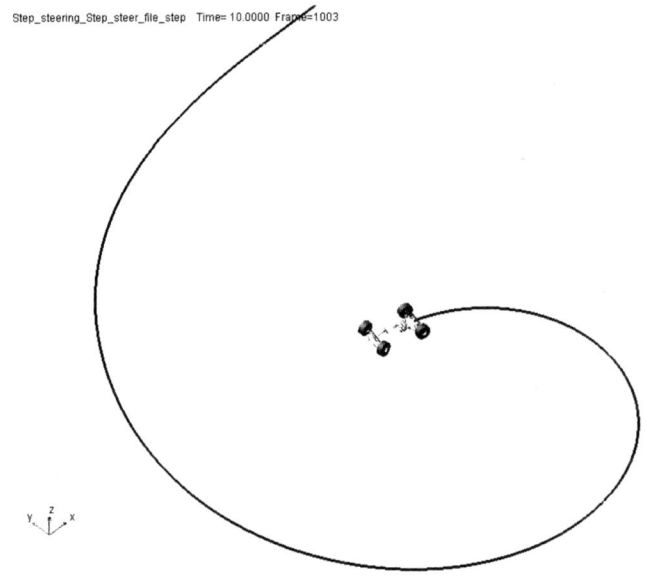

图 19-15 阶跃转向整车运行轨迹

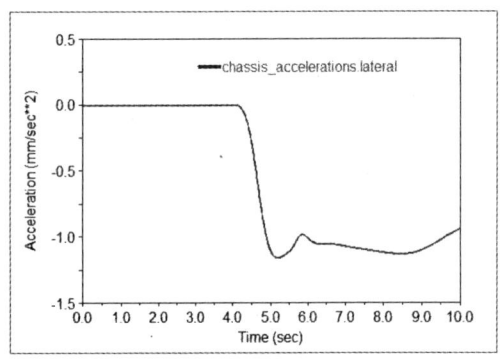

图 19-16 侧向加速度

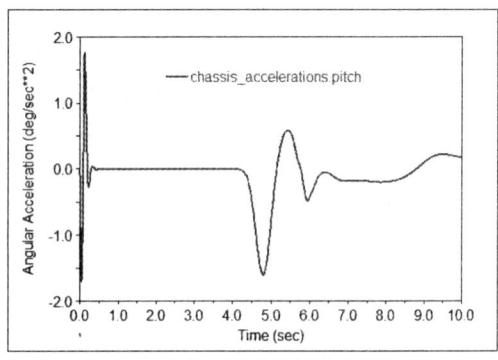

图 19-17 侧倾角加速度

第 20 章　下单横置板簧悬架模型

　　下单横置板簧悬架模型如图 20-1 所示，在之前创建的模型中完全取消下控制臂用横置板簧替代，横置板簧与车身间采用柔性衬套副连接；不同的约束副位置对应不同的板簧刚度。由于完全取消了下控制臂，避震器的设计可以通过推杆在空间内灵活布置，可以纵向布置，也可放置在上下控制臂外侧水平方向布置等。本章直接在模型_FSAE_sus_front_white.tpl 上添加其他部件完成模型的建立，完成后的前下横置板簧悬架模型：_FSAE_sus_front_leaf_down.tpl 存放在章节文件夹中，读者请自行查阅。

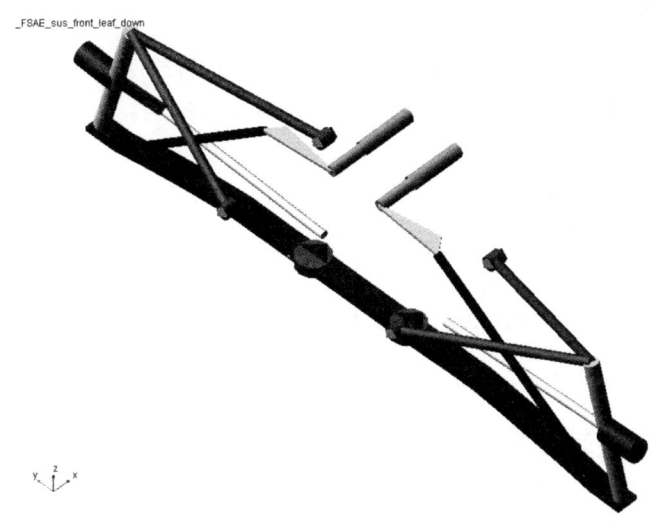

图 20-1　下单横置板簧悬架模型

学习目标

（1）了解前下单横置板簧悬架。
（2）了解后下单横置板簧悬架。
（3）了解静平衡。
（4）会准静态定半径转弯仿真。
（5）会准静态定速转弯仿真。
（6）会准静态力矩仿真。
（7）会准静态直线加速仿真。

20.1　前下单横置板簧悬架

1. 模型导入

（1）启动 ADAMS/CAR，选择专家模块进入建模界面。

（2）单击 File > Open 命令，弹出打开模板对话框，如图 20-2 所示，在 Template Name 中输入：mdids：//FASE/templates.tbl/_FSAE_sus_front_white.tpl。

（3）单击 OK 按钮，完成模型导入。

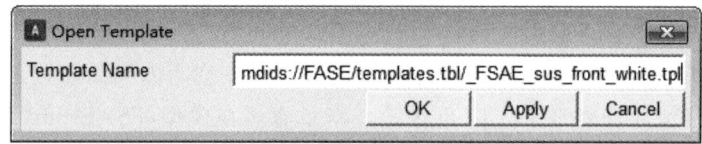

图 20-2　模板对话框

2. 删除下控制臂几何体

删除对应的上控制臂前后拉杆，但要注意不能删除上控制部件，删除后会导致模型错误，原因在于悬架的数组参数（即转向主销）由上下控制臂部件确定。

删除悬架以上控制臂前后拉杆几何体：

① ._FSAE_sus_front_white.gel_lca.gralin_lca_link_front；

② ._FSAE_sus_front_white.ger_lca.gralin_lca_link_front；

③ ._FSAE_sus_front_white.gel_lca.gralin_lca_link_rear；

④ ._FSAE_sus_front_white.ger_lca.gralin_lca_link_rear。

3. 修改部件位置

下控制臂部件位置由三个点确定，删除前后拉杆后，需要重新确定下控制臂的部件位置；将下控制臂部件放置在转向主销下位置点。

修改信息如下：

（1）单击 Build > Part > General Part > Modify 命令，在下列对话框中输入相应的数据：

- Location Dependency：Delta location from coordinate；
- Coordinate Reference（参考坐标）：._FSAE_sus_front_white.ground.hpl_lca_outer；
- Location：0，0，0。

（2）其余信息保持默认，单击 OK 按钮，完成下控制臂部件的修改。

（3）单击 File > Save As 命令，弹出保存模板对话框，如图 20-3 所示，在下列对话框中输入相应的数据：

- Template Name：_FSAE_sus_front_white；
- New Template Name：._FSAE_sus_front_leaf_down；
- Major Role（主特征）：suspension；
- File Format：Binary；
- Target：Datebase/FASE。

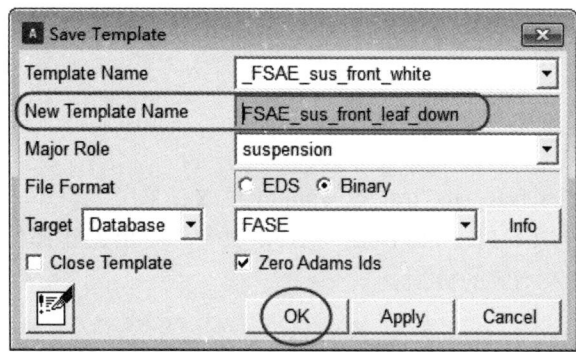

图 20-3 模型保存

（4）单击 OK 按钮，完成推杆式悬架模型模板 FSAE_sus_front_leaf_down 的保存。

4. 结构框

（1）单击 Build > Construction Frame > New 命令，弹出创建结构框对话框，如图 20-4 所示，在下列对话框中输入相应的数据：

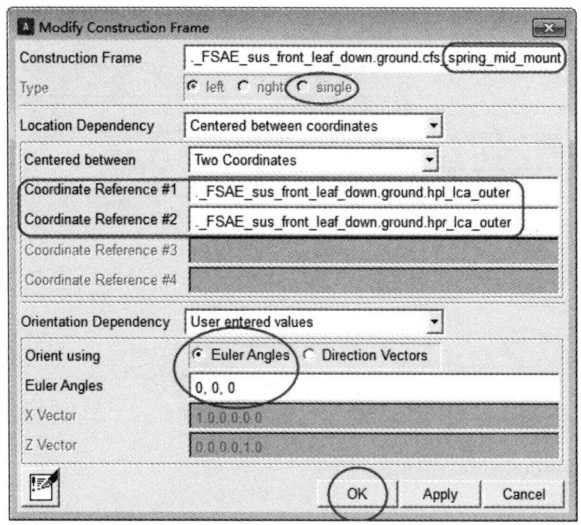

图 20-4 spring_mid_mount

- Construction Frame（结构框名称）：spring_mid_mount；
- Location Dependency：Centered between coordinates；
- Centered between：Two Coordinates；
- Coordinate Reference #1（参考坐标）：._FSAE_sus_front_leaf_down.ground.hpl_lca_outer；
- Coordinate Reference #2（参考坐标）：._FSAE_sus_front_leaf_down.ground.hpr_lca_outer；
- Orientation Dependency：User entered values；
- Orient using：Euler Angles；
- Euler Angles：0，0，0。

（2）单击 Apply 按钮，完成._FSAE_sus_front_leaf_down.ground.cfs_spring_mid_mount 结构框的创建。

5. 板簧 leafspring

板簧 MNF 中性文件：fsae_sus_leaf_down.mnf 存放在章节文件中，请读者自行调阅查看。

（1）单击 Build > Part > Flexible Body > New 命令，弹出创建部件对话框，如图 20-5 所示，在下列对话框中输入相应的数据：

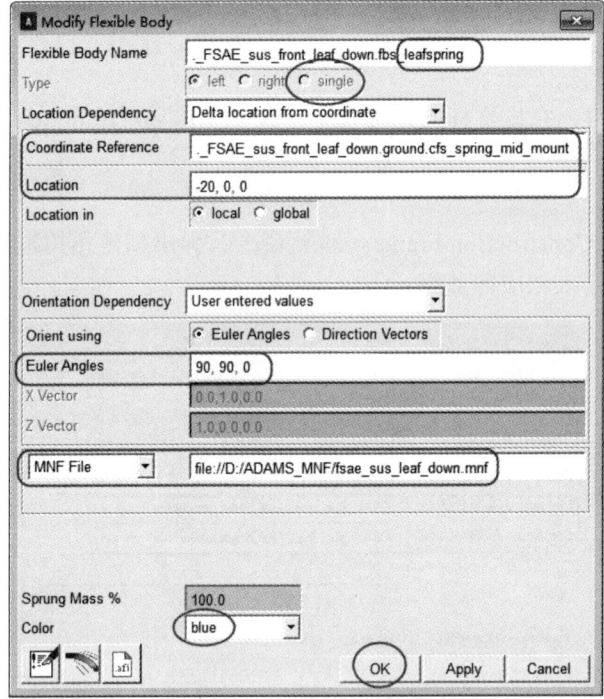

图 20-5　板簧 leafspring

- Flexible Body Name：leafspring；
- Type：single；
- Location Dependency：Delta location from coordinate；
- Coordinate Reference（参考坐标）：._FSAE_sus_front_leaf_down.ground.cfs_spring_mid_mount；
- Location：-20，0，0；
- Orientation Dependency：User entered values；
- Orient using：Euler Angles；
- Euler Angles：90，90，0；
- MNF File：file：//D：/ADAMS_MNF/fsae_sus_leaf_down.mnf；
- Color：blue。

（2）单击 OK 按钮，完成部件._FSAE_sus_front_leaf_down.fbs_leafspring 的创建。

6. 衬套连接硬点

（1）单击 Build > Hardpoind > New 命令，在下列对话框中输入相应的数据：
- Hardpoint：leaf_mount1_l；
- Type：left；
- Location：0.0，-100.0，158.2432808；此参数通过选取有限元节点获取。

（2）单击 OK 按钮，完成._FSAE_sus_front_leaf_down.ground.hps_leaf_mount1_l 硬点的创建。

7. 部件 leafspring 与 suspension_to_chassis 之间的 bushing 约束

（1）单击 Build > Attachments > Bushing > New 命令，弹出创建衬套对话框，如图 20-6 所示，在下列对话框中输入相应的数据：

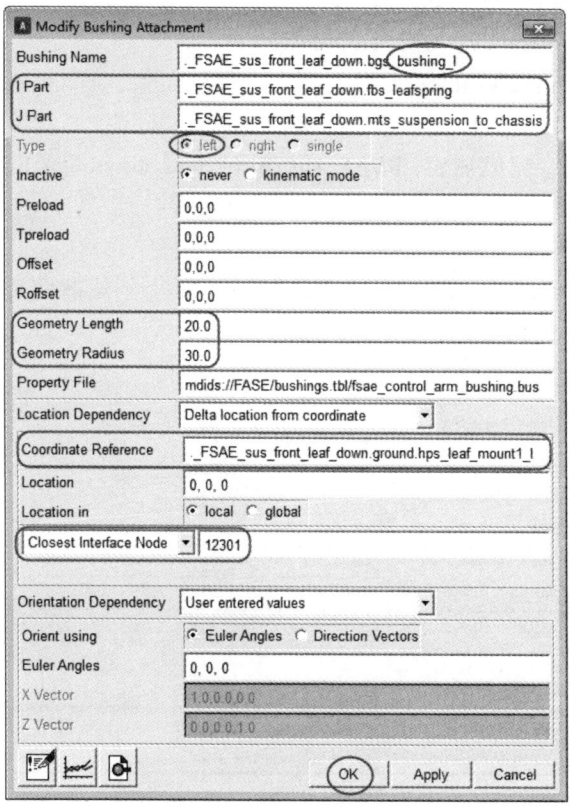

图 20-6　衬套 bushing_l

- Bushing Name（约束副名称）：bushing_l；
- I Part：._FSAE_sus_front_leaf_down.fbs_leafspring；
- J Part：._FSAE_sus_front_leaf_down.mts_suspension_to_chassis；
- Type：left；
- Inactive（抑制）：never；
- Preload：0，0，0；
- Tpreload：0，0，0；

- Offset：0，0，0；
- Roffset：0，0，0；
- Geometry Length：20；
- Geometry Radius：30；
- Property File：mdids：//FASE/bushings.tbl/fsae_control_arm_bushing.bus；
- Location Dependency：Delta location from coordinate；
- Coordinate Reference（参考坐标）：._FSAE_sus_front_leaf_down.ground.hps_lcaf_mount1_1；
- Location：0，0，0；
- Location in：local；
- Closest Interface Node：Left，12301；
- Orientation Dependency：User entered values；
- Orient using：Euler Angles；
- Euler Angles：0，0，0。

（2）单击 OK 按钮，完成轴套._FSAE_sus_front_leaf_down.bgs_bushing_1 的创建。

8. 部件 prod

（1）单击 Build > Part > General Part > New 命令，弹出创建部件对话框，如图 20-7 所示，在下列对话框中输入相应的数据：

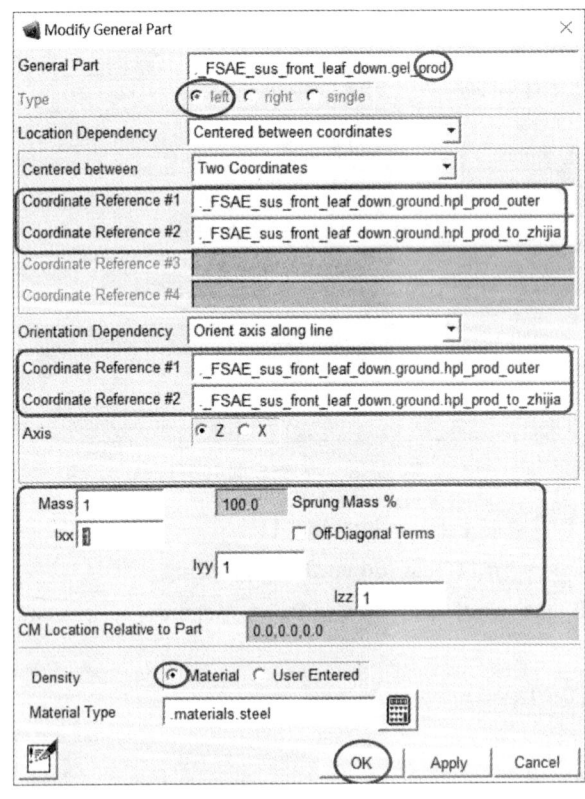

图 20-7　部件 prod

- General Part：prod；
- Type：left；
- Location Dependency：Centered between coordinates；
- Centered between：Two Coordinates；
- Coordinate Reference #1（参考坐标）：._FSAE_sus_front_leaf_down.ground.hpl_prod_outer；
- Coordinate Reference #2（参考坐标）：._FSAE_sus_front_leaf_down.ground.hpl_prod_to_bellcrank；
- Orientation Dependency：Orient axis along line；
- Coordinate Reference #1（参考坐标）：._FSAE_sus_front_leaf_down.ground.hpl_prod_outer；
- Coordinate Reference #2（参考坐标）：._FSAE_sus_front_leaf_down.ground.hpl_prod_to_bellcrank；
- Mass：1；
- Ixx：1；
- Iyy：1；
- Izz：1；
- Density：Material；
- Material Type：.materials.steel。

（2）单击 OK 按钮，完成部件._FSAE_sus_front_leaf_down.gel_prod 的创建。

9. 几何体 prod

- 单击 Build > Geometry > Link > New 命令，弹出创建几何体对话框，如图 20-8 所示，在下列对话框中输入相应的数据：

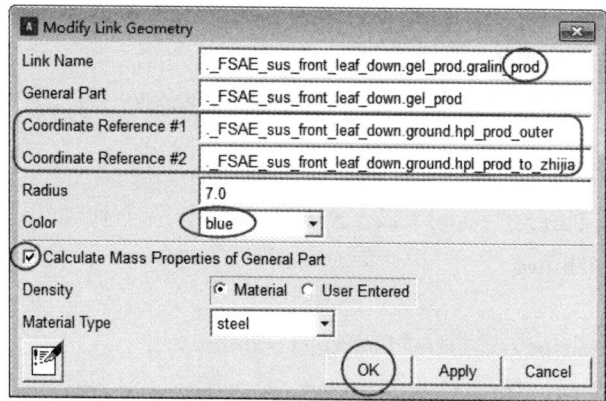

图 20-8 连杆几何体_prod

- Link Name（连杆名称）：prod；
- General Part 输入：._FSAE_sus_front_leaf_down.gel_prod；
- Coordinate Reference #1（参考坐标）：._FSAE_sus_front_leaf_down.ground.hpl_prod_

outer；

- Coordinate Reference #2（参考坐标）：._FSAE_sus_front_leaf_down.ground.hpl_prod_to_zhijia；
- Radius（半径）：7；
- Color：blue；
- 勾选 Calculate Mass Properties of General Part 复选框，当几何体建立好之后，会更新对应部件的质量和惯量参数；
- Density：Material；
- Material Type：steel。

（2）单击 OK 按钮，完成._FSAE_sus_front_leaf_down.gel_prod.gralin_prod 几何体的创建。

10. 支架摆臂硬点

（1）单击 Build > Hardpoind > New 命令，在下列对话框中输入相应的数据：

- Hardpoint：prod_to_zhijia；
- Type：left；
- Location：-20.8，-180.0，381.0。

（2）单击 Apply 按钮，完成._FSAE_sus_front_leaf_down.ground.hpl_prod_to_zhijia 硬点的创建。同理，创建其他硬点。

- Hardpoint：shock_to_bellcrank；
- Type：left；
- Location：-10.0，-50.8，381.0。

（3）单击 Apply 按钮，完成._FSAE_sus_front_leaf_down.ground.hpl_shock_to_bellcrank 硬点的创建。

- Hardpoint：zhijia_pivot；
- Type：left；
- Location：5.0，-160.0，381.0。

（4）单击 OK 按钮，完成._FSAE_sus_front_leaf_down.ground.hpl_zhijia_pivot 硬点的创建。

11. 部件 zhijia

（1）单击 Build > Part > General Part > New 命令，在下列对话框中输入相应的数据：

- General Part：zhijia；
- Type：left；
- Location Dependency：Centered between coordinates；
- Centered between：Three Coordinates；
- Coordinate Reference #1（参考坐标）：._FSAE_sus_front_leaf_down.ground.hpl_prod_to_zhijia；
- Coordinate Reference #2（参考坐标）：._FSAE_sus_front_leaf_down.ground.hpl_shock_to_zhijia；
- Coordinate Reference #3（参考坐标）：._FSAE_sus_front_leaf_down.ground.hpl_

zhijia_pivot；
- Orientation Dependency：User entered values；
- Orient using：Euler Angles；
- Euler Angles：0，0，0；
- Mass：1；
- Ixx：1；
- Iyy：1；
- Izz：1；
- Density：Material；
- Material Type：.materials.steel。

（2）单击 OK 按钮，完成部件._FSAE_sus_front_leaf_down.gel_zhijia 的创建。

12. 三角臂 zhijia

（1）单击 Build > Geometry > Arm > New 命令，弹出创建三角臂几何体对话框，如图 20-9 所示，在下列对话框中输入相应的数据：

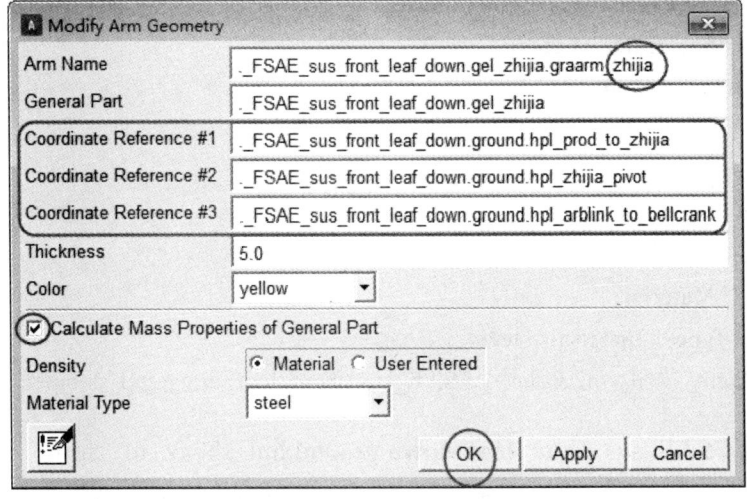

图 20-9　连杆几何体_prod

- Link Name（连杆名称）：zhijia；
- General Part：._FSAE_sus_front_leaf_down.gel_zhijia；
- Coordinate Reference #1（参考坐标）：._FSAE_sus_front_leaf_down.ground.hpl_prod_to_zhijia；
- Coordinate Reference #2（参考坐标）：._FSAE_sus_front_leaf_down.ground.hpl_zhijia_pivot；
- Coordinate Reference #3（参考坐标）：._FSAE_sus_front_leaf_down.ground.hpl_arblink_to_bellcrank；
- Thickness（三角臂厚度）：5；
- Color：yellow；

- 勾选 Calculate Mass Properties of General Part 复选框，当几何体建立好之后，会更新对应部件的质量和惯量参数；
  - Density：Material；
  - Material Type：steel。

（2）单击 OK 按钮，完成._FSAE_sus_front_leaf_down.gel_zhijia.graarm_zhijia 几何体的创建。

13. 部件 damper_chassis

（1）单击 Build > Part > General Part > New 命令，在下列对话框中输入相应的数据：
  - General Part：damper_chassis；
  - Type：left；
  - Location Dependency：Delta location from coordinate；
  - Coordinate Reference（参考坐标）：._FSAE_sus_front_leaf_down.ground.hpl_shock_to_chassis；
  - Location：0，0，0；
  - Location in：local；
  - Orientation Dependency：User entered values；
  - Orient using：Euler Angles；
  - Euler Angles：0，0，0；
  - Mass：1；
  - Ixx：1；
  - Iyy：1；
  - Izz：1；
  - Density：Material；
  - Material Type：.materials.steel。

（2）单击 Apply 按钮，完成部件._FSAE_sus_front_leaf_down.gel_damper_chassis 的创建。

14. 部件._FSAE_sus_front_leaf_down.ground.hpl_shock_to_zhijia

（1）单击 Build > Part > General Part > New 命令，在下列对话框中输入相应的数据：
  - General Part：damper_zhijia；
  - Type：left；
  - Location Dependency：Delta location from coordinate；
  - Coordinate Reference（参考坐标）：._FSAE_sus_front_leaf_down.ground.hpl_shock_to_zhijia；
  - Location：0，0，0；
  - Location in：local；
  - Orientation Dependency：User entered values；
  - Orient using：Euler Angles；
  - Euler Angles：0，0，0；
  - Mass：1；

- Ixx：1；
- Iyy：1；
- Izz：1；
- Density：Material；
- Material Type：.materials.steel。

（2）单击 OK 按钮，完成部件._FSAE_sus_front_leaf_down.gel_damper_zhijia 创建。

15．避震器

（1）单击 Build > Force > Damper > New 命令，弹出避震器创建对话框，如图 20-10 所示，在下列对话框中输入相应的数据：

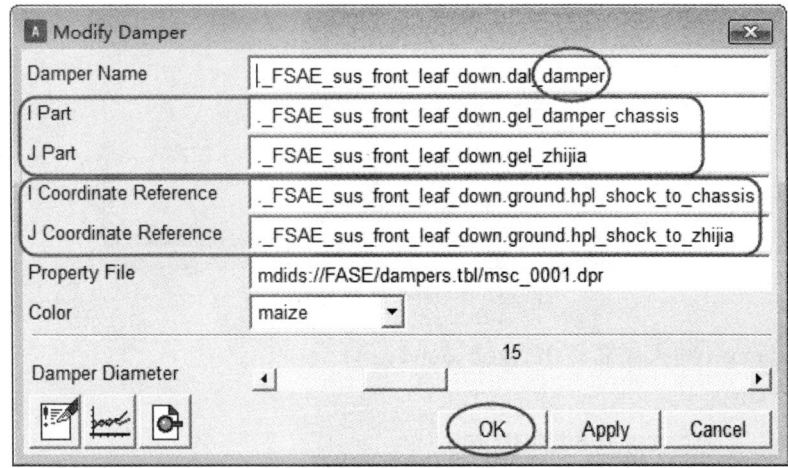

图 20-10　避震器

- Damper Name（减震器名称）：damper ；
- I Part：._FSAE_sus_front_leaf_down.gel_damper_chassis；
- J Part：._FSAE_sus_front_leaf_down.gel_zhijia；
- I Coordinate Reference（参考坐标）：._FSAE_sus_front_leaf_down.ground.hpl_shock_to_chassis；
- J Coordinate Reference（参考坐标）：._FSAE_sus_front_leaf_down.ground.hpl_shock_to_zhijia；
- Property File（属性文件）：mdids：//FASE/dampers.tbl/msc_0001.dpr；
- Damper Diameter（避震器直径）：拖动滑块选择 15 mm；
- Color：maize。

（2）单击 OK 按钮，完成避震器._FSAE_sus_front_leaf_down.dal_damper 的创建。

16．部件 uca 与 prod 之间的 spherical 约束

（1）单击 Build > Attachments > Joint > New 命令，弹出创建约束件对话框，如图 20-11 所示，在下列对话框中输入相应的数据：

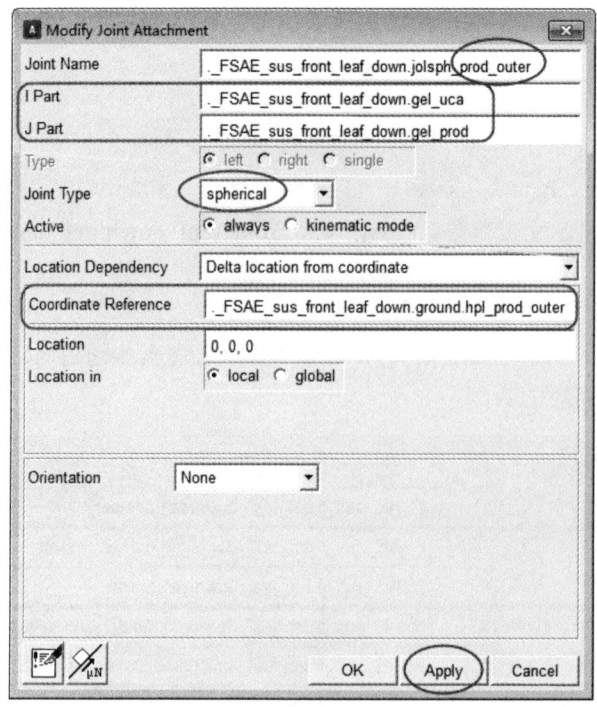

图 20-11　刚性约束-spherical

- Joint Name（约束副名称）：prod_outer；
- Type：left；
- I Part：._FSAE_sus_front_leaf_down.gel_uca；
- J Part：._FSAE_sus_front_leaf_down.gel_prod；
- Joint Type（约束副类型）：spherical；
- Active（激活）：always；
- Location Dependency：Delta location from coordinate；
- Coordinate Reference（参考坐标）：._FSAE_sus_front_leaf_down.ground.hpl_prod_outer；
- Location：0, 0, 0；
- Location in：local。

（2）单击 Apply 按钮，完成约束副._FSAE_sus_front_leaf_down.jolsph_prod_outer 的创建。

17. 部件 prod 与 zhijia 之间的 hook 约束

（1）单击 Build > Attachments > Joint > New 命令，在下列对话框中输入相应的数据：
- Joint Name（约束副名称）：prod_to_zhijia；
- Type：left；
- I Part：._FSAE_sus_front_leaf_down.gel_prod；
- J Part：._FSAE_sus_front_leaf_down.gel_zhijia；
- Joint Type（约束副类型）：hook；
- Active（激活）：always；

- Location Dependency：Delta location from coordinate；
- Coordinate Reference（参考坐标）：._FSAE_sus_front_leaf_down.ground.hpl_prod_to_zhijia；
- Location：0，0，0；
- Location in：local；
- I-Part Axis：._FSAE_sus_front_leaf_down.ground.hpl_prod_outer；
- J-Part Axis：._FSAE_sus_front_leaf_down.ground.hpl_zhijia_pivot。

（2）单击 Apply 按钮，完成约束副._FSAE_sus_front_leaf_down.jolhoo_prod_to_zhijia 的创建。

18. 部件 zhijia 与 suspension_to_chassis 之间的 revolute 约束

（1）单击 Build > Attachments > Joint > New 命令，在下列对话框中输入相应的数据：
- Joint Name（约束副名称）：zhijia_pivot；
- Type：left；
- I Part：._FSAE_sus_front_leaf_down.gel_zhijia；
- J Part：._FSAE_sus_front_leaf_down.mts_suspension_to_chassis；
- Joint Type（约束副类型）：revolute；
- Active（激活）：always；
- Location Dependency：Delta location from coordinate；
- Coordinate Reference（参考坐标）：._FSAE_sus_front_leaf_down.ground.hpl_zhijia_pivot；
- Location：0，0，0；
- Orientation Dependency：User entered values；
- Orient using：Euler Angles；
- Euler Angles：0，0，0。

（2）单击 Apply 按钮，完成约束副._FSAE_sus_front_leaf_down.jolrev_zhijia_pivot 的创建。

19. 部件 damper_zhijia 与 zhijia 之间的 hook 约束

（1）单击 Build > Attachments > Joint > New 命令，在下列对话框中输入相应的数据：
- Joint Name（约束副名称）：damper_to_zhijia；
- Type：left；
- I Part：._FSAE_sus_front_leaf_down.gel_damper_zhijia；
- J Part：._FSAE_sus_front_leaf_down.gel_zhijia；
- Joint Type（约束副类型）：hook；
- Active（激活）：always；
- Location Dependency：Delta location from coordinate；
- Coordinate Reference（参考坐标）：._FSAE_sus_front_leaf_down.ground.hpl_shock_to_zhijia；
- Location：0，0，0；

- Location in：local；
- I-Part Axis：._FSAE_sus_front_leaf_down.ground.hpl_shock_to_chassis；
- J-Part Axis：._FSAE_sus_front_leaf_down.ground.hpl_prod_to_zhijia；

（2）单击 Apply 按钮，完成约束副._FSAE_sus_front_leaf_down.jolhoo_damper_to_zhijia 的创建。

20．部件 damper_chassis 与 suspension_to_chassis 之间的 hook 约束

（1）单击 Build > Attachments > Joint > New 命令，在下列对话框中输入相应的数据：
- Joint Name（约束副名称）：damper_to_chassis；
- Type：left；
- I Part：._FSAE_sus_front_leaf_down.gel_damper_chassis；
- J Part：._FSAE_sus_front_leaf_down.mts_suspension_to_chassis；
- Joint Type（约束副类型）：hook；
- Active（激活）：always；
- Location Dependency：Delta location from coordinate；
- Coordinate Reference（参考坐标）：._FSAE_sus_front_leaf_down.ground.hpl_shock_to_chassis；
- Location：0，0，0；
- Location in：local；
- I-Part Axis：._FSAE_sus_front_leaf_down.ground.hpl_shock_to_zhijia；
- J-Part Axis：._FSAE_sus_front_leaf_down.ground.cfl_damper_chassis_orient。

（2）单击 Apply 按钮，完成约束副._FSAE_sus_front_leaf_down.jolhoo_damper_to_chassis 的创建。

21．部件 damper_chassis 与 damper_zhijia 之间的 cylindrical 约束

（1）单击 Build > Attachments > Joint > New 命令，在下列对话框中输入相应的数据：
- Joint Name（约束副名称）：damper_slide；
- Type：left；
- I Part：._FSAE_sus_front_leaf_down.gel_damper_chassis；
- J Part：._FSAE_sus_front_leaf_down.gel_damper_zhijia；
- Joint Type（约束副类型）：cylindrical；
- Active（激活）：always；
- Location Dependency：Centered between coordinates；
- Centered between：Two Coordinates；
- Coordinate Reference #1（参考坐标）：._FSAE_sus_front_leaf_down.ground.hpl_shock_to_chassis；
- Coordinate Reference #2（参考坐标）：._FSAE_sus_front_leaf_down.ground.hpl_shock_to_zhijia；
- Location Dependency：Orient axis along line；

- Coordinate Reference #1（参考坐标）：._FSAE_sus_front_leaf_down.ground.hpl_shock_to_chassis；
- Coordinate Reference #2（参考坐标）：._FSAE_sus_front_leaf_down.ground.hpl_shock_to_zhijia。

（2）单击 Apply 按钮，完成约束副._FSAE_sus_front_leaf_down.jolcyl_damper_slide 的创建。

至此，前单下横置板簧悬架模型建立完成，如图 20-1 所示。更改衬套与车身的连接位置，可以改变整个悬架的刚度，悬架与车身两点间的连接距离越大，连接段所起的横向稳定杆的作用力矩越大。

20.2　后下单横置板簧悬架

后下单横置板簧建好的模型如图 20-12 所示，建模过程不再叙述，读者请参阅模型：_FSAE_sus_rear_leaf_down.tpl；后下单横置板簧建好的模型的硬点和参数变量信息如下：

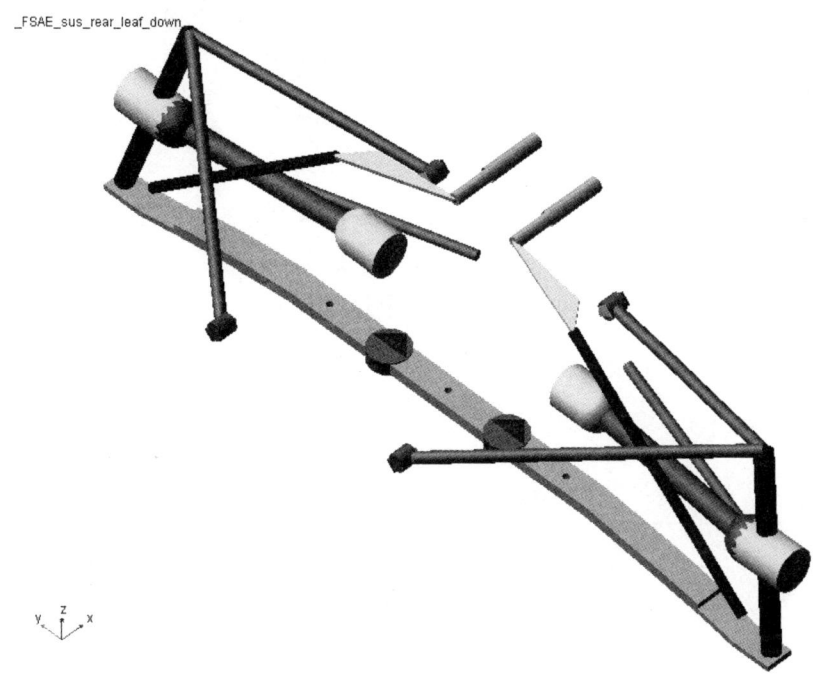

图 20-12　后下单横置板簧

```
Info for subsystem:   FSAE_sus_rear_leaf_down

File Name    :   <FASE>/subsystems.tbl/FSAE_sus_rear_leaf_down.sub
Template     :   mdids://FASE/templates.tbl/_FSAE_sus_rear_leaf_down.tpl
Comments     :   *no comments found*
```

Major Role   : suspension
Minor Role   : rear

HARDPOINTS:

hardpoint name	symmetry	x_value	y_value	z_value
global	single	1524.0	0.0	0.0
arblink_to_bellcrank	left/right	1560.5	-50.8	381.0
bellcrank_pivot	left/right	1547.8	-170.0	381.0
bellcrank_pivot_orient	left/right	1473.2	-146.05	547.3
drive_shaft_inr	left/right	1550.0	-200.0	225.0
lca_outer	left/right	1498.6	-546.1	101.6
leaf_mount	left/right	1498.6	-100.0	139.2
prod_outer	left/right	1498.6	-497.2	127.0
prod_to_bellcrank	left/right	1509.7	-200.0	381.0
shock_to_bellcrank	left/right	1560.5	-50.8	381.0
shock_to_chassis	left/right	1700.0	-50.8	381.0
tierod_inner	left/right	1676.4	-127.0	152.4
tierod_outer	left/right	1574.8	-457.2	152.4
uca_front	left/right	1270.0	-152.4	304.8
uca_outer	left/right	1549.4	-482.6	355.6
uca_rear	left/right	1625.6	-152.4	304.8
wheel_center	left/right	1524.0	-558.8	228.6

PARAMETERS:

parameter name	symmetry	type	value
driveline_active	single	integer	1
kinematic_flag	single	integer	0
camber_angle	left/right	real	-1.5
drive_shaft_offset	left/right	real	75.0
toe_angle	left/right	real	0.0

20.3 静平衡仿真

1. 静平衡

对于子系统或整车模型,静平衡是成功仿真的前提,通过静平衡可以发现车辆是否存在问题,在正确仿真前对模型进行修改,前后悬架建立完成后,对应建立其子系统,通过替换其他车辆的前后悬架得到新整车.FSAE_2020_leaf_down 模型,如图 20-13 所示。

图 20-13 整车模型.FSAE_2020_leaf_down

2. 整车静平衡

(1)单击 Simulate > Full-Vehicle Analysis > Static and Quasi-Static Maneuvers > Static Equilibrium 命令,显示静平衡仿真设置,如图 20-14 所示,在下列对话框中输入相应的数据:

- Output Prefix(输出别名):se;
- Simulation Mode(仿真类型):interactive;
- Static Set-Up:Settle;
- Road Date File:mdids: //acar_shared/roads.tbl/2d_flat.rdf;
- Gear Position:5。

(2)单击 OK 按钮,完成直线制动 FSAE_2020_leaf_up 驱动控制仿真设置并提交计算。运算完成后,弹出的命令信息窗口可以显示整车静平衡成功与否,通过静平衡仿真发现整车模型.FSAE_2020_leaf_down 静平衡成功。

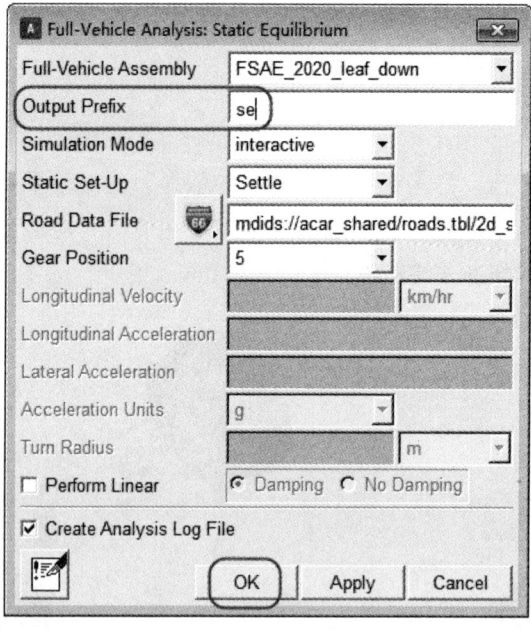

图 20-14 静平衡设置

20.4 准静态定半径转弯

（1）单击 Simulate > Full-Vehicle Analysis > Static and Quasi-Static Maneuvers > Constant Radius Cornering 命令，显示静平衡仿真设置，如图 20-15 所示，在下列对话框中输入相应的数据：

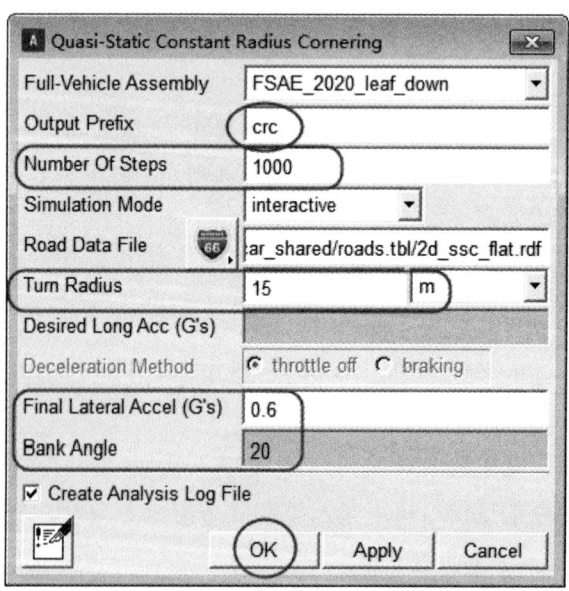

图 20-15 准静态定半径转弯仿真

- Output Prefix（输出别名）：crc；
- Number Of Steps：1000；
- Simulation Mode（仿真类型）：interactive；
- Road Date File：mdids：//acar_shared/roads.tbl/2d_ssc_flat.rdf；
- Turn Radius（单位：m）：15；
- Desired LongAcc（G'S）：0，方框为灰色可以不用设置，指在转弯过程中期望的加速度，可以为正负值，负值表示制动减速；
- Final LongAcc（G'S）：0.6；
- Bank Angle（单位：度）：20，指路面的倾斜角。

（2）单击 OK 按钮，完成直线制动 FSAE_2020_leaf_up 准静态定半径转弯设置并提交计算。运算完成后，需要采用绘图配置文件才能对计算结果绘制曲线。

（3）绘图配置文件如图 20-16 所示，导入绘图配置文件：mdids：//FASE/plot_configs.tbl/mdi_fva_ssc.plt。

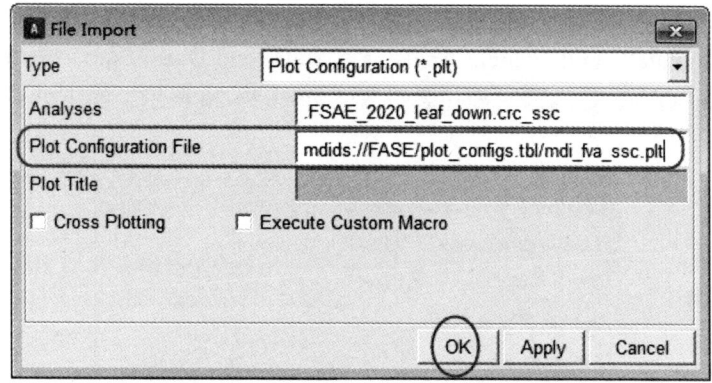

图 20-16　绘图配置文件（准静态演算）

（4）单击 OK 按钮，可以对计算结果绘制曲线图，车辆参数曲线如图 20-17～图 20-19 所示。

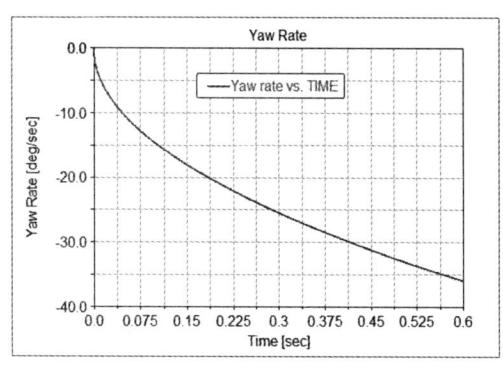

图 20-17　横摆角速度

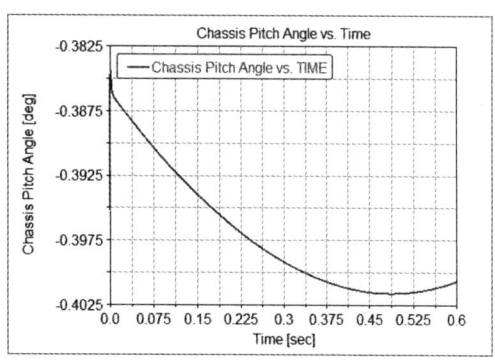

图 20-18　侧倾角速度

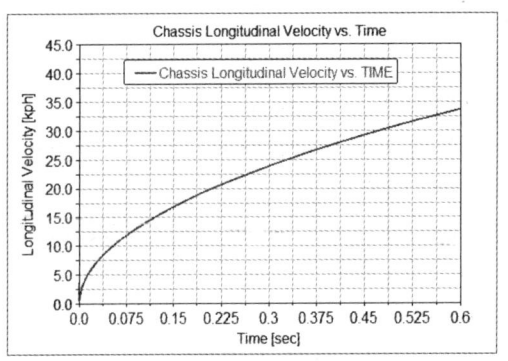

图 20-19 纵向速度

20.5 准静态定速转弯

（1）单击 Simulate > Full-Vehicle Analysis > Static and Quasi-Static Maneuvers > Constant Velocity Cornering 命令，显示静平衡仿真设置，如图 20-20 所示，在下列对话框中输入相应的数据：

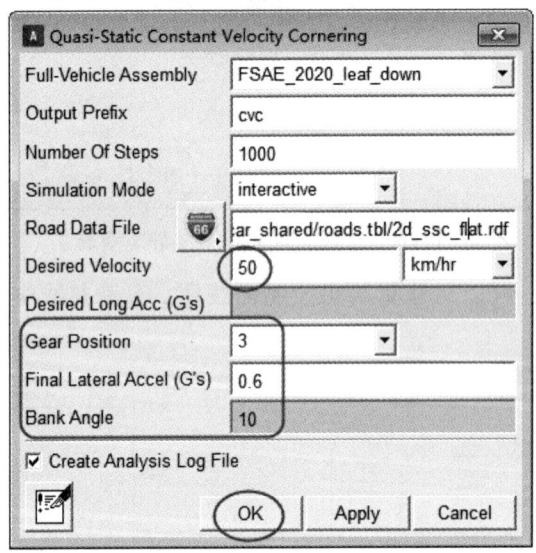

图 20-20 准静态定速转弯仿真

- Output Prefix（输出别名）：cvc；
- Number Of Steps：1000；
- Simulation Mode（仿真类型）：interactive；
- Road Date File：mdids：//acar_shared/roads.tbl/2d_ssc_flat.rdf；
- Desired Velocity（单位：km/h）：50；
- Desired LongAcc（G'S）：0，方框为灰色可以不用设置，指在转弯过程中期望的加速度，可以为正负值，负值表示制动减速；
- Gear Position：3；
- Final LongAcc（G'S）：0.6；
- Bank Angle（单位：度）：10，指路面的倾斜角。

（2）单击 OK 按钮，完成直线制动 FSAE_2020_leaf_up 准静态定速度转弯设置并提交计算。运算完成后，采用绘图配置文件绘制参数，如图 20-21～图 20-24 所示。

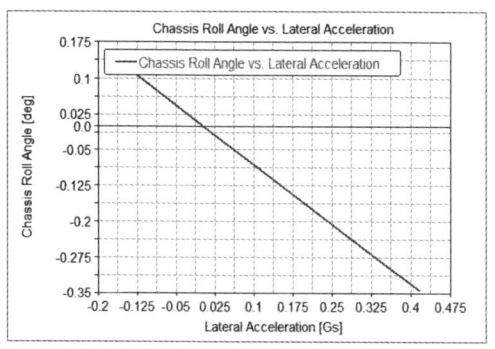

图 20-21　俯仰角与侧向加速度的关系

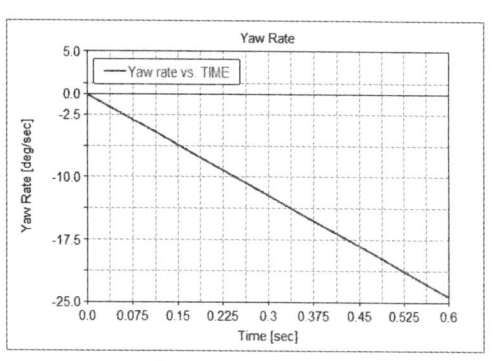

图 20-22　横摆角速度

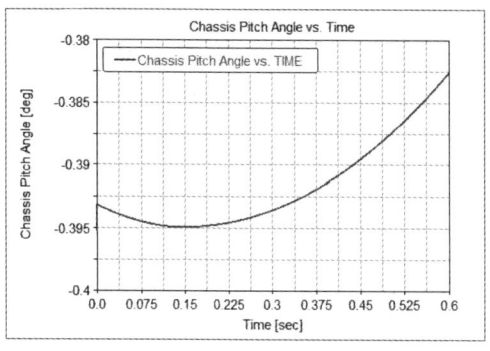

图 20-23　侧倾角

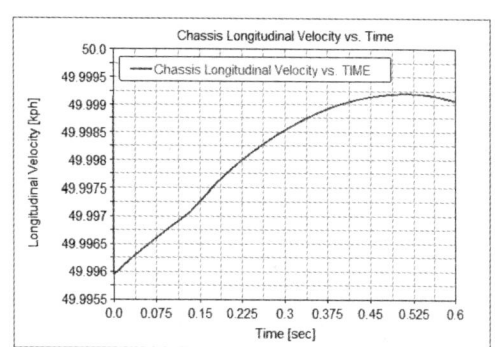

图 20-24　纵向速度

20.6　准静态力矩仿真

（1）单击 Simulate > Full-Vehicle Analysis > Static and Quasi-Static Maneuvers > Constant Velocity Cornering 命令，显示静平衡仿真设置，如图 20-25 所示，在下列对话框中输入相应的数据：
- Output Prefix（输出别名）：FMM；
- Number Of Steps：100；
- Simulation Mode（仿真类型）：interactive；
- Road Date File：mdids: //acar_shared/roads.tbl/2d_ssc_flat.rdf；
- Desired Velocity（单位：km/h）：50；

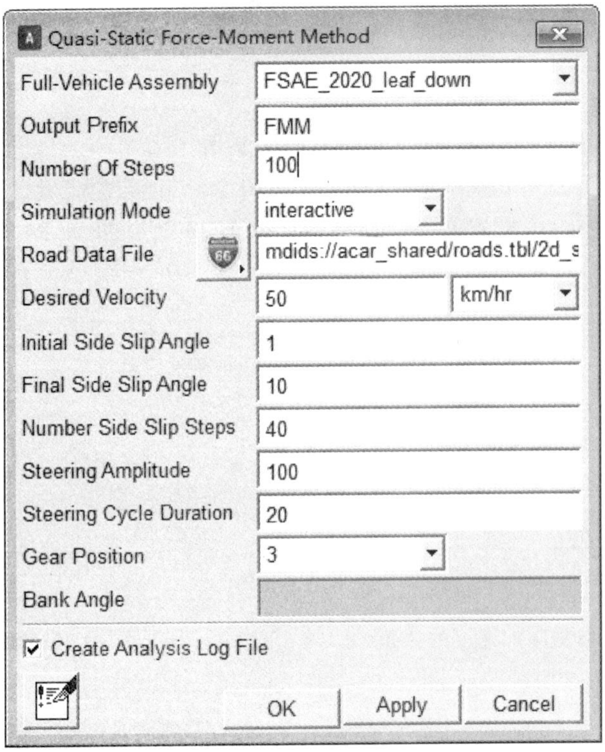

图 20-25 准静态力矩仿真设置

- Initial Side Slip Angle：1；
- Final Side Slip Angle：10；
- Number Side Slip Steps：40；
- Steering Amplitude：100；
- Steering Cycle Duration：20；
- Gear Position：3。

（2）单击 OK 按钮，完成直线制动 FSAE_2020_leaf_up 准静态力矩仿真设置并提交计算。运算完成后，采用绘图配置文件绘制参数，如图 20-26 和图 20-27 所示。

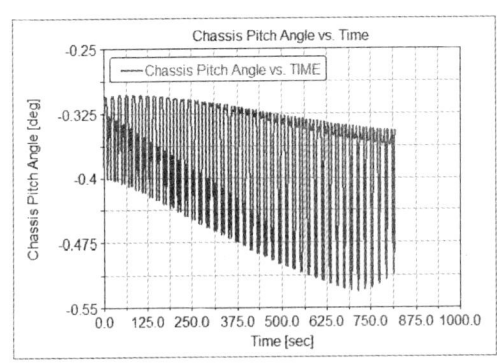

图 20-26 侧倾角

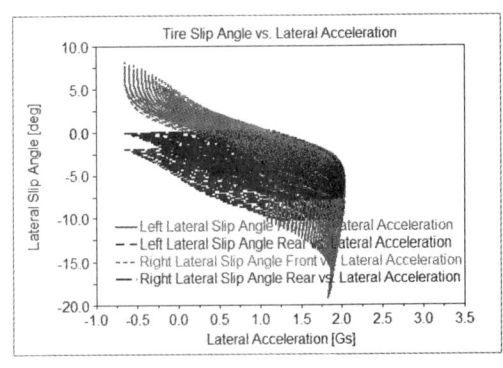

图 20-27 轮胎滑移角与侧向加速度的关系

20.7　准静态直线加速仿真

（1）单击 Simulate > Full-Vehicle Analysis > Static and Quasi-Static Maneuvers > Straight-line Acceleration 命令，显示静平衡仿真设置，如图 20-28 所示，在下列对话框中输入相应的数据：

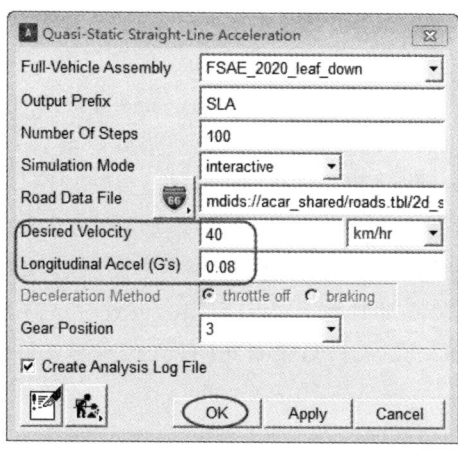

图 20-28　准静态直线加速仿真

- Output Prefix（输出别名）：SLA；
- Number Of Steps：100；
- Simulation Mode（仿真类型）：interactive；
- Road Date File：mdids：//acar_shared/roads.tbl/2d_ssc_flat.rdf；
- Desired Velocity（单位：km/h）：40；
- Longitudinal Accel（G'S）：0.08；
- Gear Position：3。

（2）单击 OK 按钮，完成直线制动 FSAE_2020_leaf_up 准静态直线加速设置并提交计算。运算完成后，采用绘图配置文件绘制参数，如图 20-29 和图 20-30 所示。

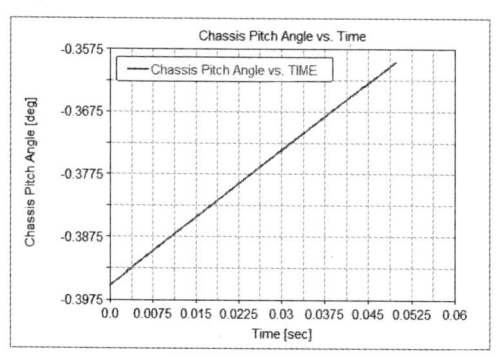

图 20-29　侧倾角

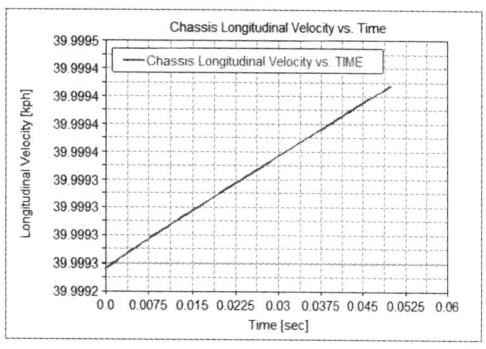

图 20-30　车身纵向速度

第 21 章　下双横置板簧悬架模型

前下双横置板簧悬架模型如图 21-1 所示，将下控制臂替换为双横置钢板弹簧。钢板弹簧的优势如下：① 可以作为弹簧，因此悬架系统不再需要螺旋弹簧；② 可以作为横向拉杆；③ 横置板簧本身会起到横向稳定杆的作用，因此可以省略稳定杆，模型更加简单。此模型中建立的板簧刚度是可以调节的，即在板簧空间的不同位置有定位孔，不同的定位孔与车身连接会导致板簧产生不同的刚度，此处板簧模型有三个刚度选项可以调节。本章直接在模型 _FSAE_sus_front_white.tpl 上添加其他部件完成模型的建立，完成后的前下双横置板簧悬架模型的：_FSAE_sus_front_doubleleaf_damper.tpl 存放在章节文件夹中，读者请自行查阅。

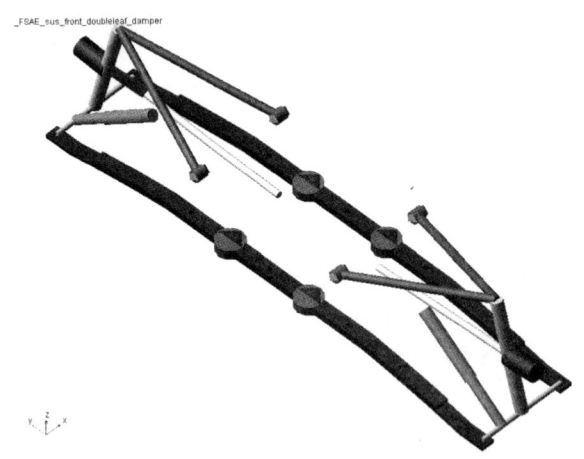

图 21-1　前下双横置板簧悬架模型

学习目标

（1）了解双片板簧 MNF。
（2）多叶片弹簧约束问题讨论。
（3）了解前下双横置板簧悬架。
（4）了解后下双横置板簧悬架。
（5）会收油门直线仿真。
（6）会稳定性参数对比。

21.1 双片板簧 MNF

制作好的双片板簧模态中性文件 MNF：fsae_sus_leaf_down_30width.mnf 存储在章节文件中；制作过程不再重复，请读者参阅"横向稳定杆Ⅱ"章节；此处的板簧提取前 10 阶模态；前 4 阶模态云图如图 21-2～图 21-5 所示，前 10 阶板簧频率如图 21-6 所示。

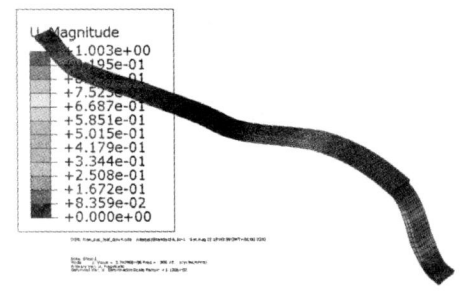

图 21-2　一阶模态

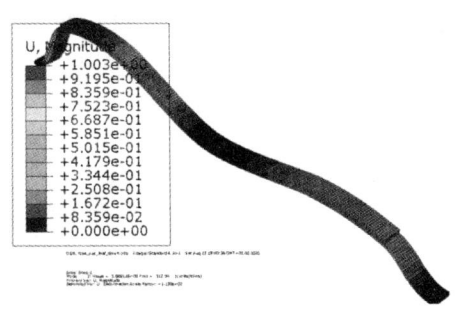

图 21-3　二阶模态

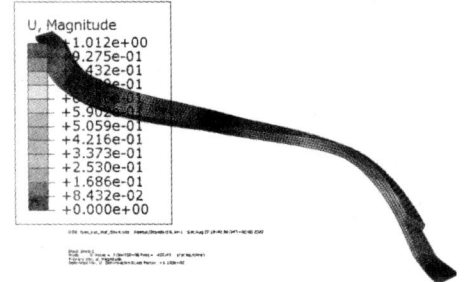

图 21-4　三阶模态

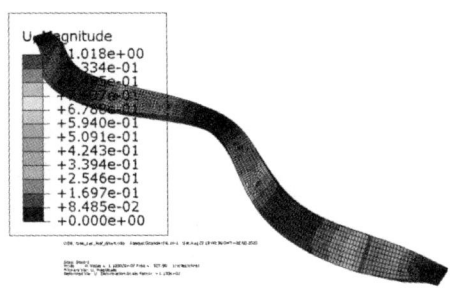

图 21-5　四阶模态

1	Mode	1: Value = 3.74796E+06 Freq =	308.12	(cycles/time)
2	Mode	2: Value = 3.86611E+06 Freq =	312.94	(cycles/time)
3	Mode	3: Value = 7.04475E+06 Freq =	422.43	(cycles/time)
4	Mode	4: Value = 1.10002E+07 Freq =	527.86	(cycles/time)
5	Mode	5: Value = 1.48146E+07 Freq =	612.58	(cycles/time)
6	Mode	6: Value = 1.57132E+07 Freq =	630.89	(cycles/time)
7	Mode	7: Value = 6.18114E+07 Freq =	1251.3	(cycles/time)
8	Mode	8: Value = 6.48691E+07 Freq =	1281.9	(cycles/time)
9	Mode	9: Value = 6.71757E+07 Freq =	1304.4	(cycles/time)
10	Mode	10: Value = 7.52697E+07 Freq =	1380.8	(cycles/time)

图 21-6　前 10 阶板簧频率

21.2 多叶片板簧约束问题

本节涉及的横置板簧为两片不等长的簧片叠加装配而成，装配体模态分析与单个零部件有很大不同，原因在于不同部件之间的约束问题。

（1）簧片之间的约束采用绑定约束；施加绑定 TIE 约束后，两簧片之间不能产生移动，此种方式在静止状态下或者少簧片装配时可以采用；如果簧片之间的滑移量过大，或者在高频动载荷下采用此种方式，这与板簧的实际工作装配是不相符的。绑定约束如图 21-7 所示，图中的粉红色与红色面采用绑定约束。

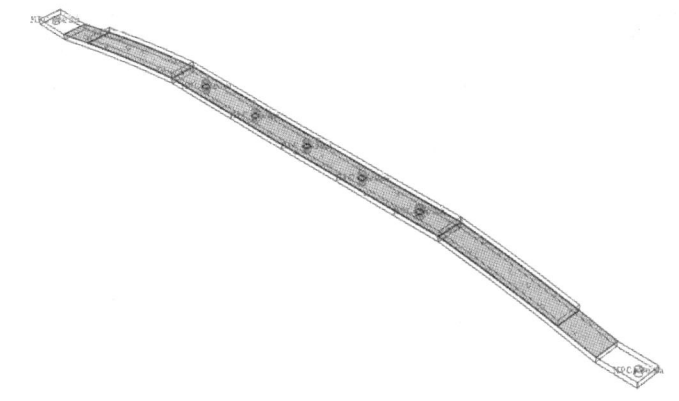

图 21-7　绑定 TIE 约束

（2）MPC 约束；在空间中建立 RP 点，RP 点与孔内的节间采用梁约束（hypermesh 中对应 RP-2 约束），此处的约束会产生过约束而导致板簧在进行模态计算时不能收敛，如图 21-8 所示；原因在于孔内第三层节点既是孔内节点，又是板簧面接触绑定的节点；装配的板簧仅有 5 个孔，孔的每个圆周有 8 个节点，因此共有 40 个节点产生过约束。此处的处理办法是先选取节点对应的 Set 集，如图 21-9 所示，在选取过程中不选择中间一层的几个节点，此时中间节点只存在绑定约束。k7 孔放大约束如图 21-10 所示；整个板簧装配约束如图 21-11 所示。

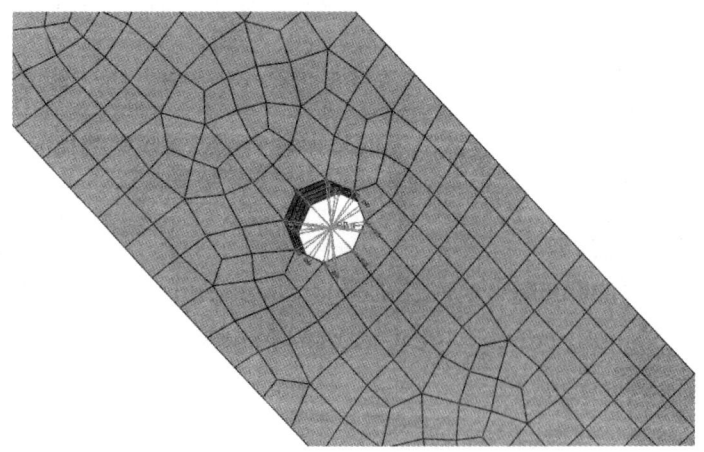

图 21-8　MPC 约束

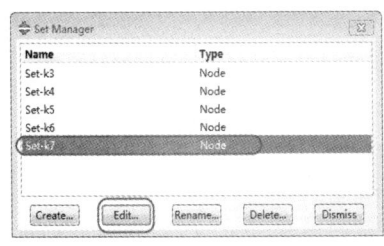 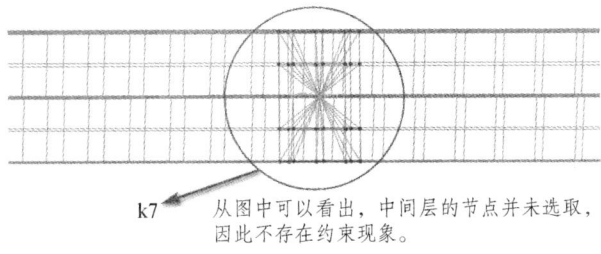

图 21-9 节点 Set 集　　　　　　　　图 21-10 k7 孔放大图

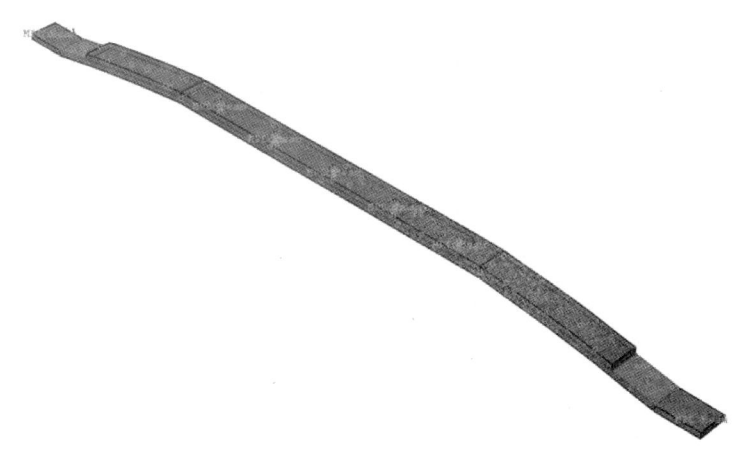

图 21-11 板簧装配体约束

21.3 前下双横置板簧悬架

1. 模型导入

（1）启动 ADAMS/CAR，选择专家模块进入建模界面。

（2）单击 File > Open 命令，弹出打开模板对话框，如图 12-12 所示，在 Template Name 中输入：mdids：//FASE/templates.tbl/_FSAE_sus_front_white.tpl。

（3）单击 OK 按钮，完成模型导入。

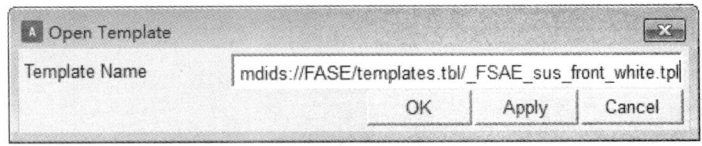

图 12-12 模板对话框

2. 删除下控制臂几何体

删除对应的下控制臂前后拉杆，但要注意不能删除下控制部件，删除后会导致模型错误，原因在于悬架的数组参数（即转向主销）由上下控制臂部件确定。

- 493 -

删除悬架以下几何体：

① ._FSAE_sus_front_white.gel_lca.gralin_lca_link_front；
② ._FSAE_sus_front_white.ger_lca.gralin_lca_link_front；
③ ._FSAE_sus_front_white.gel_lca.gralin_lca_link_rear；
④ ._FSAE_sus_front_white.ger_lca.gralin_lca_link_rear。

3．修改部件位置

下控制臂部件位置由三个点确定，删除前后拉杆后，需要重新确定下控制臂的部件位置；将下控制臂部件放置在转向主销下位置点。

修改信息如下：

（1）单击 Build > Part > General Part > Modify 命令，在下列对话框中输入相应的数据：

- Location Dependency：Delta location from coordinate；
- Coordinate Reference（参考坐标）：._FSAE_sus_front_GT_torsion_work.ground.cfl_torsion_to_chassis；
- Location：0，0，0。

（2）其余信息保持默认，单击 OK 按钮，完成下控制臂部件的修改。

（3）单击 File > Save As 命令，弹出保存模板对话框，如图 21-13 所示，在下列对话框中输入相应的数据：

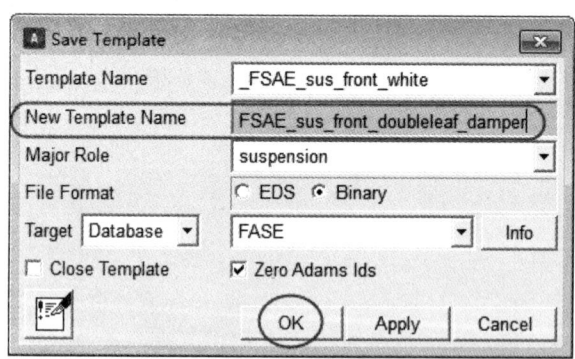

图 21-13　模型保存

- Template Name：_FSAE_sus_front_white；
- New Template Name：FSAE_sus_front_doubleleaf_damper；
- Major Role（主特征）：suspension；
- File Format：Binary；
- Target：Datebase/FASE。

（4）单击 OK 按钮，完成推杆式悬架模型模板 FSAE_sus_front_doubleleaf_damper 的保存。

4．结构框

（1）单击 Build > Construction Frame > New 命令，弹出创建结构框对话框，如图 21-14 所示，在下列对话框中输入相应的数据：

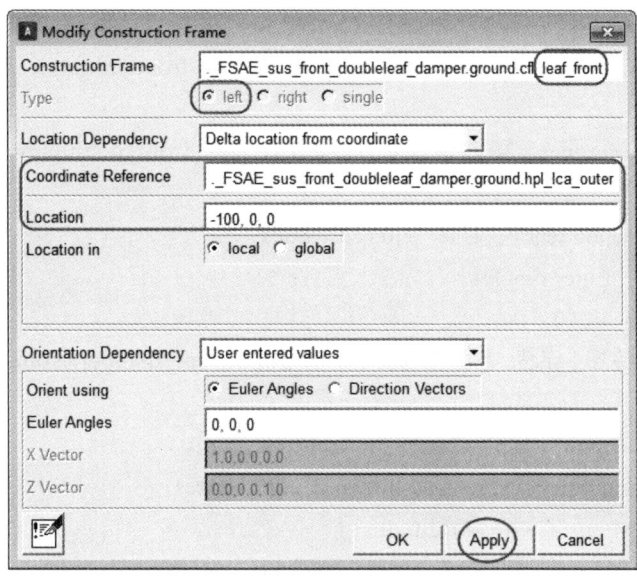

图 21-14　结构框 leaf_front

- Construction Frame（结构框名称）：leaf_front；
- Location Dependency：Delta location from coordinate；
- Coordinate Reference（参考坐标）：._FSAE_sus_front_doubleleaf_damper.ground. hpl_lca_outer；
- Location：-100，0，0；
- Location in：local；
- Orientation Dependency：User entered values；
- Orient using：Euler Angles；
- Euler Angles：0，0，0。

（2）单击 Apply 按钮，完成._FSAE_sus_front_doubleleaf_damper.ground.cfl_leaf_front 结构框的创建。同理，创建其他结构框。

- Construction Frame（结构框名称）：leaf_rear；
- Location Dependency：Delta location from coordinate；
- Coordinate Reference（参考坐标）：._FSAE_sus_front_doubleleaf_damper.ground.hpl_lca_outer；
- Location：100，0，0；
- Location in：local；
- Orientation Dependency：User entered values；
- Orient using：Euler Angles；
- Euler Angles：0，0，0。

（3）单击 Apply 按钮，完成._FSAE_sus_front_doubleleaf_damper.ground.cfl_leaf_rear 结构框的创建。

- Construction Frame（结构框名称）：leaf_front_mid；
- Location Dependency：Centered between coordinates；

- Centered between：Two Coordinates；
- Coordinate Reference #1（参考坐标）：._FSAE_sus_front_doubleleaf_damper.ground.cfl_leaf_front；
- Coordinate Reference #2（参考坐标）：._FSAE_sus_front_doubleleaf_damper.ground.cfr_leaf_front；
- Orientation Dependency：User entered values；
- Orient using：Euler Angles；
- Euler Angles：0，0，0。

（4）单击 Apply 按钮，完成._FSAE_sus_front_doubleleaf_damper.ground.cfs_leaf_front_mid 结构框的创建。

- Construction Frame（结构框名称）：leaf_rear_mid；
- Location Dependency：Centered between coordinates；
- Centered between：Two Coordinates；
- Coordinate Reference #1（参考坐标）：._FSAE_sus_front_doubleleaf_damper.ground.cfl_leaf_rear；
- Coordinate Reference #2（参考坐标）：._FSAE_sus_front_doubleleaf_damper.ground.cfr_leaf_rear；
- Orientation Dependency：User entered values；
- Orient using：Euler Angles；
- Euler Angles：0，0，0。

（5）单击 Apply 按钮，完成._FSAE_sus_front_doubleleaf_damper.ground.cfs_leaf_rear_mid 结构框的创建。

- Construction Frame（结构框名称）：damper_down；
- Location Dependency：Delta location from coordinate；
- Coordinate Reference（参考坐标）：._FSAE_sus_front_doubleleaf_damper.ground.hpl_lca_outer；
- Location：-50，0，0；
- Location in：local；
- Orientation Dependency：User entered values；
- Orient using：Euler Angles；
- Euler Angles：0，0，0。

（6）单击 Apply 按钮，完成._FSAE_sus_front_doubleleaf_damper.ground.cfl_damper_down 结构框的创建。

- Construction Frame（结构框名称）：damper_up；
- Location Dependency：Delta location from coordinate；
- Coordinate Reference（参考坐标）：._FSAE_sus_front_doubleleaf_damper.ground.cfl_damper_down；
- Location：0，200，150；
- Location in：local；

- Orientation Dependency：User entered values；
- Orient using：Euler Angles；
- Euler Angles：0，0，0。

（7）单击 Apply 按钮，完成 ._FSAE_sus_front_doubleleaf_damper.ground.cfl_damper_up 结构框的创建。

- Construction Frame（结构框名称）：damper_up_ref；
- Location Dependency：Delta location from coordinate；
- Coordinate Reference（参考坐标）：._FSAE_sus_front_doubleleaf_damper.ground.cfl_damper_up；
- Location：0，30，0；
- Location in：local；
- Orientation Dependency：User entered values；
- Orient using：Euler Angles；
- Euler Angles：0，0，0。

（8）单击 OK 按钮，完成 ._FSAE_sus_front_doubleleaf_damper.ground.cfl_damper_up_ref 结构框的创建。

5. 部件 damper_up

（1）单击 Build > Part > General Part > New 命令，弹出创建部件对话框，如图 21-15 所示，在下列对话框中输入相应的数据。

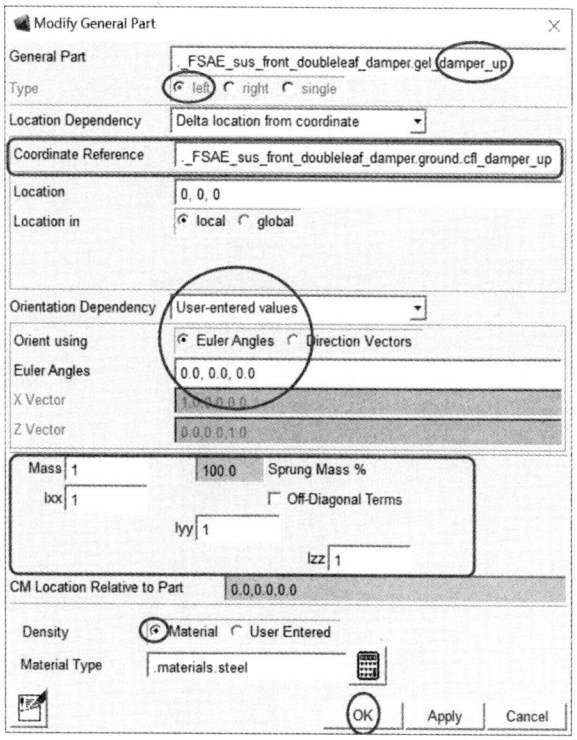

图 21-15　部件 damper_up

- General Part：damper_up；
- Type：left；
- Location Dependency：Delta location from coordinate；
- Coordinate Reference（参考坐标）：._FSAE_sus_front_doubleleaf_damper.ground.cfl_damper_up；
- Location：0，0，0；
- Location in：local；
- Orientation Dependency：User entered values；
- Orient using：Euler Angles；
- Euler Angles：0，0，0；
- Mass：1；
- Ixx：1；
- Iyy：1；
- Izz：1；
- Density：Material；
- Material Type：.materials.steel。

（2）单击 OK 按钮，完成部件 ._FSAE_sus_front_doubleleaf_damper.gel_damper_up 的创建。

6. 部件 damper_down

（1）单击 Build > Part > General Part > New 命令，在下列对话框中输入相应的数据：

- General Part：damper_down；
- Type：left；
- Location Dependency：Delta location from coordinate；
- Coordinate Reference（参考坐标）：._FSAE_sus_front_doubleleaf_damper.ground.cfl_damper_down；
- Location：0，0，0；
- Location in：local；
- Orientation Dependency：User entered values；
- Orient using：Euler Angles；
- Euler Angles：0，0，0；
- Mass：1；
- Ixx：1；
- Iyy：1；
- Izz：1；
- Density：Material；
- Material Type：.materials.steel。

（2）单击 OK 按钮，完成部件 ._FSAE_sus_front_doubleleaf_damper.gel_damper_down 创建。

7. 板簧 leaf_front

板簧 MNF 中性文件：fsae_sus_leaf_down_30width.mnf 存放在章节文件中，请读者自行调阅查看。

（1）单击 Build > Part > Flexible Body > New 命令，弹出创建部件对话框，如图 21-16 所示，在下列对话框中输入相应的数据：

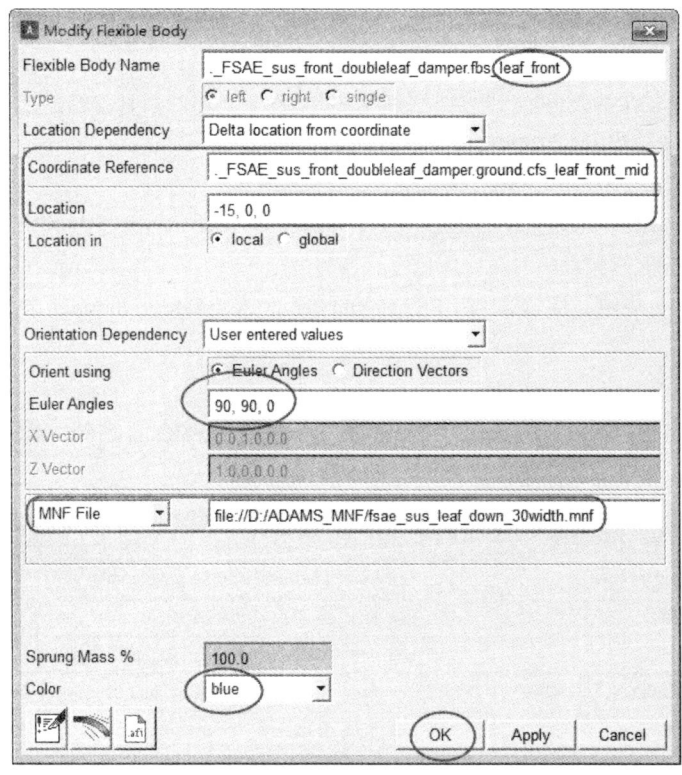

图 21-16 板簧 leaf_front

- Flexible Body Name：leaf_front；
- Type：single；
- Location Dependency：Delta location from coordinate；
- Coordinate Reference（参考坐标）：._FSAE_sus_front_doubleleaf_damper.ground.cfs_leaf_front_mid；
- Location：-15，0，0；
- Orientation Dependency：User entered values；
- Orient using：Euler Angles；
- Euler Angles：90，90，0；
- MNF File：file：//D：/ADAMS_MNF/fsae_sus_leaf_down_30width.mnf；
- Color：blue。

（2）单击 Apply 按钮，完成部件._FSAE_sus_front_doubleleaf_damper.fbs_leaf_front 的创建。

8. 板簧 leaf_rear

（1）单击 Build > Part > Flexible Body > New 命令，在下列对话框中输入相应的数据：

- Flexible Body Name：leaf_rear；
- Type：single；
- Location Dependency：Delta location from coordinate；
- Coordinate Reference（参考坐标）：._FSAE_sus_front_doubleleaf_damper.ground.cfs_leaf_rear_mid；
- Location：-15，0，0；
- Orientation Dependency：User entered values；
- Orient using：Euler Angles；
- Euler Angles：90，90，0；
- MNF File：file：//D：/ADAMS_MNF/fsae_sus_leaf_down_30width.mnf；
- Color：blue。

（2）单击 Apply 按钮，完成部件._FSAE_sus_front_doubleleaf_damper.fbs_leaf_rear 的创建。

9. 避震器

（1）单击 Build > Force > Damper > New 命令，弹出避震器创建对话框，如图 21-17 所示，在下列对话框中输入相应的数据：

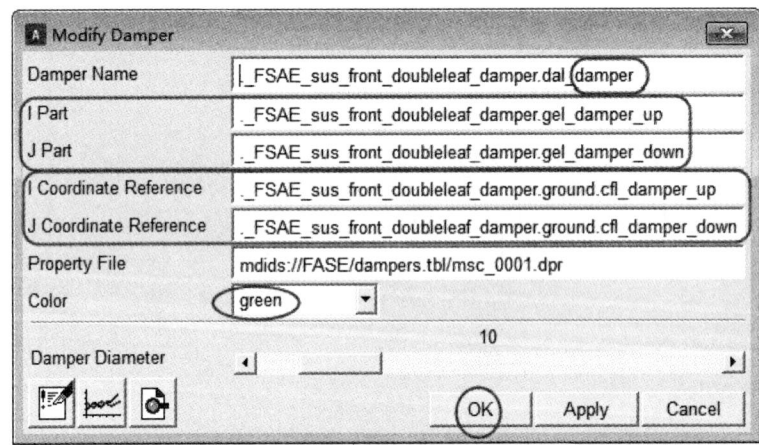

图 21-17　避震器

- Damper Name（减震器名称）：damper；
- I Part：._FSAE_sus_front_doubleleaf_damper.gel_damper_up；
- J Part：._FSAE_sus_front_doubleleaf_damper.gel_damper_down；
- I Coordinate Reference（参考坐标）：._FSAE_sus_front_doubleleaf_damper.ground.cfl_damper_up；
- J Coordinate Reference（参考坐标）：._FSAE_sus_front_doubleleaf_damper.ground.cfl_damper_down；
- Property File（属性文件）：mdids：//FASE/dampers.tbl/msc_0001.dpr；

- Damper Diameter（避震器直径）：拖动滑块选择 10 mm；
- Color：green。

（2）单击 OK 按钮，完成避震器 ._FSAE_sus_front_doubleleaf_damper.dal_damper 的创建。

10. 部件 damper_down 与 lca 之间的 convel 约束

（1）单击 Build > Attachments > Joint > New 命令，弹出创建约束件对话框，如图 21-18 所示，在下列对话框中输入相应的数据：

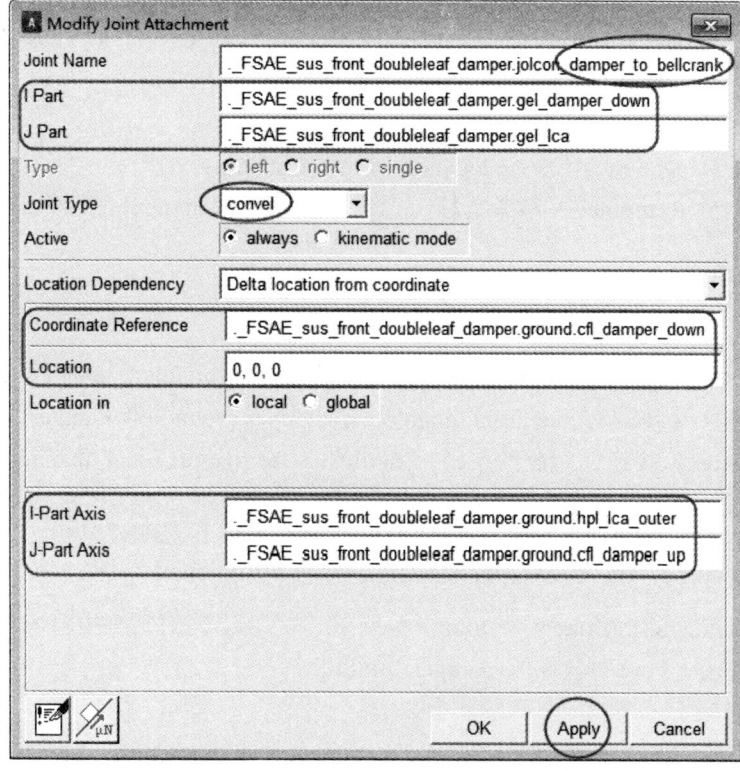

图 21-18 约束 convel

- Joint Name（约束副名称）：damper_to_bellcrank；
- I Part：._FSAE_sus_front_doubleleaf_damper.gel_damper_down；
- J Part：._FSAE_sus_front_doubleleaf_damper.gel_lca；
- Joint Type（约束副类型）：convel；
- Active（激活）：always；
- Location Dependency：Delta location from coordinate；
- Coordinate Reference（参考坐标）：._FSAE_sus_front_doubleleaf_damper.ground.cfl_damper_down；
- Location：0，0，0；
- Location in：local；
- I-Part Axis：._FSAE_sus_front_doubleleaf_damper.ground.hpl_lca_outer；

- J-Part Axis：._FSAE_sus_front_doubleleaf_damper.ground.cfl_damper_up。

（2）单击 Apply 按钮，完成约束副._FSAE_sus_front_doubleleaf_damper.jolcon_damper_to_bellcrank 的创建。

11. 部件 damper_up 与 suspension_to_chassis 之间的 convel 约束

（1）单击 Build > Attachments > Joint > New 命令，在下列对话框中输入相应的数据：
- Joint Name（约束副名称）：damper_to_chassis；
- I Part：._FSAE_sus_front_doubleleaf_damper.gel_damper_up；
- J Part：._FSAE_sus_front_doubleleaf_damper.mts_suspension_to_chassis；
- Joint Type（约束副类型）：convel；
- Active（激活）：always；
- Location Dependency：Delta location from coordinate；
- Coordinate Reference（参考坐标）：._FSAE_sus_front_doubleleaf_damper.ground.cfl_damper_up；
- Location：0，0，0；
- Location in：local；
- I-Part Axis：._FSAE_sus_front_doubleleaf_damper.ground.cfl_damper_down；
- J-Part Axis：._FSAE_sus_front_doubleleaf_damper.ground.cfl_damper_up_ref。

（2）单击 Apply 按钮，完成约束副._FSAE_sus_front_doubleleaf_damper.jolcon_damper_to_chassis 的创建。

12. 部件 damper_up 与 damper_down 之间的 cylindrical 约束

（1）单击 Build > Attachments > Joint > New 命令，在下列对话框中输入相应的数据：
- Joint Name（约束副名称）：damper_slide；
- Type：left；
- I Part：._FSAE_sus_front_doubleleaf_damper.gel_damper_up；
- J Part：._FSAE_sus_front_doubleleaf_damper.gel_damper_down；
- Joint Type（约束副类型）：cylindrical；
- Active（激活）：always；
- Location Dependency：Centered between coordinates；
- Centered between：Two Coordinates；
- Coordinate Reference #1（参考坐标）：._FSAE_sus_front_doubleleaf_damper.ground.cfl_damper_down；
- Coordinate Reference #2（参考坐标）：._FSAE_sus_front_doubleleaf_damper.ground.cfl_damper_up；
- Location Dependency：Orient axis along line；
- Coordinate Reference #1（参考坐标）：._FSAE_sus_front_doubleleaf_damper.ground.cfl_damper_down；
- Coordinate Reference #2（参考坐标）：._FSAE_sus_front_doubleleaf_damper.ground.cfl_

damper_up。

（2）单击 Apply 按钮，完成约束副._FSAE_sus_front_doubleleaf_damper.jolcyl_damper_slide 的创建。

13. 部件 leaf_front 与 lca 之间的 fixed 约束

（1）单击 Build > Attachments > Joint > New 命令，在下列对话框中输入相应的数据：
- Joint Name（约束副名称）：leaf_front；
- Type：left；
- I Part：._FSAE_sus_front_doubleleaf_damper.fbs_leaf_front；
- J Part：._FSAE_sus_front_doubleleaf_damper.gel_lca；
- Joint Type（约束副类型）：fixed；
- Active（激活）：always；
- Location Dependency：Delta location from coordinate；
- Coordinate Reference（参考坐标）：._FSAE_sus_front_doubleleaf_damper.ground.cfl_leaf_front；
- Location：0，0，0；
- Location in：local；
- Closest Interface Node：left/9496，right/9496（9496 指有限元软件中创建的接口点）。

（2）单击 Apply 按钮，完成约束副._FSAE_sus_front_doubleleaf_damper.jolfix_leaf_front 的创建。

14. 部件 leaf_rear 与 lca 之间的 fixed 约束

（1）单击 Build > Attachments > Joint > New 命令，在下列对话框中输入相应的数据：
- Joint Name（约束副名称）：leaf_rear；
- Type：left；
- I Part：._FSAE_sus_front_doubleleaf_damper.fbs_leaf_rear；
- J Part：._FSAE_sus_front_doubleleaf_damper.gel_lca；
- Joint Type（约束副类型）：fixed；
- Active（激活）：always；
- Location Dependency：Delta location from coordinate；
- Coordinate Reference（参考坐标）：._FSAE_sus_front_doubleleaf_damper.ground.cfl_leaf_rear；
- Location：0，0，0；
- Location in：local；
- Closest Interface Node：left/9496，right/9496（9496 指有限元软件中创建的接口点）。

（2）单击 OK 按钮，完成约束副._FSAE_sus_front_doubleleaf_damper.jolfix_leaf_rear 的创建。

15. 衬套连接硬点

（1）单击 Build > Hardpoind > New 命令，在下列对话框中输入相应的数据：
- Hardpoint：leaf_front_mount；
- Type：left；
- Location：-100.0，-100.0，158.2432808，通过鼠标选取节点位置获取参数；

（2）单击 Apply 按钮，完成._FSAE_sus_front_doubleleaf_damper.ground.hpl_leaf_front_mount 硬点的创建。同理，创建其他硬点。
- Hardpoint：leaf_rear_mount；
- Type：left；
- Location：100.0，-100.0，158.2432808。

（3）单击 OK 按钮，完成._FSAE_sus_front_doubleleaf_damper.ground.hpl_leaf_rear_mount 硬点的创建。

16. 部件 leaf_front 与 suspension_to_chassis 之间的 bushing 约束

（1）单击 Build > Attachments > Bushing > New 命令，弹出创建衬套对话框，如图 21-19 所示，在下列对话框中输入相应的数据：

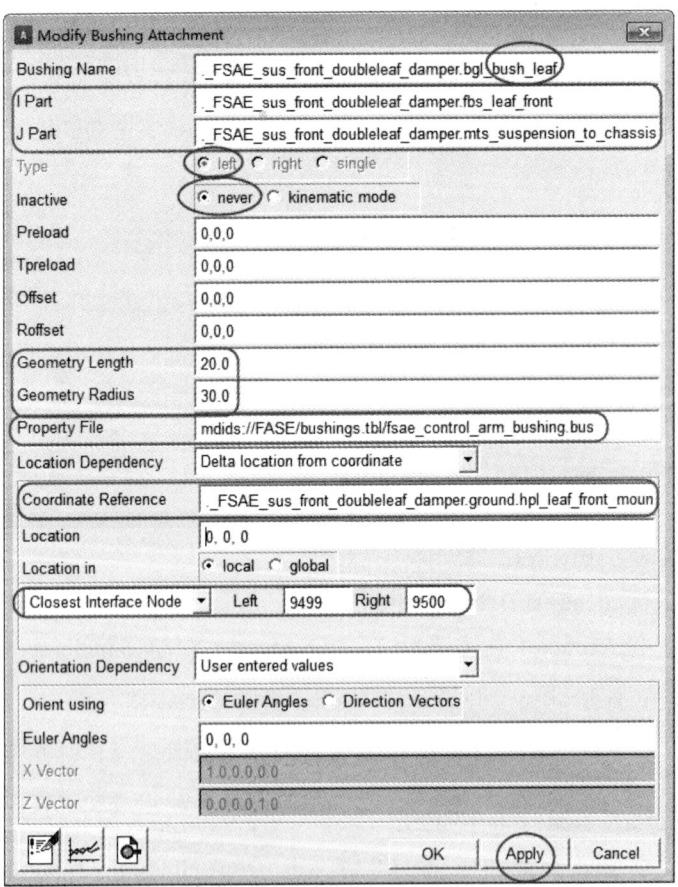

图 21-19　衬套-bush_leaf

- Bushing Name（约束副名称）：bush_leaf；
- I Part：._FSAE_sus_front_doubleleaf_damper.fbs_leaf_front；
- J Part：._FSAE_sus_front_doubleleaf_damper.mts_suspension_to_chassis；
- Inactive（抑制）：never；
- Preload：0，0，0；
- Tpreload：0，0，0；
- Offset：0，0，0；
- Roffset：0，0，0；
- Geometry Length：20；
- Geometry Radius：30；
- Property File：mdids：//FASE/bushings.tbl/fsae_control_arm_bushing.bus；
- Location Dependency：Delta location from coordinate；
- Coordinate Reference（参考坐标）：._FSAE_sus_front_doubleleaf_damper.ground.hpl_leaf_front_mount；
- Location：0，0，0；
- Location in：local；
- Closest Interface Node：Left：9499 / Right：9500；
- Orientation Dependency：User entered values；
- Orient using：Euler Angles；
- Euler Angles：0，0，0。

（2）单击 Apply 按钮，完成轴套._FSAE_sus_front_doubleleaf_damper.bgl_bush_leaf 的创建。

17. 部件 leaf_rear_mount 与 suspension_to_chassis 之间的 bushing 约束

（1）单击 Build > Attachments > Bushing > New 命令，在下列对话框中输入相应的数据：
- Bushing Name（约束副名称）：leaf_rear_mount；
- I Part：._FSAE_sus_front_doubleleaf_damper.fbs_leaf_rear；
- J Part：._FSAE_sus_front_doubleleaf_damper.mts_suspension_to_chassis；
- Inactive（抑制）：never；
- Preload：0，0，0；
- Tpreload：0，0，0；
- Offset：0，0，0；
- Roffset：0，0，0；
- Geometry Length：20；
- Geometry Radius：30；
- Property File：mdids：//FASE/bushings.tbl/fsae_control_arm_bushing.bus；
- Location Dependency：Delta location from coordinate；
- Coordinate Reference（参考坐标）：._FSAE_sus_front_doubleleaf_damper.ground.hpl_

leaf_rear_mount；
- Location：0，0，0；
- Location in：local；
- Closest Interface Node：Left：9499 / Right：9500；
- Orientation Dependency：User entered values；
- Orient using：Euler Angles；
- Euler Angles：0，0，0。

（2）单击 OK 按钮，完成轴套._FSAE_sus_front_doubleleaf_damper.bgl_leaf_rear_mount 的创建。

18．几何体 lca

（1）单击 Build > Geometry > Link > New 命令，弹出创建几何体对话框，如图 21-20 所示，在下列对话框中输入相应的数据：

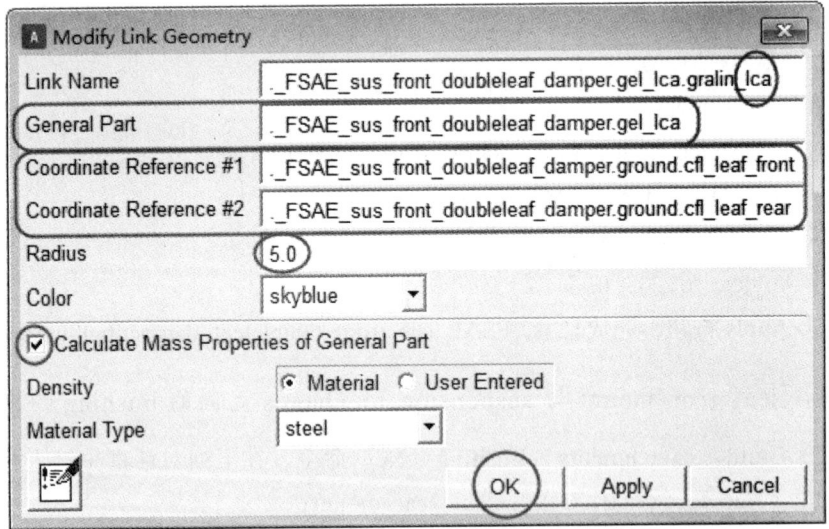

图 21-20　连杆几何体_lca

- Link Name（连杆名称）：lca；
- General Part：._FSAE_sus_front_doubleleaf_damper.gel_lca；
- Coordinate Reference #1（参考坐标）：._FSAE_sus_front_doubleleaf_damper.ground.cfl_leaf_front；
- Coordinate Reference #2（参考坐标）：._FSAE_sus_front_doubleleaf_damper.ground.cfl_leaf_rear；
- Radius（半径）：5；
- Color：skyblue；
- 勾选 Calculate Mass Properties of General Part 复选框，当几何体建立好之后，会更新对应部件的质量和惯量参数；
- Density：Material；

- Material Type：steel；

（2）单击 OK 按钮，完成._FSAE_sus_front_doubleleaf_damper.gel_lca.gralin_lca 几何体的创建。

19. 单轮跳动仿真

至此，悬架模型建立完成，如图 21-1 所示；切换到专家模板，建立子系统；装配悬架，悬架与振动台装配好后，修改轮胎的半径为 200 mm；单轮跳动仿真如图 21-21 所示，上下跳动距离为 30 mm；四轮定位参数如图 21-22 ~ 图 21-25 所示。

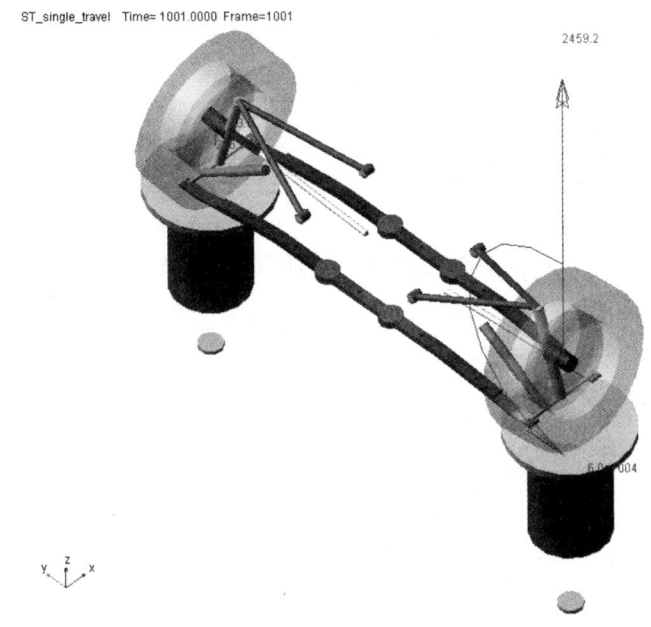

图 21-21 单轮跳动仿真

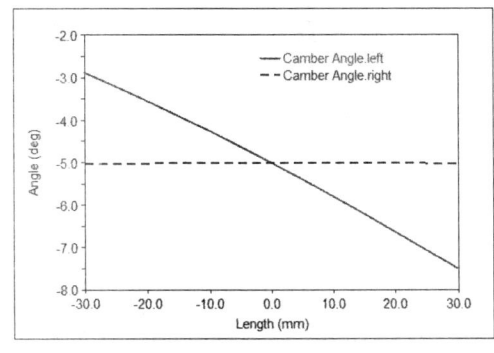

图 21-22 外倾角

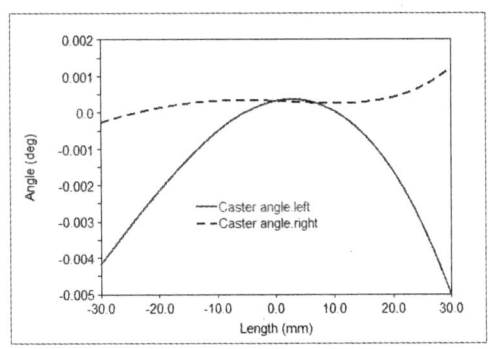

图 21-23 主销后倾角

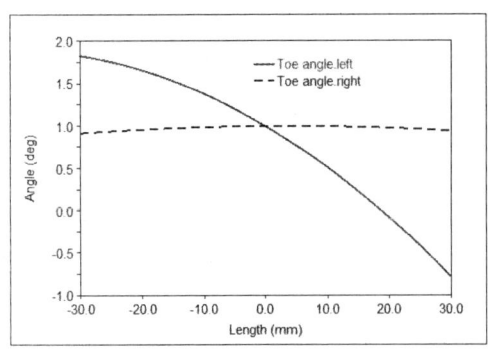
图 21-24　前束角

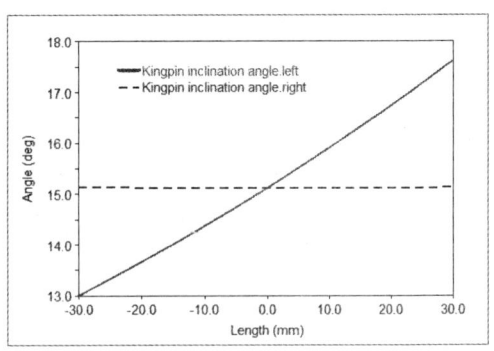
图 21-25　主销内倾角

21.4　后下双横置板簧悬架

后悬架模型建模不再展开，请读者参阅前下双横置板簧悬架模型建立；建立好的后下双横置板簧悬架如图 21-26 所示。相关信息如下：

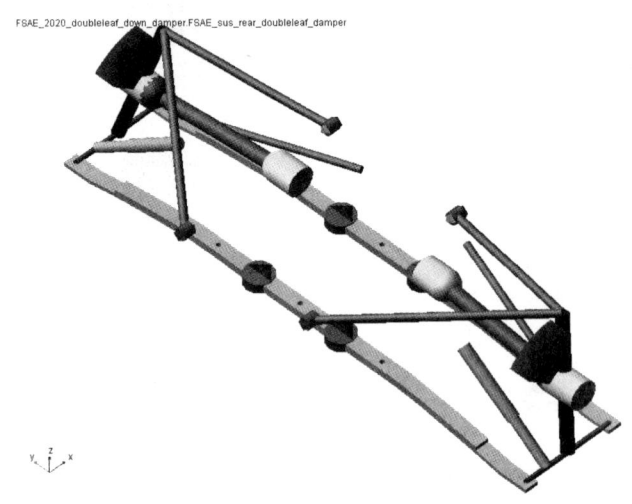

图 21-26　后下双横置板簧悬架模型

后下双横置板簧悬架模型硬点与变量信息如下：

Info for subsystem:　FSAE_sus_rear_doubleleaf_damper

File Name　　　:　<FASE>/subsystems.tbl/FSAE_sus_rear_doubleleaf_damper.sub

Template　　　 :　mdids://FASE/templates.tbl/_FSAE_sus_rear_doubleleaf_damper.tpl

Comments　　　:　*no comments found*

Major Role　　 :　suspension

| Minor Role | : | rear | | |

HARDPOINTS:

hardpoint name	symmetry	x_value	y_value	z_value
global	single	1524.0	0.0	0.0
bellcrank_pivot	left/right	1547.8	-170.0	381.0
drive_shaft_inr	left/right	1550.0	-200.0	225.0
lca_outer	left/right	1498.6	-546.1	101.6
leaf_front_mount	left/right	1398.6	-100.0	139.2
leaf_rear_mount	left/right	1598.6	-100.0	139.2
prod_outer	left/right	1498.6	-497.2	127.0
prod_to_bellcrank	left/right	1509.7	-200.0	381.0
tierod_inner	left/right	1676.4	-127.0	182.4
tierod_outer	left/right	1574.8	-457.2	152.4
uca_front	left/right	1270.0	-152.4	304.8
uca_outer	left/right	1549.4	-482.6	355.6
uca_rear	left/right	1625.6	-152.4	304.8
wheel_center	left/right	1524.0	-558.8	228.6

PARAMETERS:

parameter name	symmetry	type	value
driveline_active	single	integer	1
kinematic_flag	single	integer	0
camber_angle	left/right	real	-1.5
drive_shaft_offset	left/right	real	75.0
toe_angle	left/right	real	0.0

21.5 收油门直线仿真

收油门直线仿真指的是在直线行驶中突然抬起（释放）油门踏板，检查汽车保持运动轨迹的性能；通过替换前后悬架建立好的整车模型如图 21-27 所示。

图 21-27　整车模型

整车信息如下（子系统包含硬点信息省略）：

**************************** 　TESTRIG INFO　 ****************************

　　Testrig Name　 : 　__MDI_SDI_TESTRIG

**************************** 　ASSEMBLY INFO　 ****************************

　　Assembly Name　 : 　FSAE_2020_doubleleaf_down_damper
　　Assembly Class : 　full_vehicle
　　File Name　　　 : 　<FASE>/assemblies.tbl/FSAE_2020_doubleleaf_down_damper.asy

SUBSYSTEM NAME	MAJOR ROLE	MINOR ROLE
front_tire	wheel	front
rear_tire	wheel	rear
brake	brake_system	any
FSAE_Body_2017	body	any
FSAE_steering_2017	steering	front
powertrain_fsae_2017	powertrain	rear
FSAE_sus_front_doubleleaf_damper	suspension	front
FSAE_sus_rear_doubleleaf_damper	suspension	rear

（1）单击 Simulate > Full-Vehicle Analysis > Straight-line Events >Power-Off Straight Line 命令，弹出阶跃仿真对话框，如图 21-28 所示，在下列对话框中输入相应的数据：

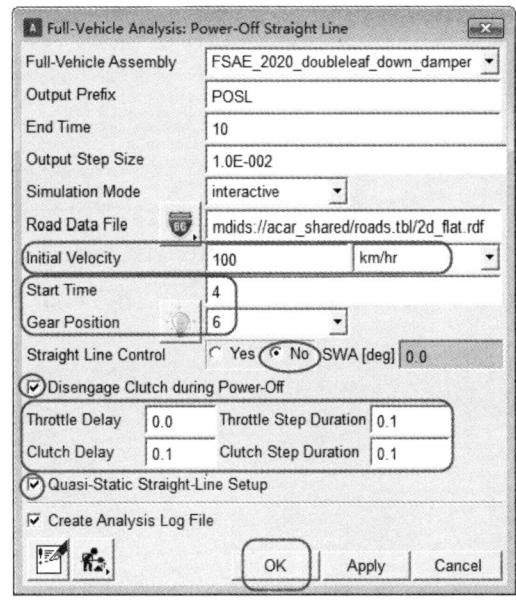

图 21-28　收油门直线仿真设置

- Full-Vehicle Assembly：FSAE_sus_front_doubleleaf_damper；
- Output Prefix：POSL；
- End Time：10；
- Output Step Sizes：0.01；
- Mode of Simulation：interactive；
- Road Date File：mdids：//acar_shared/roads.tbl/2d_flat.rdf；
- Initial Velocity（单位：km/h）：100；
- Start Time：4；
- Gear Position：6；
- Straight Line Control：No，在释放油门的时候方向盘可以自由转动；
- 勾选 Disengage Clutch during Power-Off，离合器不中断；
- Throttle Delay：0，释放油门延迟量；
- Throttle Step Duration：0.1，释放油门时间；
- Clutch Delay：0.1，释放离合器延迟量；
- Clutch Step Duration：0.1，释放离合器时间；
- 勾选 Quasi-Static Straight-Line Setup。

（2）单击 Apply 按钮，完成 FSAE_sus_front_doubleleaf_damper.赛车收油门直线仿真设置并提交运算。运算完成后，车辆纵向加速度及侧向偏移量如图 21-29 和图 21-30 所示。

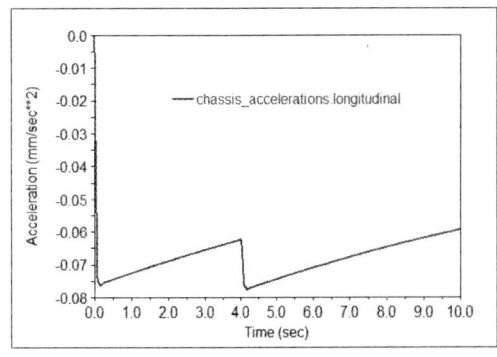
图 21-29 纵向加速度（制动加速度）/X 轴

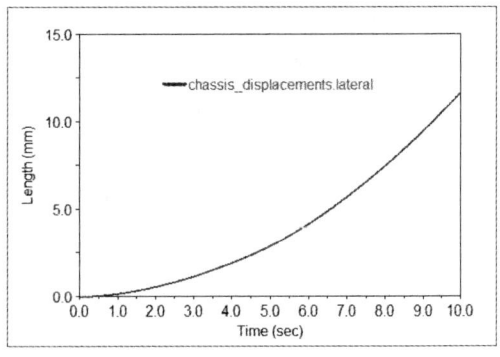
图 21-30 侧向偏移量/Y 轴

第 22 章　优化设计实验

对于一个动力学模型，设计越复杂，影响设计的因素也就越多。由于各个参数之间是相互影响的，所以每次改变一个参数很难提高设计的性能。如果同时改变多个参数，需要大量的仿真计算，并产生庞大的仿真数据，这样很难判断到底哪个参数是主要的，哪个是次要的。利用 ADAMS/Insight，工程师们可以对虚拟样机和物理样机进行系统的研究、深入的分析，并可以与整个团队分享自己的成果；研究策略可以应用于部件或子系统，或者扩展到评估多层次问题中，实现跨部门的设计方案优化。ADAMS/Insight 可以通过网页或者数据表格实现数据交换，从而使设计人员、研究人员以及项目管理人员能够直接参与到"如果?—怎样?"的研究中，而不需要接触到实际的仿真模型。

ADAMS/Insight 的特点如下：

① 研究策略：设计研究、蒙特卡罗法研究、设计实验、扫描研究、周期研究、单目标和多目标优化；

② 支持用户自定义策略或将已有策略应用于其他模型；

③ 响应曲面法（Response Surface Methods）是通过对试验数据进行数学回归分析的方法，帮助工程师更好地理解产品的性能和系统内部各个参数之间的相互关系；

④ 可综合考虑各种制造因素的影响（例如：公差、装配误差、加工精度等）；

⑤ 对拥有共同输入的不同域的实验进行综合分析；

⑥ 将实验结果与解算结果进行综合比较，以便更深入地研究；

⑦ 网络发布实验结果；

⑧ 可输出为 Excel、MATLAB 以及 Visual Basic 文件格式；

⑨ 既可与其他 MSC.Adams 模块联合使用，也可脱离 MSC.Adams 环境单独使用。

学习目标

（1）会双 A 臂悬架前束角优化。
（2）会推杆式悬架模型外倾角优化。

22.1 双 A 臂悬架前束角优化数据库导入模型

（1）启动 ADAMS/View，保持界面为默认设置。

（2）单击 File > Import 命令，弹出导入模型界面，如图 22-1 所示，在下列对话框中输入相应的数据：

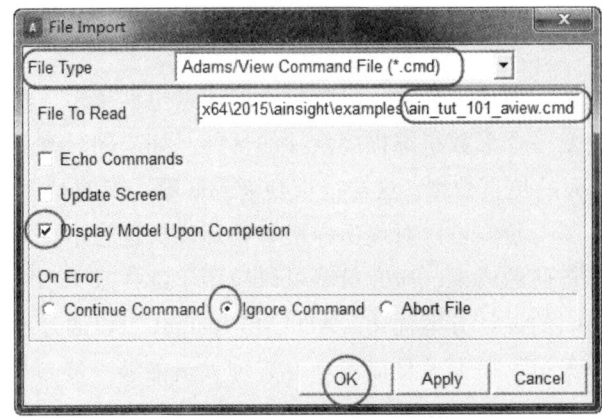

图 22-1　导入 CMD 模型

- File Type：ADAMS/View Command File（*.cmd）；
- File To Read：D：\MSC.Software\Adams_x64\2015\ainsight\examples\ain_tut_101_aview.cmd；
- 勾选 Display Model Upon Completion。

（3）单击 OK 按钮，完成模型导入，如图 22-2 所示。图 22-2 为双 A 臂悬架概念模型，模型中的部件特性、约束、驱动等参数请读者自行查看学习，此处不做详细叙述。

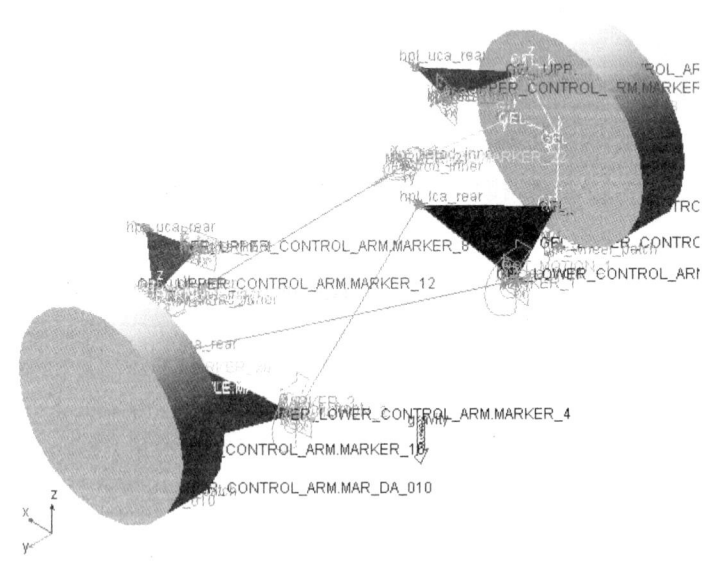

图 22-2　双 A 臂悬架概念模型

1. 运行仿真设定

(1) 单击 Simulation > Simulate 命令，弹出仿真设定对话框，如图 22-3 所示，在下列对话框中输入相应的数据：

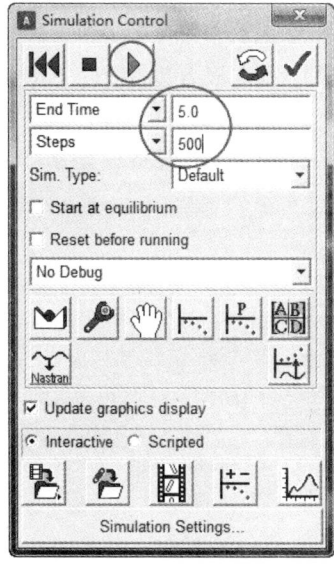

图 22-3　仿真设定

- End Time：5；
- Steps：500；
- 其余保持默认设置。

(2) 单击 Start simulation 按钮，运行完成仿真。

2. 优化设计实验

(1) 单击 Design Exploration > Adams/Insight Export Dialog box 命令，弹出优化输出界面对话框，如图 22-4 所示，在下列对话框中输入相应的数据：

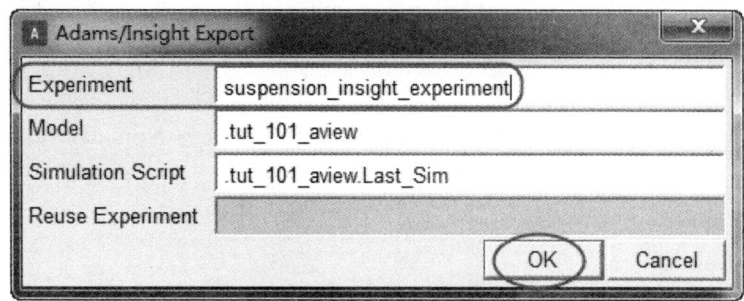

图 22-4　ADAMS/Insight 输出界面

- Experiment：suspension_insight_experiment；
- Model：tut_101_aview；
- Simulation Script：.tut_101_aview.Last_Sim。

（2）单击 OK 按钮，完成输出接口的设置，此时 ADAMS/View 界面消失，弹出 ADAMS/Insight 界面，此界面较为简单，包含菜单栏、工具条、模型树及显示窗口四部分；实验矩阵的设置、分析等将在此界面完成。

3．创建设计矩阵（优化变量选取）

（1）选取优化变量，按如下顺序依次展开模型树：Factors > Candidates > tut_101_aview > ground > hpl_tierod_outer。

（2）选择：ground.hpl_tierod_outer.x，此时视窗显示如图 22-5 所示。

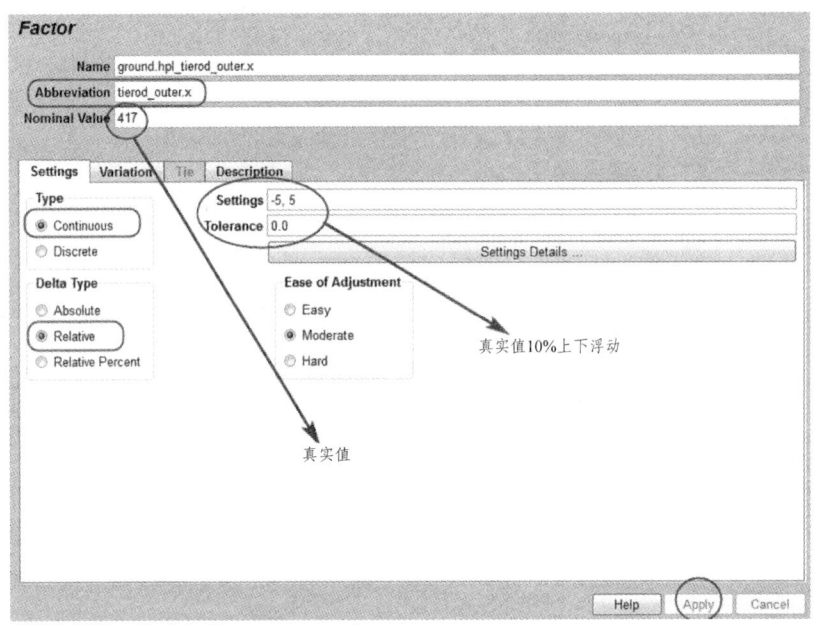

图 22-5　设计变量因素

（3）单击 Promote to inclusion，将设计因素：ground.hpl_tierod_outer.x 提升为设计变量，在下列对话框中输入相应的数据：

- Abbreviation（简称）：ground.hpl_tierod_outer.x；
- Nominal Value：417，此数值为转向横拉杆在 X 方向的真实值；
- Type：Continuous；
- Delta Type：Relative，相对值，一般允许其变化范围为 Nominal Value 的 10%；
- Settings：-5，5；在 Nominal Value 的基础减少/增加 5，即允许数值的变化范围为[412，422]；
- 切换到 Description 菜单；
- Units：mm。

（4）单击 Apply 按钮，完成 ground.hpl_tierod_outer.x 参数的设定。

（5）重复上述步骤，完成以下设计因素的设定：

① ground.hpl_tierod_outer.y；

② ground.hpl_tierod_outer.z。

4. 优化目标

(1) 选取优化目标，按如下顺序依次展开模型树：Responses > Candidates > tut_101_aview > toe_left_REQ（此优化目标模型导入之前已经创建好，因此不需要创建，如果是自建模型，则需要根据优化的任务创建自己所希望的优化目标）。

(2) 选择：toe_left_REQ，此时视窗显示如图 22-6 所示。

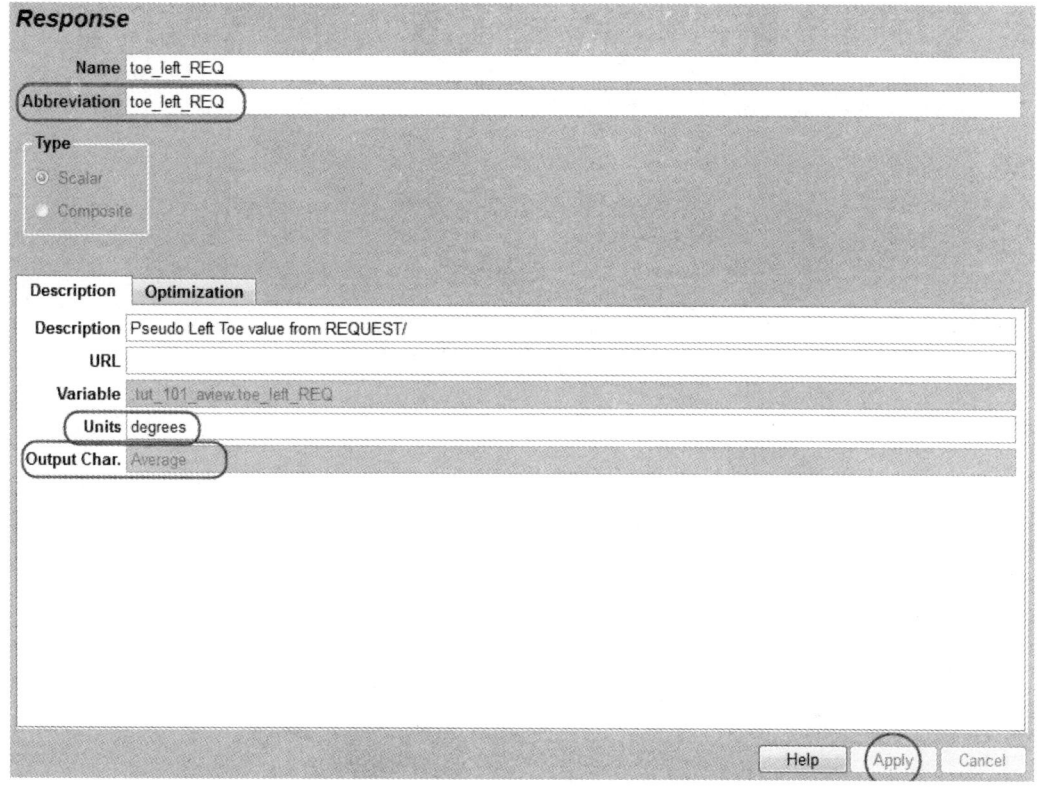

图 22-6 优化目标

(3) 单击 Promote to inclusion，将响应因素：ground.hpl_tierod_outer.x 提升为响应目标（优化目标），在下列对话框中输入相应的数据：
- Abbreviation（简称）：toe_left_REQ；
- Units：degrees。

(4) 单击 Apply 按钮，完成 toe_left_REQ 优化目标的设定。

(5) 重复上述步骤，完成 toe_right_REQ 优化目标的设定。

5. 设置设计规范

(1) 单击 Define > Experiment Design > Set Design Specification，弹出设计规范界面，如图 22-7 所示，在下列对话框中输入相应的数据：
- Investigation Strategy：DOE Screening（2 Level）；
- Model：Linear；
- DOE Design Type：Full Factorial。

（2）单击 Apply 按钮，完成设计规范中参数设定。

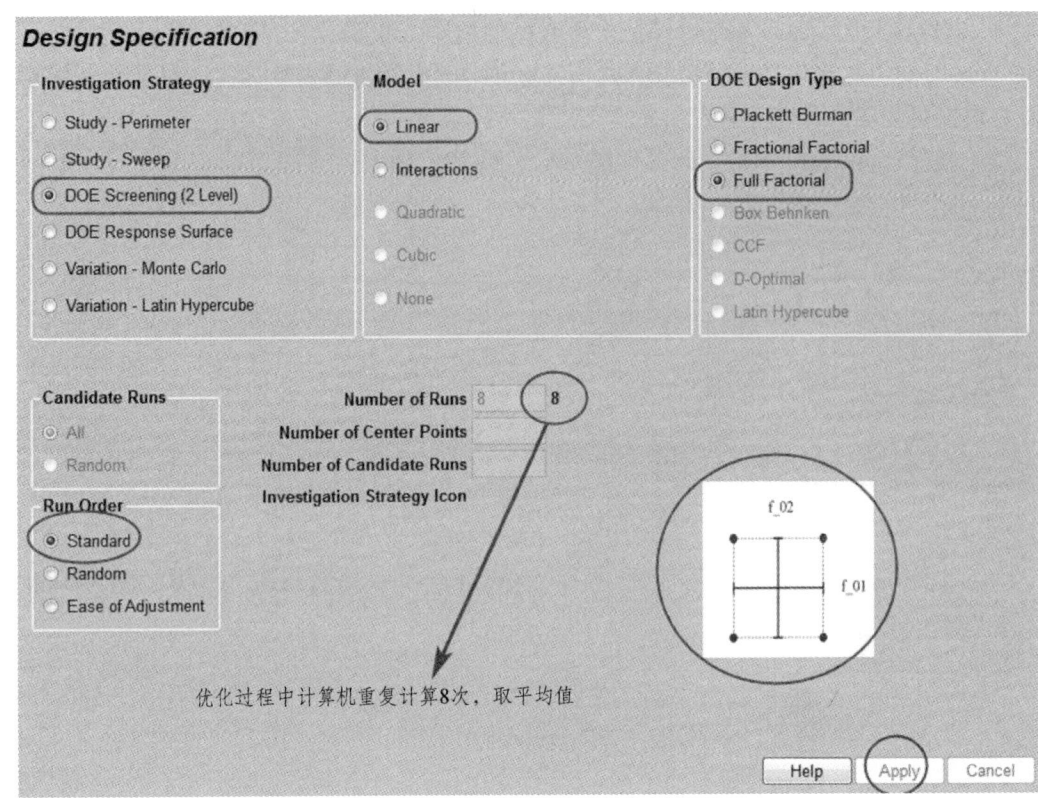

图 22-7　设计规范

6. 创建设计与工作空间

（1）单击 Define > Experiment Design > Create Design Space。
（2）单击 Define > Experiment Design > Create Work Space。
（3）在模型树上单击 Work Space，弹出视图窗口，如图 22-8 所示，在图 22-8 中，可以看到上述所选取的优化变量及优化目标参数，优化变量通过不同的组合共有 8 种不同的组合，提交计算后计算机需要重复计算 8 次。

Trial	tierod outer.y	tierod outer.z	tierod outer.x	toe right REQ	toe left REQ
Trial 1	-755	325	412		
Trial 2	-755	325	422		
Trial 3	-755	335	412		
Trial 4	-755	335	422		
Trial 5	-745	325	412		
Trial 6	-745	325	422		
Trial 7	-745	335	412		
Trial 8	-745	335	422		

图 22-8　工作空间

7. 提交计算

（1）单击 Simulation > Build-Run-Load > All，ADAMS/View 打开并运行由实验定义的仿真。ADAMS/View 状态栏显示模拟进度的消息，消息窗口也会显示有关关节位置的警告（在本教程中可以忽略这些警告）。

（2）计算完成后，界面显示如图 22-9 和图 22-10 所示；图 22-9 所示参数保持恒定值并没有变化，原因在于左右车轮为独立悬架，相互不影响，同时优化变量选取的是右侧横向拉杆外侧点 X、Y、Z 三个方向的参数，因此优化目标输出变化仅为左侧车轮的前束角。

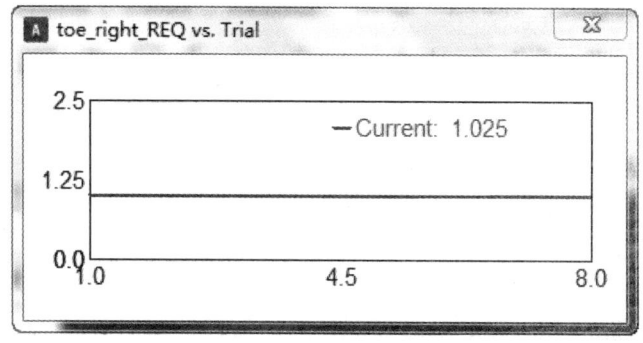

图 22-9　右轮前束角

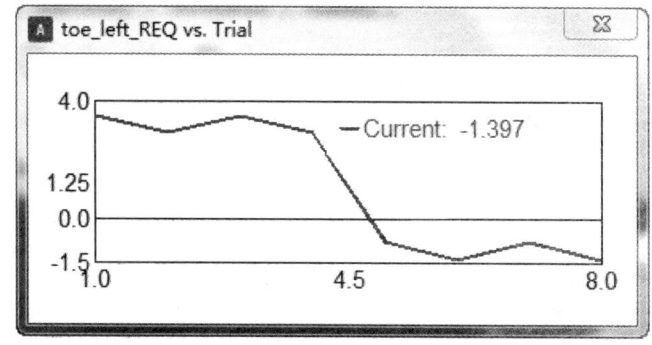

图 22-10　左轮前束角

8. 优化结果

（1）单击 Simulation > Adams/Insight > Display 命令，显示如图 22-11 所示的界面。

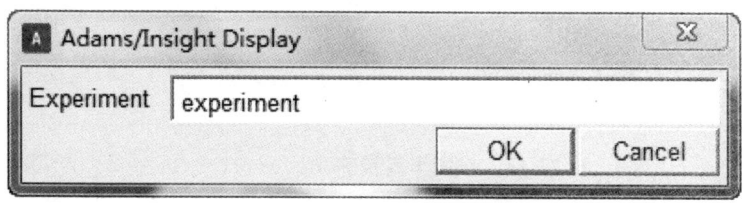

图 22-11　优化结果输出界面

（2）在模型树下单击 Work Space，计算优化出的结果如图 22-12 所示，从结果中可以看出，每种不同的组合左侧车轮前束角的值完全不同，根据设计要求，可以选取最符合目标的一组值。

Work Space					
Trial		tierod outer.y	tierod outer.z	tierod outer.x	toe right REQ	toe left REQ
1	Trial 1	-755	325	412	1.02452	3.5103
2	Trial 2	-755	325	422	1.02452	2.90266
3	Trial 3	-755	335	412	1.02452	3.5103
4	Trial 4	-755	335	422	1.02452	2.90266
5	Trial 5	-745	325	412	1.02452	-0.777356
6	Trial 6	-745	325	422	1.02452	-1.39724
7	Trial 7	-745	335	412	1.02452	-0.777356
8	Trial 8	-745	335	422	1.02452	-1.39724

图 22-12　优化结果

9. 拟合结果

Adams/View 已经完成了工作空间矩阵中定义的测试，接下来可以使用 Adams/Insight 将结果拟合到多项式或响应曲面。

（1）单击 Tools > Fit New Model 命令。

（2）单击 Regression > toe_left_REQ 命令。

（3）Display 界面可以选择需要显示的参数，此处选择 Fit，显示结果如图 22-13 所示。

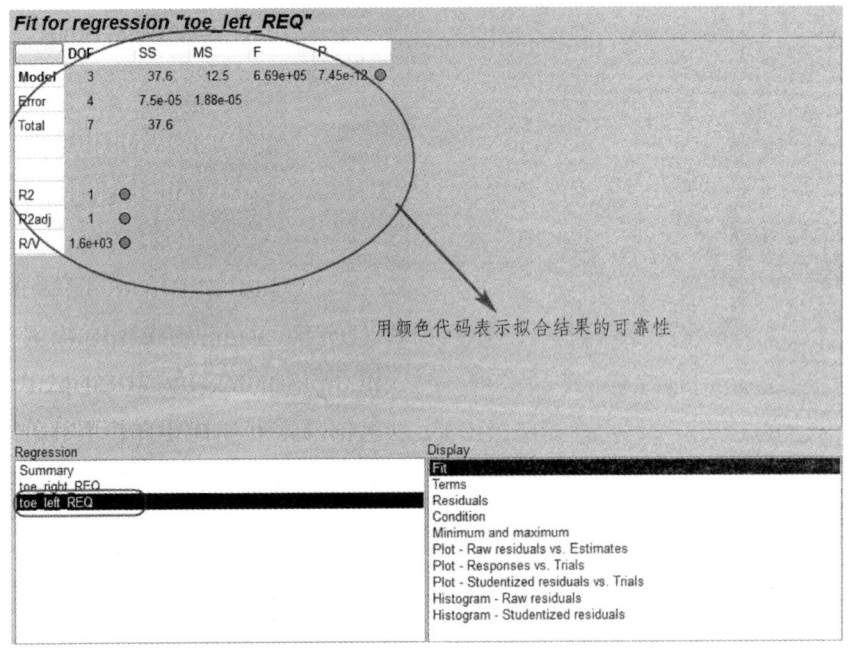

图 22-13　左侧车轮前束角回归拟合分析

① 绿色表示所有拟合标准均满足或超过最高拟合阈值；
② 黄色表示适合标准，可能需要调查；
③ 红色表示应调查拟合标准。

（4）Display 界面可以选择 Plot-Responses vs.Trials，显示结果如图 22-14 所示。图 22-14 与图 22-12 中的数值相同。

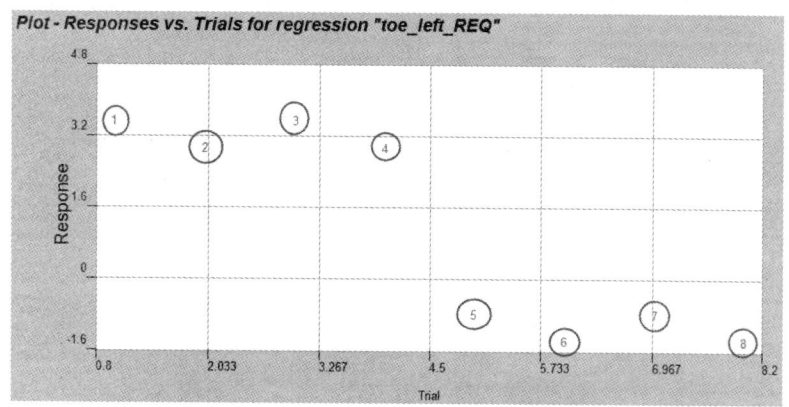

图 22-14 响应（Responses vs.Trials）

10. 刷新因素设定

可以使用 Adams / Insight 执行单目标和多目标优化。单目标优化旨在标量响应；多目标优化涉及多个标量响应。

（1）单击 Tools > Optimize Model 命令，显示如图 22-15 所示的界面。

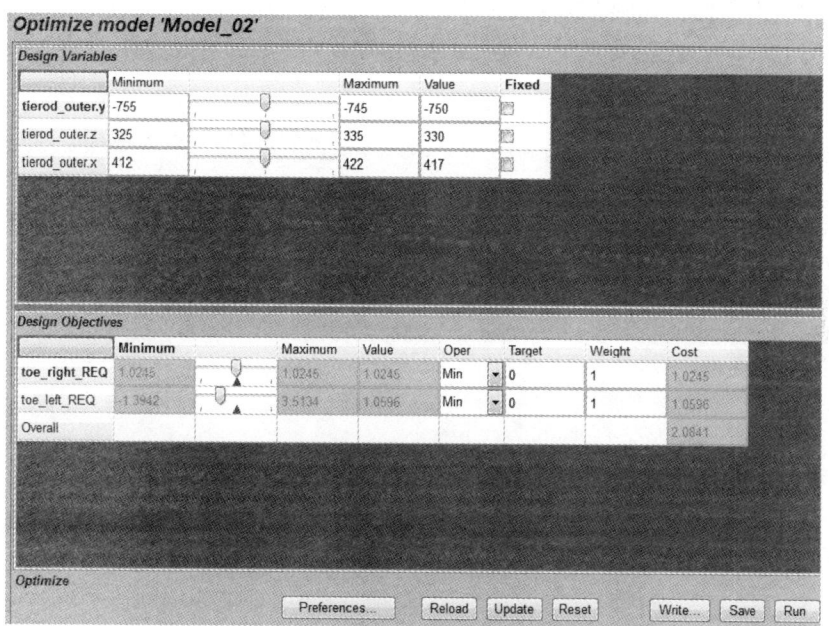

图 22-15 优化模型界面

（2）在优化模型界面中，可以通过滑动条幅修改参数 tierod_outer.x\ tierod_outer.y\ tierod_outer.z 的值，修改完成后，设计目标值会通过刷新按钮更新。

（3）单击 Analysis > Model_01 > Export to Web > Model_01 命令，显示如图 22-16 所示的界面。

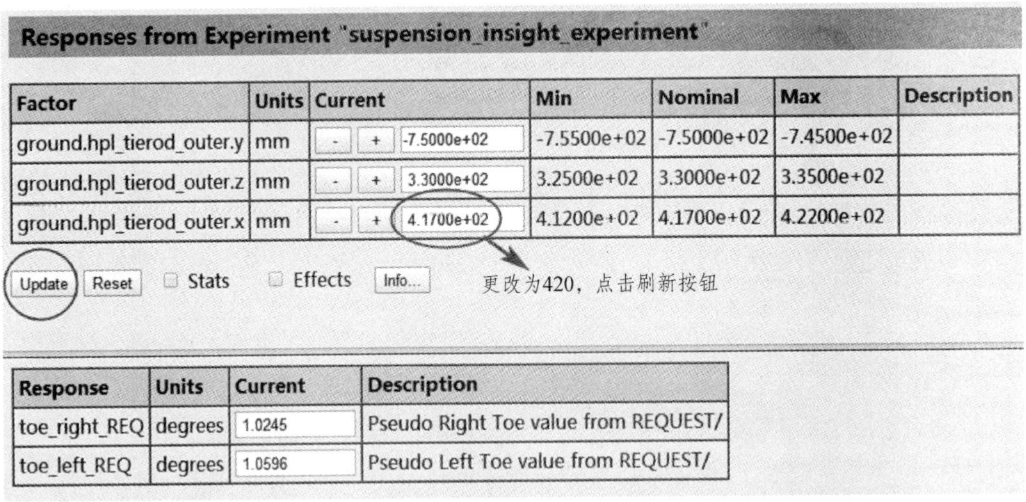

图 22-16 响应输出（网页模式）

（4）将第一个因子 hpl_tierod_outer.x 的值从 417 更改为 420，然后选择 Update。目标响应会进行调整以反映新的因子值。注意，只有一个响应中的 toe_left_REQ 发生了变化；因为模型是一个独立的悬架，其中右拉杆未与左拉杆相连，用户所做的因子值更改仅影响悬架的左侧；左车轮前束角变化如图 22-17 所示。

（5）勾选 Effects，优化变量灵敏度显示如图 22-18 所示，可以看出，Y 方向的数值对前束角的影响最大。

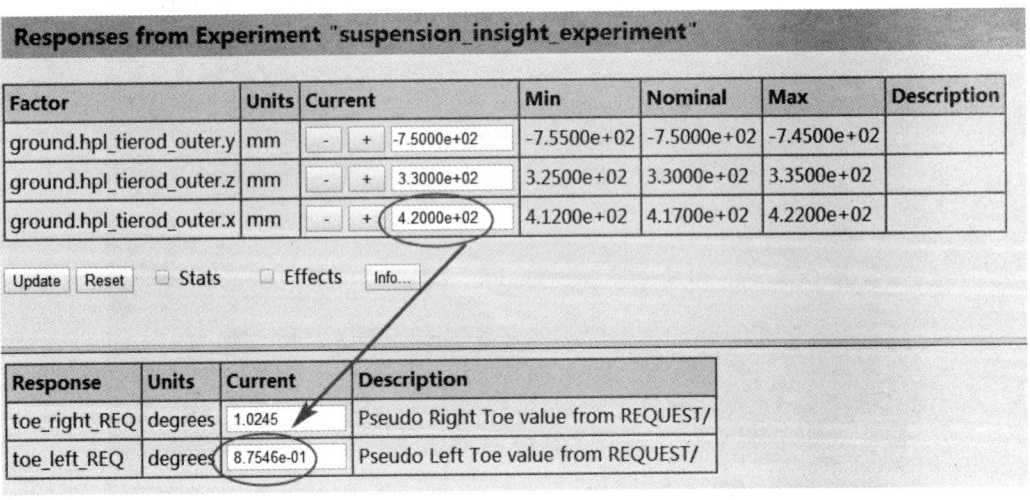

图 22-17 响应输出（网页模式，改变数值为 420）

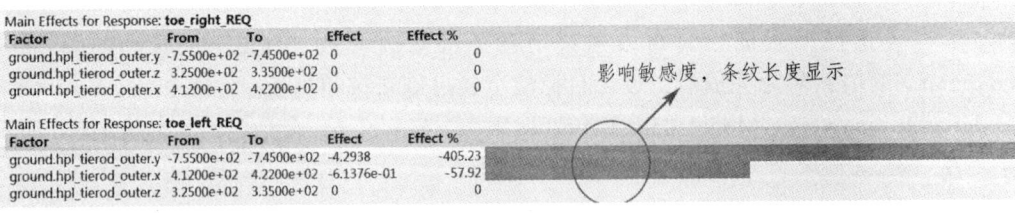

图 22-18 参数灵敏度分析

22.2 推杆式悬架外倾角优化——ACAR

1. 悬架装配

（1）单击 File > New > Suspension Assembly 命令，弹出推杆式悬架装配对话框，如图 22-19 所示，在下列对话框中输入相应的数据：

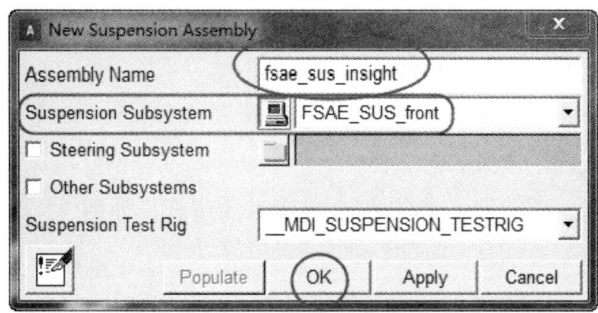

图 22-19　推杆式悬架装配

- Assembly Name（系统名称）：fsae_sus_insight；
- Suspension Subsystem（模板路径）：FSAE_SUS_front。

（2）单击 OK 按钮，完成推杆式悬架的装配。

2. 仿真设置

（1）单击 Simulate > Suspension Analysis > Opposite Travel 命令，弹出双轮同向激振对话框，如图 22-20 所示，在下列对话框中输入相应的数据。

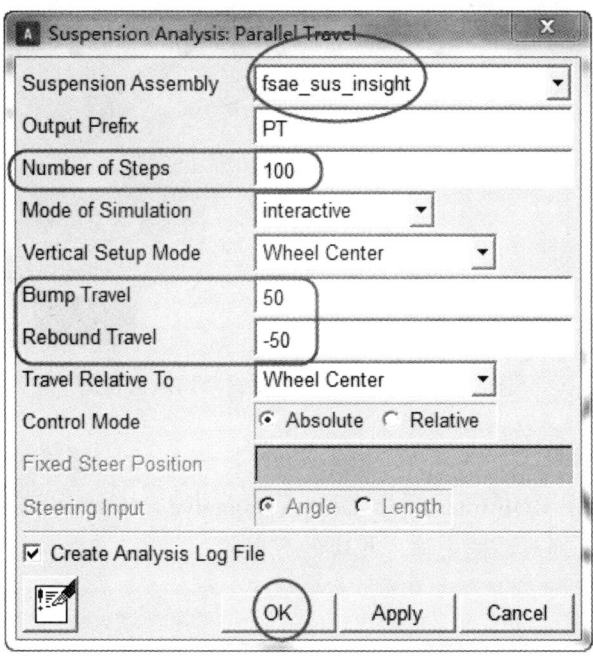

图 22-20　车轮同向激振仿真

- Output Prefix：PT；
- Number of Steps（仿真步数）：100；
- Mode of Simulation：interactive；
- Vertical Setup Mode：Wheel Center；
- Bump Travel：50；
- Rebound Travel：-50；
- Travel Relative To：Wheel Center；
- Control Mode：Absolute；
- Coordinate System：Vehicle。

（2）单击 OK 按钮，完成耦合悬架在 C 模式下的仿真，如图 22-21 所示。

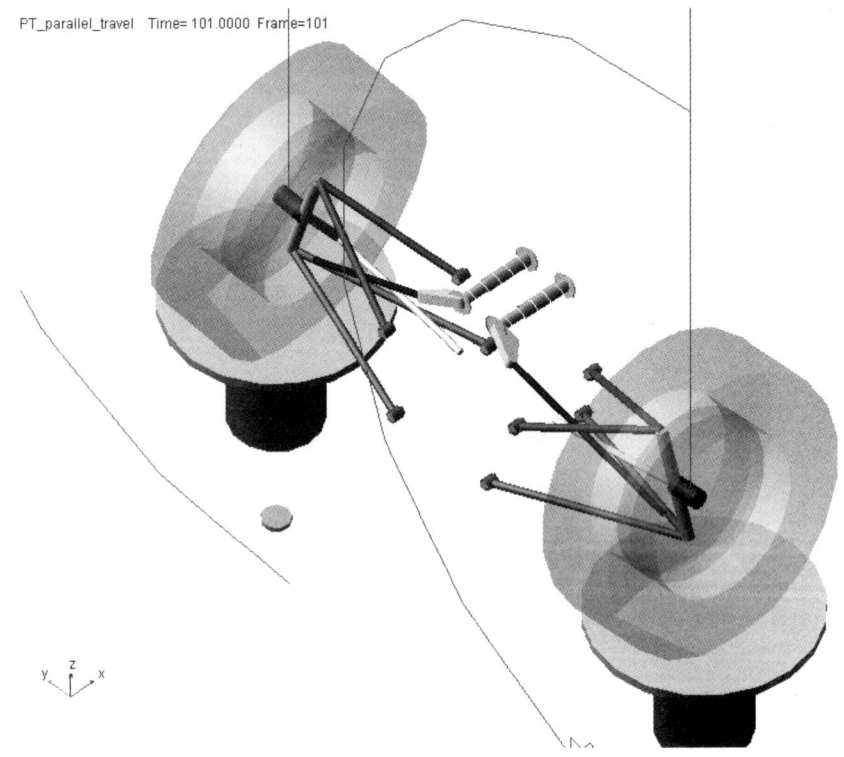

图 22-21 推杆式悬架车轮同向激振仿真

3. 创建优化目标

（1）单击 Simulate > DOE Interface > Design Objective > New 命令，弹出设计目标对话框，如图 22-22 所示，在下列对话框中输入相应的数据：

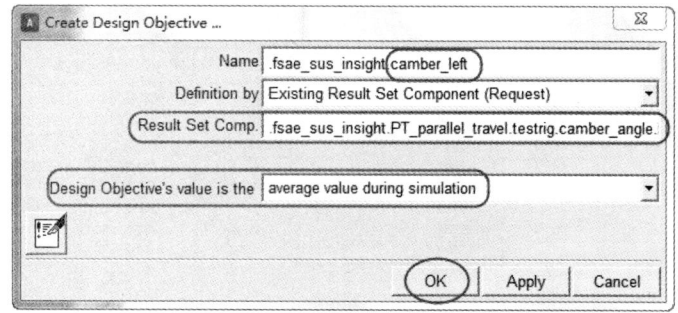

图 22-22　设计目标对话框-Camber_left

- Definition by：Existing Result Set Component（Request）；
- Result Set Comp：.fsae_sus_insight.PT_parallel_travel.testrig.camber_angle.left；
- Design Objective's value is the：average value during simulation。

（2）单击 Apply 按钮，完成 camber_angle.left 优化目标的创建。同理，创建其他优化目标。

- Result Set Comp：.fsae_sus_insight.PT_parallel_travel.testrig.camber_angle.right；
- Design Objective's value is the：average value during simulation。

（3）单击 OK 按钮，完成 camber_angle. right 优化目标的创建。

4．优化设计实验

（1）单击 Simulate > DOE Interface > Adams/Insight > Export 命令，弹出优化输出界面，如图 22-23 所示，在下列对话框中输入相应的数据：

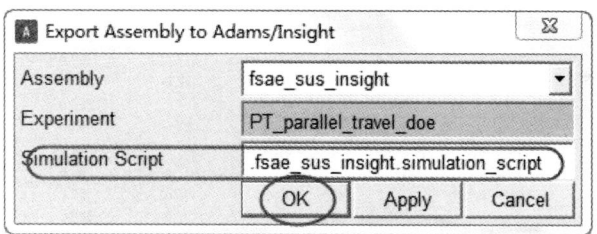

图 22-23　Adams/Insight 输出接口

- Assembly：fsae_sus_insight；
- Experiment：PT_parallel_travel_doe；
- Simulation Script：fsae_sus_insight.simulation_script。

（2）单击 OK 按钮，完成输出接口的设置，此时 Adams/View 界面消失，弹出 Adams/Insight 界面。

5．优化结果

- 后续具体操作请参考前束角优化，此处不再详细叙述。计算机共运行 64 次，优化结果如图 22-24 所示；优化变量敏感度如图 22-25 所示。

Work Space

Trial	hpl uca outer.x	hpl uca outer.y	hpl uca outer.z	hpr uca outer.x	hpr uca outer.y	hpr uca outer.z	camber angle left	camber angle right
Trial 1	-5	-487.6	350.6	-5	477.6	350.6	-7.64193	-7.71144
Trial 2	-5	-487.6	350.6	-5	477.6	360.6	-7.64193	-7.94597
Trial 3	-5	-487.6	350.6	-5	487.6	350.6	-7.64193	-7.64193
Trial 4	-5	-487.6	350.6	-5	487.6	360.6	-7.64193	-7.87602
Trial 5	-5	-487.6	350.6	5	477.6	350.6	-7.64193	-7.7873
Trial 6	-5	-487.6	350.6	5	477.6	360.6	-7.64193	-8.0239
Trial 7	-5	-487.6	350.6	5	487.6	350.6	-7.64193	-7.7165
Trial 8	-5	-487.6	350.6	5	487.6	360.6	-7.64193	-7.95269
Trial 9	-5	-487.6	360.6	-5	477.6	350.6	-7.87602	-7.71144
Trial 10	-5	-487.6	360.6	-5	477.6	360.6	-7.87602	-7.94597
Trial 11	-5	-487.6	360.6	-5	487.6	350.6	-7.87602	-7.64193
Trial 12	-5	-487.6	360.6	-5	487.6	360.6	-7.87602	-7.87602
Trial 13	-5	-487.6	360.6	5	477.6	350.6	-7.87602	-7.7873
Trial 14	-5	-487.6	360.6	5	477.6	360.6	-7.87602	-8.0239
Trial 15	-5	-487.6	360.6	5	487.6	350.6	-7.87602	-7.7165
Trial 16	-5	-487.6	360.6	5	487.6	360.6	-7.87602	-7.95269
Trial 17	-5	-477.6	350.6	-5	477.6	350.6	-7.71144	-7.71144
Trial 18	-5	-477.6	350.6	-5	477.6	360.6	-7.71144	-7.94597
Trial 19	-5	-477.6	350.6	-5	487.6	350.6	-7.71144	-7.64193
Trial 20	-5	-477.6	350.6	-5	487.6	360.6	-7.71144	-7.87602
Trial 21	-5	-477.6	350.6	5	477.6	350.6	-7.71144	-7.7873
Trial 22	-5	-477.6	350.6	5	477.6	360.6	-7.71144	-8.0239
Trial 23	-5	-477.6	350.6	5	487.6	350.6	-7.71144	-7.7165
Trial 24	-5	-477.6	350.6	5	487.6	360.6	-7.71144	-7.95269
Trial 25	-5	-477.6	360.6	-5	477.6	350.6	-7.94597	-7.71144
Trial 26	-5	-477.6	360.6	-5	477.6	360.6	-7.94597	-7.94597
Trial 27	-5	-477.6	360.6	-5	487.6	350.6	-7.94597	-7.64193
Trial 28	-5	-477.6	360.6	-5	487.6	360.6	-7.94597	-7.87602
Trial 29	-5	-477.6	360.6	5	477.6	350.6	-7.94597	-7.7873
Trial 30	-5	-477.6	360.6	5	477.6	360.6	-7.94597	-8.0239
Trial 31	-5	-477.6	360.6	5	487.6	350.6	-7.94597	-7.7165
Trial 32	-5	-477.6	360.6	5	487.6	360.6	-7.94597	-7.95269
Trial 33	5	-487.6	350.6	-5	477.6	350.6	-7.7165	-7.71144
Trial 34	5	-487.6	350.6	-5	477.6	360.6	-7.7165	-7.94597
Trial 35	5	-487.6	350.6	-5	487.6	350.6	-7.7165	-7.64193
Trial 36	5	-487.6	350.6	-5	487.6	360.6	-7.7165	-7.87602
Trial 37	5	-487.6	350.6	5	477.6	350.6	-7.7165	-7.7873
Trial 38	5	-487.6	350.6	5	477.6	360.6	-7.7165	-8.0239
Trial 39	5	-487.6	350.6	5	487.6	350.6	-7.7165	-7.7165
Trial 40	5	-487.6	350.6	5	487.6	360.6	-7.7165	-7.95269
Trial 41	5	-487.6	360.6	-5	477.6	350.6	-7.95269	-7.71144
Trial 42	5	-487.6	360.6	-5	477.6	360.6	-7.95269	-7.94597
Trial 43	5	-487.6	360.6	-5	487.6	350.6	-7.95269	-7.64193
Trial 44	5	-487.6	360.6	-5	487.6	360.6	-7.95269	-7.87602
Trial 45	5	-487.6	360.6	5	477.6	350.6	-7.95269	-7.7873
Trial 46	5	-487.6	360.6	5	477.6	360.6	-7.95269	-8.0239
Trial 47	5	-487.6	360.6	5	487.6	350.6	-7.95269	-7.7165
Trial 48	5	-487.6	360.6	5	487.6	360.6	-7.95269	-7.95269
Trial 49	5	-477.6	350.6	-5	477.6	350.6	-7.7873	-7.71144
Trial 50	5	-477.6	350.6	-5	477.6	360.6	-7.7873	-7.94597
Trial 51	5	-477.6	350.6	-5	487.6	350.6	-7.7873	-7.64193
Trial 52	5	-477.6	350.6	-5	487.6	360.6	-7.7873	-7.87602
Trial 53	5	-477.6	350.6	5	477.6	350.6	-7.7873	-7.7873
Trial 54	5	-477.6	350.6	5	477.6	360.6	-7.7873	-8.0239
Trial 55	5	-477.6	350.6	5	487.6	350.6	-7.7873	-7.7165
Trial 56	5	-477.6	350.6	5	487.6	360.6	-7.7873	-7.95269
Trial 57	5	-477.6	360.6	-5	477.6	350.6	-8.0239	-7.71144
Trial 58	5	-477.6	360.6	-5	477.6	360.6	-8.0239	-7.94597
Trial 59	5	-477.6	360.6	-5	487.6	350.6	-8.0239	-7.64193
Trial 60	5	-477.6	360.6	-5	487.6	360.6	-8.0239	-7.87602
Trial 61	5	-477.6	360.6	5	477.6	350.6	-8.0239	-7.7873
Trial 62	5	-477.6	360.6	5	477.6	360.6	-8.0239	-8.0239
Trial 63	5	-477.6	360.6	5	487.6	350.6	-8.0239	-7.7165
Trial 64	5	-477.6	360.6	5	487.6	360.6	-8.0239	-7.95269

图 22-24 优化结果

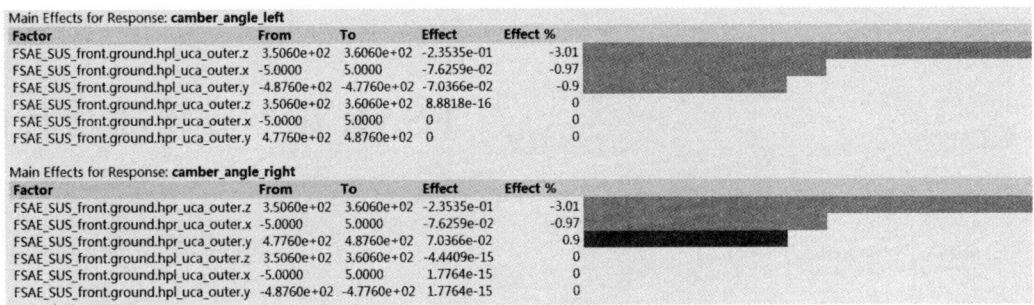

图 22-25 外倾角优化变量敏感度分析

第23章 FSAE赛车后轮随动转向

通过理论分析建立后轮瞬态随动转向系统数学模型，分析得出摆臂旋转角度及摆臂与车身连接衬套刚度是影响随动转向的主要因素；为验证理论模型的正确性，用ADAMS软件建立包含后轮随动转向特性的FSAE整车模型，后轮随动悬架模型设计为扭力梁悬架，衬套刚度通过动静刚度试验机获取。柔性扭转梁通过ABAQUS输出模态中性文件获取。反向车轮激振仿真表明：左右车轮中心可以获取随动转向位移，与理论数模模型对比，误差仅为1.7%。整车弯道仿真表明：车辆入弯时为过度转向，出弯时为不足转向，整车兼顾平顺性与操控性。扭力梁安装位置C值变动时，随着C值的增加，不足转向特性趋势减小，整车稳定性能变差；衬套安装角度θ增加时，不足转向特性趋势减小，整车稳定性能提升。后轮随动转向FSAE整车模型如图23-1所示。

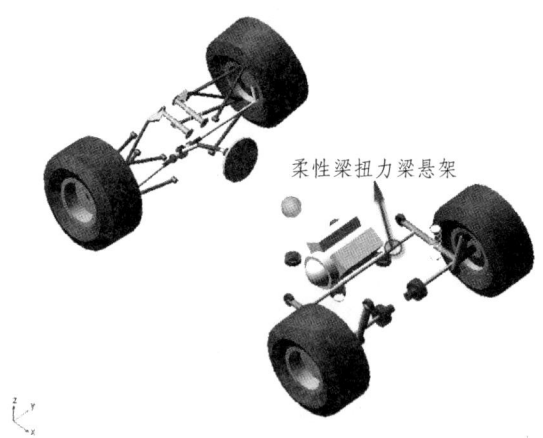

图23-1 后轮随动转向FSAE模型

学习目标

（1）了解随动转向数学模型。
（2）了解随动转向物理模型。
（3）会衬套实验。
（4）了解扭转梁MNF。
（5）会反向激振仿真。
（6）了解扭力梁位置因素。
（7）了解衬套安装角度。

23.1 随动转向数学模型

后轮随动转向特性在一定程度上可以改善整车行驶平顺性并兼顾稳定性,主要体现在低速模式下转弯半径小,高速模式下入弯半径小、出弯半径大。后轮随动转向特性相关文献较少,检索相关文献主要集中在两方面:① 商用车后轮随动转向特性设计,商用车后驱动轮及挂车随动转向以减小整车及汽车列车转向半径为目的,不考虑整车瞬时转向特性及整车稳定性;② 采用控制算法设计控制后轮转向,文献研究的重心主要在于验证算法,并没有从理论上及结构模型特性上对随动转向特性进行系统分析,此类文献研究主旨本质上是四轮转向。FSAE 赛车属于小型赛车,设计定位的方向是以操控稳定性为主导。适宜的后轮随动转向特性对提升 FSAE 赛车稳定性具有促进作用。要实现后轮随动转向特性,后悬架需设计成半独立式扭转梁悬架或独立式扭转梁悬架;从设计效果上看,独立式扭转梁悬架瞬时转向特性效果更好,但两根独立的扭转梁在车身底部占用较大空间,扭转梁的作用是替代了螺旋弹簧,同时独立式扭转梁悬架与车身之间需要安装四个不同刚度的柔性衬套,才能实现较好的转向特性。因此,选择较为简单半独立式扭转梁较为适宜,其占用空间小,结构简单,成本低,同时半独立式扭转梁悬架存在旋转臂(拖拽臂),在整车长度一定的前提下可以进一步减少车身的长度,降低车身质量,提升稳定性。

FSAE 赛车向左转向时,车身向右侧倾斜,右侧车身向下压缩,即右侧轮胎向上跳动,左侧车轮向下跳动,后悬架车轮跳动模型简化如图 23-2 所示;FSAE 赛车在静止状态时后悬架摆臂与车身之间的安装位置高于车轮中心时摆臂角为正,反之为负;当摆臂角为正时,转向瞬间整车为过度转向状态,后轮转向与前轮方向相反,FSAE 整车瞬态转向模型简化如图 23-3 所示。后轮瞬态转向数学模型如下:

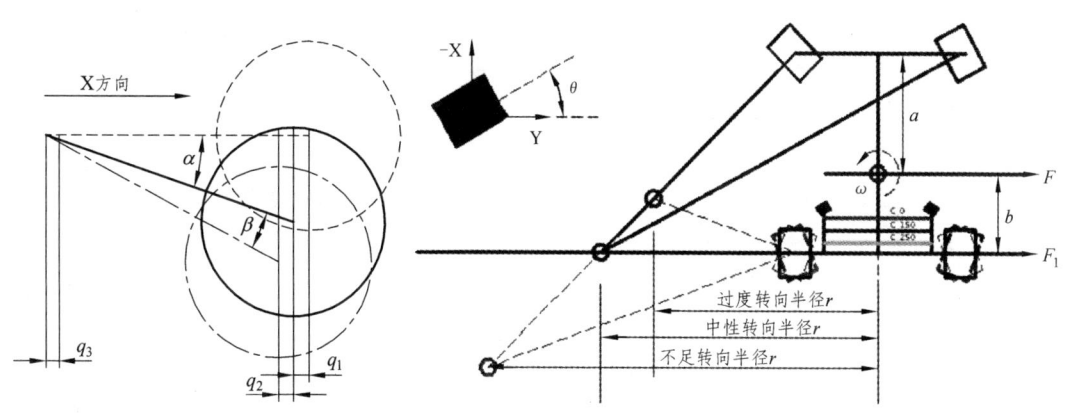

图 23-2 后悬架车轮跳动模型　　图 23-3 FSAE 整车瞬态转向模型

右后车轮正 X 方向运动位移:

$$l - l\cos\alpha = q_1 \quad (23\text{-}1)$$

左后车轮负 X 方向运动位移:

$$l - l\cos\beta = q_2 \tag{23-2}$$

FSAE 赛车离心力：

$$F = mr\omega^2 \tag{23-3}$$

忽略轮胎侧偏力及轮胎本身的迟滞特性，后左右轮胎侧向受力总和为

$$F_1 = F\frac{a}{a+b} \tag{23-4}$$

右后车轮摆臂衬套 X 方向的位移量为

$$F_1\sin\theta = 2k\frac{q_3}{\cos\theta} \tag{23-5}$$

衬套有增加过度转向的趋势，后悬架左右车轮在 X 方向之间的总位移为

$$q_1 + q_2 + q_3 = q \tag{23-6}$$

将式（23-1）~式（23-5）代入式（23-6）并整理得：

$$l(2 - \cos\alpha - \cos\beta) + \frac{mr\omega^2 a\sin\theta}{2k(a+b)}\cos\theta = q \tag{23-7}$$

摆臂角为负时，后轮转向与前轮保持同方向，整车弯道转向为不足转向状态，同时衬套有减小不足转向的趋势：

$$q_1 + q_2 - q_3 = q \tag{23-8}$$

将式（23-1）~式（23-5）代入式（23-8）并整理得：

$$l(2 - \cos\alpha - \cos\beta) - \frac{mr\omega^2 a\sin\theta}{2k(a+b)}\cos\theta = q \tag{23-9}$$

式中：l 为摆臂长度；α 为右侧车轮摆臂旋转角度；β 为左侧车轮摆臂旋转角度；q_1 为右侧车轮中心移动距离；q_2 为左侧车轮中心移动距离；q_3 为摆臂衬套 X 方向的位移量；q 为左右车轮中心偏移总距离；m 为整车质量；r 为车辆转向半径；ω 为车身横摆角速度；a 为车辆质心距前轴距离；b 为车辆质心距后轴距离；θ 为摆臂与车身连接衬套的安装角度；k 为衬套径向刚度。

从公式中可以看出，影响瞬态转向的因素很多，其中摆臂旋转角度及摆臂与车身连接衬套的刚度是影响随动转向的主要因素。当摆臂角为正时，即后轮与前轮转向相反，较小的衬套刚度会加大转向的力度，此时整车舒适性变好但稳定性变差；当摆臂角为负时，即后轮与前轮转向相同，较小的衬套刚度会抵消车轮的同向偏转，因此必须增加衬套刚度，增大不足转向特性，提升整车的稳定性。

23.2 随动转向物理模型

赛车对稳定性要求较高,因此摆臂角应为负值,此时在结构上存在如下问题:悬架与车身之间没有空间安装减震器与螺旋弹簧。借鉴雪铁龙系列车型的特点,螺旋弹簧可以改为双横置扭杆弹簧,减震器为大角度斜置,但由于 FSAE 赛车为中置后驱底盘布置形式,这种设计较难实现,因此选择半独立式扭力梁悬架设计后轮随动转向特性。

衬套刚度采用衬套动静刚度试验机获取,实验前需要将衬套嵌套在夹具内,然后把夹具安装在丝杠上;实验之前要确保接线准确并进行预热 15 min,以保证传感器的稳定性;实验过程中,上下限位块的位移需要合理设定,防止过大位移导致夹具和试样损坏。X 方向的位移设定为 9 mm,扭转角度设定为 10°;Y 方向的位移设定为 6 mm,扭转角度设定为 10°;Z 方向的位移设定为 15 mm,扭转角度设定为 15°。衬套实验如图 23-4 所示,各方向刚度如图 23-5 和图 23-6 所示。

图 23-4 衬套刚度实验

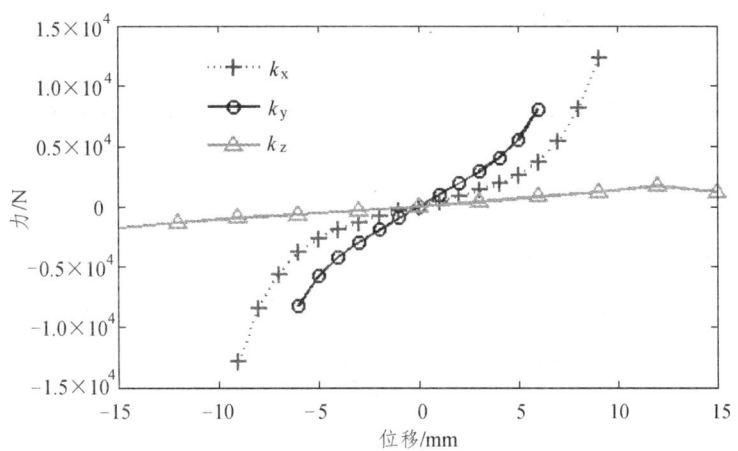

图 23-5 衬套 X/Y/Z 垂向刚度

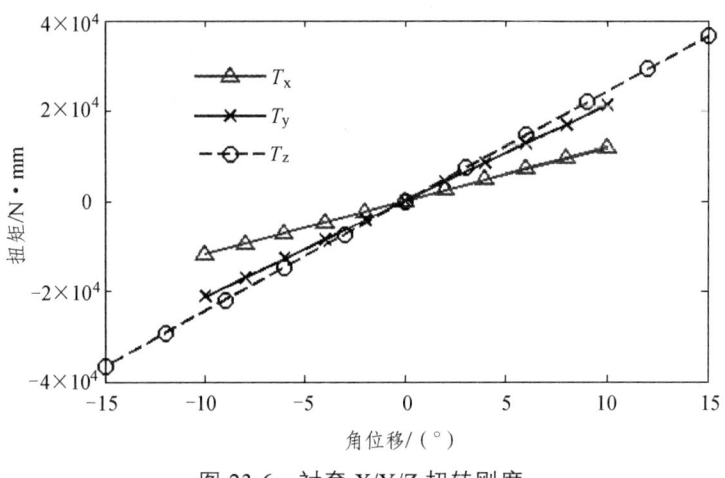

图 23-6 衬套 X/Y/Z 扭转刚度

23.3 柔性扭转梁

扭力梁物理模型建立过程的核心是扭转梁柔性化处理，采用 ABAQUS 软件输出扭转杆模态中性文件，将文件导入到 ADAMS 中建立扭力梁悬架模型，建模过程中要保证悬架通信器与发动机总成及车身正确匹配。ABAQUS 创建扭力梁模态中性文件程序如下：

子结构数据块生成程序：

Substructure Generate，overwrite，type=Z1，recovery matrix=YES，MASS MATRIX=YES

柔性体转换程序：

FLEXIBLE BODY，TYPE=ADAMS

应力应变输出程序：

ELEMENT RECOVERY MATRIX，POSITION=AVERAGED AT NODES

S,

E,

模态中性文件 MNF 生成程序：

abaqus adams job=torsion_beam substructure_sim= torsion_beam _Z1 model_odb= torsion_beam length=mm mass=tonne time=sec force=N

扭转梁模态中性文件转换完后共 20 阶，其中 4 阶、6 阶、8 阶位移变化如图 23-7 ~ 图 23-9 所示。

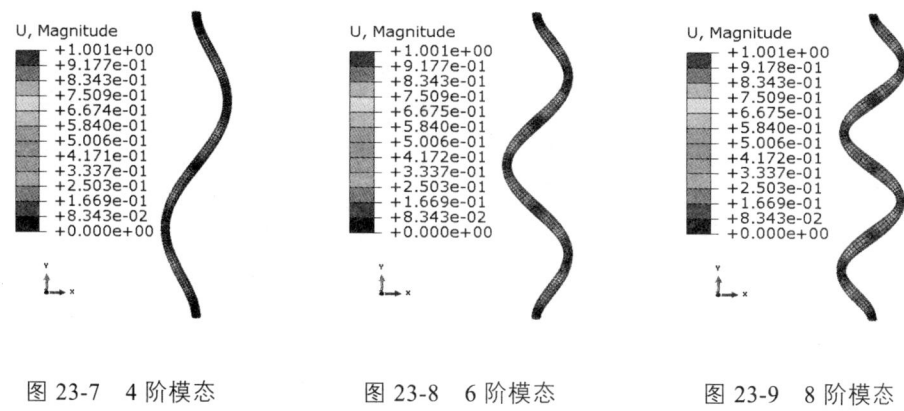

图 23-7　4 阶模态　　　图 23-8　6 阶模态　　　图 23-9　8 阶模态

23.4　反向激振仿真

扭力梁物理悬架模型建立好后对其进行车轮反向激振仿真，如图 23-10 所示；车轮反向激振实验的目的是获取车轮上下跳动过程中车轮中心在纵向的位移偏移量，计算结果如图 23-11 所示。左侧车轮下跳最大位移为 10.60 mm，平衡状态位移为 0.89 mm，上跳最大位移为 13.45 mm；右侧车轮位移变化量与左侧数值大小相反。左转向过程中，左侧车轮跳动量较小，测试左右车辆之间最大偏移量为 12.56 mm，可以实现后轮瞬态转向特性。

图 23-10　车轮反向激振实验

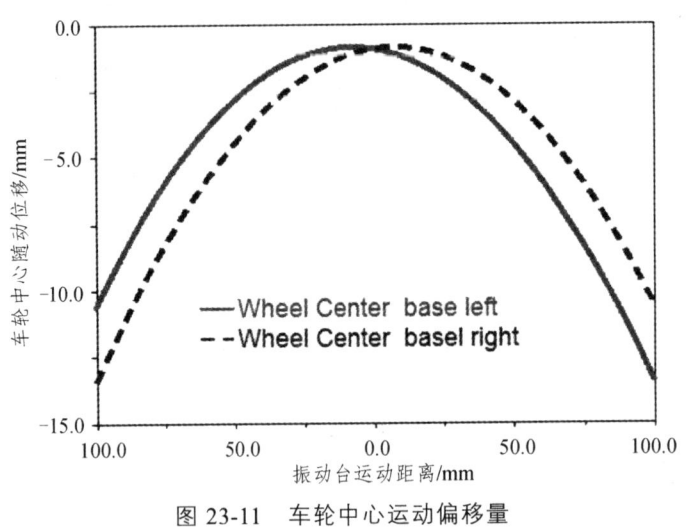

图 23-11 车轮中心运动偏移量

悬架摆臂的长度为 375.63 mm，车轮上跳 100 mm，实测车轮上调角度为 14.907°，车轮反向激振实验不考虑衬套安装角度，将参数代入公式（23-1）、（23-2），计算出左右车辆之间最大偏移量为 12.77 mm，误差仅为 1.7%。

23.5 弯道仿真

构建包含扭力梁悬架的整车模型，当扭力悬架中的扭转梁为刚体时，整车模型为 58 个自由度；将刚性扭转梁替换为柔性扭转梁时，整车为 84 个自由度。通过阶跃转向弯道仿真，对比后扭力梁悬架在整车环境模式下瞬时随动转向特性对整车性能的影响。仿真时间为 5 s，方向盘向左转 180°，转向时刻在 0 s，即仿真初始状态已经开始转向，转向时间持续 2 s，初始速度为 50 km/h，挡位为 2 挡。整车模型如图 23-1 所示。

图 23-12 中实线为整车在刚性扭转梁悬架模式下的运动轨迹，虚线为将刚性梁替换为柔性扭转梁后整车的运动轨迹。通过提取 0 s 时刻的数据，刚性扭力梁整车模型初始位移为 0 mm，柔性扭力梁整车模型初始位移为 5.5×10^{-15} mm，说明在仿真开始时后轮随动转向已经起作用。从图中可以看出，柔性扭力梁的运动轨迹整体比刚性梁略大，在 4 s 之前，重合度较高，即整车在入弯时进入到瞬时过度转向，转向半径减小，4 s 之后整车在出弯道时进入不足转向状态，转向半径增大，整车稳定性能提升。

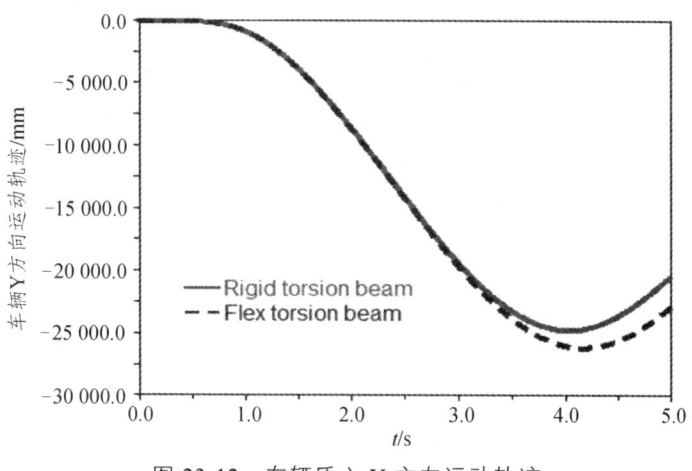

图 23-12　车辆质心 Y 方向运动轨迹

为进一步验证整车的稳定性是否提升，需要计算车身质心处的侧向速度与横摆角速度，如图 23-13 和图 23-14 所示，从图中可以看出，柔性梁悬架模式下整车的侧向速度及横摆角速度变化范围均比刚性扭转梁小，横摆角速度 RMS 值分别为：54.29、50.95，性能提升 6.15%；车身侧向速度最大值分别为：1.24、1.06，性能提升 14.52%。

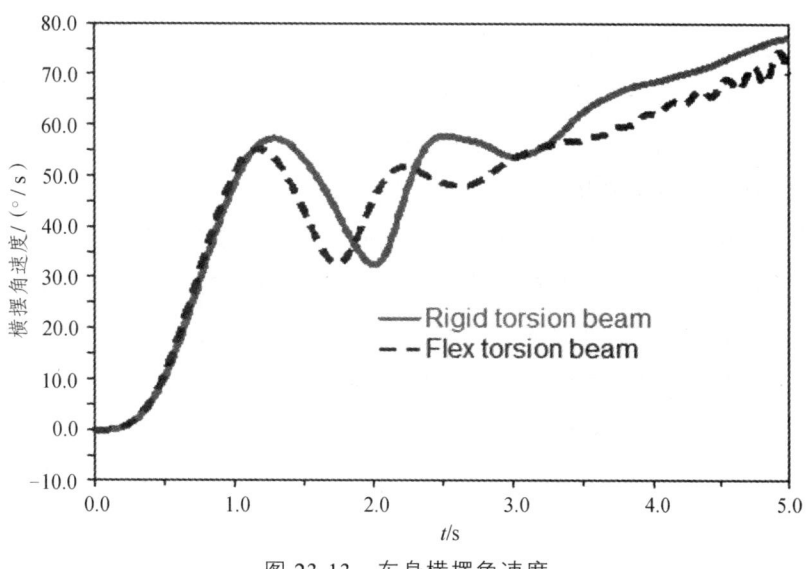

图 23-13　车身横摆角速度

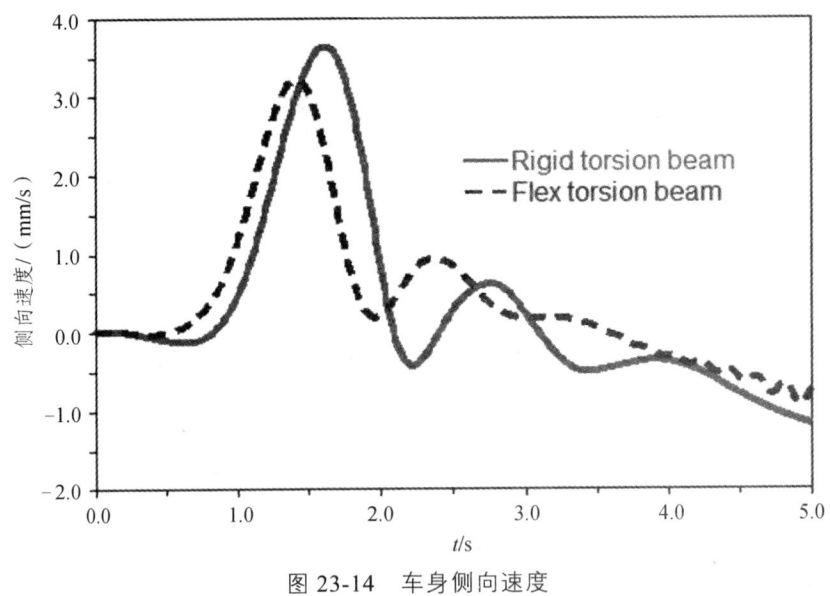

图 23-14　车身侧向速度

23.6　扭力梁位置因素

扭力梁悬架中扭转梁的安装位置不同会导致车轮中心随动位移大小产生变化。为系统研究此问题，分别将柔性扭转梁安装在摆臂上不同的 3 个位置：C_0 指原有柔性梁安装位置；C_150 指在原有位置上向正 X 方向移动 150 mm；C_250 指在原有位置上向正 X 方向移动 250 mm。图 23-15 所示的柔性扭转梁安装的位置为 C_150，计算结果如图 23-16 所示，随着柔性梁偏移距离的增加，后轮随动转向特性逐渐减小；C_0 的 RMS 值为 24 591.33，C_150 的 RMS 值为 24 423.17，C_250 的 RMS 值为 24 257.48。从图 23-17 和图 23-18 中可以看出，随着柔性梁偏移距离的增加，车身侧向速度、横摆角速度幅值逐渐增加，车辆稳定性变差。车身侧向速度：C_0 的 RMS 值为 1.06，C_150 的 RMS 值为 1.00，C_250 的 RMS 值为 1.07；车身横摆角速度：C_0 的 RMS 值为 64.99，C_150 的 RMS 值为 65.62，C_250 的 RMS 值为 66.07。

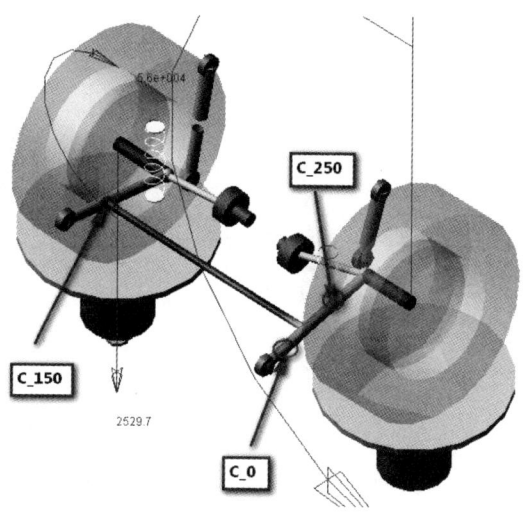

图 23-15　扭转梁安装位置

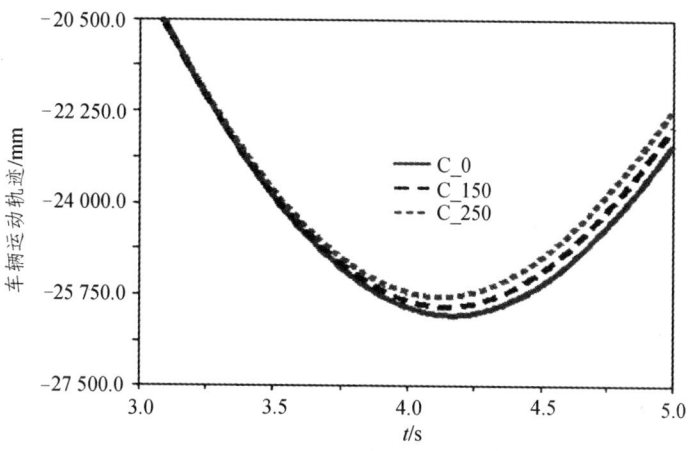

图 23-16　车辆质心 Y 方向运动轨迹

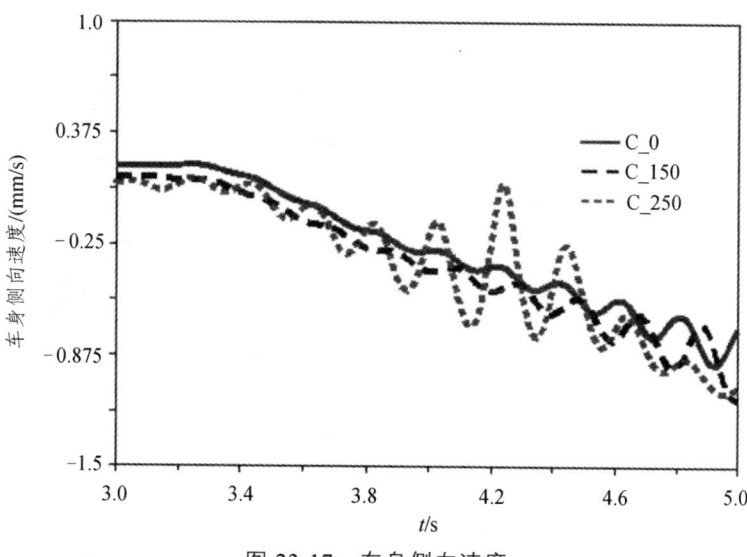

图 23-17　车身侧向速度

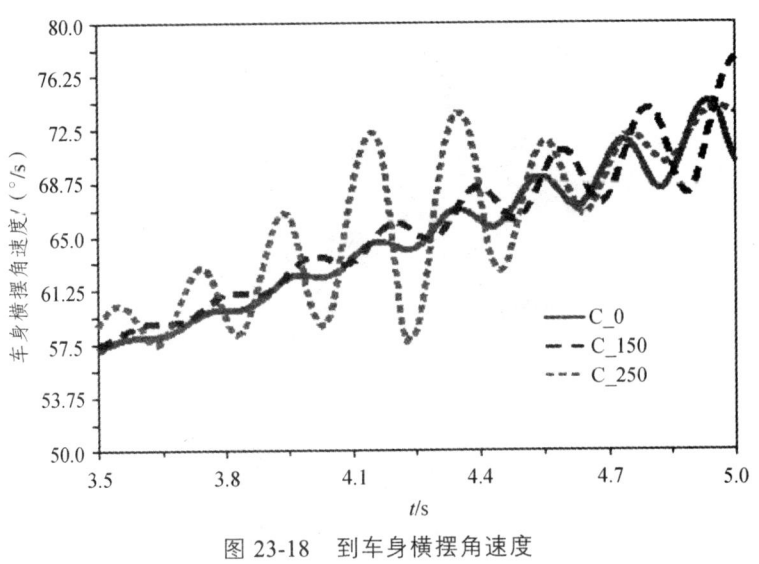

图 23-18 到车身横摆角速度

23.7 衬套安装角度

柔性衬套安装角度会影响整车的稳定性，如果采用与扭转梁轴向平行的布置形式，会导致后悬架整体横向偏移量过大，此时需要添加横向连杆保证侧向位移，这对于 FSAE 赛车后悬架安装空间来说不可行。由于衬套的各方向刚度不同，采用不同的安装角度会抑制侧向偏移，但衬套受力较大，图 23-19 中衬套安装角度为 45°。

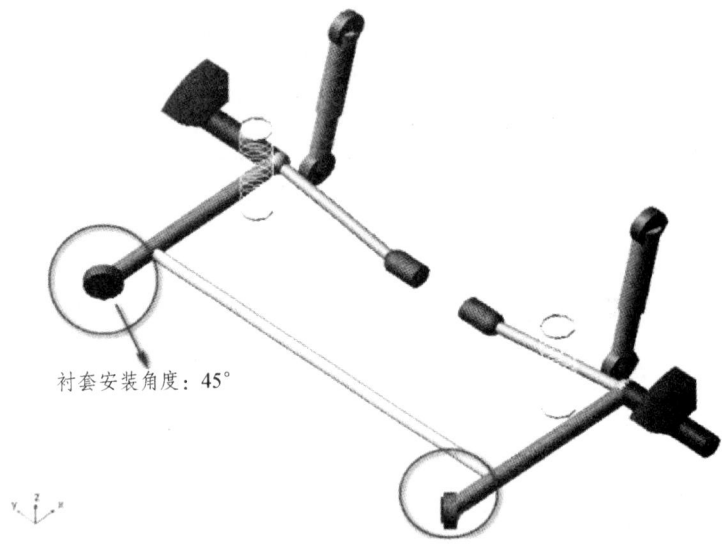

图 23-19 衬套安装角度

图 23-20 所示为不同安装角度衬套力，从图中可以看出，45°角安装衬套受力最大，最大值为 760.78 N，RMS 值为 626.10；30°角安装衬套受力次之，最大值为 703.76 N，RMS 值为 593.18；0°角安装衬套受力最小，最大值为 531.0 N，RMS 值为 538.88。

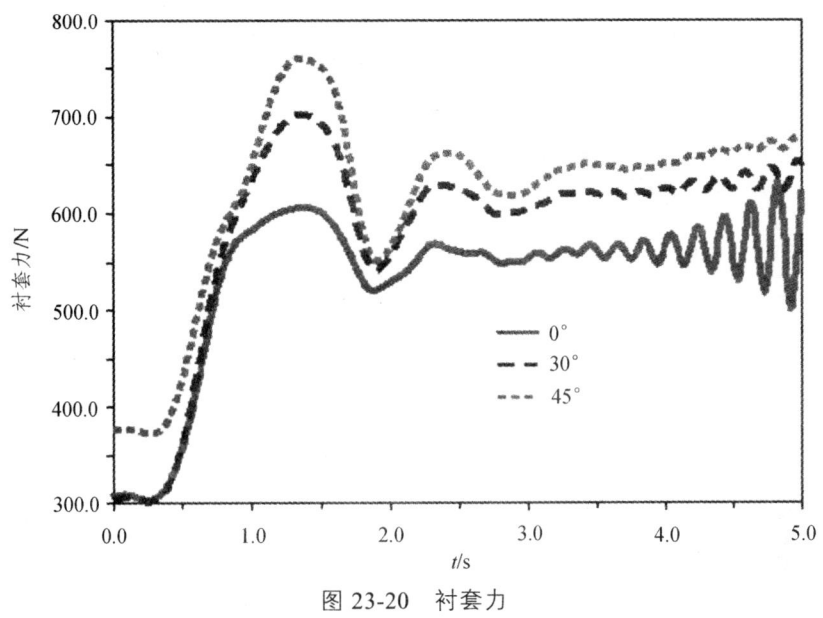

图 23-20 衬套力

从图 22-21 中可以看出,随着衬套安装角度的逐步增加,整车不足转向有减小的趋势,衬套 0°安装时 RMS 值为 25 597.84;衬套 30°安装时 RMS 值为 25 479.06;衬套 45°安装时 RMS 值为 25 369.80;从图 23-22 中可以看出,随着衬套安装角度的逐步增加,整车车身横摆角速度逐步减小,整车稳定性能提升。

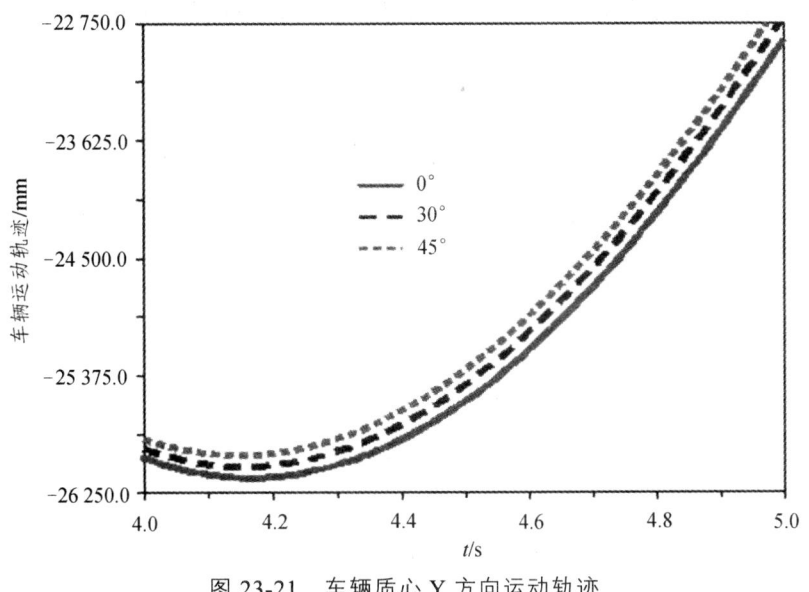

图 23-21 车辆质心 Y 方向运动轨迹

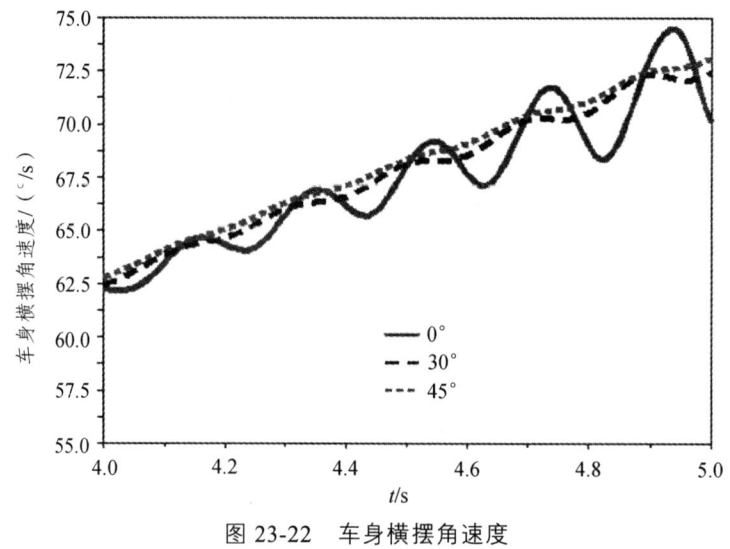

图 23-22 车身横摆角速度

23.8 总　结

（1）摆臂旋转角度及摆臂与车身连接处的衬套刚度是影响瞬态转向的两个最主要因素。

（2）与刚性扭力梁悬架对比，柔性扭力梁悬架入弯半径小，出弯半径大，兼顾平顺性与操纵稳定性。

（3）扭转梁位置影响随动转向特性，随着安装位置距离的增加，不足转向趋势减小，整车稳定性变差。

（4）衬套安装角度逐步增加时，整车不足转向特性趋势减小，整车稳定性能提升。

第 24 章 变刚度悬架特性研究

通过变刚度横置板簧悬架模型设计，板簧刚度可实现 9 倍范围变化，FSAE 赛车可实现 16 种变刚度底盘特性组合。弯道仿真实验表明，FSAE 赛车前后悬架不同刚度组合均可降低整车质心高度，提升稳定性，其中 BB 刚度组合性能改善最为明显，车身高度降低 81.18 mm，整车横摆角速度指标提升 35.39%，侧向加速度指标提升 57.50%；横置板簧与车身及下控制连接处添加衬套后，可有效改善实验初期震荡现象；板簧结构优化后，质量减少 24.00%，稳定性指标微幅提升，但实验过程伴有较小的震荡现象。建立好的整车模型如图 24-1 所示。

图 24-1 FSAE 整车

学习目标

（1）会板簧有限元前处理。
（2）了解板簧模态。
（3）了解板簧刚度。
（4）了解 FSAE 整车。
（5）会定常半径转向转弯值。
（6）会板簧结构优化。

24.1 横置板簧悬架

FSAE 赛车设计关注的重点是整车的操控稳定性，其对整车的底盘要求较高。纵观国内近些年赛事，绝大多数 FSAE 赛车前后悬架均采用推杆式双横臂悬架，有少数车辆采用双横臂悬架。推杆式悬架与双横臂悬架对整车稳定的提升均有改善作用，但推杆式悬架最大的缺点是安装时需要占用较大的车身空间。FSAE 赛车驱动采用中置后驱模式，发动机、传动系统、车身附属装置及悬架系统均布置在后轮附近空间，集成度较高，同时导致整车质心后移，稳定性变差。针对此问题提出横置板簧悬架模型设计，旨在去掉推杆及螺旋弹簧部件，减少空间的占用，同时车轮及非簧载质量减轻，固有频率提升，车辆振动减小；横置板簧既起到螺旋弹簧的作用，同时又起到横向拉杆的作用。在横置板簧悬架模型的基础上，通过改变板簧与车身的固定位置即可以改变板簧横向力臂，最终可以分段调节板簧的刚度特性，同时板簧两端与下控制臂连接，安装在悬架最底部，不占用空间，可以进一步降低整车质心高度，这对提升操控稳定性极为有利。

国内 FSAE 赛车前后悬架均为推杆式双 A 臂悬架，此悬架的优点是悬架空气动力学性能较好，阻尼效率高，缺点是悬架的整体质量增幅较大，占用较多的安装空间。FSAE 赛车为中后置后轮驱动，后悬架附近需要安装发动机、变速器、传动机构及悬架等附属装置，系统部件布置空间小且后轴系偏中；针对此问题提出一种横置板簧悬架模型设计，以增大后轴布置空间，降低车身质心，进一步提升整车稳定性。

24.1.1 变刚度板簧模型

变刚度板簧模型如图 24-2 所示。钢板宽度为 20 mm，厚度为 5 mm，长度为 730 mm；板簧长度中心线上设计出 9 个孔，孔径为 5 mm。此板簧有 4 种刚度：RP-5 为板簧长度的中心，固定 RP-5 时，单侧臂 RP-5 与 RP-1 之间的刚度为 A，单侧臂 RP-5 与 RP-9 之间的刚度为 A；RP-4 与 RP-6 关于 RP-5 对称，固定 RP-4 与 RP-6 时，单侧臂 RP-4 与 RP-1 之间的刚度为 B，单侧臂 RP-6 与 RP-9 之间的刚度为 B；RP-3 与 RP-7 关于 RP-5 对称，固定 RP-4 与 RP-6 时，单侧臂 RP-3 与 RP-1 之间的刚度为 C，单侧臂 RP-7 与 RP-9 之间的刚度为 C；RP-2 与 RP-8 关于 RP-5 对称，固定 RP-2 与 RP-8 时，单侧臂 RP-2 与 RP-1 之间的刚度为 D，单侧臂 RP-8 与 RP-9 之间的刚度为 D；RP-1 与 RP-9 与下控制臂刚性固定连接。

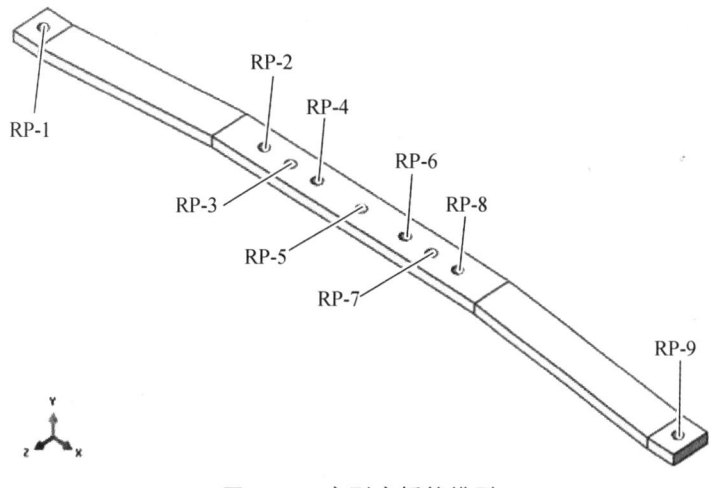

图 24-2 变刚度板簧模型

24.1.2 板簧模态

在 ABAQUS 软件中计算板簧前 20 阶模态并输出板簧中性模态文件 MNF 到 ADAMS 软件中构建横置板簧悬架模型。在 RP-1 至 RP-9 孔中分别建立 MPC 多点约束,输出模态中固定约束 RP-1、RP-5、RP-9;网格划分为六面体,共 2 808 个单元,单元类型为 C3D8R;计算并生成子数据块,其中 5、6、7 阶模态如图 24-3 ~ 图 23-5 所示。

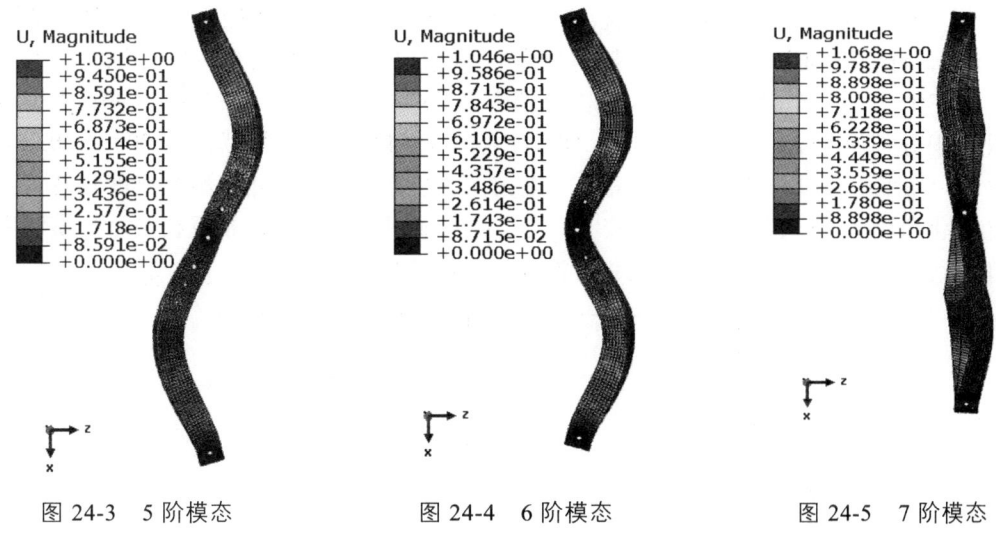

图 24-3 5 阶模态　　　　图 24-4 6 阶模态　　　　图 24-5 7 阶模态

24.1.3 板簧刚度

板簧子数据块完成计算后通过转换命令生成板簧中性文件 MNF,在 ADAMS 中导入中性文件添加约束、驱动计算板簧刚度。单侧臂刚度测试过程如下:RP-9 处添加与 Y 轴平行的移动副,在移动副上添加驱动位移,每秒运动 20 mm,分别固定约束 RP-5、RP-6、RP-7、

RP-8 计算出刚度 A、B、C、D，如图 24-6 所示；刚度 A 为 26.10 N/mm、刚度 B 为 56.04 N/mm、刚度 C 为 107.54 N/mm、刚度 D 为 232.55 N/mm。从计算结果可以看出，同一片钢板弹簧，通过改变力臂大小，刚度实现了约 9 倍范围的变化。

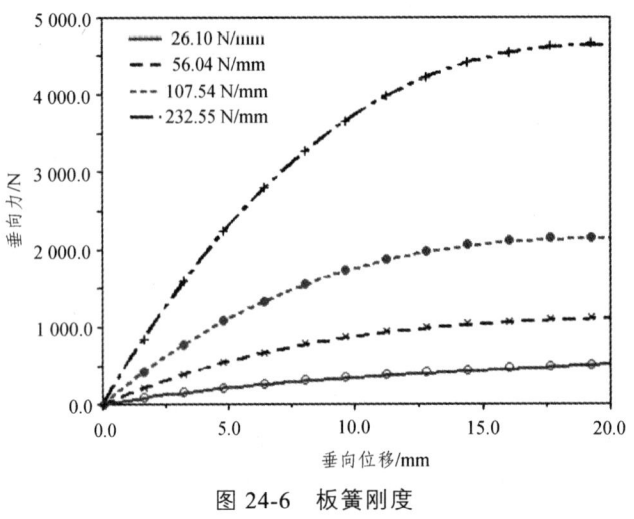

图 24-6　板簧刚度

24.1.4　参数测试

前横置板簧悬架模型前束角设置为 1°，车轮外倾设置为 -5°，外倾角为负且角度较大有利于提升整车的稳定性。对前横置板簧悬架模型进行同向车轮激振实验，车轮跳动距离为 50 mm，计算出推杆式双横臂悬架与横置板簧悬架的前束角变化范围分别为 -1.17 ~ 2.85、-1.30 ~ 2.75；车轮外倾角变化范围分别为 -2.53 ~ -7.83、-2.57 ~ -7.90；主销内倾角变化范围分别为 12.66 ~ 17.97、12.70 ~ 18.04；主销后倾角变化范围分别为 0.004 9 ~ 0.084 3、-0.01 ~ 6.8 × 10^{-5}。从计算结果可以看出：前束角、外倾角、内倾角曲线重合度较高，主销后倾角变化角度小，但相对变化范围较大，变化趋势如图 24-7 所示。横置板簧悬架模型后倾角在车轮跳动中变化范围不大，性能相对推杆式双横臂悬架有所提升。

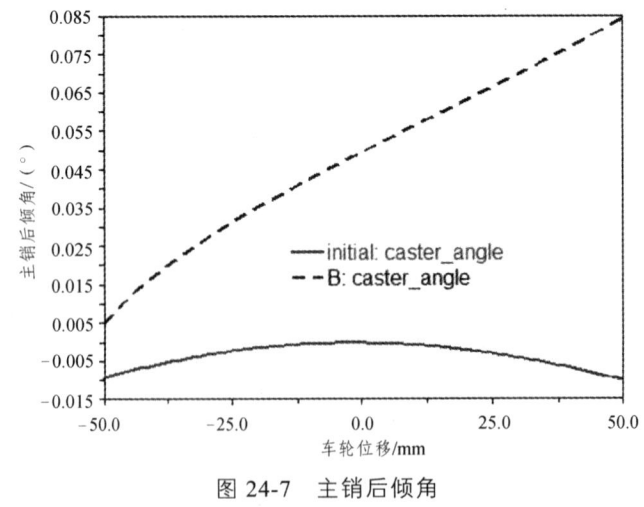

图 24-7　主销后倾角

24.2 定常半径弯道仿真

用前后横置板簧悬架模型完成 FSAE 整车模型的建立，如图 24-1 所示，整车共 196 个自由度。前后横置板簧均有 A、B、C、D 四种刚度，通过前后悬架刚度组合共有 16 组刚度可调，刚度组合如表 24-1 所示。表中 G_1 表示前轴悬架板簧刚度，G_2 表示后轴板簧刚度。板簧刚度 C、D 相对于刚度 A 大很多，接近于刚性连接，同时由于后悬架刚度相对于前悬架刚度一般会略大或者相同，在表 24-1 所示的刚度组合表中，具有实际研究意义的刚度组合为 AA、BA、BB、CC、DD。如果变刚度板簧的刚度增量变化较小，表中 16 种刚度组合均具有研究意义。AA、BA、BB、CC、DD 五种刚度组合中，AA 组合整车静平衡发散，原因在于后悬架板簧刚度为 A 时变形量过大，BA、BB 两种刚度组合计算结果相近，说明后轴刚度对整车稳定性具有主导作用，因此选取 BB、CC、DD 三种刚度组合与采用推杆式悬架 FSAE 整车进行对比分析。对整车进行定常值半径转弯仿真，相同工况下测试整车横摆角速度、侧向加速度稳定性参数，车辆转向半径为 15 m，初始速度为 10 km/h，最终速度为 50 km/h，发动机变速器均为 3 挡工况，运行时间为 10 s。

表 24-1 刚度组合表

G_2	G_1			
	A	B	C	D
A	AA	AB	AC	AD
B	BA	BB	BC	BD
C	CA	CB	CC	CD
D	DA	DB	DC	DD

图 24-8～图 24-10 表示推杆式双横臂悬架 FSAE 整车模型的计算结果。从图 24-8 中可以看出，推杆式双横臂悬架整车模型车身高度为 348.17 mm，采用横置板簧悬架后整车的车身高均有降低，BB 刚度组合后车身高度为 302.75 mm，降低 45.42 mm；CC 组合后车身高度为 310.88 mm，降低 37.29 mm；采用 DD 组合后车身高度为 314.37 mm，降低 33.8 mm。图 24-9 为横摆角速度变化曲线，initial 为 29.39°/s；BB 刚度组合最大值为 24.50°/s，性能提升 16.64%；CC 刚度组合最大值为 26.65°/s，性能提升 9.32%；DD 刚度组合最大值为 27.48°/s，性能提升 6.50%。图 24-10 为侧向加速度，initial 为 -0.40 mm/s^2；BB 刚度组合最大值为 -0.28 mm/s^2，性能提升 30.00%；CC 刚度组合最大值为 -0.33 mm/s^2，性能提升 17.50%；DD 刚度组合最大值为 -0.35 mm/s^2，性能提升 12.50%。

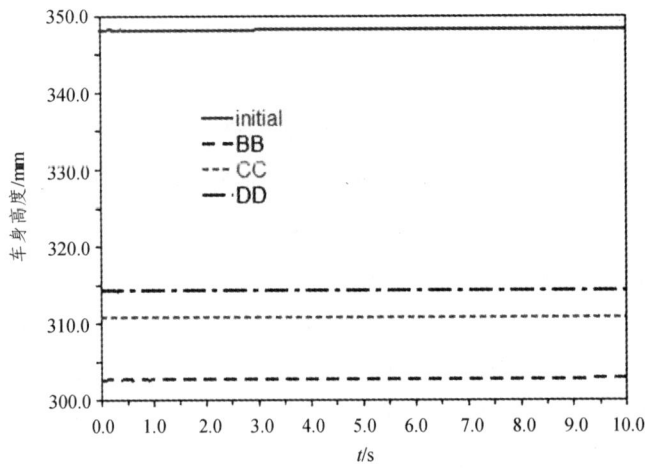

图 24-8　车身高度

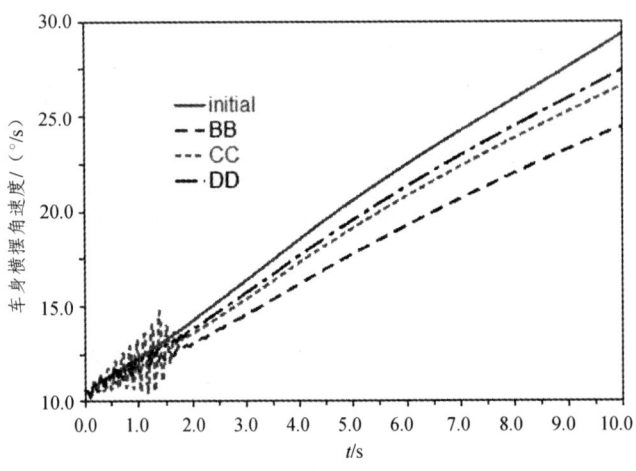

图 24-9　横摆角速度

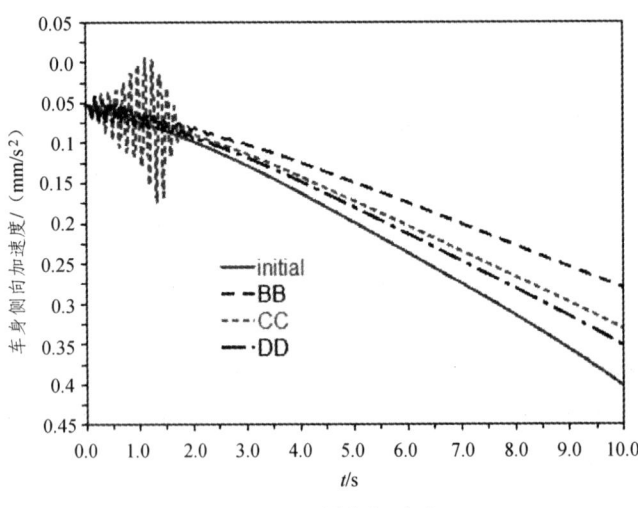

图 24-10　侧向加速度

24.3 板簧优化

通过定值半径转弯计算，发现采用横置板簧悬架的整车稳定性均具有提升，其中 BB 刚度组合的整车性能最佳。但 BB、CC 刚度组合在实验初始都伴有较大的震荡现象，针对此问题继续对板簧进行优化。

24.3.1 板簧衬套

针对实验初期存在的震荡现象，当刚度组合为 BB 时，在板簧孔 RP-4、RP-6 与 RP-1、RP-9 处添加柔性衬套；当刚度组合为 CC 时，在板簧孔 RP-3、RP-7 与 RP-1、RP-9 处添加柔性衬套；当刚度组合为 CC 时，添加衬套，如图 24-11 所示。

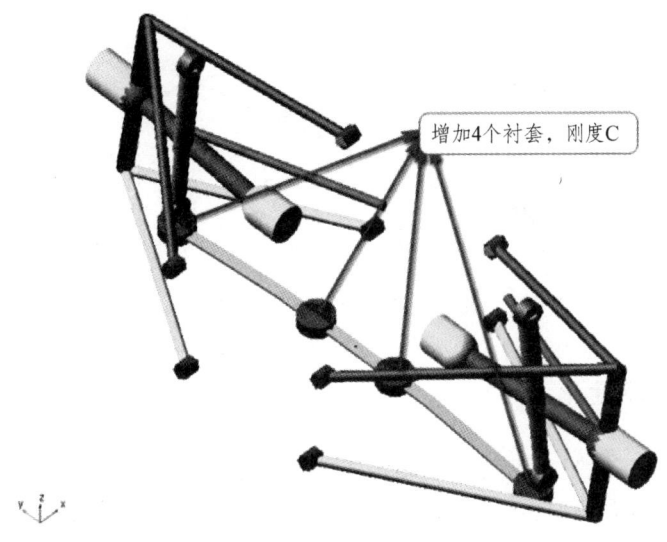

图 24-11　横置板簧衬套

对优化后的横置板簧悬架整车模型进行计算，当刚度组合为 CC 时，图 24-12 中显示整车实验初期震荡现象改善明显，同时稳定性指标参数进一步提升，其中横摆角度最大值降低为 25.44°/s，性能相对添加衬套前提升 4.54%，相对于推杆式双横臂悬架提升 13.44%；侧向加速度最大降低为 $-0.30 mm/s^2$，性能相对添加衬套前提升 9.10%，相对于推杆式双横臂悬架提升 25.00%。

当刚度组合为 BB 时，图 24-13 中显示整车稳定性指标参数进一步提升，其中横摆角度最大值降低为 18.99°/s，性能相对添加衬套前提升 22.49%，相对于推杆式双横臂悬架提升 35.39%；侧向加速度最大降低为 $-0.17 mm/s^2$，性能相对添加衬套前提升 39.29%，相对于推杆式双横臂悬架提升 57.50%；车身高度降低为 266.99 mm，相对于推杆式悬架整车模型降低 81.18 mm。

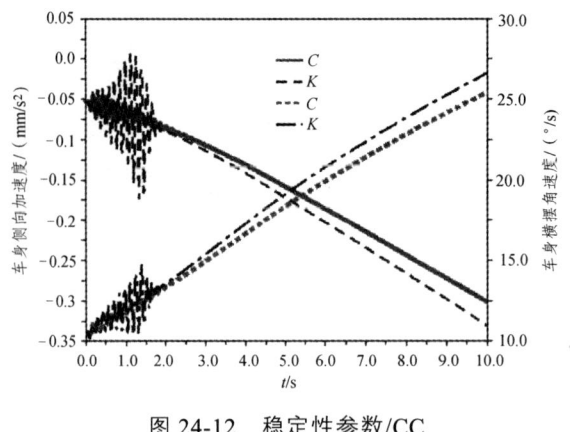

图 24-12 稳定性参数/CC

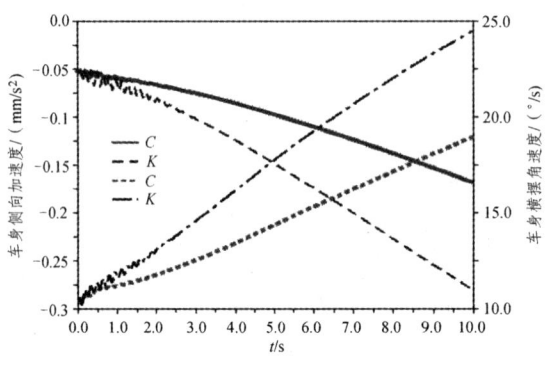

图 24-13 稳定性参数/BB

24.3.2 板簧轻量化

对 FSAE 赛车右后轮处板簧与下控制臂连接点测量受力,如图 24-14 所示,其中侧向力最大为 4 650 N,垂向力为 183 N;板簧纵向力微小,可以忽略,从计算结果可以看出横置板簧主要承受侧向力与垂向力,其作用相当于在悬架上增加了一根横向拉杆,同时起到弹簧作用,这也是采用横置板簧悬架整车模型弯道模式下侧向加速度参数大幅降低的最主要原因,整车纵向力主要由悬架上下控制臂承受。

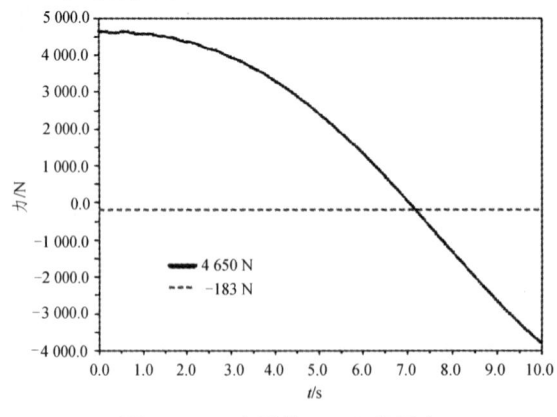

图 24-14 右后轮 RP-1 点受力

横置板簧材料：60Si2Mn；弹性模量：2.06×10^5 MPa；泊桑比：0.29；密度：7.74×10^{-9} t/mm^3；抗拉强度 1 270 MPa。对板簧进行有限元分析，单侧臂应力变化如图 24-15 所示，最大应力为 449.7 MPa，最大应变为 28.3 mm。图中绿色区域承受应力较小，远小于抗拉强度，对此区域进行拓扑优化，循环计算 30 次，在拓扑优化的基础上输出板簧几何图形，并对其进行形貌与尺寸修正，最终板簧单侧几何体如图 24-16 所示。对优化后的板簧再次进行分析，约束与载荷条件相同，计算结果显示最大应力为 555.90 MPa，最大应变为 28.10 mm，与优化之前对比最大应力增加，最大应变减小，最大应力依然小于抗拉强度，符合设计要求。

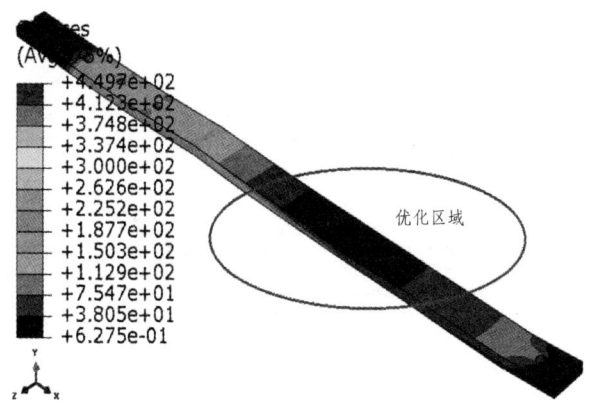

图 24-15　单侧板簧应力

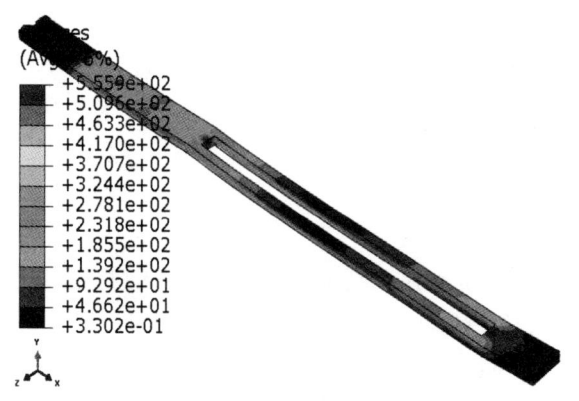

图 24-16　优化后单侧板簧应力

优化前横置板簧质量为 0.595 7 kg，优化后质量为 0.452 7 kg，质量减少 24.0%，优化效果较明显。对优化后的板簧进行刚度测试，计算显示优化后的板簧刚度为 50.65 N/mm，如图 24-17 所示，相对优化前刚度 56.04 N/mm 有所减小；横置板簧替换后通过仿真实验计算整车稳定性参数，如图 24-18 所示，整车侧向加速度与横摆角速度分别为 0.157 7 mm/s^2、18.32°/s，性能相对优化前整车 BB 刚度组合提升 8.05%、4.58%，但运行过程中伴有微小震荡现象。

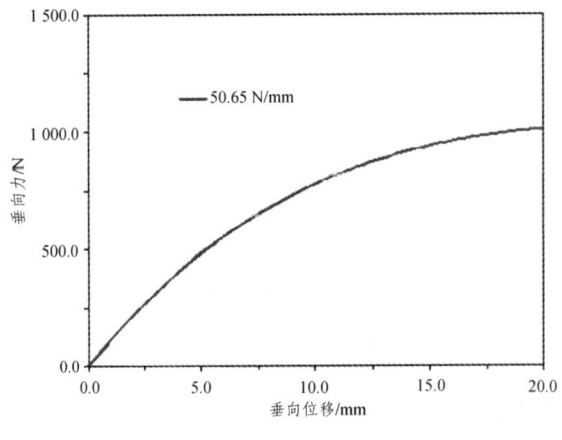

图 24-17 优化后板簧刚度

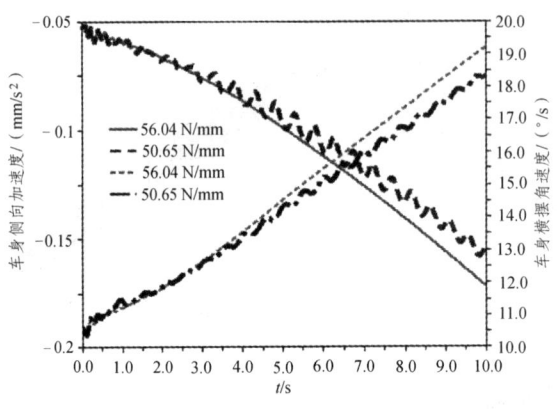

图 24-18 优化后稳定性参数

24.4 总 结

（1）变刚度横置板簧可以实现弹簧刚度在 9 倍范围内的变化，最小刚度为 26.10 N/mm，最大刚度为 232.55 N/mm，FSAE 整车模型可以实现 16 种可变刚度组合底盘模型。

（2）FSAE 采用横置板簧后，相对于推杆式双横臂悬架整车车身高度均有所降低，侧向加速度、横摆角速度等稳定性参数均有提升，其中 BB 刚度组合性能提升最为明显，分别为 16.64%、30.00%。

（3）横置板簧与下控制臂、车身连接处添加柔性衬套后，仿真实验初期震荡现象改善明显，同时整车稳定性指标进一步提升，车身高度共降低 81.18 mm。

（4）板簧结构优化后，质量减少 24%，FSAE 整车稳定性继续改善，但伴有震荡现象。

（5）该仿真对 FSAE 赛车悬架设计、底盘参数匹配及进一步对整车其他系统优化具有重要指导意义。

第 25 章　弯道制动联合仿真

相对于直线制动，高速弯道制动属于极限制动工况，对于整车的底盘设计及系统之间的匹配要求极高。大学生方程式赛车属于小型方程式赛车，其设计难点是要保证整车具有良好的动态特性。在动态测试过程中，定半径弯、发卡弯、蛇形穿桩、复合赛道、高速避障等测试项目会涉及高速弯道制动过程，如果制动系统设计不符合要求，会导致整车产生严重的侧向滑移、方向盘转向失效及翻车等严重事故，甚至危及赛车手人身安全。目前，制动系统的研究主要集中在乘用车及商用车上，对于小型方程式赛车的制动特性研究甚少，尤其是在弯道制动模式下。在研究过程中，较多文献主要以单个制动车轮模型为基础，匹配不同的制动算法，其主要目的在于验证制动控制算法的正确性，与整车制动过程实际情况不符；在整车制动过程中，各车轮的制动力大小不同，单个车轮的制动力特性也会波及整车的安全运行状态，并且制动与发动机系统相关联，需要管理并控制发动机的输出转矩与各车轮所需的制动力矩相匹配。本案例主要通过实验构建包含发动机的精确整车 FSAE 多体模型，根据滑移率产生的机理，采用双模糊控制理论算法并在前悬架上增加辅助弹簧与减震器研究赛车的制动特性，如图 25-1 所示。

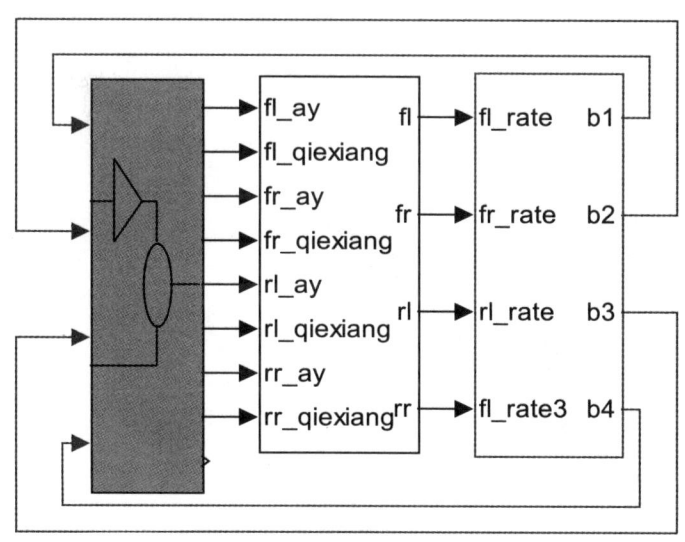

图 25-1　制动系统联合仿真模型

学习目标

（1）会制动系统设置。
（2）会函数编辑。
（3）会整车模型装配。
（4）会 ADAMS\Controls 设置。
（5）会 ADAMS 与 MATLAB 软件协同。
（6）了解双模糊控制理论。
（7）了解悬架辅助系统。
（8）会制动联合仿真。

25.1 制动系统设置

基于制动系统最优滑移率（0.2）的参数设置需要在制动系统模板上进行，在 View 模块与 Car 模块上进行联合仿真稍有不同，具体过程会在后续案例中体现。

（1）启动 ADAMS/CAR，选择专家模块进入建模界面。

（2）单击 File > Open 命令，弹出模板打开对话框，如图 25-2 所示，在模板名称中输入：mdids://FASE/templates.tbl/_brake_ABS.tpl；单击 OK 按钮；打开制动系统模板，如图 25-3 所示。

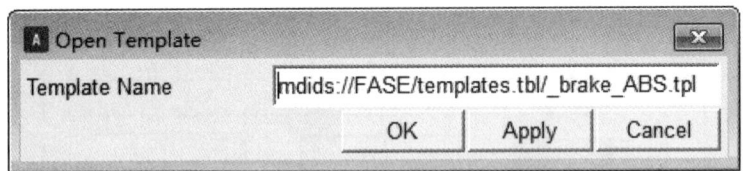

图 25-2　制动系统模板打开对话框

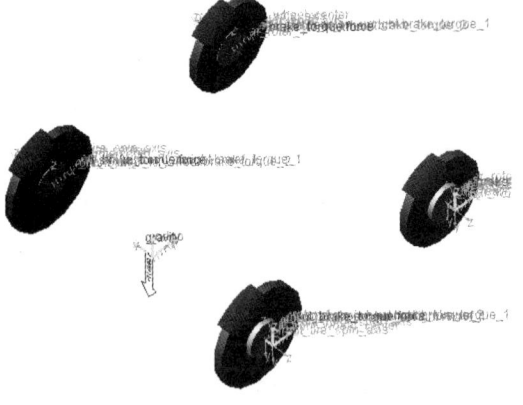

图 25-3　制动系统模板

（3）单击 Build >System Elements > State variable > New 命令，弹出创建状态变量对话框，如图 25-4 所示，在 Name（状态变量名称）中输入：abs_front_left_input，其余保持默认。

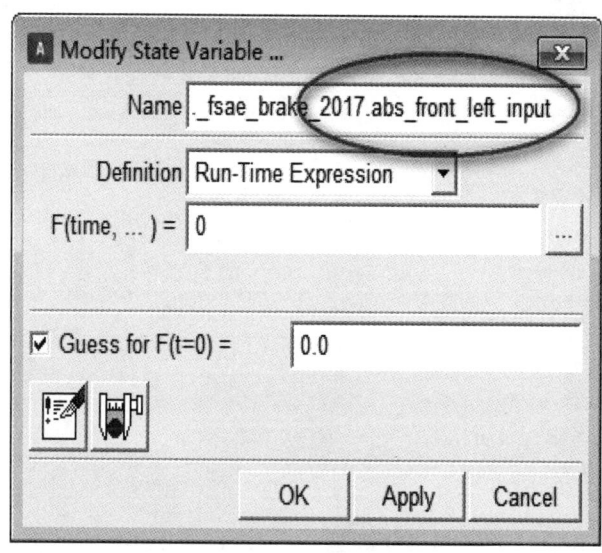

图 25-4　状态变量对话框

（4）单击 OK 按钮，完成状态变量 abs_front_left_input 的创建。

（5）重复上述步骤，依次建立状态变量：abs_front_right_input、abs_rear_left_input、abs_rear_right_input；建立好的四个状态变量分别为左前轮、右前轮、左后轮、右后轮的制动力矩变化系数；力矩变化系数由四个车轮输出的滑移率根据相应的算法计算得到，最终目的是通过制动力矩变化系数的调节使整车的滑移率控制在理想范围内。直线制动可以较好地控制在 0.2 范围内，弯道制动控制结果较差。

（6）单击 Build >Actuator > Point Torque > Modify 命令，弹出修改制动力矩对话框，如图 24-5 所示，在下列对话框中输入相应的数据：

• Left Function/ Right Function（制动力矩函数）：分别输入左前轮、右前轮、左后轮、右后轮的制动力矩函数，制动力矩函数如下，其余保持默认。

> 制动力矩函数编写注意以下情况：制动力矩函数中有下划线部分为上述建立的制动力矩变量系数，在联合仿真中，需要把此状态变量添加到制动系统力矩的公式中。添加此状态变量后，整车模型在仿真中会出现错误，原因在于模型中不能提供制动力矩：F（time=0）。

左前轮制动力矩函数：

2.0*VARVAL（._brake_ABS.abs_front_left_input）*._brake_ABS.pvs_front_piston_area*._brake_ABS.pvs_front_brake_bias*VARVAL　（　._brake_ABS.cis_brake_demand_adams_id　）*._brake_ABS.force_to_pressure_cnvt*._brake_ABS.pvs_front_brake_mu*._brake_ABS.pvs_front_effective_piston_radius*STEP（VARVAL（._brake_ABS.left_front_wheel_omega），-10D，1，10D，-1）。

右前轮制动力矩函数：

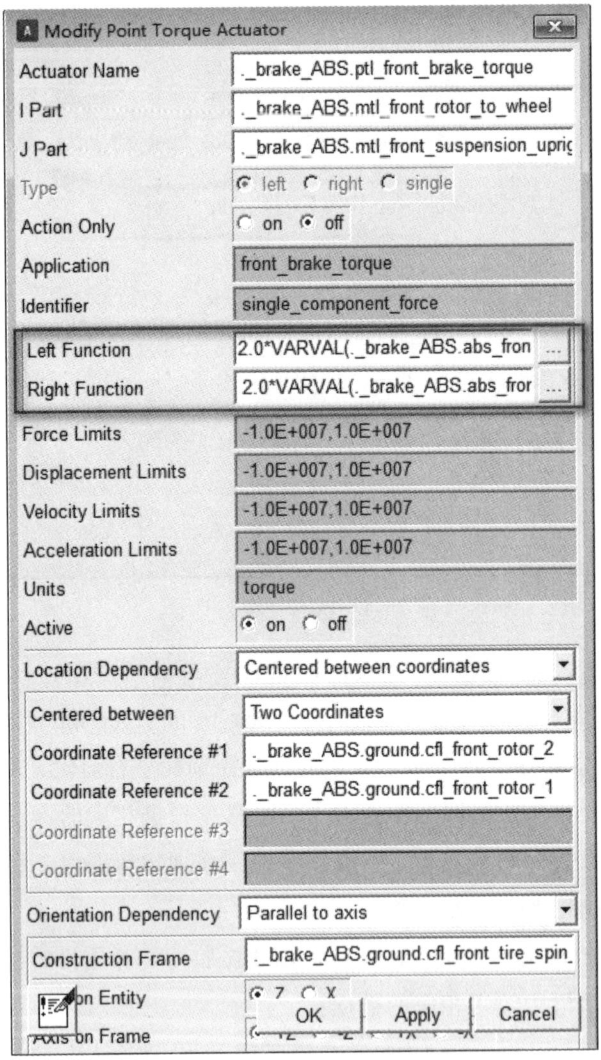

图 25-5 制动力矩函数

2.0*VARVAL（._brake_ABS.abs_front_right_input）*._brake_ABS.pvs_front_piston_area*._brake_ABS.pvs_front_brake_bias*VARVAL（._brake_ABS.cis_brake_demand_adams_id）*._brake_ABS.force_to_pressure_cnvt*._brake_ABS.pvs_front_brake_mu*._brake_ABS.pvs_front_effective_piston_radius*STEP（VARVAL（._brake_ABS.right_front_wheel_omega），-10D，1，10D，-1）。

左后轮制动力矩函数：

2.0*VARVAL（._brake_ABS.abs_rear_left_input）*._brake_ABS.pvs_rear_piston_area*（1.0-._brake_ABS.pvs_front_brake_bias）*VARVAL（._brake_ABS.cis_brake_demand_adams_id）*._brake_ABS.force_to_pressure_cnvt*._brake_ABS.pvs_rear_brake_mu*._brake_ABS.pvs_rear_effective_piston_radius*STEP（VARVAL（._brake_ABS.left_rear_wheel_omega），-10D，1，10D，-1）。

右后轮制动力矩函数：

2.0*VARVAL（._brake_ABS.abs_rear_right_input）*._brake_ABS.pvs_rear_piston_area*（1.0-._brake_ABS.pvs_front_brake_bias）*VARVAL(._brake_ABS.cis_brake_demand_adams_id）*._brake_ABS.force_to_pressure_cnvt*._brake_ABS.pvs_rear_brake_mu*._brake_ABS.pvs_rear_effective_piston_radius*STEP（VARVAL（._brake_ABS.right_rear_wheel_omega），-10D，1，10D，-1）。

以左前轮制动力矩函数为例，式中：

① ._brake_ABS.pvs_front_piston_area 为制动缸活塞有效面积；

② ._brake_ABS.pvs_front_brake_bias 为前轴系制动力分配系数；

③ VARVAL（._brake_ABS.cis_brake_demand_adams_id）为制动踏板力；

④ ._brake_ABS.force_to_pressure_cnvt 为换算系数，将制动踏板力直接转化为制动总管液体介质压强，默认 0.1；

⑤ ._brake_ABS.pvs_front_brake_mu 为制动器摩擦系数；

⑥ ._brake_ABS.pvs_front_effective_piston_radius 为制动油缸在制动盘上的作用半径；

⑦ STEP（VARVAL（._brake_ABS.left_front_wheel_omega），-10D，1，10D，-1）为阶跃函数，确保制动力矩与车轮旋转方向相反。

25.2 函数编写

25.2.1 车轮切向速度

（1）左前轮切向速度 ._brake_ABS.front_left_qiexiang：

-SQRT(VX(._brake_ABS.mtl_front_rotor_to_wheel.brake_torque_2)**2+Vy(._brake_ABS.mtl_front_rotor_to_wheel.brake_torque_2)**2)。

（2）右前轮切向速度 ._brake_ABS.front_right_qiexiang：

-SQRT(VX(._brake_ABS.mtr_front_rotor_to_wheel.brake_torque_2)**2+Vy(._brake_ABS.mtr_front_rotor_to_wheel.brake_torque_2)**2)。

（3）左后轮切向速度 ._brake_ABS.rear_left_qiexiang：

-SQRT(VX(._brake_ABS.mtl_rear_rotor_to_wheel.brake_torque_2)**2+Vy(._brake_ABS.mtl_rear_rotor_to_wheel.brake_torque_2)**2)。

（4）右后轮切向速度 ._brake_ABS.rear_right_qiexiang：

-SQRT(VX(._brake_ABS.mtr_rear_rotor_to_wheel.brake_torque_2)**2+Vy(._brake_ABS.mtr_rear_rotor_to_wheel.brake_torque_2)**2)。

25.2.2 车轮旋转速度

（1）左前轮旋转速度._brake_ABS.front_left_av_y：
WY（._brake_ABS.mtl_front_rotor_to_wheel.brake_torque_2）*243.65。
（2）右前轮旋转速度._brakc_ABS.front_right_av_y：
WY（._brake_ABS.mtr_front_rotor_to_wheel.brake_torque_2）*243.65。
（3）左后轮旋转速度._brake_ABS.rear_left_av_y：
WY（._brake_ABS.mtl_rear_rotor_to_wheel.brake_torque_2）*243.65。
（4）右后轮旋转速度._brake_ABS.rear_right_av_y：
WY（._brake_ABS.mtr_rear_rotor_to_wheel.brake_torque_2）*243.65。

25.2.3 状态变量

（1）单击 Build >System Elements > State variable > New 命令，弹出创建状态变量对话框，如图 24-6 所示，在下列对话框输入相应的数据：

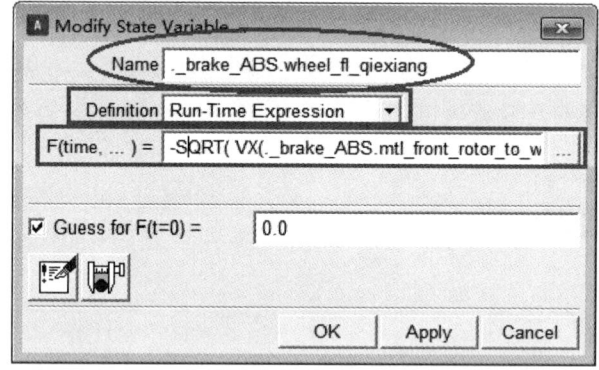

图 25-6　状态变量创建对话框

- Name（状态变量名称）：wheel_fl_qiexiang；
- Definition：Run-Time Expression；
- F（time=0）：输入之前创建的左前轮切向速度函数【-SQRT（VX（._brake_ABS.mtl_front_rotor_to_wheel.brake_torque_2）**2+Vy（._brake_ABS.mtl_front_rotor_to_wheel.brake_torque_2）**2）】。

（2）单击 OK 按钮，完成状态变量 wheel_fl_qiexiang 的创建。

（3）重复上述步骤，依次完成右前轮切向速度._brake_ABS.front_right_qiexiang、左后轮切向速度._brake_ABS.rear_left_qiexiang、右后轮旋转速度._brake_ABS.rear_right_av_y、左前轮旋转速度._brake_ABS.front_left_av_y、右前轮旋转速度._brake_ABS.front_right_av_y、左后轮旋转速度._brake_ABS.rear_left_av_y、右后轮旋转速度._brake_ABS.rear_right_av_y 状态变量的建立；创建好的状态变量主要用于机械模型系统的输出，后续用于滑移率的计算；滑移率的计算也可以直接编写函数。

（4）单击 Build >Date Elements > Plant Input > New 命令，弹出创建输入集对话框，如图

25-7 所示，在下列对话框中输入相应的数据：

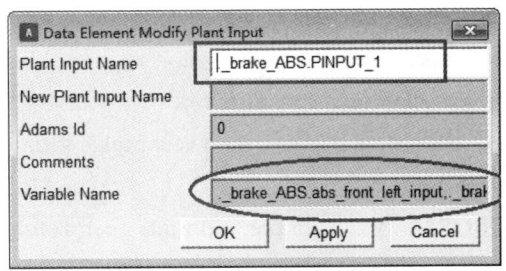

图 25-7 输入集对话框

- Variable Name（变量名称，输入之前建立好的状态变量）：._brake_ABS.abs_front_left_input，._brake_ABS.abs_front_right_input，._brake_ABS.abs_rear_left_input，._brake_ABS.abs_rear_right_input。

（5）单击 OK 按钮，完成输入集._brake_ABS.POUTPUT_1 的创建。

（6）单击 Build >Date Elements > Plant Output > New 命令，弹出创建输出集对话框，如图 25-8 所示，在下列对话框中输入相应的数据：

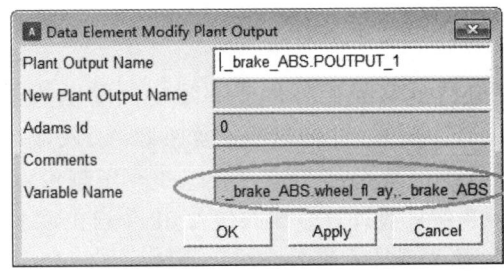

图 25-8 输出集对话框

- Variable Name（变量名称，输出之前建立好的状态变量）：._brake_ABS. wheel_fl_ay，._brake_ABS.wheel_fl_qiexiang，._brake_ABS.wheel_fr_ay，._brake_ABS.wheel_fr_qiexiang，._brake_ABS.wheel_rl_ay，._brake_ABS.wheel_rl_qiexiang，._brake_ABS.wheel_rr_ay，._brake_ABS.wheel_rr_qiexiang。

（7）单击 OK 按钮，完成输出集._brake_ABS.POUTPUT_1 的创建。

25.2.4 车轮滑移率

（1）左前轮滑移率._brake_ABS.front_left_slip_rate：

（-SQRT（VX（._brake_ABS.mtl_front_rotor_to_wheel.brake_torque_2）**2+Vy（._brake_ABS.mtl_front_rotor_to_wheel.brake_torque_2）**2）-WY（._brake_ABS.mtl_front_rotor_to_wheel.brake_torque_2）*243.65）/-SQRT（ VX（._brake_ABS.mtl_front_rotor_to_wheel.brake_torque_2）**2+Vy（._brake_ABS.mtl_front_rotor_to_wheel.brake_torque_2）**2）。

（2）右前轮滑移率._brake_ABS.front_right_slip_rate：

（-SQRT（VX（._brake_ABS.mtr_front_rotor_to_wheel.brake_torque_2）**2+Vy（._brake_

ABS.mtr_front_rotor_to_wheel.brake_torque_2）**2)-WY（._brake_ABS.mtr_front_rotor_to_wheel.brake_torque_2）*243.65）/-SQRT（VX（._brake_ABS. mtr_front_rotor_to_wheel.brake_torque_2）**2+Vy（._brake_ABS.mtr_front_rotor_to_wheel.brake_torque_2）**2）。

（3）左后轮滑移率._brake_ABS.rear_left_slip_rate：

（-SQRT（VX（._brake_ABS.mtl_rear_rotor_to_wheel.brake_torque_2）**2+Vy（._brake_ABS.mtl_rear_rotor_to_wheel.brake_torque_2）**2）-WY（._brake_ABS.mtl_rear_rotor_to_wheel.brake_torque_2）*243.65）/-SQRT（VX（._brake_ABS.mtl_rear_rotor_to_wheel.brake_torque_2）**2+Vy（._brake_ABS.mtl_rear_rotor_to_wheel.brake_torque_2）**2）。

（4）右后轮滑移率._brake_ABS.rear_right_slip_rate：

（-SQRT（VX（._brake_ABS.mtr_rear_rotor_to_wheel.brake_torque_2）**2+Vy（._brake_ABS.mtr_rear_rotor_to_wheel.brake_torque_2）**2）-WY（._brake_ABS.mtr_rear_rotor_to_wheel.brake_torque_2）*243.65）/-SQRT（VX（._brake_ABS.mtr_rear_rotor_to_wheel.brake_torque_2）**2+Vy（._brake_ABS.mtr_rear_rotor_to_wheel.brake_torque_2）**2）。

25.3 整车模型装配

制动模板设置并保存好之后转换到标准界面建立制动系统子系统，子系统名称为：brake_ABS.sub。此前整车已经装配好，此处只需要替换FSAE赛车原有的制动子系统即可（也可以重新装配）。装配好的整车模型如图25-9所示，整车模型包含前后推杆式悬挂、前后轮胎、中舵转向系统、发动机系统；经计算整车共包含58个自由度。对于整车精确建模，悬架与车身连接处的橡胶衬套、悬置系统刚度等应尽可能详细，衬套刚度对整车性能影响不可忽略。对于实验条件有限的情况，可以采用主流有限元软件（例如ABAQUS）分析橡胶衬套的刚度等。

图25-9 整车模型

在进行整车制动系统仿真时，应尽量建立包含发动机的整车模型，原因在于制动过程中由于路面条件各不相同，发动机输出的转矩通过传动系统及变速箱传递到车轮上，四个车轮的输出力矩存在差异。

单击 File > Info> Subsystem 命令，在 Subsystem name 菜单栏中下拉选中 brake_ABS 子系统，如图 25-10 所示，单击 OK 按钮显示出制动系统的相关信息。

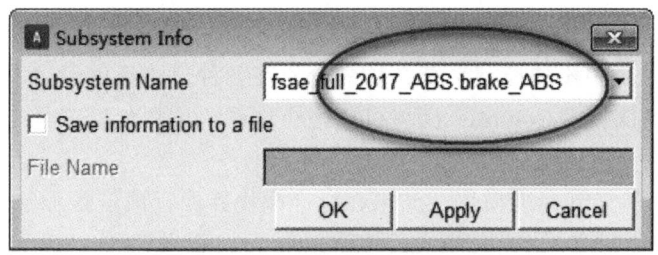

图 25-10 子系统信息对话框

制动系统的详细信息如下。可以通过模板设置不同的参数，使其与真实的车辆相符，同时在编写制动力矩函数时，根据模型的精确程度可以适当地增加或者减少某些项，越精确的制动系统，涉及的因素越多。

```
Info for subsystem    : brake_ABS
    File Name         : <FASE>/subsystems.tbl/brake_ABS.sub
    Template          : mdids://FASE/templates.tbl/_brake_ABS.tpl
    Comments          :
    Template          : 4 Wheel Disk Brake System
    Subsystem         : *no subsystem comments found*
    Major Role        : brake_system
    Minor Role        : any
    PARAMETERS：

        parameter name                  symmetry    type        value
        ---------------                 --------    ----        -----
        kinematic_flag                  single      integer     0
        front_brake_bias                single      real        0.6
        front_brake_mu                  single      real        0.4
        front_effective_piston_radius   single      real        135.0
        front_piston_area               single      real        2500.0
        front_rotor_hub_wheel_offset    single      real        25.0
        front_rotor_hub_width           single      real        40.0
        front_rotor_width               single      real        -25.0
        max_brake_value                 single      real        100.0
        rear_brake_mu                   single      real        0.4
        rear_effective_piston_radius    single      real        120.0
        rear_piston_area                single      real        2500.0
        rear_rotor_hub_wheel_offset     single      real        25.0
        rear_rotor_hub_width            single      real        40.0
        rear_rotor_width                single      real        -25.0
```

25.4　ADAMS\Controls 设置

运用多体动力学分析软件 ADAMS 建立各个子系统及组装后的整车模型，然后在 ADAMS\Controls 模块中添加控制系统，仿真分析在各种道路条件激励下，所得到的汽车操纵稳定性的响应。ADAMS\Controls 模块可以将机械系统仿真分析工具与控制仿真软件有机地链接起来。

（1）按 F9 键，转换到标准模块，如果在标准模块界面，该步骤可忽略。

（2）在 D 盘中建立文件夹 brake_cosimulation，设置 ADAMS 的工作路径为 D:\brake_cosimulation。

（3）单击 Control> Plant Export 命令，弹出控制接口输出对话框，如图 25-11 所示，在 Initial Static Analysis（初始静态分析）中选择 No。

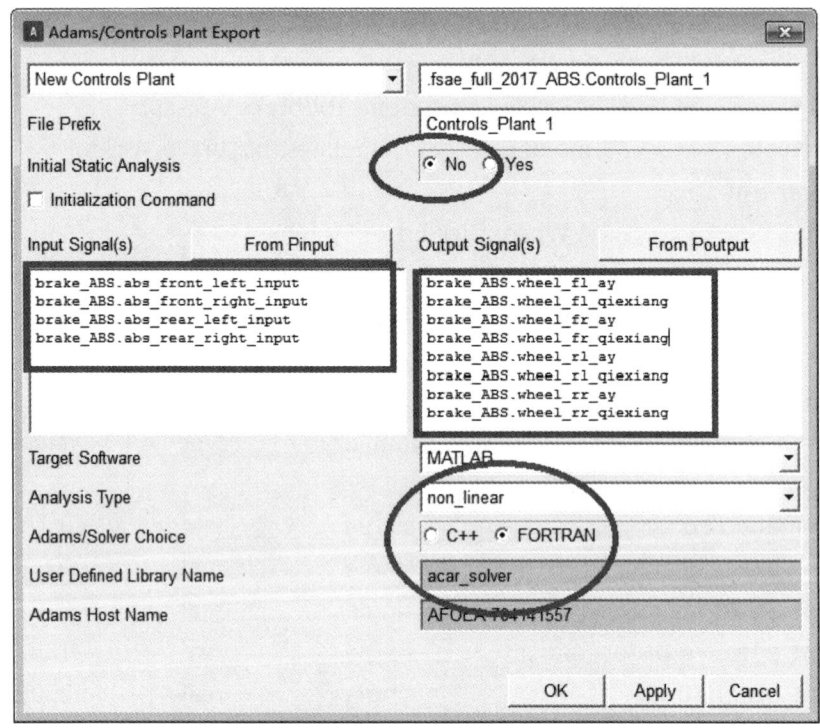

图 25-11　控制接口输出对话框

（4）单击 From Pinput，在弹出的数据命令窗口中双击_brake_ABS 下的 PINPUT1。

（5）单击 From Poutput，在弹出的数据命令窗口中双击_brake_ABS 下的 POUTPUT1。

（6）在 Target Software（目标软件或者对接软件）中选择 MATLAB。

（7）在 Analysis Type（分析类型）中选择非线性 non_linear，整车模型存在轮胎模型等非线性因素。

（8）在 Adams/Solver Choice 中选择 C++，选择 FORTRAN 语言也可以；如果是非线性计算，推荐选择 C++，所见案例较多选择 FORTRAN 语言。

（9）其余保持默认，单击 OK 按钮，完成 ADAMS\Controls 模块下的输入输出集的创建。

> ADAMS\Controls 模块下的输入输出集也可以继续添加其他系统的状态变量，例如可以增加方程式赛车车身横摆角加速度输出：fsae_full_2017_ABS.FSAE_Body_2017.state_wdtz，其他变量根据模型需要可进行相应修改，后续控制系统可以根据横摆角加速度判断整车的稳定性情况。

25.5 ADAMS 与 MATLAB 软件协同

（1）同时启动 ADAMS 与 MATLAB 软件，路径统一设置为 D：\brake_cosimulation。

（2）单击 Simulate > Full-Vehicle > Cornering Event > Braking-In-Turn 命令，弹出弯道制动仿真对话框，如图 25-12 所示，在下列对话框中输入相应的数据：

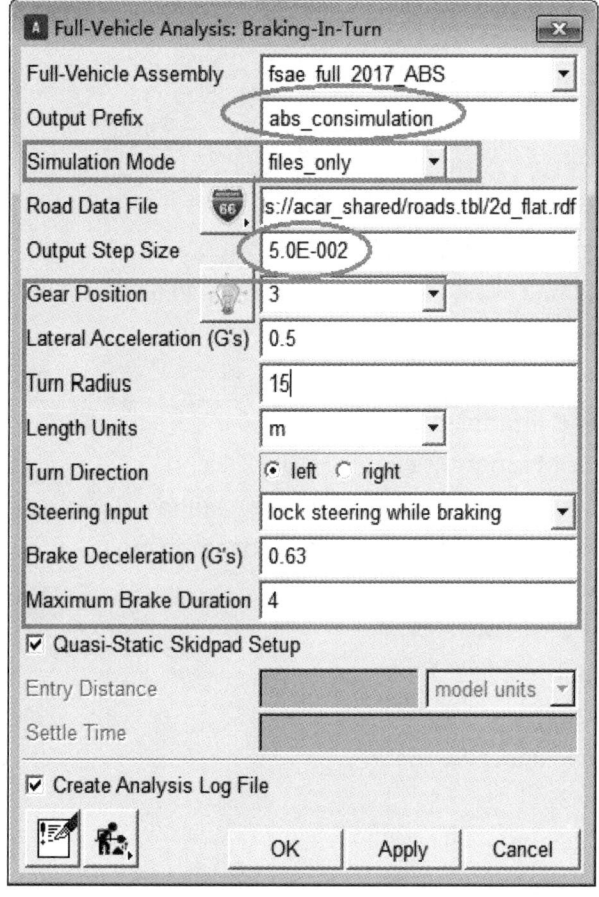

图 25-12　弯道制动仿真

- Prefix（输出别名）：abs_consimulation；
- Simulation Mode（仿真类型）：files_only；
- Road Date File：mdids://acar_shared/roads.tbl/2d_flat.rdf，路面为共享数据库中路面，此处可以选择其他路面模型或者编写路面模型，包括对开路面、对接路面等；

- Output Step Size（计算步长）：5.0E-002；
- Gear Position（挡位）：3 挡；
- Lateral Acceleration（G's）（制动时侧向加速度）：0.5；
- Turn Radius（转弯半径）：15；
- Length Units（长度单位）：m；
- Steering Input（转向输入）：lock steering while braking，转向时保持转向锁定；
- Brake Deceleration（G's）（制动时减速度）：0.63；
- Maximum Brake Duration（制动时间）：4。

（3）单击 OK 按钮，完成弯道制动设置并提交软件进行计算。

ADAMS 软件计算完成后，在目标 D：\brake_cosimulation 文件夹中存在 Controls_Plant_1.m、abs_bit.m、abs_bit.acf 三个文件。在 View 模块与 Car 模块中进行联合仿真稍有不同，采用 Car 模块进行联合仿真时，需要对文件中的参数稍做修改，仿真时采用两种方案均可，推荐采用方案一。

25.5.1 方案一

用记事本打开文件 Controls_Plant_1.m：

（1）修改 ADAMS_prefix = 'abs_bit';

（2）修改 ADAMS_init = 'file/command=Controls_Plant_1_controls.acf 为 ADAMS_init= 'file/command=file/command=abs_bit_controls.acf。

具体操作过程如下，程序修改部分用下划线区别：

```
% Adams / MATLAB Interface - Release 2014.0.0
system（'taskkill /IM scontrols.exe /F >NUL'）;clc;
global ADAMS_sysdir; % used by setup_rtw_for_adams.m
global ADAMS_host; % used by start_adams_daemon.m
machine=computer;
datestr（now）
if strcmp（machine，'SOL2'）
    arch = 'solaris32';
elseif strcmp（machine，'SOL64'）
    arch = 'solaris32';
elseif strcmp（machine，'GLNX86'）
    arch = 'linux32';
elseif strcmp（machine，'GLNXA64'）
    arch = 'linux64';
elseif strcmp（machine，'PCWIN'）
    arch = 'win32';
elseif strcmp（machine，'PCWIN64'）
```

```
        arch = 'win64';
    else
        disp（'%%% Error: Platform unknown or unsupported by Adams/Controls.'）;
        arch = 'unknown_or_unsupported';
        return
    end
    if strcmp（arch, 'win64'）
        [flag, topdir]=system（'adams2014_x64 -top'）;
    else
        [flag, topdir]=system（'adams2014 -top'）;
    end
    if flag == 0
      temp_str=strcat（topdir, '/controls/', arch）;
      addpath（temp_str）
      temp_str=strcat（topdir, '/controls/', 'matlab'）;
      addpath（temp_str）
      temp_str=strcat（topdir, '/controls/', 'utils'）;
      addpath（temp_str）
      ADAMS_sysdir = strcat（topdir, ''）;
    else
      addpath（'D: \MSC~1.SOF\ADAMS_~1\2014\controls/win64'）;
      addpath（'D: \MSC~1.SOF\ADAMS_~1\2014\controls/win32'）;
      addpath（'D: \MSC~1.SOF\ADAMS_~1\2014\controls/matlab'）;
      addpath（'D: \MSC~1.SOF\ADAMS_~1\2014\controls/utils'）;
      ADAMS_sysdir = 'D: \MSC~1.SOF\ADAMS_~1\2014\';
    end
    ADAMS_exec = 'acar_solver';
    ADAMS_host = 'AFOEA-704141557';
    ADAMS_cwd ='D: \brake_cosimulation'  ;
    ADAMS_prefix = 'abs_bit';
    ADAMS_static = 'no';
    ADAMS_solver_type = 'C++';
    if exist（[ADAMS_prefix, '.adm']）== 0
        disp（''）;
        disp（'%%% Warning: missing ADAMS plant model file（.adm）for Co-simulation
or Function Evaluation.'）;
        disp（'%%% If necessary, please re-export model files or copy the exported plant
model files into the'）;
```

```
        disp( '%%% working directory.  You may disregard this warning if the Co-simulation/Function Evaluation' );
        disp( '%%% is TCP/IP-based( running Adams on another machine ), or if setting up MATLAB/Real-Time Workshop' );
        disp( '%%% for generation of an External System Library.' );
        disp( '' );
    end
    ADAMS_init = 'file/command=abs_bit_controls.acf' ;
    ADAMS_inputs                                                                      = 'brake_ABS.abs_front_left_input!brake_ABS.abs_front_right_input!brake_ABS.abs_rear_left_input!brake_ABS.abs_rear_right_input' ;
    ADAMS_outputs

                     = 'brake_ABS.wheel_fl_ay!brake_ABS.wheel_fl_qiexiang!brake_ABS.wheel_fr_ay!brake_ABS.wheel_fr_qiexiang!brake_ABS.wheel_rl_ay!brake_ABS.wheel_rl_qiexiang!brake_ABS.wheel_rr_ay!brake_ABS.wheel_rr_qiexiang!FSAE_Body_2017.state_av_x!FSAE_Body_2017.state_av_y!FSAE_Body_2017.state_qiexiang!FSAE_Body_2017.state_wdtz' ;
    ADAMS_pinput = 'Controls_Plant_1.ctrl_pinput' ;
    ADAMS_poutput = 'Controls_Plant_1.ctrl_poutput' ;
    ADAMS_uy_ids   = [
                        290
                        291
                        292
                        293
                        306
                        310
                        307
                        311
                        308
                        312
                        309
                        313
                        255
                        256
                        261
                        260
                     ];
```

```
    ADAMS_mode    = 'non-linear' ;
    tmp_in   = decode（ADAMS_inputs）;
    tmp_out = decode（ADAMS_outputs）;
    disp（' '）;
    disp（'%%% INFO：ADAMS plant actuators names：'）;
    disp（[int2str（[1: size（tmp_in, 1）]'）, blanks（size（tmp_in, 1））', tmp_in]）;
    disp（'%%% INFO：ADAMS plant sensors    names：'）;
    disp（[int2str（[1: size（tmp_out, 1）]'）, blanks（size（tmp_out, 1））', tmp_out]）;
    disp（' '）;
    clear tmp_in tmp_out ;
% Adams / MATLAB Interface - Release 2014.0.0
```

25.5.2　方案二

用记事本打开文件 abs_bit.m，如下参数与 Controls_Plant_1.m 文件对应的参数相同，可以把 Controls_Plant_1.m 中对应的参数复制粘贴保存即可。

（1）修改 ADAMS_outputs = '\\\\' ；

（2）修改 ADAMS_poutput = '\\\\';

（3）修改 ADAMS_uy_ids　 = [\\\\] 。

具体操作过程如下，程序修改部分用黑斜体区别，黑斜体与 Controls_Plant_1.m 文件对应的参数相同：

```
% Adams / MATLAB Interface - Release 2014.0.0
system（'taskkill /IM scontrols.exe /F >NUL'）;clc;
global ADAMS_sysdir; % used by setup_rtw_for_adams.m
global ADAMS_host; % used by start_adams_daemon.m
machine=computer;
datestr（now）
if strcmp（machine，'SOL2'）
    arch = 'solaris32';
elseif strcmp（machine，'SOL64'）
    arch = 'solaris32';
elseif strcmp（machine，'GLNX86'）
    arch = 'linux32';
elseif strcmp（machine，'GLNXA64'）
    arch = 'linux64';
elseif strcmp（machine，'PCWIN'）
    arch = 'win32';
```

```
        elseif strcmp ( machine, 'PCWIN64' )
            arch = 'win64';
        else
            disp ( '%%% Error: Platform unknown or unsupported by Adams/Controls.' );
            arch = 'unknown_or_unsupported';
            return
        end
        if strcmp ( arch, 'win64' )
            [flag, topdir]=system ( 'adams2014_x64 -top' );
        else
            [flag, topdir]=system ( 'adams2014 -top' );
        end
        if flag == 0
           temp_str=strcat ( topdir, '/controls/', arch );
           addpath ( temp_str )
           temp_str=strcat ( topdir, '/controls/', 'matlab' );
           addpath ( temp_str )
           temp_str=strcat ( topdir, '/controls/', 'utils' );
           addpath ( temp_str )
           ADAMS_sysdir = strcat ( topdir, '' );
        else
           addpath ( 'D: \MSC~1.SOF\ADAMS_~1\2014\controls/win64' );
           addpath ( 'D: \MSC~1.SOF\ADAMS_~1\2014\controls/win32' );
           addpath ( 'D: \MSC~1.SOF\ADAMS_~1\2014\controls/matlab' );
           addpath ( 'D: \MSC~1.SOF\ADAMS_~1\2014\controls/utils' );
           ADAMS_sysdir = 'D: \MSC~1.SOF\ADAMS_~1\2014\';
        end
        ADAMS_exec = 'acar_solver';
        ADAMS_host = '';
        ADAMS_cwd ='D: \brake_cosimulation'  ;
        ADAMS_prefix = 'abs_bit';
        ADAMS_static = 'no';
        ADAMS_solver_type = 'C++';
        if exist ( [ADAMS_prefix, '.adm'] ) == 0
            disp ( '' );
            disp ( '%%% Warning: missing ADAMS plant model file ( .adm ) for Co-simulation or Function Evaluation.' );
            disp ( '%%% If necessary, please re-export model files or copy the exported plant
```

model files into the') ;
        disp（ '%%% working directory. You may disregard this warning if the Co-simulation/Function Evaluation' ） ;
        disp（ '%%% is TCP/IP-based（ running Adams on another machine ）, or if setting up MATLAB/Real-Time Workshop' ） ;
        disp（ '%%% for generation of an External System Library.' ） ;
        disp（ '' ） ;
    end
    ADAMS_init = 'file/command=abs_bit_controls.acf' ;
    ADAMS_inputs = 'brake_ABS.abs_front_left_input!brake_ABS.abs_front_right_input!brake_ABS.abs_rear_left_input!brake_ABS.abs_rear_right_input' ;
    ADAMS_outputs
      =
'brake_ABS.wheel_fl_ay!brake_ABS.wheel_fl_qiexiang!brake_ABS.wheel_fr_ay!brake_ABS.wheel_fr_qiexiang!brake_ABS.wheel_rl_ay!brake_ABS.wheel_rl_qiexiang!brake_ABS.wheel_rr_ay!brake_ABS.wheel_rr_qiexiang!FSAE_Body_2017.state_av_x!FSAE_Body_2017.state_av_y!FSAE_Body_2017.state_qiexiang!FSAE_Body_2017.state_wdtz' ;
    ADAMS_pinput = 'Controls_Plant_1.ctrl_pinput' ;
    ADAMS_poutput = 'Controls_Plant_1.ctrl_poutput' ;
    ADAMS_uy_ids = [
                    290
                    291
                    292
                    293
                    306
                    310
                    307
                    311
                    308
                    312
                    309
                    313
                    255
                    256
                    261
                    260
                ];

```
ADAMS_mode    = 'non-linear' ;
tmp_in   = decode ( ADAMS_inputs   ) ;
tmp_out = decode ( ADAMS_outputs  ) ;
disp ( '' ) ;
disp ( '%%% INFO: ADAMS plant actuators names: ' ) ;
disp ( [int2str ( [1: size ( tmp_in, 1 ) ]' ), blanks ( size ( tmp_in, 1 ))', tmp_in] ) ;
disp ( '%%% INFO: ADAMS plant sensors    names: ' ) ;
disp ( [int2str ( [1: size ( tmp_out, 1 ) ]' ), blanks ( size ( tmp_out, 1 ))', tmp_out] ) ;
disp ( '' ) ;
clear tmp_in tmp_out ;
% Adams / MATLAB Interface - Release 2014.0.0
```

在 MATLAB 软件命令窗口中输入：Controls_Plant_1；单击 Enter 键，此时命令窗口显示出如下信息；信息包含输入输出集信息。

```
命令窗口显示信息：
Controls_Plant_1

13-Jun-2018 11:22:17
%%% INFO: ADAMS plant actuators names:
1 brake_ABS.abs_front_left_input
2 brake_ABS.abs_front_right_input
3 brake_ABS.abs_rear_left_input
4 brake_ABS.abs_rear_right_input
%%% INFO: ADAMS plant sensors    names:
 1 brake_ABS.wheel_fl_ay
 2 brake_ABS.wheel_fl_qiexiang
 3 brake_ABS.wheel_fr_ay
 4 brake_ABS.wheel_fr_qiexiang
 5 brake_ABS.wheel_rl_ay
 6 brake_ABS.wheel_rl_qiexiang
 7 brake_ABS.wheel_rr_ay
 8 brake_ABS.wheel_rr_qiexiang
 9 FSAE_Body_2017.state_wdtz
```

运行 adams_sys，调出 adams_plant 对话框，如图 25-13 所示。

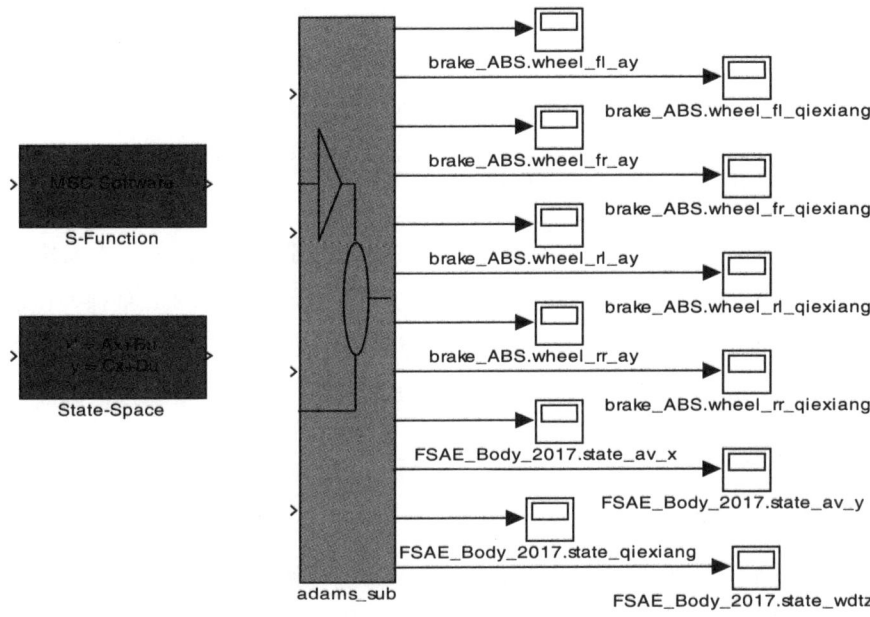

图 25-13 adams_plant 对话框

25.6 双模糊理论

1. 制动系统双模糊控制算法

实际制动过程是直线制动与弯道制动的混合模式，不存在严格意义上单一的直线制动或者弯道制动；弯道制动又可以分为低速弯道制动与高速弯道制动。如果继续细分，弯道制动又可以划分为不同车速弯道制动状态，不同车速在弯道制动中所占的权重不同，基于此提出制动系统连续模糊控制定义公式为

$$B = k_{11} \cdot b_{11} + k_{12} \cdot b_{12} + \cdots + k_{1n} \cdot b_{1n} \tag{25-1}$$

针对方程式赛车的特殊性，只考虑直线制动、低速弯道制动与高速弯道制动三种模式，左前轮、右前轮、左后轮、右后轮四个不同车轮的制动力矩定义为式（25-2）~式（25-5）。三种模式中直线制动为常态制动模式，因此公式（25-2）~式（25-5）定义为制动系统双模糊控制算法。

$$B_1 = k_{11} \cdot b_{11} + k_{12} \cdot b_{12} + k_{13} \cdot b_{13} \tag{25-2}$$

$$B_2 = k_{21} \cdot b_{21} + k_{22} \cdot b_{22} + k_{23} \cdot b_{23} \tag{25-3}$$

$$B_3 = k_{31} \cdot b_{31} + k_{32} \cdot b_{32} + k_{33} \cdot b_{33} \tag{25-4}$$

$$B_4 = k_{41} \cdot b_{41} + k_{42} \cdot b_{42} + k_{43} \cdot b_{43} \tag{25-5}$$

整理式（25-2）~式（25-5），改写成矩阵形式：

$$\begin{bmatrix} B_1 \\ B_2 \\ B_3 \\ B_4 \end{bmatrix} = \begin{bmatrix} k_{11} & k_{12} & k_{13} \\ k_{21} & k_{22} & k_{23} \\ k_{31} & k_{32} & k_{33} \\ k_{41} & k_{42} & k_{43} \end{bmatrix} \begin{bmatrix} b_{11} & b_{21} & b_{31} & b_{41} \\ b_{12} & b_{22} & b_{32} & b_{42} \\ b_{13} & b_{23} & b_{33} & b_{43} \end{bmatrix} \qquad (25\text{-}6)$$

式中：B_i（$i=1、2、3、4$）为车轮总制动力矩，按顺序分别对应左前轮、右前轮、左后轮、右后轮；k_{i1}、k_{i2}、k_{i3}（$i=1、2、3、4$）为方程式赛车在直线制动模式、低速弯道制动模式、高速弯道制动模式中的权系数，权系数大，制动力矩输出以对应的制动模式为主，其他制动模式为辅；b_{i1}、b_{i2}、b_{i3}（$i=1、2、3、4$）为制动过程中直线制动、低速弯道制动、高速弯道制动输出的制动力矩。

2. 双模糊控制规则

方程式赛车以车身横摆角加速度对制动力权系数进行调节，即以横摆角加速度对直线制动模式、低速弯道制动模式、高速弯道制动模式进行识别，合理分配权系数。制动力权系数模糊控制规则如表 25-1 所示；直线制动模式下模糊控制规则如表 25-2 所示，弯道制动模式下模糊控制规则如表 25-3 所示。

表 25-1　制动力权系数模糊控制规则

$\ddot{\varphi}_c$	－3	－2	－1	0	1	2	3
k_{i1}	0.1	0.2	0.5	0.8	0.5	0.2	0.1
k_{i2}	0.2	0.3	0.3	0.1	0.3	0.3	0.2
k_{i3}	0.7	0.5	0.2	0.1	0.2	0.5	0.7

表 25-2　直线制动模糊控制规则

E_2	EC_2		
	-3	0	3
－3	3	2	2
－2	2	2	1
0	1	－1	－1
1	－1	－1	－2
2	－2	－2	－3
3	－2	－3	－3

表 25-3　弯道制动模糊控制规则

E_2	EC_3						
	-3	-2	-1	0	1	2	3
-3	3	3	2	2	1	1	-1
-2	3	3	2	1	-1	-1	-2
0	3	2	1	-1	-1	-1	-2
1	3	2	1	-1	-1	-2	-3
2	2	1	1	-2	-2	-3	-3
3	1	1	-1	-3	-3	-3	-3

打开 simulink，根据双模控制理论及模糊控制规则搭建系统，如图 25-14 所示。

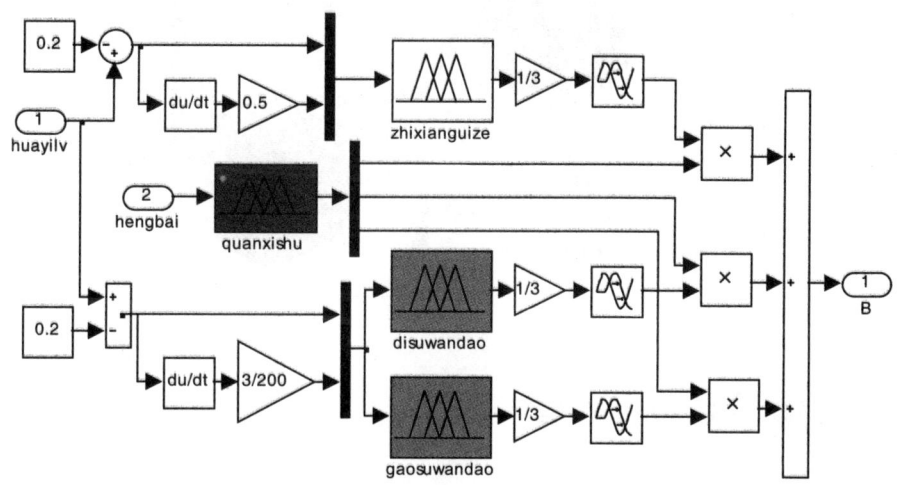

图 25-14　双模糊控制系统图

25.7　悬架辅助系统

1. 整车行驶过程中产生滑移的原因

弯道制动导致的纵侧向滑移偏大及整车失稳因素较多，且不同因素之间又互为矛盾体，设计过程中较多采用折中方案。整车行驶过程中产生滑移的原因主要有 4 点：① 轮胎的非线性因素；② 发动机输出扭矩的合理控制；③ 重心偏移及轴系侧倾刚度；④ 基于最优滑移率控制策略。针对轮胎非线性因素，可塑性不大，通过增加轮胎断面宽度可以有效改变弯道稳定性，但整车燃油经济性及动力输出造成浪费，跑车大多采用此方案；发动机扭矩输出匹配是大多文献忽略的因素，整车包含准确的发动机模型是制动仿真精确的关键因素，过大的功率输出使整车加速能力强,但轮胎磨损严重,过弯时侧向滑移更加严重;重心偏移及轴系刚度不同会导致内外侧轮胎的附着力不同,滑移严重,主动可控悬架可以有效改变此现象,但系统复杂,成本较高;基于最优滑移率控制是大多文献的研究方向,可塑性强,控制架构及策略有多种形式可以探讨。

2. 前轴推杆式悬架设计

针对 FSAE 方程式赛车的特殊性,只在赛道上做定半径弯、发卡弯、蛇形穿桩等特定行驶工况,结合第 3 点因素,提出在赛车前轴系增加辅助弹簧与避震器,改善高速过弯时产生的重心偏移及侧倾刚度过低的现象。前轴推杆式悬架设计方案如图 25-15 所示,在下控制臂与钢架车身之间增加辅助避震器与弹簧。重新组装整车模型,其余保持不变,模型另存为 fsae_full_2017_fuzhu.asy;系统输入输出集接口、软件协同等同前述内容,此处不再述。

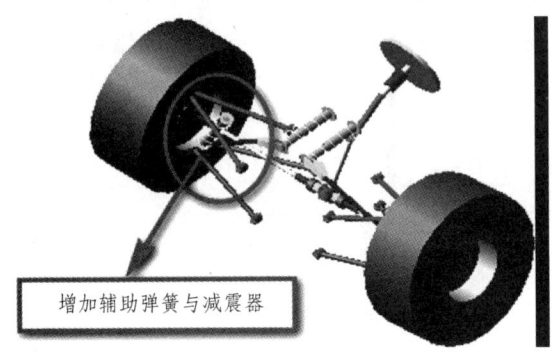

图 25-15　推杆式悬架设计方案

25.8　制动联合仿真模型

根据双模糊理论及制动系统模糊控制规则搭建双模糊控制架构,如图 25-16 所示,基于此建立联合仿真模型。计算整车在辅助悬架、双模糊控制器下 FSAE 整车在弯道运行的状态,计算结果如图 27-17 ~ 图 25-19 所示。左前轮滑移在改进前产生抱死现象,满足方程式赛车设计要求,在制动时有瞬间抖动现象;通过增加辅助避震器与弹簧后,前轴车轮的滑移率在 1.1 ~ 1.15 s 有伴随瞬间抖动现象,在 1.2 ~ 2.25 s 时显著降低,2.3 s 开始产生抱死,依然符合赛车制动系统设计要求,滑移改善明显;采用双模糊控制后,在制动系统的整个制动过程中,滑移率变化非常平稳,但增加趋势比采用辅助避震器方案大。右后轮滑移率在制动过程中均产生抱死现象;其中改进后滑移率从 1.35 s ~ 2.25 s 降低明显;采用双模糊控制后滑移率变化平稳,但有增加趋势;通过对比前后轴滑移率,前轴滑移率整体偏大,原因在于前轴制动力分配系数大;采用双模糊控制后,前后轴均未抱死。采用辅助避震器与弹簧后,车身横摆角加速度最大值从 3.024 rad/s^2 降低至 1.741 rad/s^2,性能提升 42.4%;采用双模控制后,车身横摆角加速度最大值仅为 0.509 rad/s^2,性能提升 83.2%,整车稳定性提升极为显著。

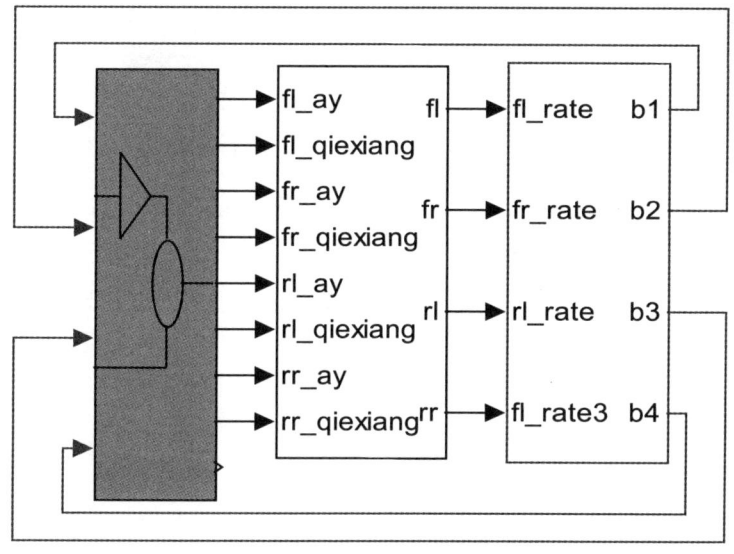

图 25-16 双模糊控制器架构

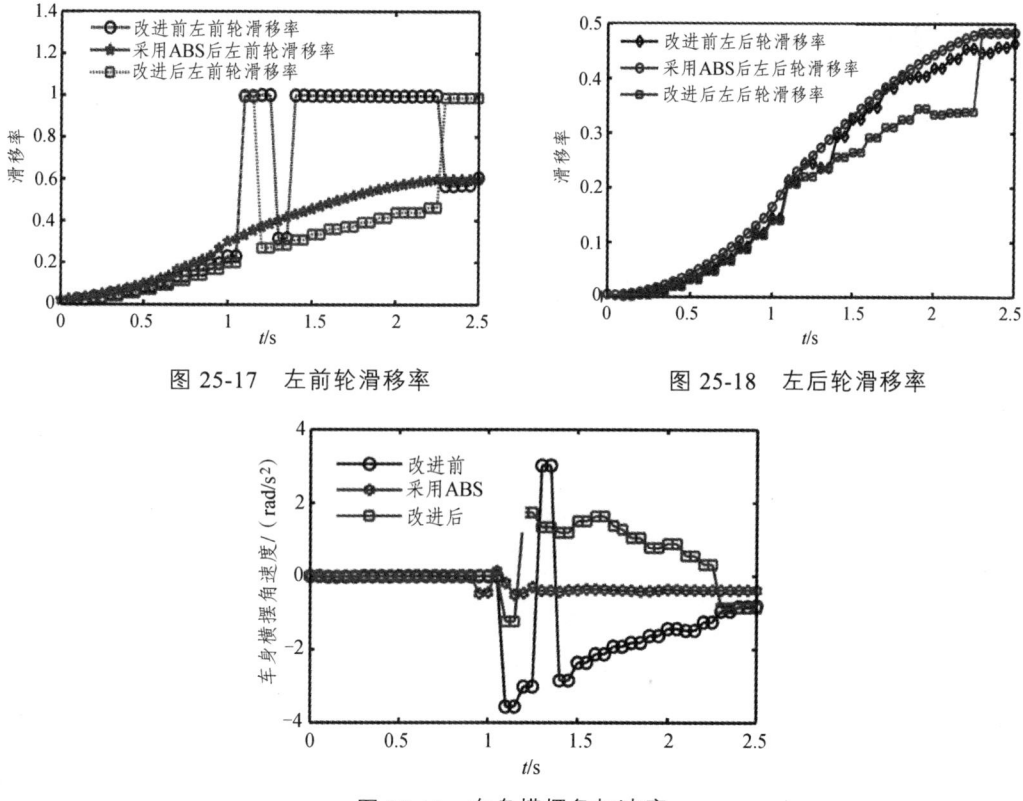

图 25-17 左前轮滑移率　　　　图 25-18 左后轮滑移率

图 25-19 车身横摆角加速度

（1）单击 File > Info> Assembly 命令，在 Assembly Name 菜单栏中下拉选中 fsae_full_2017_fuzhu 装配模型，如图 25-20 所示。

图 25-20 整车装配模型信息对话框

（2）单击 OK 按钮，显示出整车信息。子系统信息及整车模型信息是了解整车包含子系统、参数、路面、衬套、硬点等最直接的方式，在学习过程中，包含通信器在内，经常需要在整车信息中查询。

第 26 章 整车平顺性仿真

平顺性研究车身在垂向（Z 轴）方向的振动特性，ADAMS 通过虚拟的四柱实验台架与整车装配完成振动实验，路面的位移及频率特性可以通过振动产生。振动台与车轮接触，可以约束车轮在振动台架上的平面移动。四柱实验台可以同时振动，也可以前后、左右、对角测实验台柱振动并设置时间延迟振动特性。平顺性仿真需要通过插件工具调入仿真平台。平顺性仿真有两种装配方式：① 子系统与台架统一装配完成；② 先建立整车模型，然后通过添加试验台完成。推荐采用第二种方式，实验完成后，实验台可以移去或者替换成其他试验台。已经建立好的平顺性实验台架与整车如图 26-1 所示。

图 26-1 平顺性仿真台架

学习目标

（1）会整车装配。
（2）会平顺性仿真。

26.1 整车装配

1. 调入平顺性插件

(1) 单击 Tools > Plugin Manage 命令,弹出插件管理器对话框,如图 26-2 所示,勾选 Adams Car Ride 后对应的 Load,Load at Startup。

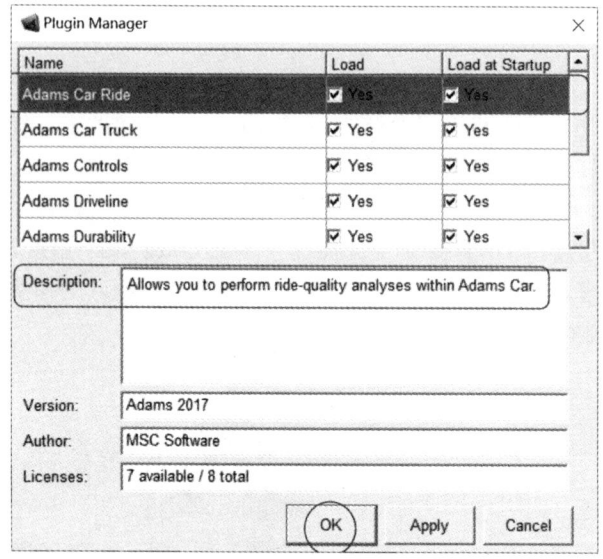

图 26-2 调入平顺仿真平台

(2) 单击 OK 按钮,完成平顺性仿真插件的导入。

2. 装配方法(1)

(1) 单击 File > Full-Vehicle Assembly 命令,弹出整车装配对话框,如图 26-3 所示,在下列对话框中输入相应的数据:

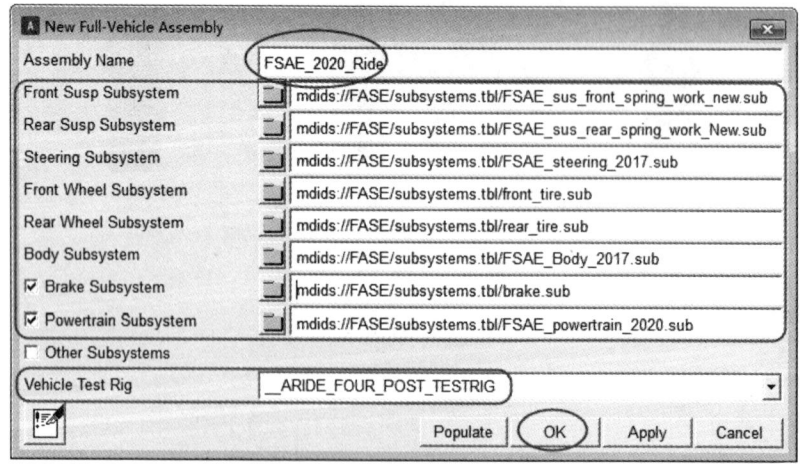

图 26-3 整车装配对话框

- Assembly Name：FSAE_2020_Ride；
- Front Susp Subsystem：mdids：//FASE/subsystems.tbl/FSAE_sus_front_spring_ work_new.sub；
- Rear Susp Subsystem：mdids：//FASE/subsystems.tbl/FSAE_sus_rear_spring_work_New.sub；
- Steering Subsystem：mdids：//FASE/subsystems.tbl/FSAE_steering_2017.sub；
- Front Wheel Subsystem：mdids：//FASE/subsystems.tbl/front_tire.sub；
- Rear Wheel Subsystem：mdids：//FASE/subsystems.tbl/rear_tire.sub；
- Body Subsystem：mdids：//FASE/subsystems.tbl/FSAE_Body_2017.sub；
- 勾选 Brake Subsystem：mdids：//FASE/subsystems.tbl/FSAE_powertrain_2020.sub；
- 勾选 Powertrain Subsystem：mdids：//FASE/subsystems.tbl/FSAE_powertrain.sub；
- Vehicle Test Rig（整车实验台架）：_ARIDE_FOUR_POST_TESTRIG。
- 单击 OK 按钮，完成整车模型 FSAE_2020_Ride 的装配，装配好的模型如图 26-1 所示。

3. 装配方法（2）

（1）启动 ADAMS/CAR，选择专家模块进入建模界面。

（2）单击 File > Open > Assembly 命令，弹出打开装配模板对话框，如图 26-4 所示，在 Assembly Name 中输入：mdids：//FASE/assemblies.tbl/FSAE_2020_spring_work_New_arb.asy。

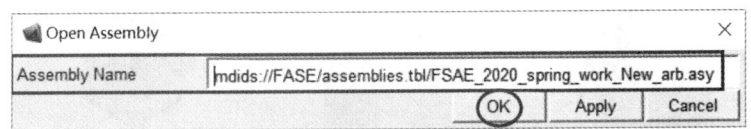

图 26-4　模板对话框

（3）单击 OK 按钮，完成整车模型的导入，如图 26-5 所示。

图 26-5　整车模型（包含横向稳定杆）

（4）单击 File > Manage Assemblies > Add Testing 命令，添加实验台架，如图 26-6 所示，在下列对话框中输入相应的数据：

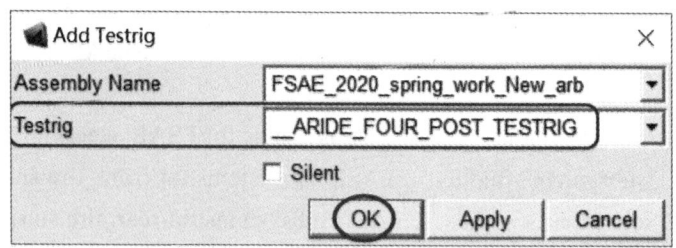

图 26-6　添加四柱实验台架

- Assembly Name：FSAE_2020_spring_work_New_arb；
- Testrig：_ARIDE_FOUR_POST_TESTRIG。

（5）单击 OK 按钮，完成实验台架的添加，此时整车模型与实验台架如图 26-7 所示。相关信息如下：

图 26-7　整车模型（包含横向稳定杆）

> 整车添加四柱实验台架命令窗口显示如下信息：
> Copying the .__ARIDE_FOUR_POST_TESTRIG testrig into assembly '.FSAE_2020_spring_work_New_arb'...
> Assembling subsystems...
> Assigning communicators...
> WARNING：The following input communicators were not assigned during assembly：
> testrig.cil_outside_wheel_center_rear
> testrig.cir_outside_wheel_center_rear
> FSAE_brake.cis_brake_demand

```
FSAE_body.cis_std_tire_ref
FSAE_powertrain_2020.cis_throttle_demand
FSAE_powertrain_2020.cis_transmission_demand
FSAE_powertrain_2020.cis_clutch_demand
FSAE_powertrain_2020.cis_initial_engine_rpm
FSAE_powertrain_2020.cis_sse_diff1
FSAE_sus_rear_spring_work_New_arb.cil_droplink_to_bellcrank
FSAE_sus_rear_spring_work_New_arb.cir_droplink_to_bellcrank
Assignment of communicators completed.
Assembly of subsystems completed.
Assembly '.FSAE_2020_spring_work_New_arb' ready.
```

26.2 平顺性仿真

（1）单击 Ride > Full-Vehicle Analysis > Four Poster Testrig 命令，弹出弹簧平顺性仿真设置，如图 26-8 所示，在下列对话框中输入相应的数据：

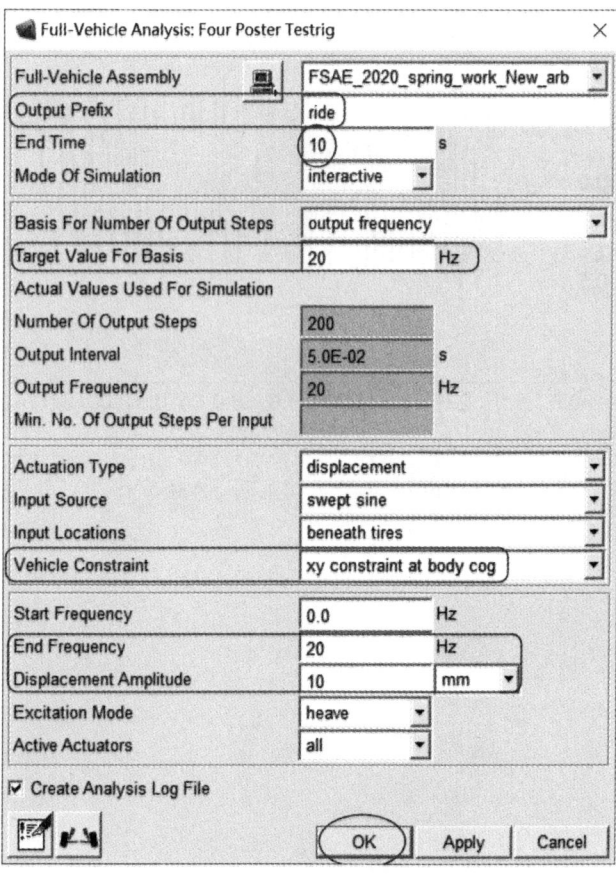

图 26-8　平顺仿真参数设置

- Full-Vehicle Assembly：FSAE_2020_spring_work_New_arb；
- Output Prefix：ride；
- End Time：10；
- Mode Of Simulation：interactive；
- Basis For Number Of Output Steps：output frequency；
- Target Value For Basis（基准频率）：20 Hz；
- Actuation Type：displacement；
- Input Source：swept sine；
- Input Locations：beneath tires；
- Vehicle Constraint：xy constraint at body cog；
- Start Frequency：0 Hz；
- End Frequency：20 Hz；
- Displacement Amplitude：10 mm。

（2）单击 OK 按钮，完成平顺参数设置并提交运算；平顺性计算结果如图 26-9 所示，根据 ISO 评价体系，加速度小于 315 mm/s^2，平顺性极好。

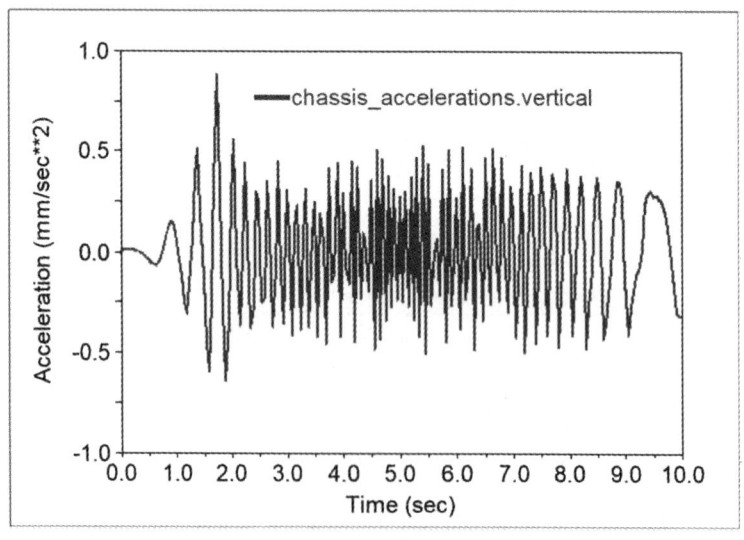

图 26-9　车身垂向振动（Z 轴）

附 录

附录 A

将模型 FSAE_sus_rear_rigid_ben 中的刚性横向稳定杆替换成柔性梁后,完成对应约束施加并另存为 FSAE_sus_rear_flexible_ben,如图 A-1 所示。torsion_10p200.mnf 中性文件及模型存在附录电子文件中,请读者自行查阅。将悬架 FSAE_sus_rear_flexible_ben 替换到整车上,整车另存为 FSAE_2020_flexible_Ben,如图 A-2 所示。

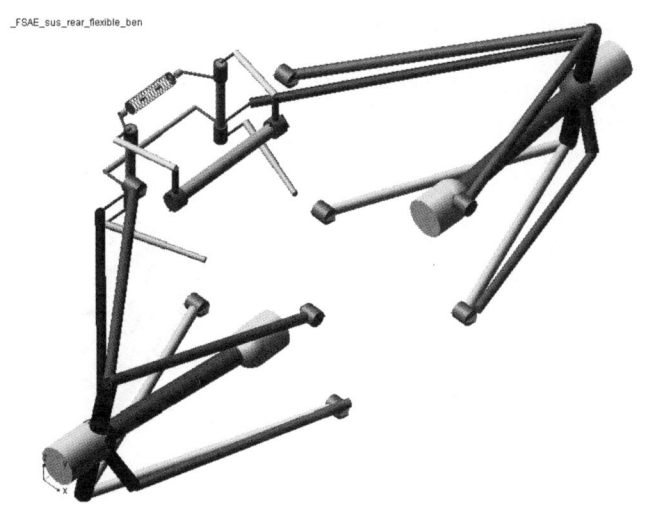

图 A-1 后悬架 FSAE_sus_rear_flexible_ben

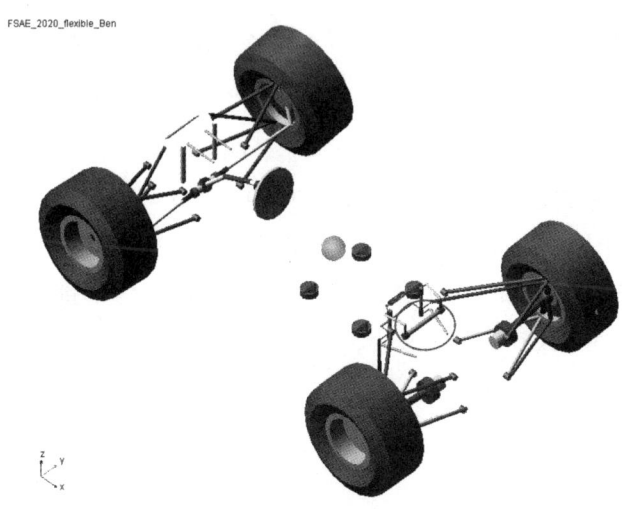

图 A-2 整车.FSAE_2020_flexible_Ben

附录 B

图 B-1 悬架模型 FSAE_sus_front_complex_colink 与图 B-2 悬架模型 FSAE_sus_front_complex_colink_flex 存放在附录电子文件中。图 B-1 中推杆悬架中的稳定杆（联动左右支架摆臂）为刚性体，当一侧车轮上下运动时，可以带动另外一侧车轮按相同的方向上下同向运动，可降低弯道侧倾角，提升整车弯道稳定性。此种布置方式被用在方程式赛车上，且结构较为简单。图 B-1 中的刚性推杆悬架不能进行车轮反向跳动仿真（仅从仿真计算的角度看），实际上方程式赛车为提升车辆性能几乎也是不允许车轮有反向跳动（即使悬架有此项功能，大多会采用第三个避震器，例如前面章节所讨论的通过横向避震器加以限制，允许悬架有运动的趋势，但是在很小的范围内又通过避震器阻止这种运动趋势）；在反向运动方向增加避震器是一种非常有效且简单的方法，当然前提条件是空间允许，例如家用轿车避震器与弹簧均为垂向布置，在反方向没有安装空间。

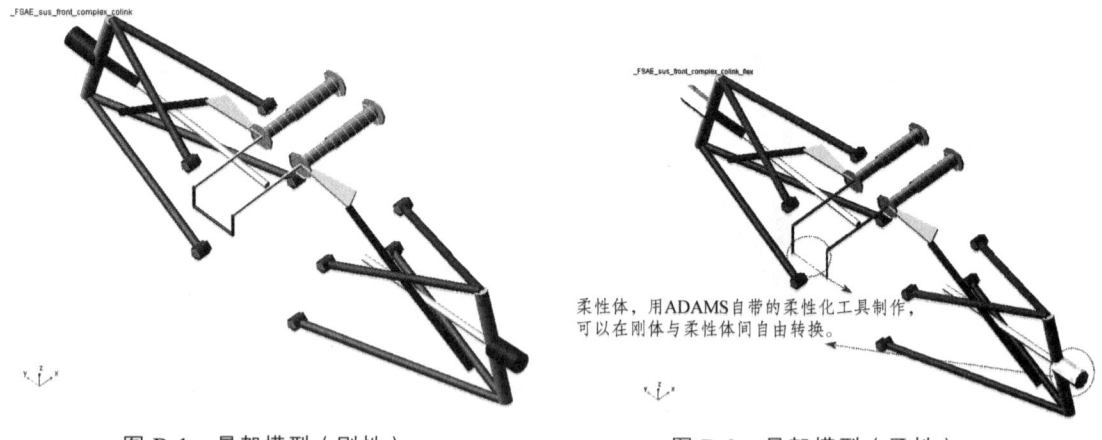

图 B-1　悬架模型（刚性）　　　　　图 B-2　悬架模型（柔性）

针对不能双轮反向跳动仿真的问题，通过 ADAMS 软件自动柔性化工具对扭杆进行柔性化，柔性化后如图 B-2 所示，此时可以进行反向车轮跳动仿真（此处扭杆横截面过小，读者可以通过增加扭杆横截面再次柔性化）。ADAMS 中的柔性化与刚体之间可以自由转换，同时可以把柔性体转化成刚体，软件中的柔性化单元均为四面体（与 Marc 协同），单元划分质量较差，推荐采用第三方软件例如 ABAQUS、Hypermesh、ANSYS 等进行六面体网格划分，并制作 MNF 模态中性文件导入到 ADAMS 中，本书所有的柔性均通过 ABAQUS 制作。

附录 C

双横置叶片弹簧悬架模型如图 C-1 所示，下控制用双横置板簧替代，横置摆臂既起到弹簧的作用，也起到横向拉杆的作用，还起到横向稳定杆的作用。因此采用横置板簧后，不用单独考虑横向稳定杆。图 C-1 中的悬架模型中的板簧刚度是可调的。

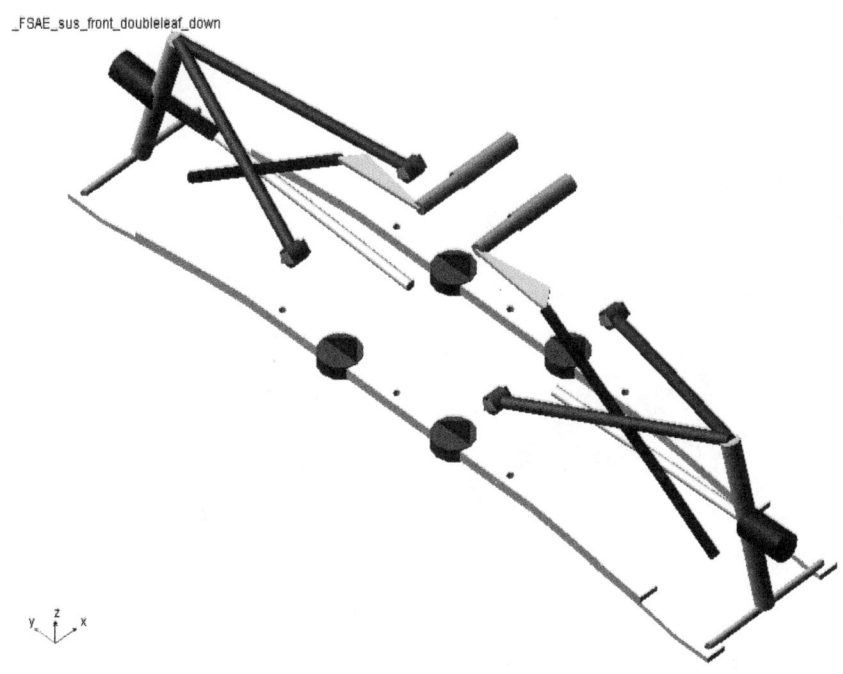

图 C-1 双横置叶片弹簧悬架模型

附录 D

考虑横向稳定杆的空间垂置扭杆弹簧拉杆式解耦悬架模型如图 D-1 所示,此悬架模式采用扭杆弹簧与螺旋弹簧两种工作模式,可以根据整车速度特性切换弹簧工作模式,具体建模不详细叙述,请读者参阅"空间斜置扭杆弹簧推杆式解耦悬架模型"章节;整车稳定性指标与平顺性指标完全解耦。

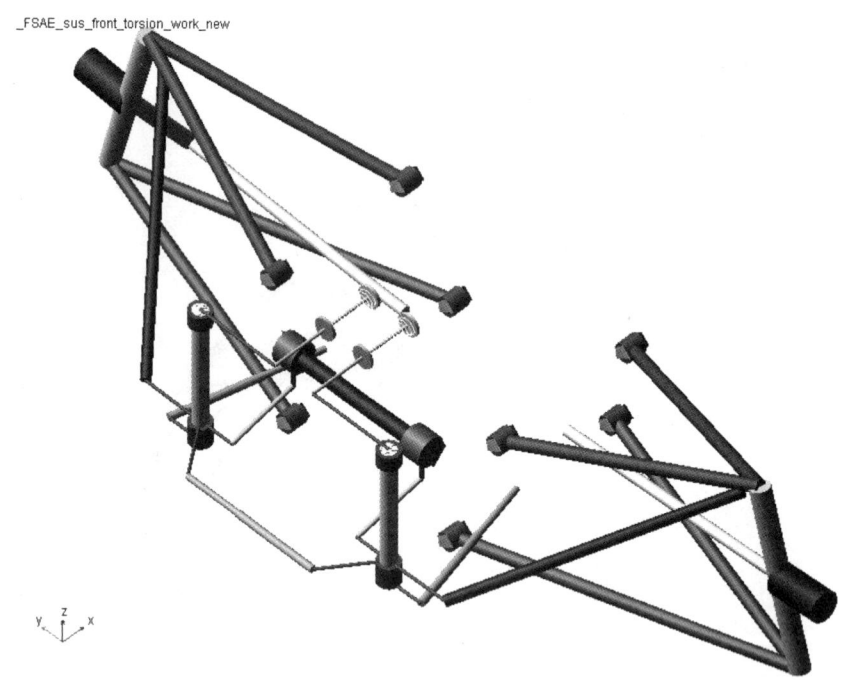

图 D-1 考虑横向稳定杆的空间垂置扭杆弹簧拉杆式解耦悬架模型

参考文献

[1] 邓景怀，林海英，刘明众. FSC 赛车碳纤维悬架制造工艺优化[A]. 2020 中国汽车工程学会年会论文集（8）[C]. 北京：中国汽车工程学会，2020.

[2] 柴源. 基于 ADAMS 的 FSAE 赛车悬架仿真优化与评价[D]. 太原：中北大学，2020.

[3] 赖锦雄，周绍鹏，刘诗汉，等. FSAE 方程式赛车悬架系统设计与仿真研究[J]. 汽车实用技术，2020（07）：52-57.

[4] 王孝鹏,刘建军. 弯道制动模式下 FSAE 赛车稳定性研究[J]. 机械设计与制造,2019(10)：110-114.

[5] 刘慧，任峰，李龙海. 方程式赛车悬架设计[J]. 时代汽车，2019（15）：102-103.

[6] 岳湘棣，唐琦军，彭才望. 基于 ADAMS 的方程式赛车前悬架仿真与优化[J]. 农业装备与车辆工程，2019，57（07）：103-106.

[7] 洪聪. 大学生方程式赛车轮胎模型参数辨识及整车操稳性仿真分析[D]. 西安：长安大学，2019.

[8] 石剑英. 大学生方程式赛车悬架和转向系统优化设计与仿真分析[D]. 锦州：辽宁工业大学，2019.

[9] 高青云. 双电机独立驱动方程式赛车稳定性与再生制动控制研究[D]. 锦州：辽宁工业大学，2019.

[10] 陆海. FSC 赛车操纵稳定性的试验评价与优化设计[D]. 南宁：广西大学，2018.

[11] 王贤民. 基于 Hyperworks 的 FSEC 赛车车架轻量化研究[D]. 成都：西华大学，2018.

[12] 徐小康. 基于前后悬架刚度匹配的 FSAE 赛车操稳性研究[D]. 合肥:合肥工业大学,2018.

[13] 胡苏楠. 纯电动大学生方程式赛车制动能量回收系统研究及优化[D]. 合肥：合肥工业大学，2018.

[14] 王乐，朱建军，田宇，等. 大学生方程式赛车多连杆悬架设计及优化[J]. 机械设计与制造，2018（02）：9-12.

[15] 罗凤，杨忠炯，周立强. 变节距悬架弹簧对方程式赛车平顺性的影响[J]. 中国机械工程，2017，28（24）：2971-2975.

[16] 杨振. 基于 ADAMS 的 FSAE 赛车操纵稳定性主观与客观评价及优化分析[D]. 合肥：合肥工业大学，2017.

[17] 张宁. FSEC 电动方程式赛车底盘设计及操纵稳定性仿真分析[D]. 锦州：辽宁工业大学，2017.

[18] 黄通尧,陈圳艳,侯占峰. 基于 ADAMS/CAR 的某方程式赛车双横臂悬架刚度计算[J]. 内蒙古农业大学学报（自然科学版），2016，37（06）：90-95.

[19] 李素华. 基于 Adams 与 MATLAB 联合仿真的赛车操纵稳定性研究[D]. 长沙：湖南大学，2016.

[20] 智淑亚，骆阳，张劼. 基于 ADAMS 赛车前悬架多柔体模型优化设计[J]. 哈尔滨理工大学学报，2015，20（01）：80-84.

[21] 胡溧，施耀贵，杨啟梁. 基于有限元法的某型大学生方程式赛车车架优化设计[J]. 武汉科技大学学报，2015，38（01）：31-34.

[22] 王振刚，杨世文. 悬架布置形式对操纵稳定性影响的仿真试验分析[J]. 液压气动与密封，2014，34（10）：50-52.

[23] 于晓燕. 平衡悬架在 FSAE 赛车上的应用[D]. 太原：中北大学，2014.

[24] 王行，阳林，彭仁杰，等. 基于 ADAMS 的 FSAE 赛车前悬架优化设计[J]. 广东工业大学学报，2013，30（03）：105-108.

[25] 白世鹏，王国权，段卫洁. FSAE 方程式赛车的操纵稳定性仿真[J]. 北京信息科技大学学报（自然科学版），2013，28（04）：46-51.

[26] 王行. 方程式赛车操纵稳定性研究[D]. 广州：广东工业大学，2013.

[27] 周东玉. FSAE 赛车总布置、悬架设计及整车操稳性分析[D]. 西安：长安大学，2013.

[28] 姜立嫚. FSAE赛车车架结构动态特性分析与优化设计[D].北京:北京信息科技大学,2013.